Ordinary Differential Equations and Special Functions

Scrivener Publishing
100 Cummings Center, Suite 541J
Beverly, MA 01915-6106

Publishers at Scrivener
Martin Scrivener (martin@scrivenerpublishing.com)
Phillip Carmical (pcarmical@scrivenerpublishing.com)

Preface

The theory of *Ordinary Differential Equations and Special Functions* is the most important branch of pure mathematics, applied mathematics, and engineering sciences. Differential equations can be used to represent a wide range of issues that arise in engineering, all scientific disciplines, and the medical sciences. The author has taken into consideration, in writing this book, the needs of those branches. Differential equations have become essential mathematical tools for the solutions of many physical problems. There are not many books covering the topics in detail with sufficient examples. The theory of differential equations has originated many new functions. We call them special functions. Most of the special functions are the series solutions of differential equations. The series solutions of certain differential equations give rise to special functions. This present book on "Ordinary Differential Equations and Special Functions" has been written to meet the requirements of the students of post-graduate and research courses in mathematics. My goal has been to provide a basic, methodical overview of the topic. All findings are supported by theorems and meticulous proofs, and sufficient examples are provided to make them simple to understand.

We hope that this study is of interest to researchers from a variety of mathematics, science, and engineering backgrounds and that Master-level students will also be able to readily study more advanced concepts such as general properties of equations and their solutions.

This book has two parts. Part I contains seven chapters dealing with the Ordinary Differential Equations. In the first chapter, we deal with some preliminaries (I) necessary for the treatment of the materials in the remaining six chapters of this part. In the second chapter, we discuss Picard's approximation method, and the third deals with matrix operations in systems of linear equations. Chapters 4 to 6 deals with homogeneous and non-homogeneous linear equations, and the last chapter deals with part-I, which contains Strum Liouville problems. Part II contains nine chapters which are devoted to the theory of special functions such as hypergeometric functions, generalized hypergeometric functions, Bessel, Legendre, Laguerre, Hermite, Chebyshev and Jacobi polynomials and their properties. A large number of problems have been completely solved, and the exercise at the end of each chapter will further help the student.

I would like to thank Professor Rashmi Bhardwaj, Professor Rajkumar Roy Choudhury, and Professor Valentina Emilia Balas who gave me the scope and encouragement, with support in various ways, to write this book. Special thanks are extended to Dr. Tarun Maiti, who helped me with every aspect in writing this book. I am also supported by books gathered by my daughter Dipanjana. Also, I would like to thank my wife Suranjana for all her support.

In the end, I wish to record my appreciation to the publishers Scrivener Publishing, Wiley, and their printers for their care and full cooperation in bringing out this study in the present nice form book. Suggestions for further improvements of the book will be gratefully acknowledged.

Dr. Dipankar De
Agartala, India
11 February 2025

Part I

ORDINARY DIFFERENTIAL EQUATIONS

Remark 1.2.1. Independence of the n arbitrary constant a in a general solution, means that the solution cannot be reduced to a form containing (n−1) or fewer constants.

Let $y = c_1e^{x+c_2}$

be a solution of the equation $\frac{d^2y}{dx^2} - y = 0$, containing two arbitrary constant c_1 and c_2.

But, it is not a general solution, since it can be written as

$$y = c_1e^{x+c_2} = c_1e^xe^{c_2} = c_1e^{c_2}e^x = ce^x, \textit{ where } c = c_1e^{c_2},$$

which contains only one constant c.

Example 1.2.4. Consider the differential equation

$$\frac{d^2y}{dx^2} = 4y \tag{i}$$

Suppose, by the same means, we obtain from it the relation

$$y = e^{2x} \tag{ii}$$

It is a solution of Equation (i). For putting $y = e^{2x}$, we get $\frac{d^2y}{dx^2} = 4e^{2x} = 4y$. We found that Equation (1.1) is identically satisfied. But it may not be the only solution if we verify that $y = e^{-2x}$ is also solution. Generally, the relation $y = Ae^{2x} + Be^{2x}$ also satisfies Equation (i): A, B are independent, arbitrary constants.

The solution of a differential equation, which contains the same number of independent arbitrary constants as the order of the differential equation, is called its general solution (*primitive or complete integral or complete*).

Any solution obtained from the general solution by giving particular values to the arbitrary constants is called a *particular solution*.

The general solution does not always include all possible solutions of a differential equation. Solutions may exist, which are not deducible from the general solution by giving particular values to the constants. Such solutions are called *singular solutions*.

The equation

$$y = px + \sqrt{a^2p^2 + b^2} \ (p = \frac{dy}{dx}) \tag{iii}$$

where a and b are constants, is satisfied by

$$y = cx + \sqrt{a^2c^2 + b^2} \qquad \text{(iv)}$$

where c is an arbitrary constant. This is a general solution of Equation (iii).
But the relation

$$\frac{x^2}{a^2} + \frac{y^2}{b^2} = 1 \qquad \text{(v)}$$

also satisfies Equation (iii) but not deducible from Equation (iv). Hence, Equation (v) is a *singular solution* of Equation (iii).

Since the general solution of an nth-order equation contains n independent arbitrary constants, the maximum number of independent conditions, which can be imposed on a solution, is n. If these conditions pertain to the behavior of the solution at one specific point, they are called *initial conditions*, and the differential equation with such conditions is called an *initial value problem* (IVP). On the other hand, a differential equation with a condition prescribed at more than one point is called a boundary value problem (BVP).

For example: $\frac{d^2y}{dx^2} + 4y = 0$, y(0)=0 and $\frac{dy}{dx}(0) = 1$ is an example of IVP and $y = sin2x$ is the solution of the problem.

For BVP, we consider $\frac{d^2y}{dx^2} + y = 0$, y(0)=1, $y(\frac{\pi}{2}) = 2$.

It can be seen that $y = cosx + 2sinx$ is a solution of the above BVP.

1.3 Family of Curves Represented by Ordinary Differential Equations

The primitive of an ordinary differential equation of first order is a relation between the two variables x and y and a parameter c. For each numerical value of c, we get a separate curve, and hence, the primitive gives a *one-parameter family of curves.* The differential equation is then said to represent a one-parameter family of curves. Each curve of family is called an *integral curve.*

The equation of first order and first degree is

$$\frac{dy}{dx} = f(x,y).$$

As we have seen earlier, a general solution of the nth order differential equation is

$$f(x,y,\ldots\ldots,y^{(n)}) = 0 \tag{1.1}$$

If it exists, it can be written in the form

$$F(x,y,c_1,c_2,\ldots.c_n) = 0 \tag{1.2}$$

$c_1, c_2, \ldots c_n$ *are arbitrary constants,*
which represents an n-parameter family of curves.

For the nth parameter of the above family of curves Equation (1.1), if we differentiate it n times successively with respect to x, treating y as an implicit function of *x*, every differentiation would be a new relation among x, y, the derivatives of y, and the parameters $c_1, c_2, \ldots. c_n$. The n relations with Equation (1.2) will give an nth order differential equation, which is able to eliminate n parameters from $(n+1)$ relations. This *n*th order differential equation represents an n-parameter family of curves.

Example 1.3.1 The Equation

$$y = ax^2 + b \tag{i}$$

represents a family of parabolas.

Differentiating Equation (i), we get $\frac{dy}{dx} = 2ax \quad \frac{d^2y}{dx^2} = 2a.$

We have

$$\frac{dy}{dx} = x\frac{d^2y}{dx^2} \tag{ii}$$

Thus, Equation (ii) is the differential equation of the family represented by Equation (i).

Example 1.3.2. The equation $x^2 + y^2 = a^2$ represent a family of circles.

Differentiating, we get $2x + 2y\dfrac{dy}{dx} = 0$

or,
$$\frac{y}{x}\frac{dy}{dx} = -1.$$

We see that the radius vector and the tangent have gradients whose product is equal to −1. That means that they are perpendicular to each other.

Thus, integral curves are, therefore, circles of any radius with the origin in center. It implies that the locus of the differential equation is of all circles that have the origin as the center.

Example 1.3.3. Considering the family of all ellipses with foci at $(-1,0)$ and $(1,0)$, we know that the sum of the focal distances of a point on the ellipse is a constant and equal to the length of its major axis.

With this consideration, we get the equation $\dfrac{x^2}{a^2} + \dfrac{x^2}{a^2 - 1} = 1, a > 1$, which is a one-parameter family with parameter a. After differention, we get the differential equation of this family $(x + yy')(xy' - y) = y'$.

Example 1.3.4. The primitive of the equation $\dfrac{dy}{dx} = \dfrac{y}{x}$ is $y = cx$. Find the equation of the integral curve through (2,3) and (0,0).

Solution: When $x = 2, y = 3, c = \dfrac{3}{2}$, and the required equation is $2y = 3x$.

When $x = 0, y = 0, c$ is not determined. All the integral curves pass through $(0,0)$. It is observed that $f(x, y) = y/x$ is not continuous at the origin, *i.e.*, $(0,0)$ is a singular point, and hence, it occurred such behavior at that point.

The geometrical concept of $\dfrac{dy}{dx} = y/x$ is that the tangent and radius vector have the same gradient. It implies that all straight lines have to pass through the origin. The general solution represents a family of curves passing through the origin.

1.4 Equation of the First Order and First Degree

We shall discuss some special methods of solutions, which are applied to the following types of equations:

1. Equations solvable by separation of variables
2. Homogeneous Equations
3. Linear Equations
4. Exact Equations

1.4.1 Variable Seperable

If in an equation it is possible to collect all functions of $x \textit{ and } dx$ on one side and all the functions of $y \textit{ and } dy$ on the other side, then the variables are said to be separable. Generally, for practice, when in the equation

$Mdx + Ndy = 0, M$ is a function of x alone, and N is a fuction of y *alone*; then, the equation $f_1(x)dx + f_2(y)dy = 0$ is said to have *separated variable.*

Thus, the general form of such an equation is

$$f(y)dy = g(x)dx.$$

Integrating both sides, we get $\int f(y)dy = \int g(x)dx + c$, where c is a constant.

Example 1.4.1. Solve the equation

$$(x+y)^2 \frac{dy}{dx} = a^2.$$

Solution: Put $x + y = v, \textit{then } 1 + \frac{dy}{dx} = \frac{dv}{dx}.$

It becomes $v^2\left(\frac{dv}{dx} - 1\right) = a^2$

or,
$$\frac{dv}{dx} = \frac{a^2}{v^2} + 1 = \frac{a^2 + v^2}{v^2}$$

or, $$\frac{v^2}{a^2+v^2}dv = dx.$$

Integrating, $\int \frac{v^2}{a^2+v^2} dv = \int dx + c$

or, $$\int 1 - \frac{a^2}{a^2+v^2} dv = \int dx + c\phi\phi$$

or, $$\int (1 - \frac{1}{1+\frac{v^2}{a^2}}) dv = \int dx + c$$

or, $$v - a\,tan^{-1}\frac{v}{a} = x + c$$

or, $$x + y - atan^{-1}\frac{x+y}{a} = x + c,$$

which is the required solution.

1.4.2 Homogeneous Equations

This equation is in the form of

$$\frac{dy}{dx} = \frac{f(x,y)}{\varphi(x,y)},$$

where $f(x,y)$ *and* $\varphi(x,y)$ are homogeneous of the same degree in *and* y.

To solve a homogeneous equation, take the following two steps:

1. Put $y = vx$, then $\frac{dy}{dx} = v + x\frac{dv}{dx}$.
2. Separate x and v and then integrate.

Definition 1.4.1. Homogeneous equation: If a function $f(x,y)$ can be in the form $x^n\varphi\left(\frac{y}{x}\right)$ or in the form $y^n\varphi\left(\frac{x}{y}\right)$, then $f(x,y)$ is said to be a homogeneous function of degree n in x *and* y. Otherwise, a function $f(x,y)$ *is homogeneous of degree* n if it can be in the form $x^n\psi(\frac{y}{x})$.

Note $\frac{dy}{dx} = f(x,y)$ is a homogeneous equation, if

$$f(zx,zy) = f(x,y) \text{ for all } z.$$

Example 1.4.2. Solve

$$(x^2 - y^2)dx - xydy = 0 \tag{i}$$

Solution: Here, $\frac{dy}{dx} = \frac{x^2 - y^2}{xy}$, which is a homogeneous equation in x and y.

Put $y = vx$ *then* $\frac{dy}{dx} = v + x\frac{dv}{dx}$.

(i) It becomes $v + x\frac{dv}{dx} = \frac{1 - v^2}{v}$

or, $$x\frac{dv}{dx}=\frac{1-v^2}{v}-v=\frac{1-2v^2}{v}$$

or, $$\frac{vdv}{1-2v^2}=\frac{dx}{x}$$

Integrating both sides, $\int\frac{vdv}{1-2v^2}=\int\frac{dx}{x}+logc$

$$-\frac{1}{4}\log(1-2v^2)=logcx.$$

Taking the antilog, $(1-2v^2)^{-\frac{1}{4}}=cx$

or, $$\frac{1}{1-2v^2}=hx^4$$

or, $$x^4\left(1-2\frac{y^2}{x^2}\right)=k$$

or, $$x^2(x^2-2y^2)=k_1,$$

where c, h, k, k_1 are constants.

Example 1.4.3. Solve $(3y+2x+4)dx-(4x+6y+5)dy=0$.
Solution:

$$\frac{dy}{dx}=\frac{(3y+2x+4)}{(4x+6y+5)} \tag{i}$$

Put $3y+2x=t$ *so that* $2+3\frac{dy}{dx}=\frac{dt}{dx}$.

(i) It becomes $\frac{1}{3}\left(\frac{dt}{dx}-2\right)=\frac{t+4}{2t+5}$

or, $$\frac{dt}{dx}=\frac{3t+12}{2t+5}+2=\frac{7t+22}{2t+5}$$

or, $$\frac{2t+5}{7t+22}dt=dx$$

Integrating, we get $\int\frac{2t+5}{7t+22}dt=\int dx+c$

or, $$\int\left(\frac{2}{7}-\frac{9}{7}\frac{1}{7t+22}\right)dt=x+c$$

or, $$\frac{2}{7}t-\frac{9}{49}\log(7t+22)=x+c$$

or, $$\frac{2}{7}(3y+2x)-\frac{9}{49}\log\{7(3y+2x)+22\}=x+c$$

or, $$14(3y+2x)-9\log\{7(3y+2x)+22\}=49x+k,$$

which is the required solution.

1.4.3 Linear Equation

A differential equation of the form

$$dy/dx + Py = Q \tag{1.3}$$

where P and Q are functions of x or constants but do not involve y, is called a *Linear Equation.*

To solve the equation, multiply both sides by $e^{\int Pdx}$ so that we get

$$\frac{dy}{dx} e^{\int Pdx} + yPe^{\int Pdx} = Qe^{\int Pdx}$$

or,

$$\frac{d}{dx}\left(ye^{\int Pdx}\right) = Qe^{\int Pdx}.$$

Integrating both sides, we get $ye^{\int Pdx} = \int Qe^{\int Pdx}dx + c$, which is the required solution.

Definition 1.4.2. The factor $e^{\int Pdx}$, by which, when multiplying, the left-hand side of (i) becomes the differential coefficient of a single function, is called *the integrating factor (I.F.)* of the linear Equation (1.3).

Here, I.F. $= e^{\int Pdx}$.

The solution is $y(I.F.) = \int Q(I.F.)dx + c$.

Example 1.4.4. Solve

$$1 + y^2 + \left(x - e^{-tan^{-1}y}\right)\frac{dy}{dx} = 0.$$

Solution: The equation can be written as

$$\frac{dx}{dy} + \frac{x}{1+y^2} = \frac{e^{-tan^{-1}y}}{1+y^2},$$

which is the form of a linear equation.

$$Integrating\ Factor(I.F) = e^{\int \frac{dy}{1+y^2}} = e^{tan^{-1}y}$$

Hence, the solution is $x(I.F.) = \int Q(I.F.)dy + c$

or, $$xe^{tan^{-1}y} = \int \frac{e^{-tan^{-1}y}}{1+y^2} e^{tan^{-1}y} y + c = \int \frac{dy}{1+y^2} + c = tan^{-1}y + c$$

or, $$\therefore xe^{tan-1y} = tan^{-1}y + c$$

1.4.4 Bernoulli's Equation

The Equation

$$\frac{dy}{dx} + Py = Qy^n \tag{1.4}$$

where P and Q are functions of x, is reducible to the Leibnitz's linear equation and is usually called the *Bernoulli's Equation n*[1]

Put $y^{n-1} = z$.

The differential equation becomes

$$\frac{dz}{dx} + P(1-n)z = Q(1-n) \tag{1.5}$$

[1] The Swiss mathematician *Jacob Bernoulli* (1654–1705) who is known for his basic work in probability and elastic theory. He was professor at Basel, and his son, Daniel Bernoulli (1700–1782), is known for his contributions to the kinetic theory of gases and fluid flow.

which is Leibnitz's linear in z and can be solved. The general solution of the equation is as follows:

$$y^{1-n} = e^{(n-1)\int Pdx}\left[(1-n)\int Q(x)e^{(1-n)\int Pdx}dx + c\right].$$

Example 1.4.5. Solve $\frac{dz}{dx} + \left(\frac{z}{x}\right)logz = \frac{z}{x}(logz)^2$.

Solution: By dividing z, the given equation becomes

$$\frac{1}{z}\frac{dz}{dx} + \left(\frac{1}{x}\right)logz = \frac{1}{x}(logz)^2 \qquad \text{(i)}$$

Put $=t$, differentiating with respect to x, so that $\frac{1}{z}\frac{dz}{dx} = \frac{dt}{dx}$,

(i) It becomes $\frac{dt}{dx} + \left(\frac{t}{x}\right) = \frac{1}{x}(t)^2$

or,
$$\frac{1}{t^2}\frac{dt}{dx} + \left(\frac{1}{x}\right)\frac{1}{t} = \frac{1}{x} \qquad \text{(ii)}$$

This would be Bernoulli's equation. Put $\frac{1}{t} = v$ so that Equation (ii) becomes $\left[\therefore \frac{dv}{dx} = -\frac{1}{t^2}\frac{dt}{dx}\right]$

$$-\frac{dv}{dx} + \left(\frac{1}{x}\right)v = \frac{1}{x}.$$

This is Leibnitz's linear in v.

$$\therefore I.F. = e^{-\int e^{\frac{1}{X}dX}} = \frac{1}{x}.$$

The solution is $v\frac{1}{x}=-\int\frac{1}{x}\frac{1}{x}dx+c=\frac{1}{x}+c.$

Replacing $v=\frac{1}{logz}$, we get

$$(xlogz)^{-1}=\frac{1}{x}+c$$

or, $$(logz)^{-1}=1+cx.$$

Example 1.4.6. Solve $\frac{dy}{dx}(x^2y^3+xy)=1.$

Solution: The equation is $\frac{dx}{dy}-xy=x^2y^3$

or, $$\frac{1}{x^2}\frac{dx}{dy}-\frac{1}{x}y=y^3;$$

putting $-\frac{1}{x}=z$ and $\frac{dz}{dy}=\frac{dx}{dy}\frac{d}{dx}\left(-\frac{1}{x}\right)=\frac{1}{x^2}\frac{dx}{dy}$, the equation becomes $\frac{dz}{dy}+yz=y^3$, which is a linear differential equation.

Its I.F. $=e^{\int ydy}=e^{\frac{y^2}{2}}$.

Hence, the required solution is $ze^{\frac{y^2}{2}}=\int e^{\frac{y^2}{2}}.y^3dy+c$

or, $$-\frac{1}{x}e^{\frac{y^2}{2}}=\frac{1}{2}\int e^{\frac{t}{2}}.tdt \qquad [\text{let } y^2=t, dt=2ydy]$$

or, $$-\frac{1}{x}e^{\frac{y^2}{2}} = \frac{1}{2}\left[2te^{\frac{t}{2}} - \int 2e^{\frac{t}{2}}dt\right] = te^{\frac{t}{2}} - 2e^{\frac{t}{2}} = e^{\frac{y^2}{2}}(y^2-2)+c$$

$$\therefore \frac{1}{x} = Ae^{-\frac{y^2}{2}} - (y^2-2).$$

1.4.5 Exact Differential Equation

A differential equation of the form $M(x,y)dx + N(x,y)dy = 0$ is said to be *exact* if its left-hand member is the exact differential of some function $u(x.y)$, i.e., $du = Mdx + Ndy = 0$. Its solution is: $u(x,y) = c$.

Necessary Condition: Let $u(x,y)$ have partial derivatives. Then, the total differentiation of $u(x,y)$ (It may be exact) is

$$du = \frac{\partial u}{\partial x}dx + \frac{\partial u}{\partial x}dy.$$

The given differential equation is $M(x,y)dx + N(x,y)dy = 0$, where $M(x,y), N(x,y)$ are continuous and have continuous partial derivatives. From above equations we can write

$$du = \frac{\partial u}{\partial x}dx + \frac{\partial u}{\partial y}dy = M(x,y)dx + N(x,y)dy.$$

$$\therefore M(x,y) = \frac{\partial u}{\partial x} \text{ and } N(x,y) = \frac{\partial u}{\partial y}.$$

Hence, we have $\frac{\partial}{\partial x}(\frac{\partial u}{\partial x}) = \frac{\partial}{\partial y}(\frac{\partial u}{\partial x}) \Rightarrow \frac{\partial M}{\partial x} = \frac{\partial N}{\partial y}$.
which is the necessary condition for exactness.

It can be easily showed

If $\frac{\partial M}{\partial x} = \frac{\partial N}{\partial y}$ then $M(x,y)\,dx + N(x,y)\,dy = 0$.
which is sufficient part.

The necessary and sufficient condition for the differential equation $Mdx + Ndy = 0$ to be exact is

$$\frac{\partial M}{\partial y} = \frac{\partial N}{\partial x}.$$

Example 1.4.7. Solve: $x\frac{dy}{dx} - y = x\sqrt{x^2 + y^2}$.

Solution: $\frac{xdy - ydx}{x^2} = \sqrt{1+\left(\frac{y}{x}\right)^2}\,dx$

or, $$\frac{d\left(\frac{y}{x}\right)}{\sqrt{1+\left(\frac{y}{x}\right)^2}} = dx$$

or, $$sinh^{-1}\left(\frac{y}{x}\right) = x + c$$

or, $$y = xsinh(x + c).$$

Sometimes, a differential equation, which is not exact, can be made so, on multiplication, by a suitable factor, called *Integrating Factor.*

Example 1.4.8. Solve $(1+xy)ydx + (1-xy)xdy = 0$.

Solution: Here, $M = (1+xy)y, \quad N = (1-xy)x$.

$$I.F. = \frac{1}{Mx - Ny} = \frac{1}{(1+xy)yx - (1-xy)xy} = \frac{1}{2x^2y^2}.$$

or,
$$\frac{vdx.}{\sqrt{1+v^2}[1-\sqrt{1+v^2}]}=\frac{dx}{x}.$$

Now, putting $\sqrt{1+v^2}=t,$ we get $\dfrac{vdv}{\sqrt{1+v^2}}=dt.$

The equation reduces to $\dfrac{dt}{1-t}=\dfrac{dx}{x}.$

By integrating, we get $-\log(1-t)=logx-lggc$

or,
$$(1-t)=\frac{c}{x}.$$

$$\therefore\left(1-\frac{c}{x}\right)^2=t^2=1+v^2=1+\frac{y^2}{x^2}.$$

Therefore, $(x-c)^2=x^2+y^2$, or $y^2+2cx=c^2$.

1.5 Equations of the First Order and Higher Degree

The most general form of a differential equation of the first order but of nth degree is

$$p^n+P_1p^{n-1}+P_2p^{n-2}+\cdots+P_n=0,$$

where p denotes $\dfrac{dy}{dx}$ *and* $P_1, P_2 \ldots\ldots P_n$ are the functions of x *and* y.

Example 1.5.1. Solve $p^2+2pycotx=y^2$.

Solution: We have $p^2+2pycotx+(ycotx)^2=y^2+(ycotx)^2$

or,
$$(p+ycotx)^2=y^2(1+cot^2x)=y^2cosec^2x$$

or, $$p + y\cot x = \pm y\,\mathrm{cosec}\,x.$$

$$\therefore\ p = y(-\cot x + \mathrm{cosec}\,x) \quad \text{(i)}$$

and

$$p = y(-\cot x - \mathrm{cosec}\,x) \quad \text{(ii)}$$

From Equation (i), $\frac{dy}{y} = (-\cot x + \mathrm{cosec}\,x)dx.$

Integrating, $\log y = -\log\sin x + \log\tan\frac{x}{2} + \log c = \log\frac{c\tan\frac{x}{2}}{\sin x}$

or, $$y = \frac{c\tan\frac{x}{2}}{\sin x} = \frac{c}{2\cos^2\frac{x}{2}}$$

or, $$y(1+\cos x) = c.$$

From Equation (ii), $p = y(-\cot x - \mathrm{cosec}\,x)$

or, $$\frac{dy}{y} = -(\cot x + \mathrm{cosec}\,x)dx.$$

Integrating, we get $\log y = -\log\sin x - \log\tan\frac{x}{2} + \log c = \log\frac{c}{\sin x\tan\frac{x}{2}}$

or,
$$y = \frac{c}{sinx tan\frac{x}{2}} = \frac{c}{2sin^2\frac{x}{2}} = \frac{c}{(1-cosx)}$$

or,
$$y(1+cosx) = c.$$

The required general solution is

$$y(1 \pm cosx) = c.$$

If the given equation, on solving for y, is in the form

$$y = f(x, p),$$

then, the differentiation with respect to x gives the equation in the form

$$\frac{dy}{dx} = p = \frac{\partial f}{\partial x} + \frac{\partial f}{\partial p}\frac{dp}{dx} = \phi(x, p, \frac{dp}{dx}).$$

Example 1.5.2. Solve $y = yp^2 + 2px.$

Solution: $y = \frac{2px}{1-p^2}$

$$\frac{dy}{dx} = \frac{2p}{1-p^2} + x.\frac{2(1-p^2)-(-2p)2p}{(1-p^2)^2} = \frac{2p}{1-p^2} + x.\frac{2(1+p^2)}{(1-p^2)^2}\frac{dp}{dx}$$

or,
$$p - \frac{2p}{1-p^2} = x.\frac{2(1+p^2)}{(1-p^2)^2}\frac{dp}{dx}$$

or, $$\frac{-p(1+p^2)}{1-p^2} = x.\frac{2(1+p^2)}{(1-p^2)^2}\frac{dp}{dx}$$

or, $$\frac{dx}{x} = \frac{-2}{p(1-p^2)}dp = -\left(\frac{1}{p+1}+\frac{1}{p-1}-\frac{2}{p}\right)dp.$$

Integrating, $-logx + logc = \log(1+p) + \log(p-1) - logp^2$

or, $$c/x = \frac{(p+1)(p-1)}{p^2} = \frac{p^2-1}{p^2}$$

$$\therefore x = \frac{cp^2}{p^2-1}$$

Example 1.5.3 Solve $y = y^2p^3 + 2px.$

Solution: Here, $x = \frac{y - y^2p^3}{2p}.$

Differentiating with respect to y,

$$\frac{dx}{dy} = \frac{1}{p} = \frac{1}{2}\frac{\left(1-2yp^3 - y^2.3p^2\frac{dp}{dy}\right)p - (y-y^2p^3)\frac{dp}{dy}}{p^2}$$

$$2p = p - 2yp^4 - 3y^2p^3\frac{dp}{dy} - y\frac{dp}{dy} + y^2p^3\frac{dp}{dy}$$

or, $$p + 2yp^4 + y\frac{dp}{dy} + 2y^2p^3\frac{dp}{dy} = 0$$

or, $$p(1+2yp^3) + y\frac{dp}{dy}(1+2yp^3) = 0$$

or, $$(1+2yp^3)\left(p+y\frac{dp}{dy}\right)=0,$$

which gives $p+y\dfrac{dp}{dy}=0 \quad or, \dfrac{d}{dy}(yp)=0.$

Integrating, $yp=c.$

Given this value of p in the given equation, we get

$$y=y^2\frac{c^3}{y^3}+2\frac{c}{y}x$$

or $y^2=c^3+2cx,$ which is the required solution.

1.5.1 Clairaut's Equation

An equation of the form

$$y=px+f(p) \tag{1.6}$$

is known as *Clairaut's Equation.*

Differentiating with respect to x, we get $p=p+x\dfrac{dp}{dx}+f'(p)\dfrac{dp}{dx}$

or, $$\frac{dp}{dx}\left(x+f'(p)\right)=0$$

$$\therefore \frac{dp}{dx}=0,$$

which gives us $p=c.$

Putting this values in Equation (1.3), we get $y=cx+f(c),$ which is the general solution of Equation (i).

Example 1.5.4. Solve $y = y^2p^3 + 2px$.
Solution:
Let $z = y^2$, then $\frac{dz}{dx} = 2yp$.
Then, the given equation can be written as

$$y^2 = y^3p^3 + 2pyx,$$

which can be transformed into $z = \frac{dz}{dx}x + (\frac{1}{2}\frac{dz}{dx})^3$.
Putting $\frac{dz}{dx} = P$, the equation becomes

$$z = Px + \frac{1}{8}P^3, \tag{i}$$

which is Clairaut's equation.
Differentiating with respect to x, we have

$$\frac{dz}{dx} = P + x\frac{dP}{dx} + \frac{3}{8}P^2\frac{dP}{dx}$$

or,
$$\frac{dP}{dx}\left(x + \frac{3}{8}P^2\right) = 0.$$

Using the above method, i.e., $\frac{dP}{dx} = 0 \; or \; P = c$, put in Equation (i). Then,

the general solution is $z = cx + \frac{1}{8}c^3$

or,
$$y^2 = cx + \frac{1}{8}c^3.$$

If $x + \frac{3}{8}P^2 = 0$, *then* $P^2 = -\frac{8}{3}x$.
From Equation (i), $z^2 = P^2(x + \frac{1}{8}P^2)^2$

or, $$y^4 = -\frac{8}{3}x(x - \frac{1}{3}x)^2$$

or, $$y^4 = -\frac{8}{3}x(\frac{2}{3}x)^2$$

or, $$y^4 = -\frac{32}{27}x^3$$

or, $$27y^4 + 32x^3 = 0,$$

which is a singular solution.

1.6 Linear Differential Equation

1.6.1 The General Form

The general form of a linear differential equation of the nth order is

$$\frac{d^n y}{dx^n} + p_1\frac{d^{n-1}y}{dx^{n-1}} + p_2\frac{d^{n-2}y}{dx^{n-2}} + \ldots\ldots + p_n y = X \tag{1.7}$$

where $p_1, p_2 \ldots p_n$ *and* X are constants or functions of x only.

In this chapter, we shall assume that linear differential equations with constant coefficients, i.e., the coefficients $p_1, p_2 \ldots p_n$ are constants.

We shall consider two forms of Equation (1.7). First, we consider the form in which the right-hand side, i.e., X is zero, and then, we consider the form in which X is a function of x only or constant.

Here, we assume that the coefficients $p_1, p_2 \ldots p_n$ are constants here.

Equation of First Order: Let $n = 1, X = 0$ in Equation (1.6), then

$$\frac{dy}{dx} + p_1 y = 0.$$

We easily get, $y = Ae^{-p_1 x}$, where A is an arbitrary constant.
Equation of Second Order: If $n = 2$, $X = 0$, then Equation (1.6) becomes

$$\frac{d^2y}{dx^2} + p_1\frac{dy}{dx} + p_2 y = 0 \tag{1.8}$$

The above solution suggests $y = Ae^{mx}$, where m is a constant, which may satisfy Equation (1.7).

By with this $y = Ae^{mx}$, Equation (1.10) reduces to

$$Ae^{mx}(m^2 + p_1 m + p_2) = 0.$$

If m is a root of

$$m^2 + p_1 m + p_2 = 0 \tag{1.9}$$

$y = Ae^{mx}$ is a solution of Equation (1.10) whatever may be the value of A.

Let Equation (1.11) have two distinct real roots α *and* β. Then, we have two solutions of Equation (1.6):

$y = Ae^{\alpha x}$ and $y = Be^{\beta x}$ where A and B are arbitrary constants.

Now, we substitute

$y = Ae^{\alpha x} + Be^{\beta x}$ in the left side of Equation (1.6). We obtain

$Ae^{\alpha x}(\alpha^2 + p_1\alpha + p_2) + Be^{\beta x}(\beta^2 + p_1\beta + p_2)$, which must be zero since roots α *and* β satisfies the equation, which shows that α *and* β are its roots. Thus,

$y = Ae^{\alpha x} + Be^{\beta x}$ is a solution of Equation (1.8).

Since it contains two arbitrary constants A and B, it is regarded as the *general solution or complete primitive of the equation.*

Two Cases

(i) When the auxiliary equation has imaginary roots.

Let it have two imaginary roots of the form

$$\alpha + i\beta \text{ and } \alpha - i\beta.$$

Then, the general solution is:

$$\begin{aligned} y &= Ae^{(\alpha+i\beta)x} + Be^{(\alpha-i\beta)x} \\ &= e^{\alpha x}\left(Ae^{i\beta x} + be^{-i\beta x}\right) \\ &= e^{\alpha x}[Mcos\beta x + Nsin\beta x), \end{aligned}$$

where M and N are constants.

(ii) When the roots are equal: Let the roots be $\alpha \text{ and } \alpha$

The solution becomes $y = Ae^{\alpha x} + Be^{\alpha x} = (A+B)e^{\alpha x}$.

It cannot be not regarded as the most general solution since it has two roots. It will be

$$y = (A + Bx)e^{\alpha x}.$$

Theorem 1.6.1. If $y_1 \text{ and } y_2$ are the only two solutions of the equation

$$\frac{d^n y}{dx^n} + k_1 \frac{d^n y}{dx^n} + k_2 \frac{d^{n-2} y}{dx^{n-2}} + \cdots + k_n y = 0, \tag{1.10}$$

Then, $c_1 y_1 + c_2 y_2 (= u)$ is also its solution.

If $y = y_1, y = y_2, \ldots y = y_n,$ are integrals of the equation, then $y = C_1 y_1 + C_2 y_2 + \cdots + C_n y_n$ are also integrals of the equation where $C_1, C_1, .. C_n$ are arbitrary constants.

Theorem 1.6.2. If $y = f(x)$ is the general solution of the equation

$$\frac{d^n y}{dx^n} + p_1 \frac{d^n y}{dx^n} + p_2 \frac{d^{n-2} y}{dx^{n-2}} + \cdots + p_n y = 0,$$

and $y = g(x)$ is a solution of the equation

$$\frac{d^n y}{dx^n} + p_1 \frac{d^{n-1} y}{dx^{n-1}} + p_2 \frac{d^{n-2} y}{dx^{n-2}} + \cdots + p_n y = X,$$

then, $y = f(x) + g(x)$ is the general solution of Equation (1.9).

1.6.2 Rules for Finding the Complementary Function

To solve the equation,

$$\frac{d^n y}{dx^n} + k_1 \frac{d^n y}{dx^n} + k_2 \frac{d^{n-2} y}{dx^{n-2}} + \cdots + k_n y = 0, \tag{1.11}$$

where k's are constants.

Equation (i) in symbolic form is

$$(D^n + k_1 D^{n-1} + k_2 D^{n-2} + \cdots + k_n) y = 0.$$

Denoting $\frac{d}{dx}, \frac{d^2}{dx^2}, \frac{d^3}{dx^3}$, etc., by D, D^2, D^3, etc.,

$$D^n + k_1 D^{n-1} + k_2 D^{n-2} + \cdots + k_n = 0 \tag{1.12}$$

is called an *auxiliary equation.*

Let $m_1, m_2, \ldots m_n$ be its roots.

Case I. If all roots are real and different, then Equation (ii) is equivalent to

$$(D - m_1)(D - m_2) \ldots (D - m_n) y = 0 \tag{1.13}$$

The solution of any one of the equations

$$(D - m_1) y = 0, \ (D - m_2) y = 0, \ldots, (D - m_n) y = 0 \tag{1.14}$$

is also a solution of Equation (1.13).

Let $(D - m_n) y = 0$ and its solution $y = c_n e^{m_n x}$.

Similarly, $(D - m_1) y = 0 \ (D - m_2) y = 0, \ldots, (D - m_n) y = 0$, and its solution be $y = c_1 e^{m_1 x}, y = c_2 e^{m_2 x}, \ \ldots, y = c_n e^{m_n x}$.

Thus, the complete solution of Equation (1.13) is

$$y = (u) = c_1 e^{m_1 x} + c_2 e^{m_2 x} + \cdots + c_n e^{m_n x} \tag{1.15}$$

4. When $X = e^{ax}V$, *where V is a function of* x,

$$\frac{1}{f(D)} e^{ax} V = e^{ax} \frac{1}{f(D+a)} V.$$

Example 1.6.1. $\frac{d^3y}{dx^3} - 3\frac{d^2y}{dx^2} + 4\frac{dy}{dx} - 2y = e^x + cosx$

Solution: The equation is

$$(D^3 - 3D^2 + 4D - 2)y = e^x + cosx.$$

The auxiliary equation is $m^3 - 3m^2 + 4m - 2 = 0$.
The complimentary function is $y = Ae^x + (Bcosx + C\,sinx)e^x$

$$[(m^3 - 3m^2 + 4m - 2 = (m-1)(m^2 - 2m + 2) = 0,\ m = 1, 1 \pm i].$$

The particular integral is

$$P.I. = \frac{1}{D^3 - 3D^2 + 4D - 2}(e^x + cosx) = \frac{1}{D^3 - 3D^2 + 4D - 2} e^x + \frac{1}{D^3 - 3D^2 + 4D - 2} cosx.$$

$$\begin{aligned} P.I. &= \frac{1}{(D-1)(D^2 - 2D + 2)} e^x + \frac{1}{(D-1)(D^2 - 2D + 2)} cosx \\ &= \frac{1}{(D-1)(1^2 - 2.1 + 2)} e^x + \frac{1}{(D-1)(-1^2 - 2D + 2)} cosx \\ &= e^x \frac{1}{(D+1-1).1}.1 + \frac{1}{-2D^2 + 3D - 1} cosx \\ &= e^x \frac{1}{D}.1 + \frac{1}{-2(-1) + 3D - 1} cosx = e^x x + \frac{3D-1}{9D^2 - 1} cosx \\ &= e^x x + \frac{3D-1}{9(-1) - 1} cosx = e^x x - \frac{1}{10}(3D-1)cosx \\ &= e^x x + \frac{3}{10} sinx + \frac{1}{10} cosx. \end{aligned}$$

The solution is

$$y = Ae^x + (B\cos x + C\ \sin x)e^x + e^x x + \frac{3}{10}\sin x + \frac{1}{10}\cos x.$$

Example 1.6.2. Solve $\frac{d^3y}{dx^3} + a^2 y = \sec ax.$

Solution: The auxiliary equation is $m^2 + a^2 = 0$

or,
$$(m + ai)(m - ai) = 0.$$

The roots are $ai, -ai.$
The complimentary equation is $y = A\cos ax + B\sin ax.$
For P.I.:

$$P.I. = \frac{1}{D^2 + a^2}\sec ax = \frac{1}{(D + ai)(D \pm ai)}\sec ax$$
$$= \frac{1}{2ai}\left(\frac{1}{D - ai} - \frac{1}{D + ai}\right)\sec ax.$$

Now, $\frac{1}{D - ai}\sec ax = e^{iax}\int e^{-iax}\sec ax\, dx$

$$= e^{iax}\int \frac{(\cos ax - i\sin ax)}{\cos ax}dx = e^{iax}[x + \frac{i}{a}\log\cos x]$$
$$= (\cos ax + i\sin ax)(x + \frac{i}{a}\log\cos x)$$
$$= \left(x\cos ax - \frac{1}{a}\sin ax\log\cos x\right) + i(x\sin ax + \frac{1}{a}\cos ax\log\sin ax).$$

Similarly,

$$\frac{1}{D + ai}\sec ax = \left(x\cos ax - \frac{1}{a}\sin ax\log\cos x\right) - i(x\sin ax + \frac{1}{a}\cos ax\log\sin ax).$$

$$P.I. = \frac{1}{2ai} 2i\left(xsinax + \frac{1}{a} cosaxlogsinax \right) = \frac{1}{a} xsinax$$

$$+ \frac{1}{a^2} cosaxlogsinax.$$

Therefore, the solution is

$$y = CF + PI = Acosax + Bsinax + \frac{1}{a} xsinax + \frac{1}{a^2} cosaxlogsinax.$$

Example 1.6.3. Solve $\frac{d^2y}{dx^2} - 3\frac{dy}{dx} + 2y = xe^{3x} + sin2x.$

Solution: The symbolic form is $(D^2 - 3D + 2)y = xe^{3x} + sin2x.$

The auxiliary equation is $D^2 - 3D + 2 = 0$ $or, (D-2)(D-1) = 0$ *where* $D = 2,1.$

Thus, $C.F. = A\,e^x + Be^{2x}.$

Now, $$P.I. = \frac{1}{D^2 - 3D + 2}[xe^{3x} + sin2x]$$

$$= \frac{1}{D^2 - 3D + 2} xe^{3x} + \frac{1}{D^2 - 3D + 2} sin2x$$

$$= e^{3x} \frac{1}{(D+3)^2 - 3(D+3) + 2} x + \frac{1}{-(2)^2 - 3D + 2} \sin 2x$$

$$= e^{3x} \frac{1}{D^2 + 3D + 2} x - \frac{3D - 2}{9D^2 - 4} \sin 2x$$

$$= \frac{e^{3x}}{2}(1 + \frac{3D + D^2}{2})^{-1} x - \frac{3D - 2}{-9.2^2 - 4} \sin 2x$$

$$= \frac{e^{3x}}{2}\left(1 - \frac{3D + D^2}{2} + ..\right) x + \frac{1}{40}(3D - 2)sin2x$$

$$= \frac{e^{3x}}{2}(x - \frac{3}{2}) + \frac{1}{20}(3cos2x - sin2x).$$

Hence, the complete solution is:

$$y = Ae^x + Be^{2x} + \frac{e^{3x}}{2}(x - \frac{3}{2}) + \frac{1}{20}(3cos2x - sin2x).$$

Example 1.6.4. Solve $\frac{d^2y}{dx^2} - 2\frac{dy}{dx} + y = xe^x sinx$

Solution: The equation in symbolic form is:

$$(D^2 - 2D + 1)y = xe^x sinx.$$

The roots are 1,1.
Therefore, C.F. is $(A + Bx)e^x$.

$$P.I. = \frac{1}{(D^2 - 2D + 1)} xe^x sinx$$

$$= e^x \frac{1}{(D - 1 + 1)^2} xsinx$$

$$= e^x \frac{1}{D^2} xsinx = e^x \frac{1}{D} \int xsinx\, dx = e^x \frac{1}{D}(x. - cosx + \int cosxdx)$$

$$= e^x \frac{1}{D}(x. - cosx + sinx) = e^x(-cosx - xsinx - cosx)$$

$$= -e^x(xsinx + 2cosx).$$

The complete solution is

$$y = (A + Bx)e^x - e^x(xsinx + 2cosx).$$

Example 1.6.5. Solve $\frac{d^4y}{dx^4} + 2\frac{d^2y}{dx^2} + y = x^2 cosx.$

Solution: The equation in symbolic form is:

$$(D^4 + 2D^2 + 1)y = x^2 cosx.$$

The auxiliary equation is $m^4 + 2m^2 + 1 = 0$. $\therefore$ m $= \pm i, \pm i$.
Therefore, C.F. is $(a + bx)cosx + (c + dx)sinx$.

$$P.I. = \frac{1}{D^4+2D^2+1}x^2 cosx$$

$$= Real\ part\ of \frac{1}{D^4+2D^2+1}x^2 e^{ix}$$

$$= Real\ part\ of\ e^{ix}\frac{1}{(D+i)^4+2(D+i)^2+1}x^2$$

$$= Real\ part\ of\ e^{ix}\frac{1}{\{D^2+2Di\}^2}x^2$$

$$[\{(D+i)^2+1\}^2 = \{D^2+2Di\}^2 = 4D^2i^2(1+\frac{D}{2i})^2]$$

$$= Real\ part\ of \frac{e^{ix}}{-4}\frac{1}{D^2}(1+\frac{D}{2i})^{-2}x^2$$

$$= Real\ part\ of \frac{e^{ix}}{-4}\left[\frac{1}{D^2}+\frac{i}{D}-\frac{3}{4}-\frac{iD}{2}+\frac{5}{16}D^2\right]x^2$$

$$= Real\ part\ of \frac{e^{ix}}{-4}\left[\frac{x^4}{12}+\frac{ix^3}{3}-\frac{3}{4}x^2-ix+\frac{5}{8}\right]$$

$$= -\frac{1}{4}\left[\frac{x^4}{12}cosx-\frac{x^3}{3}sinx-\frac{3}{4}x^2cosx+xsinx+\frac{5}{8}cosx\right].$$

1.7 Other Methods of Finding P.I.

1.7.1 Method of Variation of Parameters

This method applies to the equations of the form

$$y'' + py' + qy = X,$$

where $p, q,$ *and* X *are the functions* of x,

and $\text{P.I.} = -y_1\int\frac{y_2X}{W}dx + y_2\int\frac{y_1X}{W}dx$, where y_1 *and* y_2 *are the solutions of* $y''+py'+qy=0$, and $W = \begin{vmatrix} y_1 & y_2 \\ y_2' & y_2' \end{vmatrix}$ is called the Wronskian of y_1 *and* y_2.

Example 1.7.1. Using the method of variation of parameters, solve

$$\frac{d^2y}{dx^2} + 4y = tan2x.$$

Solution: The symbolic form of the equation is $(D^2+4)y = tan2x$.

The auxiliary equation is $D^2+4=0$, *roots are* $2i, -2i$.

The C.F. is $y = Acos2x + B\,sin2x$.

To find P.I.:

Here, $y_1 = cos2x$, $y_2 = sin2x$ *and* $X = tan2x$.

$$W = \begin{vmatrix} y_1 & y_2 \\ y_1' & y_2' \end{vmatrix} = \begin{vmatrix} cos2x & sin2x \\ -2sin2x & 2cos2x \end{vmatrix} = 2.$$

$$\begin{aligned}
P.I. &= -y_1 \int \frac{y_2 X}{W} dx + y_2 \int \frac{y_1 X}{W} dx \\
&= -cos2x \int \frac{sin2xtan2x}{2} dx + sin2x \int \frac{cos2xtan2x}{2} dx \\
&= -\frac{1}{2} cos2x \int (sec2x - cos2x)dx + \frac{sin2x}{2} \int sin2xdx \\
&= -\frac{1}{4} cos2x[\log(sex2x + tan2x) - sin2x] - \frac{sin2xcos2x}{4} \\
&= -\frac{1}{4} cos2xlog(sex2x + tan2x).
\end{aligned}$$

Hence, the required solution is

$$y = Acos2x + B\,sin2x - \frac{1}{4} cos2xlog(sex2x + tan2x).$$

Example 1.7.2. Using the method of variation of parameters, solve

$$\frac{d^2y}{dx^2}-2\frac{dy}{dx}+y=e^x logx.$$

Solution: The auxiliary equation is $D^2-2D+1=0;$ *the roots are* 1,1.

The C.F. is $y=(A+Bx)e^x$.

Find P.I. Here, $y_1=e^x, y_2=xe^x$ *and* $X=e^x logx.$

$$W=\begin{vmatrix} y_1 & y_2 \\ y_1' & y_2' \end{vmatrix}=\begin{vmatrix} e^x & xe^x \\ e^x & (1+x)e^x \end{vmatrix}=e^{2x}.$$

$$P.I.=-y_1\int\frac{y_2X}{W}dx+y_2\int\frac{y_1X}{W}dx=-e^x\int\frac{xe^xe^x logx.}{e^{2x}}dx+xe^x\int\frac{e^xe^x logx}{e^{2x}}dx=$$

$$=-e^x\int xlogxdx+xe^x\int logxdx=\frac{1}{4}x^2e^x(2logx-3).$$

Therefore, the complete solution is

$$y=(A+Bx)e^x+\frac{1}{4}x^2e^x(2logx-3).$$

1.7.2 Cauchy's Homogeneous Linear Equation

An equation is of the form

$$x^n\frac{d^ny}{dx^n}+k_1x^{n-1}\frac{d^ny}{dx^{n-2}}+k_2x^{n-2}\frac{d^{n-2}y}{dx^{n-2}}+\cdots+k_{n-1}x\frac{dy}{dx}+k_ny=X.$$

Here, let $x=e^z$ *or* $z=logx$ and $\frac{dz}{dx}=\frac{1}{x}$.

Then, $\frac{dy}{dx}=\frac{dy}{dz}\frac{dz}{dx}=\frac{1}{x}\frac{dy}{dz}$ $\therefore x\frac{dy}{dx}=\frac{dy}{dz}$, *i.e.,* $\boldsymbol{x\frac{dy}{dx}=Dy}$, *where* $D=\frac{d}{dz}$

$$\frac{d^2y}{dx^2}=\frac{d}{dx}\frac{dy}{dx}=\frac{d}{dx}\left(\frac{1}{x}\frac{dy}{dz}\right)=-\frac{1}{x^2}\frac{dy}{dz}+\frac{1}{x}\frac{d}{dx}\left(\frac{dy}{dz}\right)$$
$$=-\frac{1}{x^2}\frac{dy}{dz}+\frac{1}{x}\frac{d}{dz}\left(\frac{dy}{dz}\right)\frac{dz}{dx}=\frac{1}{x^2}\left(\frac{d^2y}{dz^2}-\frac{dy}{dz}\right).$$
$$\therefore x^2\frac{d^2y}{dx^2}=(D^2-D)y=D(D-1)y.$$

Similarly, $x^3\frac{d^3y}{dx^3}=D(D-1)(D-2)y$ and so on.

Substituting these in the given equation, the equation with constant coefficients can be easily solved as before.

Example 1.7.3. Solve $x^2\frac{d^2y}{dx^2}+x\frac{dy}{dx}+y=logxsin(logx)$.

Solution: Let $x=e^z$ or $z=logx$ and $\frac{dz}{dx}=\frac{1}{x}$.

Now, the equation is in the symbolic form

$$[D(D-1)+D+1]y=zsinz.$$

The auxiliary equation is $D^2+1=0,$ $roots\ are\ \pm i.$

C.F. is $Acosz+Bsinz.$

$$\text{P.I.}=\frac{1}{D^2+1}zsinz=\frac{1}{D^2+1}z(imz.part\ \ of\ e^{iz})$$
$$=IP\ of\ e^{iz}\frac{1}{(D+i)^2+1}z=IP\ of\ e^{iz}\frac{1}{D^2+2Di}z$$
$$=IP\ of\ \frac{1}{2i}e^{iz}\frac{1}{D}(1-\frac{iD}{2})^{-1}z=IP\ of\ \frac{1}{2i}e^{iz}\frac{1}{D}\left(1+\frac{Di}{2}\right)z$$
$$=IP\ of\ \frac{1}{2i}e^{iz}(\frac{z^2}{2}+\frac{iz}{2})=IP\ of\ e^{iz}\left(-\frac{z^2i}{4}+\frac{z}{4}\right)$$
$$=I.P\ of\ (cosz+isinz)\left(-\frac{z^2i}{4}+\frac{z}{4}\right)=-\frac{z^2cosz}{4}+\frac{zsinz}{4}.$$

The required solution is

$$y = Acosz + Bsinz - \frac{z^2}{4cosz} + \frac{zsinz}{4}$$

$$= Acos(logx) + Bsin(logx) - \frac{(logx)^2 coz(logx)}{4} + \frac{logxsin(logx)}{4}.$$

Example 1.7.4. Solve $(D^2 + 1)y = sinx$

Solution: Here, C.F. is $D^2 + 1 = 0;\quad the\ roots\ are \pm i.$

Therefore, C.F. $= c_1 cosx + c_2 sinx.$

We assume a trail solution such as $a_1 cosx + a_2 sinx.$

Since these terms appear in the C.F., we multiply it by x and assume the trail P.I. as

$$y = x(a_1 cosx + a_2 sinx)$$

$$Dy = (a_1 + a_2 x)cosx + (a_2 - a_1 x)sinx$$

$$D^2 y = (2a_2 - a_1 x)cosx - (2a_1 + a_2 x)sinx.$$

Substituting in the equation, we get

$$2a_1 cosx - 2a_2 sinx = sinx.$$

Equating the coefficients: $2a_1 = 0, 2a_2 = -1, so\ that\ a_1 = 0\ and\ a_2 = -\frac{1}{2}.$

Therefore, P.I. $= -\frac{1}{2} xsinx.$

The required solution is $y = c_1 cosx + c_2 sinx - \frac{1}{2} xsinx.$

1.8 Differential Equation of Other Types

1.8.1 Equations Whose One Solution is Known

Consider the differential equation

$$\frac{d^2 y}{dx^2} + P\frac{dy}{dx} + Q = X,$$

where P, Q, and X are functions of x only. If $y = u(x)$ is a known solution of this equation, then put $y = uv$ in it. It reduces the equation to one of first order in $\frac{dv}{dx}$, which can be solved easily.

One Integral belonging to the C.F., can be found by inspection as follows:

1. If $1+P+Q=0$, *then* $y=e^{x}$ is a solution.
2. If $1-P+Q=0$, *then* $y=e^{-x}$ is a solution.
3. If $P+Qx=0$, *then* $y=x$ is a solution.

1.8.2 Reduction to Normal Form by Removing the First Derivative

Consider the differential equation

$$\frac{d^2y}{dx^2}+P\frac{dy}{dx}+Q=X \tag{1.18}$$

We assume $y=uv$,
where both u,v are functions of x, and none of u *and* v is a part of the complementary function of Equation (1.18).

Then, Equation (1.18) would be

$$u\frac{d^2v}{dx^2}+u\left(P+\frac{2}{u}\frac{du}{dx}\right)\frac{dv}{dx}+\left(\frac{d^2u}{dx^2}+P\frac{du}{dx}+Qu\right)v=X.$$

Removing $\frac{dv}{dx}$, we choose u in such a manner that

$$P+\frac{2}{u}\frac{du}{dx}=0$$

or,

$$u=e^{-\frac{1}{2}\int P dx}.$$

We get the equation reduced to the *normal form* as

$$\frac{d^2v}{dx^2}+Iv=S \tag{1.19}$$

where $I = Q - \frac{1}{4}P^2 - \frac{1}{2}\frac{dP}{dx}$ *and* $S = Xe^{\int Pdx}$.

If the value of I is a constant or a constant divided by x^2, then Equation (ii) can easily be solved.

1.8.3 Change of Independent Variable

Consider the differential equation

$$\frac{d^2y}{dx^2} + P\frac{dy}{dx} + Qy = X \tag{1.20}$$

To change the independent variable x *to* z, *let* $z = f(x)$.

Then, $\frac{dy}{dx} = \frac{dy}{dz}\frac{dz}{dx}$ and $\frac{d^2y}{dx^2} = \frac{dy}{dz}\frac{d^2z}{dx^2} + (\frac{dz}{dx})^2\frac{d^2y}{dx^2}$.

Substituting in Equation (i), we get

$$\frac{d^2y}{dz^2} + P_1\frac{dy}{dx} + Q_1 y = X_1 \tag{1.21}$$

where $P_1 = \frac{\frac{d^2z}{dx^2} + P\frac{dz}{dx}}{(\frac{dz}{dx})^2}$, and $Q_1 = \frac{Q}{(\frac{dz}{dx})^2}$ and $X_1 = \frac{X}{(\frac{dz}{dx})^2}$

can be solved by taking $Q_1 = constant$.

1.8.4 Total Differential Equations

(i) An ordinary differential equation of the first order and first degree involving three variables is in the form:

$$P + Q\frac{dy}{dx} + R\frac{dz}{dx} = 0 \tag{1.22}$$

where P, Q, R are function of x, y, z, and x is the independent variable.

In terms of differentials, Equation (i) can be written as

$$Pdx + Qdy + R\,dz = 0,$$

which is integrable only if

$$P\left(\frac{\partial Q}{\partial z} - \frac{\partial R}{\partial y}\right) + Q\left(\frac{\partial R}{\partial x} - \frac{\partial P}{\partial z}\right) + R\left(\frac{\partial P}{\partial y} - \frac{\partial Q}{\partial x}\right) = 0.$$

(ii) To solve, $Pdx + Qdy + R\,dz = 0.$
If the condition of integrability is satisfied, consider one of the variable, say z, as a constant so that $dz = 0.$

Then, $Pdx + Qdy = 0.$

Replace the arbitrary constant by $\varphi(x)$. Now, differentiate the integral just obtained with respect to $x, y, z.$ Finally, compare this result with the given differential equation to determine $\varphi(z)$.

Example 1.8.1. Solve $x\frac{d^2y}{dx^2} - (2x-1)\frac{dy}{dx} + (x-1)y = 0.$

Solution: It can be written as

$$\frac{d^2y}{dx^2} - \left(2 - \frac{1}{x}\right)\frac{dy}{dx} + (1 - \frac{1}{x})y = 0 \qquad \text{(i)}$$

Here, $1 + P + Q = 1 - \left(2 - \frac{1}{x}\right) + \left(1 - \frac{1}{x}\right) = 0.$

$\therefore\ y = e^x$ is part of the C.F. of (i).

Let $y = e^x v$

so that $\frac{dy}{dx} = e^x v + e^x\frac{dv}{dx}$ and $\frac{d^2y}{dx^2} = e^x v + 2e^x\frac{dv}{dx} + e^x\frac{d^2v}{dx^2}.$

Substituting it in Equation (i), we get

$$x(e^x v + 2e^x\frac{dv}{dx} + e^x\frac{d^2v}{dx^2}) - (2x-1)(e^x v + e^x\frac{dv}{dx}) + (x-1)e^x v = 0$$

or, $$x(v+2\frac{dv}{dx}+\frac{d^2v}{dx^2})-(2x-1)(v+\frac{dv}{dx})+(x-1)v=0$$

or, $$x\frac{d^2v}{dx^2}+\frac{dv}{dx}=0 \quad or \quad x\frac{dp}{dx}+p=0,\text{ where } p=\frac{dv}{dx}.$$

Integrating, we get $logp=-logx+logc$

or, $$p=\frac{c}{x} \quad or, \frac{dv}{dx}=\frac{c}{x}$$

or, $$v=clogx+k.$$

The complete solution is $y=e^x(clogx+k)$.

Example 1.8.2. Solve $\frac{d^2y}{dx^2}-\frac{2}{x}\frac{dy}{dx}+(a^2+\frac{2}{x^2})y=0$ by reducing to normal form.

Solution: Comparing with $\frac{d^2y}{dx^2}+P\frac{dy}{dx}+Qy=X,$ we get

$$P=-\frac{2}{x},\ Q=a^2+\frac{2}{x^2} \ and\ X=0.$$

We take $u=e^{-\frac{1}{2}\int Pdx}=e^{-\frac{1}{2}\int -\frac{2}{x}dx}=x.$
Now, substituting $y=uv=xv,$

the equation becomes $\frac{d^2v}{dx^2}+(Q-\frac{1}{4}P^2-\frac{1}{2}\frac{dP}{dx})v=0$

or,
$$\frac{d^2v}{dx^2}+(a^2+\frac{2}{x^2}-\frac{1}{4}\frac{4}{x^2}-\frac{1}{2}\frac{2}{x^2})v=0$$

or,
$$(D^2+a^2)v=0,\ where\ D=\frac{d}{dx}.$$

Its solution is $v=c_1cosax+c_2sinax$.
The required solution is $y=uv=x(c_1cosax+c_2sinax)$.

Example 1.8.3. Solve by changing the independent variable,

$$x\frac{d^2y}{dx^2}-\frac{dy}{dx}+4x^3y=x^5.$$

Solution: The equation is

$$\frac{d^2y}{dx^2}-\frac{1}{x}\frac{dy}{dx}+4x^2y=x^4 \qquad \text{(i)}$$

Here, $P=-\frac{1}{x}$, $Q=4x^2$ and $X=x^4$.
Choose z so that $\frac{Q}{(\frac{dz}{dx})^2}=constant\ or\ (\frac{dz}{dx})^2=4x^2$

or,
$$\frac{dz}{dx}=2x\ \ or\ z=x^2.$$

$$\therefore\frac{d^2y}{dz^2}+P_1\frac{dy}{dz}+Q_1y=X_1 \qquad \text{(ii)}$$

where $P_1 = \dfrac{\dfrac{d^2z}{dx^2} + P\dfrac{dz}{dx}}{(\dfrac{dz}{dx})^2} = \dfrac{[2+(-x^{-1})2x]}{4x^2} = 0,$

$$Q_1 = \frac{Q}{(\frac{dz}{dx})^2} = \frac{4x^2}{4x^2} = 1, \text{ and } X_1 = \frac{X}{(\frac{dz}{dx})^2} = \frac{x^4}{4x^2} = \frac{x^2}{4} = \frac{z}{4}.$$

(iii) Take the form $\dfrac{d^2y}{dz^2} + y = \dfrac{z}{4}$

or,
$$(D^2+1)y = \frac{z}{4}.$$

Its A.E. is $D^2+1=0, or\ D=\pm i.$
C.F.: $y = c_1cosz + c_2sinz.$

$$\text{P.I.} = \frac{1}{D^2+1}\frac{z}{4} = \frac{1}{4}(D^2+1)^{-1}z = \frac{1}{4}z.$$

Hence, the complete solution is

$$y = c_1cosz + c_2sinz + \frac{1}{4}z$$
$$= c_1cosx^2 + c_2sinx^2 + \frac{1}{4}x^2.$$

Example 1.8.4. Solve $P\left(\dfrac{\partial Q}{\partial z} - \dfrac{\partial R}{\partial y}\right) + Q\left(\dfrac{\partial R}{\partial x} - \dfrac{\partial P}{\partial z}\right) + R\left(\dfrac{\partial P}{\partial y} - \dfrac{\partial Q}{\partial x}\right)$

$= dx + dy + dz = 0.$

Solution: Here, $P = y+z, Q = R = 1.$

Therefore, $P\left(\frac{\partial Q}{\partial z}-\frac{\partial R}{\partial y}\right)+Q\left(\frac{\partial R}{\partial x}-\frac{\partial P}{\partial z}\right)+R\left(\frac{\partial P}{\partial y}-\frac{\partial Q}{\partial x}\right)$

$$=(y+z).0+1.(-1)+1.(1)=0.$$

Hence, it is integrable.
The given equation can be written as

$$Pdx+Qdy=0$$

or, $$(y+z)dx+d(y+z)=0$$

or, $$dx+\frac{d(y+z)}{y+z}=0.$$

Integrating, $x+\log(y+z)=C.$
This is the complete solution.

1.9 Orthogonal Trajectories

Two families of curves, such that every member of either family cuts each member of other family at right angles, are called *orthogonal trajectories* of each other.

To find the orthogonal trajectories of the family of curves $F(x,y,c)=0$:

1. Form the differential equation in the form $f\left(x,y,\frac{dy}{dx}\right)=0$ by eliminating c.
2. Replace from the differential equation $\frac{dy}{dx}$ *by* $-\frac{dx}{dy}$.
3. Solve the differential equation of orthogonal trajectories: $f\left(x,y,-\frac{dx}{dy}\right)=0.$

1.10 Examples

Example 1.10.1. Solve $\frac{dy}{dx} + x sin2y = x^3 cos^2 y.$

Solution: Dividing both sides by $cos^2 y,$ we get

$$sec^2 y \frac{dy}{dx} + 2x tany = x^3$$

[General equation reducible to *Leibnitz's linear:* $f'(y)\frac{dy}{dx} + Pf(y) = Q$ *where* P, Q are functions of x.]

Put $tany = z$ on the above equation,

$\frac{dz}{dx} = sec^2 y \frac{dy}{dx}$. Thus, we get $\frac{dz}{dx} + 2xz = x^3$.

This is Leibnitz's equation in z.

$$\text{I.F.} = e^{\int 2x dx} = e^{x^2}.$$

Hence, the solution is $ze^{x^2} = \int x^3 e^{x^2} dx + c = \frac{1}{2}(x^2 - 1)e^{x^2} + c$

$$\left[\int x^3 e^{x^2} dx = \frac{1}{2}\int te^t dt = \frac{1}{2}\left(t\, e^t - e^t\right) = \frac{1}{2}e^t (t-1) = \frac{1}{2}e^t (t-1),\right.$$

$put\ t = x^2,\ dt = 2xdx$

or, $$tany\, e^{x^2} = \frac{1}{2}(x^2 - 1)e^{x^2} + c$$

or, $$tany = \frac{1}{2}(x^2 - 1) + c\, e^{-x^2}.$$

Example 1.10.2. Solve $\left\{y\left(1 + \frac{1}{x}\right) + cosy\right\}dx + (x + logx - xsiny)dy = 0.$

Solution: Here, $M = y\left(1+\frac{1}{x}\right)+cosy,\ and\ N = x + logx - xsiny$

$$\frac{\partial M}{\partial y} = 1+\frac{1}{x} - siny = \frac{\partial N}{\partial x}.$$

Therefore, the equation is exact, and its solution is

$$\int_{(y\,const)} Mdx + \int (terms\ of\ N\ not\ containing\ x)dy = c$$

or, $\int (y+\frac{y}{x}+cosy)dx + 0 = c \quad or,\ yx + ylogx + xcosy = c.$

I.F. found by inspection: When $Mdx + Ndy = 0,\ and$

if $\dfrac{\frac{\partial M}{\partial y} - \frac{\partial N}{\partial x}}{N}$ be a function of x only$= f(x)say,$ then

$$\text{I.F.} = e^{\int f(x)dx}.$$

Example 1.10.3. Solve $\left(xy^2 - e^{\frac{1}{x^3}}\right)dx - x^2ydy = 0.$

Solution: Here, $M = xy^2 - e^{\frac{1}{x^3}},\ N = -x^2y.$

Now, $\dfrac{\frac{\partial M}{\partial y} - \frac{\partial N}{\partial x}}{N} = \dfrac{2xy + 2xy}{-x^2y} = -\dfrac{4}{x}$, a function of x only.

$$\therefore \text{I.F.} = e^{\int f(x)dx} = e^{\int -\frac{4}{x}dx} = e^{-4logx} = \frac{1}{x^4}.$$

Multiplying the equation by $\frac{1}{x^4}$, *we get*

$$\frac{1}{x^4}\left(xy^2 - e^{\frac{1}{x^3}}\right)dx - \frac{1}{x^4}x^2 y dy = 0$$

or $\left(\frac{y^2}{x^3} - \frac{e^{\frac{1}{x^3}}}{x^4}\right)dx - \frac{y}{x^2}dy = 0,$ which is an exact equation.

The solution is $\int_{(y\,const)} Mdx + \int (terms\ of\ N\ not\ containing\ x)dy = c$

or, $\int\left(\frac{y^2}{x^3} - \frac{e^{\frac{1}{x^3}}}{x^4}\right)dx = c \quad , \quad or -\frac{1}{2}y^2\frac{1}{x^2} + \frac{1}{3}e^{\frac{1}{x^3}} = c$

$[Let\ \frac{1}{x^3} = z, -3\frac{1}{x^4}dx = dz,\ or -\frac{1}{x^4}dx = \frac{1}{3}dz, \int -\frac{e^{\frac{1}{x^3}}}{x^4}dx$

$= \frac{1}{3}\int e^z dz = \frac{1}{3}e^{\frac{1}{x^3}}].$

Example 1.10.4. Solve $y - 2px = tan^{-1}(xp^2)$.
Solution: The equation is

$$y - 2px = tan^{-1}(xp^2) \tag{i}$$

Differentiating with respect to x, we get

$$p - 2p - 2x\frac{dp}{dx} = \frac{p^2 + 2px\frac{dp}{dx}}{1 + x^2 p^4}$$

or, $$p+2x\frac{dp}{dx}+p\left(p+2x\frac{dp}{dx}\right)\frac{1}{1+x^2p^4}=0$$

or, $$\left(p+2x\frac{dp}{dx}\right)\left(1+\frac{p}{1+x^2p^4}\right)=0.$$

This gives $\left(p+2x\frac{dp}{dx}\right)=0$

or, $$2\frac{dp}{p}+\frac{dx}{x}=0.$$

Integrating, $2logp+logx=logc$

or, $$xp^2=c \text{ or, } p=\sqrt{\frac{c}{x}}.$$

Eliminating p from Equation (i), we get $y-2\sqrt{\frac{c}{x}}x=tan^{-1}(x\frac{c}{x})$

or $y=2\sqrt{\frac{c}{x}}x+tan^{-1}(c)$, which is the general solution of Equation (i).

Example 1.10.5. Solve $(px-y)(py+x)=a^2p.$

Solution: Put $x^2=u \; and \; y^2=v \; so \; that \; 2xdx=du \; and \; 2ydy=dv$

$$p=\frac{dy}{dx}=\frac{\frac{dv}{2y}}{\frac{du}{2x}}=\frac{x}{y}P \text{ , } where \; P=\frac{dv}{du}.$$

The given equation becomes $\left(\frac{x}{y}P.x-y\right)\left(\frac{x}{y}P.y+x\right)=a^2\frac{x}{y}P$

or,
$$\left(u\frac{P}{y}-y\right)x(P+1)=a^2\frac{x}{y}P$$

or,
$$(uP-v)(P+1)=a^2P$$

or $uP-v=\frac{a^2P}{P+1}$ $\quad$ $or\ v=uP-\frac{a^2P}{P+1}$, which is Clairaut's form

$$[y^2=x^2\frac{yp}{x}-\frac{a^2\frac{yp}{x}}{\frac{yp}{x}+1}=xyp-\frac{a^2yp}{yp+x}=\frac{xy^2p^2+x^2yp-a^2yp}{yp+x}$$

$$=\frac{x^3P^2+x^3P-x^3P}{xP+x}=\frac{x^2P^2}{P+1}, or\ \frac{v}{u}].$$

Its solution is $v=uc-\frac{a^2c}{c+1}$

or,
$$y^2=cx^2-\frac{a^2c}{c+1}.$$

Example 1.10.6. Find the orthogonal trajectories of the family of confocal conics $\frac{x^2}{a^2}+\frac{y^2}{a^2+\lambda}=1$, where λ is the parameter.

Solution: Differentiating the given equation, we get

$$\frac{2x}{a^2}+\frac{2y}{a^2+\lambda}\frac{dy}{dx}=0$$

or, $$\frac{y}{a^2+\lambda}=-\frac{x}{a^2(\frac{dy}{dx})}, \text{ or } \frac{y^2}{a^2+\lambda}=-\frac{xy}{a^2(\frac{dy}{dx})}.$$

Substituting it in the given equation, we get

$$\frac{x^2}{a^2}-\frac{xy}{a^2\left(\frac{dy}{dx}\right)}=1, \quad or\ x^2\left(\frac{dy}{dx}\right)-a^2\left(\frac{dy}{dx}\right)=xy$$

or, $$[x^2-a^2]\left(\frac{dy}{dx}\right)=xy \qquad \text{(i)}$$

which is the equation of the given family.

Changing $\frac{dy}{dx}$ *to* $-\frac{dx}{dy}$ in Equation (i), we get $[a^2-x^2]\left(\frac{dx}{dy}\right)=xy$ as the differential equation of the orthogonal trajectories.

Separating the variables and integrating, we get

$$\int ydy=\int\frac{a^2-x^2}{x}dx+c$$

or, $$\frac{1}{2}y^2=a^2logx-\frac{1}{2}x^2+c_1$$

or, $$(x^2+y^2)=2a^2logx+c'$$

is the required orthogonal trajectory.

Example 1.10.7. Solve the method of variation of parameters $\frac{d^2y}{dx^2} - y = \frac{2}{1+e^x}$.

Solution: The differential equation is $(D^2-1) = \frac{2}{1+e^x}$.

A.E. is $D^2 - 1 = 0,\ D = \pm 1$ *and* $C.F. = c_1e^x + c_2e^{-x}$.

Here, $y_1 = e^x, y_2 = e^{-x}$ and $X = \frac{2}{1+e^x}$.

$$W = \begin{vmatrix} y_1 & y_2 \\ y_1' & y_2' \end{vmatrix} = \begin{vmatrix} e^x & e^{-x} \\ e^x & -e^{-x} \end{vmatrix} = -2.$$

$$P.I. = -y_1\int \frac{y_2X}{W}dx + y_2\int \frac{y_1X}{W}dx =$$

$$-e^x\int \frac{1}{-2}\cdot\frac{2e^{-x}}{1+e^x}dx + e^{-x}\int \frac{1}{-2}\cdot\frac{2e^x}{1+e^x}dx$$

$$= e^x\int.\frac{e^{-x}}{1+e^x}dx - e^{-x}\int \frac{e^x}{1+e^x}dx$$

$$= e^x\int\left(\frac{1}{e^x} - \frac{1}{1+e^x}\right)dx - e^{-x}\log(1+e^x)$$

$$= e^x[-e^{-x} - \int\left(\frac{e^{-x}}{1+e^{-x}}\right)dx] - e^{-x}\log(1+e^x)$$

$$= e^x[-e^{-x} + \log(1+e^{-x}]] - e^{-x}\log(1+e^x)$$

$$= -1 + e^x\log(1+e^{-x}] - e^{-x}\log(1+e^x).$$

The required solution is

$$y = c_1e^x + c_2e^{-x} - 1 + e^x\log(1+e^{-x}] - e^{-x}\log(1+e^x).$$

Example 1.10.8. Solve $x^2\frac{d^2y}{dx^2} + 4x\frac{dy}{dx} + 2y = e^{e^x}$.

Solution: Put $x = e^z, z = logx, \frac{dy}{dx} = \frac{dy}{dz}\frac{dz}{dx} = \frac{1}{x}\frac{dy}{dz}, x\frac{dy}{dx} = \frac{dy}{dz}$ *if* $\frac{d}{dz} = D$,

then $x\frac{dy}{dx} = Dy$ *and* $D(D-1)y = x^2\frac{d^2y}{dx^2}$, *and so on.*

$\therefore$ The equation takes the form $[D(D-1) + 4D + 2]y = e^{e^z}$.

A.E. is $D(D-1)+4D+2=0$, *or* $D^2+3D+2=0, (D+2)(D+1)=0\ \therefore D=-2.-1.$

$$\text{C.F.} = c_1e^{-2x}+c_2e^{-x}.$$

$$\text{P.I.} = \frac{1}{(D+2)(D+1)}e^{e^z} = \left[\frac{1}{D+1}-\frac{1}{D+2}\right]e^{e^z}.$$

Now, $\frac{1}{D+1}e^{e^z} = \frac{1}{D+1}e^{-z}e^{z}e^{e^z} = e^{-z}\frac{1}{D-1+1}e^{z}e^{e^z} = e^{-z}\frac{1}{D}e^{z}e^{e^z}$

$$= e^{-z}\int e^{e^z}d(e^z) = e^{-z}e^{e^z} = \frac{1}{x}e^{x}$$

$$\frac{1}{D+2}e^{e^z} = \frac{1}{D+2}e^{-2z}e^{2z}e^{e^z} = e^{-2z}\frac{1}{D-2+2}e^{2z}e^{e^z} = e^{-2z}\frac{1}{D}e^{2z}e^{e^z}$$

$$= e^{-2z}\int e^{e^z}e^{z}d(e^z)$$

$$= e^{-2z}\int e^{x}x\,dx \quad [\because e^z = x]$$

$$= \frac{1}{x^2}(xe^x - e^x).$$

Hence, the required solution is $y = c_1e^{-2x}+c_2e^{-x}+\frac{1}{x}e^x-\frac{1}{x^2}(xe^x-e^x).$

Example 1.10.9. A particle moves in a straight line toward a center of force $\frac{\mu}{(distance)^3}$ starting from rest at a distance '*a*'. from the center of force. Show that the time of reaching a point distance '*b*'. from the center of force is $\frac{\sqrt{\mu}}{ab}\sqrt{(a^2-b^2)}.$

Solution: The differential equation is:

$\frac{d^2x}{dt^2} = -\frac{\mu}{(x)^3}$, when the particle travels at a distance *x at a time t*

or $2\frac{dx}{dt}\frac{d^2x}{dt^2} = -2\frac{\mu}{(x)^3}\frac{dx}{dt}$ or integrating, we get

$$\left(\frac{dx}{dt}\right)^2 = -\mu \int \frac{1}{(x)^3} dx + c = \mu.2.\frac{1}{2}\left(\frac{1}{x^2}\right) + c = \frac{\mu}{x^2} + c.$$

Here, when $x = a, \frac{dx}{dt} = 0, \quad \therefore 0 = \frac{\mu}{a^2} + c, \ c = -\frac{\mu}{a^2}.$

$$\therefore \left(\frac{dx}{dt}\right)^2 = \frac{\mu}{x^2} - \frac{\mu}{a^2} = \mu\left(\frac{1}{x^2} - \frac{1}{a^2}\right) = \frac{\mu}{a^2}\frac{a^2 - x^2}{x^2}$$

When, at $x = b$, the velocity toward O $= \sqrt{\frac{\mu}{a^2}\frac{a^2 - x^2}{x^2}} = \frac{\sqrt{\mu}}{ab}\sqrt{a^2 - b^2}.$

$$\frac{dx}{dt} = -\frac{\sqrt{\mu}}{a}\frac{\sqrt{a^2 - x^2}}{x}.$$

Integrating $\sqrt{\mu}\int dt = -a\int \frac{xdx}{\sqrt{a^2 - x^2}} + c_1 = a\sqrt{a^2 - x^2} + c_1$

$$\sqrt{\mu}t = a\sqrt{a^2 - x^2} + c_1$$

when $t = 0, \ x = a, \ \therefore c_1 = 0.$

Hence, $\sqrt{\mu}t = a\sqrt{a^2 - x^2}. \ \therefore t = \frac{a}{\sqrt{\mu}}\sqrt{a^2 - x^2}.$

The time taken to arrive at B when $t = t, x = b$:

$$t = \frac{a}{\sqrt{\mu}}\sqrt{a^2 - b^2}.$$

Example 1.10.10. Solve $(1 - x^2)\frac{d^2y}{dx^2} + x\frac{dy}{dx} - y = x(1 - x^2)^{3/2}$ in terms of a known integral.

Solution: In standard form: $\frac{d^2y}{dx^2} + \frac{x}{1 - x^2}\frac{dy}{dx} - \frac{1}{1 - x^2}y = x(1 - x^2)^{1/2}.$

Here, $P=\frac{x}{1-x^2}$, $Q=-\frac{1}{1-x^2}$ *and* $P+Qx=0$.

Thus, $y=x$ is a part of the complementary function.
Now, for a complete solution, put

$$y=vx\text{ , so that }\frac{dy}{dx}=v+x\frac{dv}{dx}\text{ and }\frac{d^2y}{dx^2}=2\frac{dv}{dx}+x\frac{d^2v}{dx^2}.$$

Putting these in the given equation, we get

$$(1-x^2)(2\frac{dv}{dx}+x\frac{d^2v}{dx^2})+x(v+x\frac{dv}{dx})-vx=x(1-x^2)^{3/2}$$

or,
$$\frac{d^2v}{dx^2}+\left(\frac{2}{x}+\frac{x}{(1-x^2)}\right)\frac{dv}{dx}=(1-x^2)^{1/2}.$$

Let $p=\frac{dv}{dx}$. The above equation becomes

$$\frac{dp}{dx}+\left(\frac{2}{x}+\frac{x}{(1-x^2)}\right)p=(1-x^2)^{1/2}.$$

This is a linear equation of first order in p, whose integrating factor is

$$\frac{x^2}{\sqrt{1-x^2}}.$$

Thus, the solution is $p\frac{x^2}{\sqrt{1-x^2}}=\frac{1}{3}x^3+c_1$

or,
$$p=\frac{dv}{dx}=\frac{1}{3}x\sqrt{1-x^2}+\frac{c_1}{x^2}\sqrt{1-x^2}.$$

Integrating, we get

$$v=-\frac{1}{9}(1-x^2)^{\frac{3}{2}}-c_1\left(sin^{-1}x+\frac{1}{x}\sqrt{1-x^2}\right)+c_2.$$

Hence, the complete integral is

$$y = vx = -\frac{1}{9}x(1-x^2)^{\frac{3}{2}} - c_1\left(x\sin^{-1}x + \sqrt{1-x^2}\right) + c_2x.$$

Example 1.10.11. Solve $x\frac{d^2y}{dx^2} + (x-1)\frac{dy}{dx} - y = x^2$.

Solution: Here, let $D = \frac{d}{dx}$.

The equation is

$$[xD^2 + (x-1)D - 1]y = x^2 \tag{i}$$

or,
$$(xD-1)(D+1)y = x^2.$$

Now, let $(D+1)y = v$, *so that Equation (i)* becomes

$$(xD-1)v = x^2$$

or,
$$\frac{dv}{dx} - \frac{1}{x}v = x,$$

which is a linear equation of the first order in v, whose integrating factor $= e^{-\int \frac{1}{x}dx} = \frac{1}{x}$.

Thus, the equation is

$$v\frac{1}{x} = \int \frac{1}{x}xdx + c = x + c$$

or, $$v = x^2 + cx.$$

Substituting for v: $(D+1)y = v$. We get

$\frac{dy}{dx} + y = x^2 + cx$. This is a linear equation in y, whose I.F. $= e^{\int dx} = e^x$.

Hence, the solution is $xy = \int(x^2+cx)e^x dx + C$ or $y = x^2 + A(x-1) + Be^{-x}$, where A, B are constants.

1.11 Exercise

1. Solve (i) $\frac{dy}{dx} = \sin(x+y) + \cos(x+y)$

   (ii) $\frac{dy}{dx} = (4x+y+1)^2$, *if* $y(0) = 1$.

2. Solve (i) $\left(1 + e^{\frac{x}{y}}\right)dx + e^{\frac{x}{y}}\left(1 - \frac{x}{y}\right)dy = 0$

   (ii) $(x+2y)(dx - dy) = dx + dy$.
3. (i) Obtain the differential equation of all circles each of which touches the axis of x at the origin.
   (ii) Show that the differential equation of all parabolas, having their axes parallel to the $y-$ axes, is $\frac{d^3y}{dx^3} = 0$.
4. Find the solution of the equation
   $(x^2 + y^2 + 2x)dx + 2ydy = 0$ *when* $x = y = 1$.
5. Prove that $(x+y+1)^{-4}$ is an integrating factor of the equation
   $(2xy - y^2 - y)dx + (2xy - x^2 - x)dy = 0$ *and hence solve it.*
6. Solve, using the method of variation of parameters, the equations

   (i) $\frac{dz}{dx} + \frac{z}{x}\log z = \frac{z}{x^2}(\log z)^2$

   (ii) $\frac{d^2y}{dx^2} - 2\frac{dy}{dx} + y = e^x \log x$.

7. Reducethedifferentialequation $\frac{dz}{dx} = 1 - x(y-x) - x^3(y-x)^3$ to a linear form, and hence, solve it.
8. Solve $y = p^2 x + p$.
9. Using the transformations $u = x + y, v = x^2 + y^2$, reduce the equation

$$(x^2 + y^2)(1+p^2) - 2(x+y)(1+p)(x+yp) + (x+yp)^2 = 0$$

into Clairaut's form, and hence solve it.

10. Solve $\frac{d^2 y}{dx^2} + n^2 y = sec\, nx$.
11. Solve $(D^5 - D)y = e^x + sinx - x$.
12. Solve the equation $(D^2 - 2D + 1)y = xe^x$, by the method of undetermined coefficients.
13. Solve $(x^2 D^2 + 3xD + 1)y = \frac{1}{(1-x)^2}$, where $D = \frac{d}{dx}$.
14. Solve $sin^2 x \frac{d^2 y}{dx^2} = 2y$.
15. Solve $(1 + x + x^2)\frac{d^3 y}{dx^3} + (3 + 6x)\frac{d^2 y}{dx^2} + 6\frac{dy}{dx} = 0$.
16. Verify that e^x is a solution of the reduced equation of $(1-x)\frac{d^2 y}{dx^2} + x\frac{dy}{dx} - y = (1-x)^2$. Solve the equation after reducing it to a linear equation of the first order.
17. Solve $sin\, x \frac{d^2 y}{dx^2} - cos\, x \frac{dy}{dx} + 2y sin\, x = 0$.
18. Solve $y\frac{d^2 y}{dx^2} - (\frac{dy}{dx})^2 = y^2 logy$.
19. Solve $(D^4 + 2D^2 + 1)y = x^2 cos\, x$.
20. Solve $x^2 \frac{d^2 y}{dx^2} - 3x\frac{dy}{dx} + y = logx \frac{\sin(logx) + 1}{x}$.
21. Solve $x\frac{d^2 y}{dx^2} = \sqrt{\left[1 + (\frac{dy}{dx})^2\right]}$.

22. A particle is acted upon by a force $\mu(x+\frac{a^4}{x^3})$ per unit mass toward the origin, where x is the distance from the origin at time t. If it starts from a distance "a," show that it will arrive at the origin in time $\frac{\pi}{4\sqrt{\mu}}$.

23. Solve $\frac{d^2y}{dx^2} - cot\,x\frac{dy}{dx} - (1 - cotx)y = e^x sinx.$

2
Existence Theorems

Abstract
This chapter discusses theorems that guarantee the existence of solutions for initial value problems (IVPs) of ordinary differential equations. These theorems provide conditions under which a solution is guaranteed to exist, although they do not necessarily offer a method for finding the solution. Picard's existence and uniqueness theorem ensures that an IVP has a unique solution under certain conditions, and the Picard iteration method offers a way to approximate this solution. If the conditions of Picard's theorem—particularly the Lipschitz condition—are satisfied, the sequence of Picard iterates is guaranteed to converge to the unique solution of the IVP. The chapter also discusses singular solutions—cases where the IVP does not have a unique solution at certain points.

Keywords: Initial value problem, existence theorem, uniqueness theorem, Picard theorem, Lipschitz continuity, singular solution

2.1 Introduction

In the previous chapter, we have dealt with several techniques for solving first-order differential equations as well as higher-degree equations. We have largely avoided theoretical considerations about the existence and uniqueness of solutions of these differential equations. All though, we have not devised a general method, which asserts theoretically the existence and uniqueness. This chapter discusses in detail the approximation method of Picard to the solution of the initial value problem of first order non-linear differential equations of the type

$$y' = f(x.y),\, y(x_0) = y_0 \tag{i}$$

where $f(x.y)$ is some arbitrary function to be continuous in some neibourgood of (x_0, y_0).

Dipankar De. Ordinary Differential Equations and Special Functions, (67–100) © 2026 Scrivener Publishing LLC

The Picard's theorem gives the unique solution of the above initial value problem Equation (i) by the method of successive approximations, using integral equation

$$y = y_0 + \int_{x_0}^{x} f(t.y)dt.... \tag{ii}$$

We observed that Equations (i) and (ii) are equivalent. Since the Picard's theorem gives the local solution of Equation (i), we not only consider the continuation of the solution to large domain, it also to extend the theorem in the non local sense.

Example 2.1.1. Consider the initial value problem

$$\left|\frac{dy}{dx}\right| + |y| = 0, \quad y(0) = 1.$$

If possible, let $y \neq 0$. Then, divide by $|y|$, and integration leads to an absurd result. Hence, $y = 0$ *is* the only solution. Clearly, this solution does not satisfy the initial condition $y(0) = 1$, i.e., it has no solution.

Now, consider $\frac{dy}{dx} = x,\ y(0) = 1.$

Integrating, $y = \frac{x^2}{2} + c$, where c is an arbitrary constant.
Using the initial condition, $x = 0, y = 1, c = 1.$

Solution: $y = \frac{x^2}{2} + 1$ is the only one solution.

Again, consider the equation $\frac{dy}{dx} = \frac{y-1}{x},\quad y(0) = 1.$

Integrating, $logx + logc = \log(y-1)\ or\ y - 1 = cx.$

By putting $x = 0, y = 1, c$ cannot be determined. Thus, the given equation has an infinite solution. From the above three examples, it can be concluded that $\frac{dy}{dx} = f(x, y), y(x_0) = y_0$ may have one, only one, or more than one solutions.

These solutions of the equation lead us to some questions that are stated by theorems. Theorems that state such conditions are called "*Existence and Uniqueness theorems.*" It was found that the above examples are very simple, and investigation about their existence and uniqueness can be done by inspection without using any theorem. When the equation cannot be solved by standard methods, the Existence and Uniqueness theorem will play an important role.

We will devote this chapter to consider some of the above questions about existence and uniqueness.

2.2 Initial Value Problems and Boundary Value Problems

A mathematical problem consists of finding an unknown function satisfying an ordinary differential equation, which cannot be solved by one of the standard methods known so far.

We are interested in these solutions that satisfy some specified conditions. Since the general solution of an *nth* order equation contains n independent arbitrary constants, the maximum number of conditions that can be imposed on a solution is n. If these conditions pertain to the behavior of the solution at a specific point, they are called *initial value problems (IVPs).*

Example 2.2.1. The equations

(i) $\dfrac{d^2y}{dx^2}+\dfrac{dy}{dx}=0, \text{ and } y(0)=0, y'(0)=1$

(ii) $\dfrac{d^2y}{dx^2}+3\dfrac{dy}{dx}+5y=f(x), \text{ with } y=1, y'=2 \text{ at } x=0$

are initial value problems.

$y=\sin x$ is the solution of the problem.

On the other hand, a differential equation, with conditions prescribed at more than one point, is called *boundary value problem (BVP).*

The following equations are the examples of Boundary values problem.

Example 2.2.2.

(i) $\dfrac{d^2y}{dx^2}+\lambda(x)\dfrac{dy}{dx}+\mu(x)y=\gamma(x) \text{ with the conditions } y(x_0)=$
$a, y(x_n)=b$

(ii) $\dfrac{d^4y}{dx^4}+\lambda(x)y=\mu(x) \text{ with the conditions } y(x_0)=y'(x_0)=a$
$\text{and } y(x_n)=y'(x_0)=b$

(iii) $\dfrac{d^2y}{dx^2}-3\dfrac{dy}{dx}+2y=e^x, \text{ with } y=1 \text{at } x=0 \text{ and } y=2 \text{ at } x=1$

Example 2.2.3. Solve $\frac{d^2y}{dx^2}+y=0$ *with* $y(0)=0,\ y(l)=y_1,\ 0\le x\le l.$

Solution: A.E.: $D^2+1=0,\ D=\pm i.$

The general solution is $y=A\cos x+B\sin x$, where A,B are arbitrary constants.

Now, $y(0)=0$ *gives* $A=0$. The solution is $y=B\sin x$.

If $l\ne n\pi$, *where n is an integer, then*

$$y(l)=y_1 \text{ gives } y_1=B\sin l.$$

Hence, $B=\frac{y_1}{\sin l}$.

Thus, we have a unique solution $y=\frac{y_1}{\sin l}\sin x.$

If $l=n\pi$, *then* $y(l)=y_1=0$ gives $y=B\sin x$, which is a solution of BVP. In fact, we have infinite number of solutions for this problem.

If $l=n\pi$, *then* $y(n\pi)=y_1\ne 0$. *Let* $y_1=1$.

This gives no solution of BVP.

If a solution exists, then equality $1=B\sin n\pi=0$ must have been satisfied, which is not possible.

2.3 Picard's Method of Successive Approximation

We present a method due to E. Picard[1], which gives approximate solution curves of a differential equation passing through a given point. This method is also useful even when exact solutions are available.

Consider an initial value problem (first-order equation)

$$\frac{dy}{dx}=f(x,y),\ y(x_0)=y_0 \tag{2.1}$$

Integrating Equation (2.1) over the interval (x_0,x), we get

$$\int_{y_0}^{y}dy=\int_{x_0}^{x}f(x,y)dx \quad or,\quad y=y_0+\int_{x_0}^{x}f(x,y)dx \tag{2.2}$$

in which the unknown function y appears also under an integral sign.

[1]This is named after the French mathematician Emile Picard (1856–1941) who was professor in Paris since 1881 and is famous for his research in the theory of functions.

As *a first approximation* y_1 to the solution, we put $y = y_0$ in $f(x, y)$ and integrate Equation (2.2) giving

$$y_1 = y_0 + \int_{x_0}^{x} f(x, y_0)dx.$$

For a *second approximation* y_2, we put $y = y_1$ in $f(x, y)$ and integrate Equation (2.2) giving

$$y_2 = y_0 + \int_{x_0}^{x} f(x, y_1)dx.$$

Proceeding in this way, the *nth approximation* y_n is given by

$$y_n = y_0 + \int_{x_0}^{x} f(x, y_{n-1})dx \tag{2.3}$$

Thus, we arrive at a sequence of approximate solutions $y_1, y_2, \ldots\ldots y_n$. This procedure is called *Picard's method of successive approximations.*

Theorem 2.3.1. Let $f(x, y)$ *and* $\frac{\partial f}{\partial y}$ be continuous functions of x *and* y in a rectangular region R, with (x_0, y_0) at its center. There exists a unique solution of the initial value problem Equation (2.2) in some rectangular region $R' \subset R$, which the iterated approximate solution Equation (2.4) covers.

2.3.1 Picard's Method for Solving Simultaneous Differential Equations with Initial Conditions

Consider the equation

$$\frac{dy}{dx} = f(x, y, z) \quad \frac{dz}{dx} = g(x, y, z), \textit{ where } y = y_0, z = z_0 \textit{ when } x = x_0 \tag{2.4}$$

The *nth* approximation (y_n, z_n) to the initial value problem Equation (i) is given by

$$y_n = y_0 + \int_{x_0}^{x} f(x, y_{n-1}, z_{n-1})dx$$

and, $$z_n = z_0 + \int_{x_0}^{x} g(x, y_{n-1}, z_{n-1})dx.$$

Example 2.3.1. Using Picard's method of successive approximation, obtain the value of

$$\frac{dy}{dx} = y + x \text{ at } x = 0 \text{ when } y = 1 \text{ up to the third approximation.}$$

Solution: We have $y = 1 + \int_0^x (y+x)dx = 1 + yx + \frac{x^2}{2}$

First approximation: $y_1 = 1 + [yx + \frac{x^2}{2}]_{y=1} = 1 + x + \frac{x^2}{2}$

Second approximation: $y_2 = 1 + \int_0^x[(y_1 + x)dx = 1 + \int_0^x[(1 + 2x + \frac{x^2}{2})dx = 1 + x$ $+ x^2 + \frac{x^3}{6}$

Third approximation: $y_3 = 1 + \int_0^x[(y_2 + x)dx = 1 + \int_0^x[(1 + 2x + x^2 + \frac{x^3}{6})dx$

$$= 1 + (x + x^2 + \frac{x^3}{3} + \frac{x^4}{24})$$

Another process: We know that $y_n = y_0 + \int_{x_0}^x f(x, y_{n-1})dx.$

First approximation: $y_1 = y_0 + \int_{x_0}^x f(x, y_0)dx = 1 + \int_0^x f(x,1)dx = 1 + \int_0^x (1+x)$ $dx = 1 + x + \frac{x^2}{2}$

Second approximation: $y_2 = y_0 + \int_{x_0}^x f(x, y_1)dx = 1 + \int_0^x (1 + x + \frac{x^2}{2})dx = 1 + x$ $+ x^2 + \frac{x^3}{6}$

$$y_3 = y_0 + \int_{x_0}^x f(x, y_2)dx = y_0 + \int_0^x (1 + x + x^2 + \frac{x^3}{6})dx = 1 + x + \frac{x^2}{2} + \frac{x^3}{3} + \frac{x^4}{24}$$

Example 2.3.2. Using Picard's method of successive approximation, find the third approximation of the solution of the equation $\frac{dy}{dx} = x + y^2$, where $y = 0 \text{ when } x = 0.$

Solution: We know that $y_n = y_0 + \int_{x_0}^x f(x, y_{n-1})dx$

First approximation: $y_1 = y_0 + \int_{x_0}^{x} f(x, y_0)dx = 0 + \int_0^x (x+0)dx = \frac{x^2}{2}$.

Second approximation: $y_2 = y_0 + \int_{x_0}^{x} f(x, y_1)dx = \int_0^x x + (\frac{x^2}{2})^2 dx = \int_0^x x + \frac{x^4}{4}$
$dx = \frac{x^2}{2} + \frac{x^5}{20}$

Third approximation: $y_3 = \frac{x^2}{2} + \frac{x^5}{20} + \frac{x^{11}}{4400} + \frac{x^8}{160}$

Example 2.3.3. Using Picard's method of successive approximation, find the second approximation of the solution of equation $\frac{dy}{dx} = e^x + y^2$, where $y(0) = 1$.

Solution: We know that $y_n = y_0 + \int_{x_0}^{x} f(x, y_{n-1})dx$.

Here, $f(x, y) = e^x + y^2 \quad y_0 = 1, x_0 = 0$.

First approximation: $y_1 = y_0 + \int_{x_0}^{x} f(x, y_0)dx = 1 + \int_{x_0}^{x} f(x, y_0)dx = 1$
$+ \int_0^x (1 + e^x)dx = 1 + (x + e^x - 1) = x + e^x$

Second approximation: $y_2 = 1 + \int_0^x f(x, y_1)dx = x1 + \int_0^x (e^x + y_1^2)dx = 1$
$+ \int_0^x [e^x + (x + e^x)^2]dx = 1 + \int_0^x (e^x + e^{2x} + 2xe^x + x^2)dx = 1 + \frac{x^3}{3} + e^x + \frac{1}{2}e^{2x}$
$+ 2(xe^x - e^x) + \frac{3}{2}$

Example 2.3.4. Find the value of $y \ for \ x = 0.1$ by Picard's method, given that

$$\frac{dy}{dx} = \frac{y-x}{y+x}, y(0) = 1.$$

Solution: We have $y = 1 + \int_0^x \frac{y-x}{y+x}dx$.

We know that $y_n = y_0 + \int_{x_0}^{x} f(x, y_{n-1})dx$.

First approximation: $y_1 = y_0 + \int_{x_0}^{x} f(x, y_0)dx = 1 + \int_0^x \frac{1-x}{1+x}dx = 1$
$+ \int_0^x (-1 + \frac{2}{1+x})dx = 1 - x + 2\log(1+x)$

Second approximation: $y_2 = y_0 + \int_{x_0}^{x} f(x, y_1)dx = 1 + \int_0^x \frac{y_1 - x}{y_1 + x}dx = 1$
$+ \int_0^x \frac{1 - x + 2\log(1+x) - x}{1 - x + 2\log(1+x) + x}dx = 1 + \int_0^x 1 - \frac{2x}{1 - x + 2\log(1+x) + x}dx$, which is difficult to integrate.

First approximation value:

$$y_1 = 1-(0.1)+2\log[1+(0.1)] = .9+2log1.1 = .9828$$

Example 2.3.5. Using Picard's method in the equation $\frac{dy}{dx} = y$, $y(0)=1$ and show that the successive approximation tends to the limit $y = e^x$, which is the exact solution.

Solution: Here, $y_1 = 1+x$

$$y_2 = 1+\int(1+x)dx = 1+x+\frac{x^2}{2}$$

$$y_3 = 1+\int(1+x)dx = 1+\int 1+x+\frac{x^2}{2}dx = 1+x+\frac{x^2}{2!}+\frac{x^3}{3!}$$

$$y_n = 1+\int(1+x)dx = 1+\int 1+x+\frac{x^2}{2}dx = 1+x+\frac{x^2}{2!}+\frac{x^3}{3!}+\cdots+\frac{x^n}{n!} \quad \text{(i)}$$

Now, by integration, $\frac{dy}{dx} = y \ or, \frac{dy}{y} = dx$

$$y = ce^x \ , \ \therefore y = e^x.$$

Given $y(0)=1$ so that $c=1$

From Equation (i), we see that as $n \to \infty$, the successive Approximation tends toward e^x, which is the exact solution.

Example 2.3.6. Apply Picard's method of successive approximation. Find, up to the second approximation, the solution of the equation $\frac{dy}{dx} = x+z, \frac{dz}{dx} = x-y^2$ given $y=2, z=1$ *when* $x=0$.

Solution: Given $\frac{dy}{dx} = x+z, \frac{dz}{dx} = x-y^2$, given $y=2, z=1$ *when* $x=0$.

Now, $f(x,y,z) = x+z, g(x,y,z) = x-y^2$, *and* $y_0 = 2, z_0 = 1 \ x_0 = 0$.

By the iteration formula:
First approximation:

$$y_1 = y_0 + \int_{x_0}^{x} f(x, y_0, z_0)dx = 2+\int_0^x f(x,2,1)dx = 2+\int_0^x (x+1)dx = 2+x+\frac{x^2}{2}$$

$$z_1 = z_0 + \int_{x_0}^{x} g(x, y_0, z_0)dx = 1+\int_0^x (x-4)dx = \frac{x^2}{2}-4x+1$$

Second approximation:

$$y_2 = y_0 + \int_{x_0}^{x} f(x, y_1, z_1)dx = 2 + \int_0^x (x + z_1)dx = 2 + \int_0^x (\frac{x^2}{2} - 3x + 1)dx = 2 + x$$
$$- \frac{3}{2}x^2 + \frac{x^3}{6}$$

$$z_2 = z_0 + \int_{x_0}^{x} g(x, y_1, z_1)dx = 1 + \int_0^x [x - (2 + x + \frac{x^2}{2})^2]dx = 1 + \int_0^x [x - (2 + x)$$
$$- 2(2 + x)\frac{x^2}{2} - \frac{x^4}{4}]dx = 1 - 4x - \frac{3x^2}{2} - x^3 - \frac{x^4}{4} - \frac{x^5}{20}$$

Third approximation: $y_3 = y_0 + \int_{x_0}^{x} f(x, y_2, z_2)dx = 2 + \int_0^x (x + z_2)dx = 2$
$$+ x - \frac{3x^2}{2} - \frac{x^3}{2} - \frac{x^4}{4} - \frac{x^5}{20} - \frac{x^6}{100}$$

$$z_3 = z_0 + \int_{x_0}^{x} g(x, y_2, z_2)dx = 1 + \int_0^x [x - (2 + x - \frac{3}{2}x^2 + \frac{x^3}{6})^2]dx$$
$$= 1 - 4x - \frac{3}{2}x^2 + \frac{5}{3}x^3 + \frac{3}{4}x^4 - \frac{1}{6}x^4 - \frac{31}{60}x^5 + \frac{1}{12}x^6 - \frac{1}{36.7}x^7$$

Example 2.3.7. Apply Picard's method of successive approximation. Find up to the 3rd approximation of the solution of the equation $\frac{dy}{dx} = z, \frac{dz}{dx} = x^2z + x^4y$, given $y = 5, z = 1 \; when \; x = 0$.

Solution:

$$\text{Given } \frac{dy}{dx} = z, \frac{dz}{dx} = x^2z + x^4y, \;\; y = 5, z = 1 \; when \; x = 0. \qquad \text{(i)}$$

The *nth* approximation (y_n, z_n) to the initial value problem Equation (i) is given by

$$y_n = y_0 + \int_{x_0}^{x} f(x, y_{n-1}, z_{n-1})dx$$

and,
$$z_n = z_0 + \int_{x_0}^{x} g(x, y_{n-1}, z_{n-1})dx,$$

where $f(x,y,z) = z, g(x,y,z) = x^2 z + x^4 y$ *and* $y_0 = 5, z_0 = 1$ *when* $x_0 = 0$.
Hence,

$$y_n = 5 + \int_{x_0}^{x} z_{n-1}dx \quad and \quad z_n = 1 + \int_{x_0}^{x} x^2 z_{n-1} + x^4 y_{n-1}dx.$$

First approximation: $y_1 = 5 + \int_0^x z_0 dx = 5 + \int_0^x 1.dx = 5 + x$

$$z_1 = 1 + \int_0^x (x^2 z_0 + x^4 y_0)dx = 1 + \int_0^x (x^2 + 5x^4)dx = 1 + \frac{x^3}{3} + x^5$$

Second approximation: $y_2 = 5 + \int_0^x z_1 dx = 5 + \int_0^x (1 + \frac{x^3}{3} + x^5)dx = 5 + x + \frac{x^4}{12} + \frac{x^6}{6}$

$$z_2 = 1 + \int_0^x (x^2 z_1 + x^4 y_1)dx = 1 + \int_0^x [(x^2\left(1 + \frac{x^3}{3} + x^5\right) + x^4(5 + x)]dx$$

$$= 1 + \int_0^x (x^2 + 5x^4 + \frac{4}{3}x^5 + x^7)dx$$

$$= 1 + \frac{x^3}{3} + x^5 + \frac{2}{9}x^6 + \frac{x^8}{8}$$

Third approximation: $y_3 = 5 + \int_0^x z_2 dx = 5 + \int_0^x (1 + \frac{x^3}{3} + x^5 + \frac{2}{9}x^6 + \frac{x^8}{8})dx$

$$= 5 + x + \frac{x^4}{12} + \frac{x^6}{6} + \frac{2}{63}x^7 + \frac{x^9}{72}$$

$$z_3 = 1 + \int_0^x (x^2 z_2 + x^4 y_2)dx = 1 + \frac{1}{3}x^3 + \frac{2}{9}x^6 + \frac{1}{8}x^8 + \frac{11}{224}x^9 + \frac{7}{264}x^{11}$$

Example 2.3.8. Apply Picard's method of successive approximation. Find up to the third approximation of the solution of the equation $\frac{dy}{dx} = z, \frac{dz}{dx} = x^3(y + z)$, given $y = 1, z = \frac{1}{2}$ *when* $x = 0$.

Solution: Given equation: $\dfrac{dy}{dx}=z, \dfrac{dz}{dx}=x^3(y+z),\ \ y=1, z=\dfrac{1}{2}\ when\ x=0.$

The *nth* approximation (y_n, z_n) to the initial value problem Equation (i) is given by

$$y_n = y_0 + \int_{x_0}^{x} f(x, y_{n-1}, z_{n-1})dx$$

and, $$z_n = z_0 + \int_{x_0}^{x} g(x, y_{n-1}, z_{n-1})dx,$$

where $f(x,y,z)=z, g(x,y,z)=x^3(y+z)\ and\ \ y_0=1, z_0=\dfrac{1}{2}\ \ when\ x_0=0.$

First approximation: $y_1 = y_0+\int_{x_0}^{x} f(x,y_0,z_0)dx = 1+\int_0^x z_0.dx = 1+\int_0^x \frac{1}{2}.dx$ $=1+\frac{1}{2}x$

$z_1 = z_0+\int_{x_0}^{x} g(x,y_0,z_0)dx = \frac{1}{2}+\int_0^x x^3(1+\frac{1}{2})dx = \frac{1}{2}+\frac{3x^4}{8}$

Second approximation: $y_2 = y_0+\int_{x_0}^{x} f(x,y_1,z_1)dx = 1+\int_0^x z_1 dx = 1$ $+\int_0^x(\frac{1}{2}+\frac{3x^4}{8})dx = 1+\frac{1}{2}x+\frac{3x^5}{40}$

$z_2 = z_0+\int_{x_0}^{x} g(x,y_1,z_1)dx = \frac{1}{2}+\int_0^x x^3(1+\frac{1}{2}x+\frac{1}{2}+\frac{3x^4}{8})dx = \frac{1}{2}+\frac{3x^4}{8}+\frac{x^5}{10}$ $+\frac{3x^8}{64}$

Third approximation: $y_3 = y_0+\int_{x_0}^{x} f(x,y_2,z_2)dx = 1+\int_0^x z_2 dx = 1+\frac{1}{2}x$ $+\frac{3x^5}{40}+\frac{x^6}{60}+\frac{x^9}{192}$

$z_3 = z_0+\int_{x_0}^{x} g(x,y_2,z_2)dx = \frac{1}{2}+\int_0^x x^3(1+\frac{1}{2}x+\frac{3x^5}{40}+\frac{1}{2}+\frac{3x^4}{8}+\frac{x^5}{10}+\frac{3x^8}{64})dx$ $=\frac{1}{2}+\frac{3x^4}{8}+\frac{3x^5}{10}+\frac{3x^8}{64}+\frac{7x^9}{300}+\frac{x^{12}}{256}$

Example 2.3.9. Apply Picard's method of successive approximation. Find up to the third approximation of the solution of the equation $\frac{d^2y}{dx^2} = xy + 1$, given $\frac{dy}{dx} = 0,\ y = 1\ when\ x = 0$.

Solution: Let $\frac{dy}{dx} = z\ \therefore \frac{dz}{dx} = \frac{d^2y}{dx^2} = xy + 1$, given $z = \frac{dy}{dx} = 0,\ y = 1\ when$ $x = 0$.

The nth approximation (y_n, z_n) to the initial value problem Equation (i) is given by

$$y_n = y_0 + \int_{x_0}^{x} f(x, y_{n-1}, z_{n-1})dx$$

and,
$$z_n = z_0 + \int_{x_0}^{x} g(x, y_{n-1}, z_{n-1})dx.$$

Here, $\frac{dy}{dx} = f(x,y,z) = z\ and\ \frac{dz}{dx} = \frac{d^2y}{dx^2} = g(x,y,z) = xy + 1\ and\ z_0 = \frac{dy}{dx}$ $= 0, y_0 = 1, x_0 = 0\ \therefore y_1 = 1 + \int_0^x f(x, y_0, z_0)dx = 1 + \int_0^x z_0 dx = 1 + \int_0^x 0.dx = 1.$

$$z_1 = z_0 + \int_{x_0}^{x} g(x, y_0, z_0)dx = 0 + \int_{x_0}^{x} g(x,1,0)dx = \int_0^x (x.1 + 1)dx = x + \frac{x^2}{2}.$$

2.4 Lipschitz Condition

A function $f(x, y)$ is said to satisfy a *Lipschitz condition in* a region $R\ in\ the\ xy$ plane if there exists a positive constant K such that

$$|f(x, y_1) - f(x, y_2)| \le K\,|y_1 - y_2| \tag{2.5}$$

for all $(x, y_1), (x, y_2) \in R$.

The constant K is called a *Lipschitz constant* for the function $f(x, y)$.

We show that the continuity of $f(x,y)$ *and* $\frac{\partial f}{\partial y}$ in R implies that f satisfies the Lipschitz condition in R.

Theorem 2.4.1. $f(x,y)$ is a real-valued function defined on R such that $\frac{\partial f}{\partial y}$ exists, which is continuous on R and

$$\left|\frac{\partial f(x,y)}{\partial y}\right| \leq K, all(x,y) \in R \, and \text{ for some } K > 0.$$

Then, f satifies a Lipschitz condition on R with Lipchitz constant K.

Proof: Let $(x,y_1),(x,y_2)$ be in R. Then, for each fixed x,

$$\left|\frac{\partial f}{\partial y}\right| \leq K \text{ over } R.$$

$$\therefore |f(x,y_1) - f(x,y_2)| = \left|\int_{y_1}^{y_2} \frac{\partial f(x,y)}{\partial y} dy\right| = \int_{y_1}^{y_2} \left|\frac{\partial f(x,y)}{\partial y}\right| |dy| \leq K \int_{y_1}^{y_2} |dy|$$

$$= K|y_1 - y_2|.$$

$$\therefore |f(x,y_1) - f(x,y_2)| \leq |K||y_1 - y_2|.$$

Thus, f satisfies the Lipschitz condition with the Lipschitz constant $K \, in \, R.$

Example 2.4.1. Give an example to show that there exist a function that do not satisfy the Lipschitz condition.

Let $f(x,y) = y^{\frac{1}{2}}$ be defined on the rectangle.
$D = \{(x,y): |x \leq 2|, |y| \leq 2\}$. Let $y_1 > 0$.

Then, $\frac{|f(x,y_1) - f(x,0)|}{|y_1 - 0|} = \frac{y_1^{\frac{1}{2}}}{y_1} = \frac{1}{y_1^{\frac{1}{2}}}$, which is unbounded as $y_1 \to 0$.

Hence, $f(x,y)$ does not satisfy the Lipschitz condition.

Example 2.4.2. Give an example to show that the existence of a partial derivative of $f(x,y)$ is not necessary for $f(x,y)$ to be a Lipschitz function.

Let $f(x,y)=|y|$ be defined on the square $D=\{(x,y):|x|\le 1,|y|\le 1\}$.

Let the partial derivative of $f(x,y)$ *with* respect to y at the point (x',y') be defined as

$$\left(\frac{\partial f}{\partial y}\right)_{(x',y')}=\lim_{k\to 0}\frac{f(x',y'+k)-f(x',y')}{k}$$

$$=\lim_{k\to 0}\frac{f(x',0+k)-f(x',0)}{k}=\lim_{k\to 0}\frac{|k|}{k},$$

which does not exist.

Thus, $\frac{\partial f}{\partial y}$ fails to exist at $(x,0)$.

Again, $\frac{|f(x,y_1)-f(x,y_2)|}{|y_1-y_2|}=\frac{\big||y_1|-|y_2|\big|}{|y_1-y_2|}\le 1, \quad \because |y_1-y_2|\ge\big||y_1|-|y_2|\big|$

showing that $f(x,y)$ satisfies the Lipschitz condition in $x\,on\,D$ with Lipschitz constant $k=1$.

Example 2.4.3. If R is defined by the rectangle $|x|\le a,|y|\le b,$ show that the function $f(x,y)=x^2+y^2$ satisfies the Lipchitz condition and find its constant.

Solution: Here, $f(x,y)=x^2+y^2$.

By Lipschitz condition, $|f(x,y_1)-f(x,y_2)|=|x^2+y_1^2-x^2-y_2^2|=|y_1^2-y_2^2|$ $=|y_1+y_2||y_1-y_2|\le 2b\,|y_1-y_2|$.

This satisfies the Lipschitz condition,
and the constant $=2b$.

2.5 Picard's Theorem: Existence and Uniqueness Theorem

In this section, we shall consider some more conditions under which the initial value problem

$$y'=f(x,y)\quad,\quad y(x_0)=y_0\quad, \tag{2.6}$$

where $f(x,y)$ is assumed to be continuous in some neighborhood of the point $(x_0.y_0)$. We shall prove that the initial value problem of Equation (2.1) have unique solutions, which can be obtained by an approximation process, provided f satisfies an additional condition known as *Lipschitz condition*, which is defined below.

2.5.1 Picard's Theorem

Statement: Let $f(x,y)$ be continuous in a region D *of the xy* plane and M be a constant such that

$$|f(x,y)| \leq M \; in\, D \tag{2.7}$$

Let $f(x,y)$ satisfy in D, the Lipschitz condition in y, i.e.,

$$|f(x,y_1) - f(x,y_2)| \leq K\,|y_1 - y_2| \tag{2.8}$$

where K is independent of x, y_1, y_2.

Let the rectangle R defined by

$$|x - x_0| \leq h \quad , |y - y_0| \leq k \tag{2.9}$$

lie in D, where $Mh < k$. Then, for $|x - x_0| \leq h$, the differential equation $\frac{dy}{dx} = f(x,y)$ has a unique solution $y = y(x)$ for which $y(x_0) = y_0$.

Proof: Let $|x - x_0| \leq h$.

We define a sequence of functions $y_1(x), y_2(x), \ldots y_n(x)$, which are successive approximations as follows:

$$\begin{aligned} y_1(x) &= y_0 + \int_{x_0}^{x} f(x,y_0)dx \\ y_2(x) &= y_0 + \int_{x_0}^{x} f(x,y_1)dx \\ &\ldots \ldots \ldots \ldots \ldots \ldots \\ y_n(x) &= y_0 + \int_{x_0}^{x} f(x,y_{n-1})dx. \end{aligned} \tag{2.10}$$

We shall divide the following proof into the following five steps:

First Step: We prove that for $x_0 - h \leq x \leq x_0 + h$, the curve $y = y_n(x)$ lies in the rectangle R, that is to say, $y_0 - k \leq x \leq y_0 + k$.

Now, from Equation (iv), $|y_1 - y_0| = \left|\int_{x_0}^{x} f(x, y_0)dx\right| \le \left|\int_{x_0}^{x} f(x, y_0)\right| |dx|$
or $|y_1 - y_0| \le M|x - x_0| \le Mh < k$ [Using Equations (2.8) and (2.10)].
$\therefore Mh < k.$
This proves the desired result for $n = 1$.

Let $y = y_n(x)$ lie in R, and so, $f(x, y_{n-1})$ is defined and continuous and satisfies

$$|f(x, y_{n-1})| \le M \text{ on } x_0 - h, x_0 + h].$$

So, we have $|y_n - y_0| = \left|\int_{x_0}^{x} f(x, y_{n-1})dx\right| \le \left|\int_{x_0}^{x} f(x, y_{n-1})\right| |dx| \le M|x - x_0| \le Mh < k$
as before, which shows that $y_n(x)$ lies in R.

By mathematical induction, the desired result is true for all n.

Second Step: We prove by induction method that

$$|y_n - y_{n-1}| \le \frac{Mk^{n-1}}{n!}|x - x_0|^n \tag{2.11}$$

We have already verified Equation (2.11) for $n = 1$ in the first step where we have shown that

$$|y_1 - y_0| \le M|x - x_0|.$$

We suppose that Equation (2.12) is true for $n = 1$.

Let,
$$|y_{n-1} - y_{n-2}| \le \frac{Mk^{n-2}}{(n-1)!}|x - x_0|^{n-1} \tag{2.12}$$

Then,
$$|y_n - y_{n-1}| = \left|\int_{x_0}^{x} f(x, y_{n-1}) - \int_{x_0}^{x} f(x, y_{n-2})\right| |dx| \tag{2.13}$$

The Lipschitz condition Equation (2.8) gives us

$$|f(x, y_{n-1}) - f(x, y_{n-2})| \le K|y_{n-1} - y_{n-2}| \tag{2.14}$$

By Equations (2.13) and (2.14), we get

$$|y_n - y_{n-1}| = \left|\int_{x_0}^{x} f(x, y_{n-1}) - \int_{x_0}^{x} f(x, y_{n-2})\right| |dx| = \int_{x_0}^{x} |f(x, y_{n-1})$$

$$- f(x, y_{n-2})||dx| \leq K\,|y_{n-1} - y_{n-2}||dx| \leq K \frac{Mk^{n-2}}{(n-1)!}|x - x_0|^{n-1}\frac{|x - x_0|}{n}$$

$$= \frac{Mk^{n-1}}{(n)!}|x - x_0|^n$$

by Equation (2.12).

Hence, by mathematical induction, we have $|y_n - y_{n-1}| \leq \frac{Mk^{n-1}}{n!}|x - x_0|^n$, which is true for all natural numbers n.

Third Step: We shall prove that the sequence $\{y_n\}$ converges uniformly to a limit for $x_0 - h \leq x \leq x_0 + h$. For the interval under consideration, $x - x_0 \leq h$.

We get from Equation (2.1), second step, $|y_n - y_{n-1}| \leq \frac{Mk^{n-1}}{n!}h^n$.

Using the infinite series,

$$y_0 + (y_1 - y_0) + (y_2 - y_1) + \cdots \ldots . + (y_n - y_{n-1}) + \cdots \leq y_0$$

$$+ Mh + \frac{Mk}{2!}h^2 + \cdots + \frac{Mk^{n-1}}{n!}h^n + \cdots$$

$$= y_0 + \frac{M}{K}\left[(1 + hK + \frac{(hK)^2}{2!} + \cdots + \frac{(hK)^n}{n!} + \ldots \infty) - 1\right]$$

$$= y_0 + \frac{M}{K}[e^{hK} - 1] \tag{2.15}$$

This series is convergent for all values of M, K, and h. Consequently, the series Equation (2.15) is definitely convergent.

By Weierstrass M-test, the series Equation (2.15) converges uniformly on $[x_0 - h, x_0 + h]$.

Since the series is continuous functions on x,

therefore, its sum $= \lim_{n\to\infty} y_n(x) = y(x), say,$

$$[y_0 + (y_1 - y_0) + (y_2 - y_1) + \cdots \ldots . + (y_n - y_{n-1}) + \cdots = y_0$$

$$+ \sum_{n=1}^{n}(y_n - y_{n-1})\,] \tag{2.16}$$

Since the terms of Equation (2.16) are continuous, $y(x)$ must be continuous.

Fourth Step: Now, we shall prove that $y = y(x)$ satifies the differential equation $\frac{dy}{dx} = f(x, y)$.

$y_n(x)$ tends uniformly to $y(x), in\, [x_0 - h, x_0 + h]$,
and by the Lipschitz condition,

$$|f(x, y) - f(x, y_n)| \leq K|y - y_n|.$$

Now, $f(x, y_n)$ tends uniformly to $f(x, y)$.
Again, from Equation (2.11), we have $y_n(x) = y_0 + \int_{x_0}^{x} f(x, y_{n-1})dx$.
Taking $n \to \infty$, $\lim_{n\to\infty} y_n(x) = y_0 + \lim_{n\to\infty} \int_{x_0}^{x} f(x, y_{n-1})dx$.
The sequence $f(x, y_n)$ consists of continuous functions on $[x_0 - h, x_0 + h]$ and converges uniformly to $f(x, y)$ on the same interval. Hence, interchanging the limiting operation given below is valid. Using Equation (2.16), we have

$$y(x) = y_0 + \int_{x_0}^{x} \lim_{n\to\infty} f(x, y_{n-1})dx$$

$$y(x) = y_0 + \int_{x_0}^{x} [f(x, y(x)]dx \tag{2.17}$$

Since the integrand on the right-hand side of Equation (2.18) is continuous in x, we conclude that it has the derivative. Thus, the limit function $y(x)$ satisfies the differential equation $\frac{dy}{dx} = f(x, y)$ on $[x_0 - h, x_0 + h]$ and $y(x_0) = y_0$.

Fifth Step: Here, we show the uniqueness of the solution, i.e., we prove that the solution $y = y(x)$ just found the only solution at $y(x_0) = y_0$.

Let $y = \Upsilon(x)$ and $y = y(x)$ of the differential equation $\frac{dy}{dx} = f(x, y)$.

Let, $$|\Upsilon(x) - y(x)| \leq K', \text{ where } x \in [x_0 - h, x_0 + h] \tag{2.18}$$

$$|\Upsilon(x) - y(x)| = \left| \int_{x_0}^{x} \{f(x, \Upsilon) - f(x, y)\} \right| |dx|$$

or, $$|\Upsilon(x) - y(x)| \leq \int_{x_0}^{x} |\{f(x, \Upsilon) - f(x, y)\}| |dx|$$

or, $$|\Upsilon(x)-y(x)|\le K\int_{x_0}^{x}|\Upsilon(x)-y(x)||dx| \tag{2.19}$$

[by Lipschitz condition: $\because |f\{x,\Upsilon(x)\}-f\{x,y(x)\}|\le K|\Upsilon(x)-y(x)|$]

or, $$|\Upsilon(x)-y(x)|\le KK'|x-x_0| \tag{2.20}$$

using Equation (2.18).

Now, substitute it in Equation (2.19). We get

$$|\Upsilon(x)-y(x)|\le K\int_{x_0}^{x}|\Upsilon(x)-y(x)||dx| = KK'\int_{x_0}^{x}|x-x_0|\,dx$$

$$=K^2K'\frac{|x-x_0|^2}{2!} \tag{2.21}$$

Again, substituting it for the integrant in Equation (2.21), we get

$$|\Upsilon(x)-y(x)|\le K\int_{x_0}^{x}|\Upsilon(x)-y(x)||dx|\le \frac{KK^2K'}{2!}\int_{x_0}^{x}|x-x_0|^2\,dx$$
$$\le \frac{K^3K'}{3!}|x-x_0|^3.$$

Proceeding in this way, we shall get

$$|\Upsilon(x)-y(x)|\le \frac{K^nK'}{n!}|x-x_0|^n\le K'\frac{(Kh)^n}{n!},$$

where $h=|x-x_0|$.

Now, the series $\Sigma_{n=0}^{n} K'\frac{(Kh)^n}{n!}$ converges, and $\lim_{n\to\infty} K'\frac{(Kh)^n}{n!}=0.$

Thus, $|\Upsilon(x)-y(x)|$ can be less; however, it may be small and implies that

$$\Upsilon(x)-y(x)=0 \text{ ie } \Upsilon(x)=y(x).$$

This shows that the solution $y=y(x)$ is a unique solution.

Hence, the proof of the theorem is complete.

Remark 2.5.1. We have shown earlier that the continuity of $f(x,y)$ *and* $\frac{\partial f}{\partial x}$ in a rectangle R implies that f satisfies a Lipschitz condition in R. We now show in the following, with the help of an example, that there exists a function $f(x,y)$ for which $\frac{\partial f}{\partial y}$ may not exist, but even so, they may satisfy a Lipschitz condition in a rectangle, and thus, by the above theorem, the initial value problem $y' = f(x,y)$, $y(x_0) = y_0$ has a unique solution in some neighborhood of x_0.

Example 2.5.1. Prove that the continuity of $f(x,y)$ is not enough to guarantee the uniqueness of the solution of the initial value problem $\frac{dy}{dx} = f(x,y) = \sqrt{|y|}$, *where* $y(0) = 0$.

Solution: The equation is

$$\frac{dy}{dx} = \sqrt{|y|} \,,\ y(0) = 0 \tag{i}$$

Here, $f(x,y) = \sqrt{|y|}$ is continuous for all y, and the solutions are as follows:

$$[\frac{dy}{\sqrt{y}} = dx \ \ or, 2\sqrt{y} = x \ \ ie \ \ y = \frac{x^2}{4},]$$

$$y = 0 \qquad\qquad y = \begin{cases} \dfrac{x^2}{4} & , when\ x \geq 0. \\ -\dfrac{x^2}{4}, x < 0, & \end{cases}$$

Now, we show that the Lipschitz condition,

$$\frac{|f(x,y_2) - f(x,y_1)|}{|y_2 - y_1|} \leq K \tag{ii}$$

does not satisfy any region, which includes the line $y = 0$.

For, example, when $y_1 = 0$ *and* y_2 is positive, then

$$\frac{|f(x,y_2) - f(x,y_1)|}{|y_2 - y_1|} = \frac{\sqrt{y_2}}{y_2} = \frac{1}{\sqrt{y_2}} \ \text{ as } \sqrt{y_2} > 0.$$

It shows in Equation (ii) that the fraction tends to large, when y_2 is chosen to be sufficiently small, which violates the inequality in Equation (ii). So, it does not satisfy the Lipschitz condition.

Example 2.5.2. Give an example that a continuous function may not satisfy the Lipschitz condition on a rectangle.

Solution: Consider a function $f(x,y) = y^{\frac{2}{3}}$ on the rectangle

$$R: |x| \leq 1, |y| \leq 1 \quad \text{(i)}$$

Clearly, $f(x,y)$ is continuous in the rectangle R.

$$\text{Here, } \left|\frac{\partial}{\partial y} f(x,y)\right| = \left|\frac{2}{3}\frac{1}{y^{1/3}}\right| \rightarrow \infty \text{ when } y \rightarrow 0 \quad \text{(ii)}$$

Since $y = 0$ is the point inside R, Equation (ii) shows that Lipschitz condition is not satisfied in R.

Example 2.5.3. Determine the interval of existence of its solution of the initial value problem:

$$\frac{dy}{dx} = y^2 + cos^2 x, y(0) = 0,$$

given that R is the rectangle containing the origin,

$$R: \left\{(x,y): \ 0 \leq x \leq a, |y| \leq b \ a > \frac{1}{2}, b > 0\right\}.$$

Solution: Let $f(x,y) = y^2 + cos^2 x$

$\therefore \ |f(x,y)| = |y^2 + cos^2 x| \leq |y|^2 + |\cos x|^2 \leq b^2 + 1 = M \ (let)$

so that $|f(x,y)| \leq M$.

Here, $\left|\frac{\partial f}{\partial y}\right| = |2y| = 2b = K.$

Therefore, $f(x,y)$ satisfies the Lipschitz condition.

We find that $y(x)$ exists for

$$0 \leq x \leq h = \min\left(a, \frac{b}{M}\right) = \min\left(a, \frac{b}{1+b^2}\right) \quad \text{(i)}$$

Now, $\dfrac{b}{1+b^2}=\dfrac{1}{b+1/b}=\dfrac{1}{\left(\dfrac{1}{\sqrt{b}}-\sqrt{b}\right)^2+2}$. The maximum value of $\dfrac{b}{1+b^2}$ is $\dfrac{1}{2}$.

Equation (i) implies that $h=\dfrac{1}{2}$, and so, $y(x)$ exists on the interval $0\le x\le\dfrac{1}{2}$.

Example 2.5.4. Give an example to show that the existence of a partial derivative of $f(x,y)$ is not necessary for $f(x,y)$ to be a Lipschitz function.

Solution: Let $f(x,y)=|y|$ be defined in $|x|\le 1\, and\, |y|\le 1$
and by definition of partial derivative $f(x,y)\; w.r.t\; y\; at\; (x_0,y_0)$

$$\left.\frac{\partial f}{\partial y}\right]_{(x_0,y_0)}=\lim_{h\to 0}\frac{f(x_0,y_0+h)-f(x_0,y_0)}{h}=\lim_{h\to 0}\frac{|y_0+h|-|y_0|}{h}=\lim_{h\to 0}\frac{|h|}{h},$$

which does not exist, i.e., $\left.\dfrac{\partial f}{\partial y}\right]_{(x_0,y_0)}$ does not exist at (x_0,y_0).

$$\text{But } \frac{|f(x,y_1)-f(x,y_2)|}{|y_1-y_2|}=\frac{\big||y_1|-|y_2|\big|}{|y_1-y_2|}\le\frac{|y_1-y_2|}{|y_1-y_2|}\le 1\ [\because |a-b|\ge \big||a|-|b|\big|\,],$$

which shows that it satisfied the Lipschitz condition.

Note: The Cauchy–Peano theorem, also known as the Peano existence theorem, guarantees an existence solution to the initial value problem.

Statement: If a function f is continuous in a neighborhood of (k,y_0), then the initial value problem has at least one solution defined in a neighborhood of k.

The Cauchy–Peano theorem is usually used to study ordinary differential equations.

But continuity of f does not ensure the uniqueness of the solution nor does it guarantee that the Picard's iterations converge to the solution. One example is illustrated here.

Example 2.5.5. Consider the initial value problem

$$y'=y^{1/3},\ y(0)=0.$$

Here, $f(x,y)=y^{1/3}$ is a continuous function on the interval rectangle $R:|x|\le a,|y|\le b,$ where $[\alpha \ge 0]$. This means that the existence of the solution is achieved.

Now, let us verify its uniqueness.

For the Lipschitz condition, assume:

$$\left|\frac{f(x,y_1)-f(x,y_2)}{y_1-y_2}\right|\le h$$

$$\left|\frac{y_1^{1/3}-y_2^{1/3}}{y_1-y_2}\right|\le h.$$

Let $y_1=\alpha \ and \ y_2=0,$ then

$$\left|\frac{y_1^{1/3}-y_2^{1/3}}{y_1-y_2}\right|\le h\Rightarrow\left|\frac{\alpha^{1/3}}{\alpha}\right|\le h,$$

i.e., $\left|\frac{1}{\alpha^{2/3}}\right|\le h.$

It is not always true because α *is not bounded.* It means f does not satisfy the Lipchitz condition.

Thus, the function does not have a unique solution.

Here, we show that $y=0$ for all x.

By separation of variable it may be seen that

$$y=\begin{cases}\pm\left(\frac{2}{3}x\right)^{\frac{3}{2}}, x\ge 0\\ 0 \qquad x\le 0\end{cases}$$

are the solutions of the above initial value problem, impling that the uniqueness of solutions is violated. It may also be observed that

$$f(x,y)=y^{1/3}$$

does not satisfy Lipschitz condition.

Example 2.5.6. Consider the initial value problem

$$y' = 3y^{2/3}\ ,\ y(0) = 0 \tag{i}$$

which has infinitely many solutions. Indeed, one can easily check that $\phi(\mathrm{t}) = 0$ and $\psi(\mathrm{t}) = t^3, \mathrm{t} \in \mathrm{R}$, are solutions.

Now, the remaining solutions are given by, let us say, a combination of those two. Specifically, if $t_1 \le 0 \le t_2$, then the function

$$\phi(\mathrm{t}) = \begin{cases} (\mathrm{t} - t_1)^3, \text{ if } \mathrm{t} < t_1 \\ 0, \text{ if } t_1 \le \mathrm{t} \le t_2 \\ (\mathrm{t} - t_2)^3, \text{ if } \mathrm{t} > t_2 \end{cases}$$

is a solution of Equation (i). The converse is true as well: every solution of Equation (i) is one of those indicated above.

Note. The *Arzela – Ascoli theorem* establishes the existence of a solution to differential equations. It is giving necessary and sufficient conditions to decide whether every sequence of a given family of real-valued continuous functions, defined on a closed and bounded interval, has a uniformly convergent subsequence. The theorem is important because it helps establish the existence of solutions to differential equations. It is also used in other branches of mathematics such as complex analysis and harmonic analysis.

2.6 Singular Solutions

A function $\psi(x)$ is called a singular solution of a differential equation

$$F(x, y, p) = 0 \tag{2.22}$$

where $p = \dfrac{dy}{dx}$,

if the uniqueness of a solution is violated at each point of the domain of the equation. Geometrically, this indicates that more than one integral curve with a common tangent line passes through each point (x_0, y_0). A singular solution of a differential equation is not described by the general integral, i.e., it cannot be derived from the general solution for any particular value of the constant.

2.6.1 Theory of Singular Solutions

Let us consider the formation of a differential equation from the primitive

$$f(x,y,c_1,c_2\ldots c_n)=0$$

or,
$$f(x,y,c)=0 \tag{2.23}$$

It is obtained by eliminating these equations between the constants and $(n-1)$ relations [differentiation with respect to x] and the equation $\frac{\partial f}{\partial x}+\frac{\partial f}{\partial y}.\frac{dy}{dx}=0.$

But suppose c is a function of x. Then, the last equation becomes

$$\frac{\partial f}{\partial x}+\frac{\partial f}{\partial y}.\frac{dy}{dx}+\frac{\partial f}{\partial c}\frac{dc}{dx}=0.$$

It is the same as before if $\frac{\partial f}{\partial c}\frac{dc}{dx}=0.$

This will happen in two factors:

Either $\frac{dc}{dx}=0$ i.e., $c=0$.

or, $\frac{\partial f}{\partial c}=0.$

The value c from this equation is substitute in f,
which is the same method to eliminate c between the equations

$$f(x,y,c)=0 \text{ and } \frac{\partial f}{\partial c}=0.$$

This eliminant can be written as

$$Dist\ f(x,y,c)=0$$

known as *discriminant of f* with regard to c.

It will be included in the following loci:

(i) The locus of all nodal points, known as *nodal locus*.
(ii) The locus of cupsidal points, known as *cupsidal locus*

and envelop of the system of curves.

Again, let us now consider the system of curves whose differential equation is

$$\varphi(x, y, p) = 0.$$

The envelope is obtained by eliminating p between

$$\varphi(x, y, p) = 0 \text{ and } \frac{\partial\varphi}{\partial p} = 0.$$

Hence, we get *Dist* $\varphi(x, y, p) = 0$,
which is known as *discriminant of* $\varphi(x, y, p)$ with regard to p.
It will be included in the following loci:

(i) The locus of all cuspidal points known as the *cuspidal locus*
(ii) The envelope locus
(iii) The locus of all points at which two curves, which are different but not consecutive, touch, known as *tac locus.*

It is observed from the above discussion that by singular solutions of a differential equation, we mean *envelope locus* only.

Theorem 2.6.1. Suppose $F(x, y, p), \frac{\partial F(x, y, p)}{\partial y}$ and $\frac{\partial F(x, y, p)}{\partial p}$ are continuous with respect to all variables x, y *and* p in a closed neighborhood $\boldsymbol{N}$ of the point $(x_0, y_0\ p_0)$, where p_0 is a real root of

$$F(x_0, y_0\ , p) = 0 \tag{2.24}$$

If $\frac{\partial F}{\partial p} \neq 0$ in N, then there exists a unique solution
$y = y(x)$ of Equation (i) satisfying
$y_0 = y(x_0)$ and $p_0 = y'(x_0)$ in $[x_0 - h, x_0 + h]$ for some $h > 0$.

2.7 Clairaut Equation

The differential equation

$$y = xy' + f(y') \tag{2.25}$$

is called a Clairaut differential equation.

Differentiating with respect to x, we get

$$y' = xy'' + y' + f'(y')y''$$

or, $\left(x + f'(y')\right)y'' = 0.$

Now, we get $y'' = 0$

and $\left(x + f'(y')\right) = 0.$

Hence, we get $y' = c.$

Putting $y' = c$ in a given equation, we get $y = xc + f(c),$ which is the required solution and called the *integral solution.*

We eliminate y' from the given differential equation and another factor, i.e., between $y = xy' + f(y')$ and $x + f'(y') = 0.$

We get another solution, which is called the *singular solution* because it does not contain any arbitrary constant.

Remark 2.7.1. We eliminate p between the Clairaut's equation and $x + f'(p),$ which gives a singular solution.

This happens if $F(x, y, p) = f(x, p) - y = 0.$

Then, $\dfrac{\partial F}{\partial p} = \dfrac{\partial f}{\partial p}.$

Differentiating the given equation with respect to $x,$ we get

$$p = f_x + f_p \frac{dp}{dx}$$

or, $f_p . p' = p - f_x,$ which must be a factor of, f_p *and* p' i.e., it is the same as F_p in this case.

Now, eliminate the p factor and the given equation: $F(x, y, p) = 0$ *and* $F_p(x, y, p) = 0,$ which is p *discriminant.*

Thus, the branch mentioned above is a branch of p discriminant, which is the reason it is a singular solution in many cases.

Remark 2.7.2. A method of obtaining the envelope of a one parameter family of solution curves is given by

$$\varphi(x, y, c) = 0 \tag{2.26}$$

of the differential Equation (2.25).

Given a point (x, y) on the envelope, there is a particular member of the family Equation (i) touching it at the point.

This point is determined by c corresponding to that member.

Hence, we can write

$$x = x(c) \; and \; y = y(c) \tag{2.27}$$

These values satisfy =also Equation (2.26):

$$\varphi\left(x(c), y(c), c\right) = 0 \tag{2.28}$$

Differentiating with respect to c, we get

$$\frac{\partial\varphi}{\partial x}\cdot\frac{dx}{dc} + \frac{\partial\varphi}{\partial y}\cdot\frac{dy}{dc} + \frac{\partial\varphi}{\partial c} = 0 \tag{2.29}$$

Since at the point of contact $\left(x(c), y(c)\right)$, the slope of the envelope and that of the particular member

$$\varphi(x, y, c) = 0 \tag{2.30}$$

are the same.

The slope is given by

$$\frac{\partial\varphi}{\partial x} + \frac{\partial\varphi}{\partial y}\cdot\frac{dy}{dx} = 0 \tag{2.31}$$

From Equation (2.35),

$$\left(\frac{\partial\varphi}{\partial x} + \frac{\partial\varphi}{\partial y}\cdot\frac{dy}{dx}\right)\frac{dx}{dc} + \frac{\partial\varphi}{\partial c} = 0 \tag{2.32}$$

By using Equation (2.31) we get from Equation (2.32)

$$\frac{\partial\varphi}{\partial c} = 0 \tag{2.33}$$

It holds for the above point of contact and, hence, each point of the envelope.

Thus, the envelope of the family Equation (2.26), if it exists, must satisfy

$$\varphi(x, y, c) = 0 \; and \; \frac{\partial \varphi}{\partial c} = 0 \tag{2.34}$$

simultaneously.

The equation

$$\psi(x, y) = 0 \tag{2.35}$$

This equation is obtained by eliminating c from Equation (2.34), which is called *c discriminant* of the family Equation (2.26).

We have established that the envelope of the family Equation (2.32), if it exists, always satisfies the *c discriminant* Equation (2.41) of the family Equation (2.32).

Note. The *c discriminant* contains the envelope locus as a factor once, the nodal locus twice, and the cusp locus thrice.

We have seen that

(i) The envelope of the system of curves $\varphi(x, y, c) = 0$ when it exists, satisfies the equation $f(x, y, p) = 0$.

Such solution of the differential equation is known as its *singular solution*.

(ii) The nodal locus will not, in general, be a solution of the differential equation.

(iii) The cuspidal locus will not also, in general, be a solution of the differential equation.

2.8 Examples

Example 2.8.1. Solve the differential equation

$$y = xp + \frac{a}{p} \tag{i}$$

Solution: Here, $F(x,y,p) = y - xp - \frac{a}{p}$

$$\frac{\partial F}{\partial p} = -x + \frac{a}{p^2}$$

Eliminating p from $F(x,y,p) = 0$ *and* $\frac{\partial F}{\partial p} = 0$,

i.e., $y - xp - \frac{a}{p} = 0$ and $-x + \frac{a}{p^2} = 0$,

we get

$$y^2 = 4ax \qquad \text{(ii)}$$

This is the singular solution.

Now, we already know that the general solution of Equation (i) is the family of straight lines

$$y = xc + \frac{a}{c} \qquad \text{(iii)}$$

At any point (h,k) *of Equation (ii)*, the slope is given by $\frac{2a}{k}$, and we get two solutions of Equation (i) passing through the point (h,k) in the same direction, i.e., *the parabola* $y^2 = 4ax$
and the straight line through (h,k) *with slope* $\frac{2a}{k}$.

The equation is $y - k = \frac{2a}{k}(x - h)$

or, $y = \frac{2a}{k}x + k - \frac{2ah}{k} = \frac{2a}{k}x + k - \frac{k}{2}$ $[\because k^2 = 4ah \text{ or } \frac{2ah}{k} = \frac{k}{2}]$

or, $y = \frac{2a}{k}x + \frac{a}{2a/k}$.

It shows that it belongs to the family of solutions Equation (iii) for which $c = \frac{2a}{k}$.

Thus, the uniqueness of solutions in a given direction is violated at each point of Equation (ii) making it a singular solution. Since $4ah = k^2$, it belongs to the family of curves Equation (iii) for which $c = \frac{2a}{k}$. Since the uniqueness of the solution is violated at each point of Equation (ii), it is a singular solution. The figure is given [Figure 2.1].

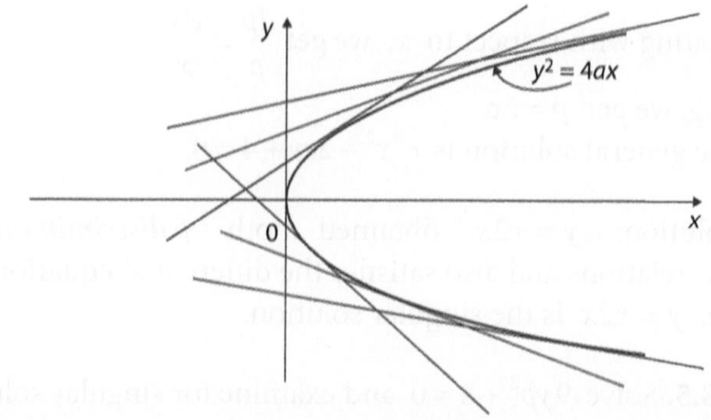

Figure 2.1

Note. We see that every member of general solution of Equation (i) *ie*, $y = xc + a/c$ touches the parabola $y^2 = 4ax$ and at any pont of the parabola it is touched by some member of $y = xc + a/c$. Thus, the singular solution is the envelope of family of lines (general solution).

Example 2.8.2. Solve the differential equation $xyp^2 + (x+y)p + 1 = 0$.

Solution. Here, p discriminant is $(x-y)^2 = 0$, which consists of a single branch $y = x$ and does not satisfy the given differential equation. Thus, the Figure 2.1

p discriminant may have branches that are not solutions at all and, hence, is not a singular solution either. The example sine p discriminant has no other branches. There is no singular solution.

Example 2.8.3. Find the singular solution, if any, of $y = 2px + p^2$.
Solution: Its p *discriminant* is

$$4x^2 + 4y = 0$$

or, $x^2 + y = 0$.

It does not satisfy the differential equation. Hence, there exists no singular solution of the equation.

Example 2.8.4. Find the general and singular solutions, if they exist, of $xp^2 - 2py + 4x = 0$.

Solution: Here, $y = \dfrac{xp^2 + 4x}{2p}$.

Differentiating with respect to x, we get $\frac{dp}{p} = \frac{dx}{x}$.

Integrating, we get $p = cx$.
Hence, the general solution is $c^2x^2 - 2cy + 4 = 0$.

Singular solution: $y = \pm 2x$ obtained both p *discriminant* and c *discriminant* relations and also satisfies the differential equation.

Therefore, $y = \pm 2x$ is the singular solution.

Example 2.8.5. Solve $9yp^2 + 4 = 0$ and examine for singular solutions and extraneous loci.

Solution: $9yp^2 + 4 = 0$

or, $y = -\frac{4}{9p^2}$.

Differentiating with respect to x:

$$p = \frac{8}{9p^3}\frac{dp}{dx}$$

or, $dx = \frac{8}{9p^4}dp$

or, $x + c = -\frac{8}{27p^3}$.

Eliminating p from this equation and given a differential equation, we get

$$y^3 + (x+c)^3 = 0.$$

p *discriminant* is $y = 0$.

c *discriminant* is $y^3 = 0$.

$y = 0$ occurs once in p *discriminant* and thrice in c *discriminant* and does not satisfy the differential equation.

So, it is a *cuspidal locus*.

The family of a semi-cubical parabola is obtained by moving $y^3 + x^2 = 0$ along the $x - axis$.

Each curve has a cusp at a point, cutting at the x *axis* and x *axis*, and is the locus of these cusps.

Example 2.8.6. Solve $(px - y)(x - py) = 2p$.

Solution: The differential equation becomes

$$p^2xy - p(x^2 + y^2 - 2) + xy = 0.$$

The p discriminant is

$$(x^2 + y^2 - 2)^2 - 4x^2y^2 = 0$$

or, $x^4 - 2x^2y^2 + y^4 - 4x^2 - 4y^2 + 4 = 0$

or, $\left(x+y+\sqrt{2}\right)\left(x+y-\sqrt{2}\right)\left(x-y+\sqrt{2}\right)\left(x-y-\sqrt{2}\right)=0.$

The primitive of the given equation is

$$(x^2 + y^2 - c(x^2 - y^2) - 1 + c^2 = 0.$$

The c discriminant is

$$(x^2 - y^2)^2 - 4(x^2 + y^2 - 1) = 0$$

or, $x^4 - 2x^2y^2 + y^4 - 4x^2 - 4y^2 + 4 = 0.$

Both p discriminant and c discriminant give the same results, and both the results satisfy the given equation. Hence, the result is a singular solution.

2.9 Exercise

1. Using Picard's method of successive approximation of the solution of the equation $\frac{dy}{dx} = x^2 - y, y = 0$, find at least the third approximation to each solution.
2. Apply Picard's method of successive approximation to the following problems, and find up to three successive approximations.

   (i) $\frac{dy}{dx} = 2xy,\ y(0) = 1$

   (ii) $\frac{dy}{dx} = 3e^x + 2y,\ y(0) = 0$

   (iii) $\frac{dy}{dx} = 1 + xy,\ y(0) = 2 =$

3. Find the third approximation of the solution of the equation $\frac{dy}{dx} = z, \frac{dz}{dx} = x^3(y+z)$ by Picard's method, where $y = 1$, $z = \frac{1}{2}$ when $x = 0$.

4. Find the third approximation of the solution of the equation $d^2y/dx^2 = x^3(y + \frac{dy}{dx}) = 0$, where $y = 1, \frac{dy}{dx} = \frac{1}{2}$ when $x = 0$.

5. If R is defined by the rectangle $|x| \le a, |y| \le b$, show that the function $f(x,y) = x sin\ y + y cos\ x$ satisfies the Lipchitz condition, and find its constant.

6. Show that the function $f(x,y) = xy^2$ satisfies the Lipschitz condition on the rectangle $R: |x| \le 1, |y| \le 1$, but does not satisfy the Lipschitz condition on the strip $S: |x| \le 1, |y| \le \infty$.

7. Consider $f(x,y) = x^3\ |y|$. Prove that f satisfies a Lipschitz condition on $R: |x| \le 2, |y| \le 2$ even though $\frac{\partial f}{\partial y}$ does not exist at $(x,0)$ if $x \ne 0$.

8 Examine the existence and uniqueness of the solution of the initial value problem $\frac{dy}{dx} = y^{\frac{1}{3}}, y(0) = 0$.

9. Examine the existence and uniqueness of the solution of the initial value problem $\frac{dy}{dx} = y^2, y(1) = -1$.

10. Verify the existence and uniqueness of the solution of the initial value problem $\frac{dy}{dx} = y^{\frac{4}{3}}, y(x_0) = y_0$.

3

System of Linear Differential Equations-I

Abstract

Physical problems often involve multiple interrelated components. Electrical circuits, mechanical systems, and similar challenges frequently exhibit this characteristic. In many such cases, the mathematical formulation consists of a system of two or more differential equations involving several variables or depending on a single independent variable. These can often be expressed as first-order differential equations.

This chapter focuses on systems of first-order linear equations with constant coefficients. It uses basic linear algebra concepts to provide a unified framework and extends the theory to matrix differential equations. The systems discussed may be homogeneous or non-homogeneous, and solution methods include elimination, matrix approaches using eigenvalues and eigenvectors.

Keywords: Matrix representation, fundamental matrix, homogeneous and non-homogeneous systems, eigenvalues, eigenvectors

3.1 Introduction

Systems of simultaneous ordinary differential equations occur naturally in problems involving multiple dependent variables, each of which is a function of a single independent variable. The independent variable will be denoted by t, and the dependent variables will be $x_1, x_2, \cdots, x_n$, which are functions of t. Differentiation with regard to t will be represented by a prime ($'$).

So, we can denote

$$x = x(t) = (x_1(t), x_2(t), \cdots, x_n(t)) \text{ and } x'(t) = (x_1'(t), x_2'(t), \cdots, x_n'(t)).$$

Let $a_{11}(t), a_{12}(t), \cdots, a_{nn}(t)$, and $f_1(t), f_2(t), \cdots, f_n(t)$ be continuous functions on the interval I.

The system of n first-order linear differential equations

Dipankar De. Ordinary Differential Equations and Special Functions, (101–140) © 2026 Scrivener Publishing LLC

$$x_1'(t) = a_{11}(t)x_1(t) + a_{12}(t)x_2(t) + \cdots + a_{1n}(t)x_n(t) + f_1(t)$$
$$x_2'(t) = a_{21}(t)x_1(t) + a_{22}(t)x_2(t) + \cdots + a_{2n}(t)x_n(t) + f_2(t)$$
$$\cdots \cdots \cdots \cdots \cdots \cdots \cdots \cdots \cdots \cdots \cdots \cdots \cdots$$
$$x_n'(t) = a_{n1}(t)x_1(t) + a_{n2}(t)x_2(t) + \cdots + a_{nn}(t)x_n(t) + f_n(t) \tag{3.1}$$

is called a *first-order linear differential system.*

The system is *homogeneous* if $f_1(t) = f_2(t) = \cdots = f_n(t) = 0$ on I. It is *non-homogeneous* if the functions $f_i(t)$ are not all identically zero on I.

This chapter extends the theory to a system of equations which gives rise to the study of matrix differential equation which will include both homogeneous and non-homogeneous type. This chapter introduces a fundamental matrix in the context of linear differential equation. It is a matrix whose columns are linearly independent solutions to a homogeneous system of differential equations and it is used for expressing the general solution of such systems and lays a vital role in solving non-homogeneous equations as well.

3.2 Matrix Form of a Linear System

Let $A(t) = \begin{pmatrix} a_{11} & a12 & \cdots & a_{1n} \\ a_{21} & a22 & \cdots & a_{2n} \\ \vdots & \vdots & \cdots & \vdots \\ a_{n1} & an2 & \cdots & a_{nn} \end{pmatrix}$ and $x(t) = \begin{pmatrix} x_1(t) \\ x_2(t) \\ \vdots \\ x_n(t) \end{pmatrix}$, $f(t) = \begin{pmatrix} f_1(t) \\ f_2(t) \\ \vdots \\ f_n(t) \end{pmatrix}$.

The first-order linear differential system can be written in the vector-matrix form

$$x'(t) = A(t)x(t) + f(t) \tag{3.2}$$

The matrix $A(t)$ is called the *matrix of coefficients* or the *coefficient matrix*. The vector $f(t)$ is called the *non-homogeneous term* or *forcing function.*

If $f(t) = 0$, we have a homogeneous first-order linear differential system $x'(t) = A(t)x(t)$.

Example 3.2.1.

(a) Write the homogeneous system $\begin{cases} x' = -5x + 5y \\ y' = -5x + y \end{cases}$ in matrix form.

(b) Write the non-homogeneous system $\begin{cases} x' = x + 2y - \sin t \\ y' = 4x - 3y + t^2 \end{cases}$ in matrix form.

Solution: (a) The homogeneous system:

$\begin{cases} x' = -5x + 5y \\ y' = -5x + y \end{cases}$ is equivalent to the system $\begin{pmatrix} x' \\ y' \end{pmatrix} = \begin{pmatrix} -5 & 5 \\ -5 & 1 \end{pmatrix} \begin{pmatrix} x \\ y \end{pmatrix}$.

The non-homogeneous system:

$\begin{cases} x' = x + 2y - \sin t \\ y' = 4x - 3y + t^2 \end{cases}$ is equivalent to the system $\begin{pmatrix} x' \\ y' \end{pmatrix} = \begin{pmatrix} 1 & 2 \\ 4 & -3 \end{pmatrix} \begin{pmatrix} x \\ y \end{pmatrix}$ $+ \begin{pmatrix} -\sin t \\ t^2 \end{pmatrix}$.

In practice, we are usually interested in solving an initial value problem; that is, we seek functions $x_1(t), x_2(t), \cdots, x_n(t)$ that satisfy not only the differential equations given by Equation (3.1) but also a set of initial conditions of the form

$$x_1(t_0) = c_1,\ x_2(t_0) = c_2, \cdots,\ x_n(t_0) = c_n. \tag{3.3}$$

where $c_1, c_2, \cdots, c_n$, and t_0 are known constants. Upon defining $C = \begin{pmatrix} c_1 \\ c_2 \\ \vdots \\ c_n \end{pmatrix}$,

it follows from the definition of $x(t)$ that $x(t_0) = \begin{pmatrix} x_1(t_0) \\ x_2(t_0) \\ \vdots \\ x_n(t_0) \end{pmatrix} = \begin{pmatrix} c_1 \\ c_2 \\ \vdots \\ c_n \end{pmatrix} = C.$

Thus, the initial conditions can be put into the matrix form

$$x(t_0) = C. \tag{3.4}$$

Definition 3.2.1. A system of differential equations is in *fundamental form* if it is given by the matrix equations

$$x'(t) = A(t)x(t) + f(t) \quad \text{and} \quad x(t_0) = C. \tag{3.5}$$

Example 3.2.2. Put the following system into fundamental form:

$$x'(t) = 2x(t) - ty(t), \quad y'(t) = t^2 x(t) - e^t \quad \text{where} \quad x(2) = 3$$
$$\text{and} \quad y(2) = 1.$$

Solution: Define $x_1(t) = x(t)$ and $x_2(t) = y(t)$.

This system is then equivalent to the matrix equations

$$\begin{pmatrix} x_1'(t) \\ x_2'(t) \end{pmatrix} = \begin{pmatrix} 2 & -t \\ t^2 & 0 \end{pmatrix} \begin{pmatrix} x_1(t) \\ x_2(t) \end{pmatrix} = \begin{pmatrix} 0 \\ e^t \end{pmatrix} \quad \text{and} \quad \begin{pmatrix} x_1(2) \\ x_2(2) \end{pmatrix} = \begin{pmatrix} 3 \\ 1 \end{pmatrix}.$$

Consequently, if we define

$$x(t) = \begin{pmatrix} x_1(t) \\ x_2(t) \end{pmatrix}, \quad A(t) = \begin{pmatrix} 2 & -t \\ t^2 & 0 \end{pmatrix}, \quad f(t) = \begin{pmatrix} 0 \\ e^t \end{pmatrix}, \quad C = \begin{pmatrix} 3 \\ 1 \end{pmatrix}$$

and $t_0 = 2$,

then, it is in fundamental form.

3.3 Reduction of an nth-Order Equation

Before seeking solutions to linear differential equations, we will first develop techniques for reducing these equations to fundamental form. In this section, we consider the initial value problems given by

$$\begin{gathered} a_n(t)\frac{d^n y}{dt^n} + a_{n-1}(t)\frac{d^{n-1}y}{dt^{n-1}} + \cdots + a_1(t)\frac{dy}{dt} + a_0(t)y = f(t) \\ x(t_0) = c_1, \quad \frac{dx(t_0)}{dt} = c_2, \cdots, \frac{d^{n-1}x(t_0)}{dt^{n-1}} = c_n \end{gathered} \tag{3.6}$$

is an nth order differential equation for $x(t)$, where $a_0(t), a_1(t), \cdots, a_n(t)$ and $f(t)$ are assumed known and continuous on some interval I containing t_0. Furthermore, we assume that $a_n(t) \neq 0$ on this interval.

A method of reduction, particularly useful for differential equations defined by a system Equation (3.6), is the following:

Step 1. Rewrite Equation (3.6) so that the nth derivative of $x(t)$ appears by itself:

$$\frac{d^n y}{dt^n} = -\frac{a_{n-1}(t)}{a_n(t)}\frac{d^{n-1}y}{dt^{n-1}} - \cdots - \frac{a_1(t)}{a_n(t)}\frac{dy}{dt} - \frac{a_0(t)}{a_n(t)}y + \frac{f(t)}{a_n(t)}. \quad (3.7)$$

Step 2. Define n new variables (the same number as the order of the differential equation), $x_1(t), x_2(t), \cdots, x_n(t)$, by the equations

$$x_1(t) = y(t), x_2(t) = \frac{dy(t)}{dt}, \; x_3(t) = \frac{d^2y(t)}{dt^2}, \cdots, \; x_{n-1}(t) = \frac{d^{n-2}y(t)}{dt^{n-2}},$$

$$x_n(t) = \frac{d^{n-1}y(t)}{dt^{n-1}}.$$

Hence, by differentiating $x_n(t)$ in the last equation, we have $x_n' = \frac{dx_n}{dt} = \frac{d^n y}{dt^n}$.

Rewriting $\frac{d^n y}{dt^n}$ in terms of the new variables $x_1, x_2, \cdots, x_n$ in Equation (3.7), we have

$$\begin{aligned} &\frac{dx_1}{dt} = x_2, \quad \frac{dx_2}{dt} = x_3, \cdots, \quad \frac{dx_{n-2}}{dt} = x_{n-1}, \quad \frac{dx_{n-1}}{dt} = x_n, \\ &\frac{dx_n}{dt} = -\frac{a_0(t)}{a_n(t)}x_1 - \frac{a_1(t)}{a_n(t)}x_2 - \cdots - \frac{a_{n-1}(t)}{a_n(t)}x_n + \frac{f(t)}{a_n(t)}, \end{aligned} \quad (3.8)$$

which can be written in matrix form as

$$x'(t)=A(t)x(t)+f(t), \tag{3.9}$$

where

$$x(t)=\begin{pmatrix} x_1 \\ x_2 \\ \vdots \\ x_n \end{pmatrix},$$

$$A(t)=\begin{pmatrix} 0 & 1 & 0 & 0 & \cdots & 0 \\ 0 & 0 & 1 & 0 & \cdots & 0 \\ 0 & 0 & 0 & 1 & \cdots & 0 \\ \vdots & \vdots & \vdots & \vdots & \vdots & \vdots \\ 0 & 0 & 0 & 0 & \cdots & 1 \\ -\frac{a_0}{a_n} & -\frac{a_1}{a_n} & -\frac{a_2}{a_n} & -\frac{a_3}{a_n} & \cdots & -\frac{a_{n-1}}{a_n} \end{pmatrix},$$

$$f(t)=\begin{pmatrix} 0 \\ 0 \\ 0 \\ \vdots \\ 0 \\ \frac{f(t)}{a_n} \end{pmatrix}.$$

Example 3.3.1. Write the equation $y''+5y+6y=\cos t$ as a system of first-order differential equations.

Solution: We let $x_1=y$ and $x_2=x_1'=y'$.

Then, $x_2' = y'' = \cos t - 6y - 5y' = \cos t - 6x_1 - 5x_2$, so the second-order equation $y'' + 5y' + 6y = \cos t$ is equivalent to the system $x_1' = x_2$, $x'2 = \cos t - 6x_1 - 5x_2$, which can be written in matrix form as

$$\begin{pmatrix} x_1 \\ x_2 \end{pmatrix}' = \begin{pmatrix} 0 & 1 \\ -6 & -5 \end{pmatrix}\begin{pmatrix} x_1 \\ x_2 \end{pmatrix} = \begin{pmatrix} 0 \\ \cos t \end{pmatrix}.$$

3.4 Matrix Preliminaries

Definition 3.4.1. The matrix $A(t) = \left[a_{ij}(t)\right], t \in I$ is said to be continuous or differentiable if every element $a_{ij}(t)$ is continuous or differentiable on I. The derivative $A'(t)$ is obtained by differentiating every element of $A(t)$ on I. That is, $A'(t) = \left[a_{ij}'(t)\right]$, $i, j = 1, 2, \ldots n$.

Definition 3.4.2. Let $A = \left[a_{ij}\right]$ denote $n \times n$ matrix, then the norm of the matrix A denoted by $|A|$ is defined as

$$|A| = \sum_{i,j=1}^{n} |a_{ij}|.$$

If $A(t) = \left[a_{ij}(t)\right], t \in I$ is continuous, then $|A(t)|$ is also continuous, since it is a sum of n^2 continuous functions.

We also get the following:

(i) $|A + B| \leq |A| + |B|$
(ii) $|AB| \leq |A|.|B|$
(iii) $|\alpha A| = |\alpha|.|A|$ for every scalar α
(iv) $|Ax| \leq |A|.|x|$ for any vector x

Definition 3.4.3. A function x, defined on some real interval I having values in C^n *or* R^n, is called a vector valued function. For each $\in I$, we associate a vector $x(t)$ *in* C^n *or* R^n, which can be written as

$$x(t) = (x_1(t), x_2(t), \ldots x_n(t)).$$

Definition 3.4.4. If $x(t)=(x_1(t),x_2(t),\dots x_n(t))$ is a vector valued function on I *to* C^n *or* R^n, then x is said to be continuous or differentiable on I, if each component $x_i(t)$ is continuous or differentiable on I.

If x is differentiable on I, then

$$x'(t)=(x_1'(t),x_2'(t),\dots x_n'(t))t\in I.$$

Definition 3.4.5. A vector valued function f *defined* for (t,x), *where* $t\in I$ *and* $x\in C^n$ *or* R^n, is said to satisfy the Lipschitz condition in t, if there exists a constant $k>0$ such that

$$|f(t,x)-f(t,y)|\le k|x-y| \text{ for all } (t,x) \textit{ in } D.$$

The constant k is called a Lipschitz constant.

Definition 3.4.6. The integral of a continuous vector valued function $x(t)=(x_1(t),x_2(t),\dots x_n(t))$ defined on an interval $I=[a,b]$ is a vector

$$\int_a^b x(t)dt=\left(\int_a^b x_1(t)dt,\int_a^b x_2(t)dt,..\int_a^b x_n(t)dt,\right).$$

Using the norm of a vector, we get

Theorem 3.4.1. $\left|\int_a^b x(t)dt\right|\le\int_a^b|x(t)|dt.$

Definition 3.4.7. A sequence (A_n) of matrices is said to be convergent, if given any $\epsilon>0$, there exists a positive integer δ such that $|A_m-A_n|<\in$ *for all* $m,n\ge\delta$.

Definition 3.4.8. A sequence (A_n) of matrices is said to tend to a limit, if given any $\epsilon>0$, there exists a positive integer δ such that $|A_n-A|<\in$ *for all* $n\ge\delta$.

Example 3.4.1. $x(t)$ is a vector valued function defined as

$$x(t)=(t,2t^2,it^3) \text{ for all } t.$$

Find (i) $x'(t)$ (ii) $\int_{-1}^{1} x(t)dt$ and verify (iii) $\left|\int_{-1}^{1} x(t)dt\right| \leq \int_{-1}^{1} |x(t)|dt$.

Solution: (i) $x(t) = (t, 2t^2, it^3)$

$$\therefore x'(t) = (1, 4t, 3it^2)$$

(ii) $\int_{-1}^{1} x(t)dt = \left(\int_{-1}^{1} tdt, \int_{-1}^{1} 2t^2dt, \int_{-1}^{1} it^3dt,\right)$

$$\Rightarrow \int_{-1}^{1} x(t)dt = \left(\left(\frac{t^2}{2}\right)_{-1}^{1}, \left(\frac{2}{3}t^3\right)_{-1}^{1}, i\left(\frac{2}{4}t^4\right)_{-1}^{1}\right) = \left(0, \frac{4}{3}, 0\right)$$

$$Hence, \left|\int_{-1}^{1} x(t)dt\right| = \frac{4}{3}$$

(iii) $|x(t)| = |t| + |2t^2| + |it^3| = |t| + 2|t^2| + |it^3|$

$$\therefore \int_{-1}^{1} |x(t)|dt = \int_{-1}^{1} \left[|t| + 2|t^2| + |it^3|\right]dt$$

$$= \int_{-1}^{0} [-t + 2t^2 - t^3]dt + \int_{0}^{1} [t + 2t^2 + t^3]dt$$

$$= \frac{1}{2} - \frac{2}{3}.1 + \frac{1}{4} + \frac{1}{2} + \frac{2}{3} + \frac{1}{4} = \frac{3}{2}$$

$$\therefore \left|\int_{-1}^{1} x(t)dt\right| \leq \int_{-1}^{1} |x(t)|\,dt, \text{ verified.}$$

At this point, given a system of ordinary differential equations, our goal is to construct either an explicit, numerical, or graphical solution of the system of equations.

3.5 Fundamental Set of Solutions

It would also seem advantageous, before trying to find the solutions, to know whether or not a given system has any solutions at all, and if it does, how many. The following subsection answers both of these questions.

3.5.1 Existence and Uniqueness

We will solve the initial value problem of the vector differential equation.

The system

$$x'(t) = A(t)x(t), \quad \text{where} \quad x(t_0) = x_0,\ t \in I$$

where $A(t)$ is an $n\times n$ matrix defined over I and $x(t)$ is a vector function on I. $x(t_0)$ is a fixed vector for a fixed t_0. We shall state and prove the following theorem in vector form.

Theorem 3.5.1. Let us consider $\mathrm{A(t)}$ as $\mathrm{n\times n}$ defined over a closed and bounded interval I. Then, the initial value problem

$$x'(t)=A(t)x(t),\quad x(t_0)=x_0,\quad t,\ t_0\in I \tag{3.10}$$

has a unique solution on I.

Proof Before proving it, we note that the IVP Equation (3.10) is equivalent to the solution of the vector integral equation

$$x(t)=x_0+\int_{t_0}^{t}A(u)x(u)du \tag{3.11}$$

given that $x(t_0)=x_0$.

By vector differentiation Equation (3.11), we get

$$x'(t)=A(t)x(t).$$

If $x(t)$ is a vector solution of Equation (3.10), then integrating, we get

$$x(t)=x(t_0)+\int_{t_0}^{t}A(u)x(u)du.$$

With the help of successive approximations, we shall show that $(x_n(t))$ is a Cauchy sequence on R^n for every $t\in I$. Since R^n is complete, $(x_n(t))$ converges uniformly to a limit $x(t)$.

Now, we prove uniqueness. Let us define the approximations by

$$x(t_0)=x_0 \text{ and } x_{n+1}=x_0+\int_{t_0}^{t}A(u)x_n(u)du,\ t\ge t_0,\ t\in I,$$

for $n=0,1,2,\ldots$

Since x_0 is a given vector, the sequence $(x_n(t))$ is well defined. First, we shall show that $(x_n(t))$ converges uniformly on I to a function $x(t)$, which is a solution of an IVP Equation (3.10), and also, we show that $(x_n(t))$ is a Cauchy sequence in R^n.

Since R^n is complete, $(x_n(t))$ converges to $x(t)$. We have to show that $x_n(t)$ is a Cauchy sequence.

Let us consider $x_0(t)+\sum_{n=1}^{\infty}\left(x_{A(u)x_n(u)n+1}(t)-x_n(t)\right)$.

Consider $x_{n+1}(t)-x_n(t)=\int_{t_0}^{t} A(u)\left[x_n(u)-x_{n-1}(u)\right]du$.

We shall construct an upper bound for $|x_{n+1}(t)-x_n(t)|$ by induction method.

Since $A(t)$ is a continuous matrix defined on I, there exists a constant M such that

$$|A(t)|\leq M. \tag{3.12}$$

Now, $x_1(t)-x_0(t)=\int_{t_0}^{t} A(u)x_0(u)du$.

So that $|x_1(t)-x_0(t)|\leq\int_{t_0}^{t}|A(u)||x_0|du$.

Using Equation (3.12) above, we get

$$|x_1(t)-x_0(t)|\leq\int_{t_0}^{t} M|x_0|(t-t_0). \tag{3.13}$$

Similarly, we have

$$x_2(t)-x_1(t)=\int_{t_0}^{t} A(u)(x_1-x_0)du,$$

$$|x_2(t)-x_1(t)|\leq\int_{t_0}^{t}|A(u)||x_1-x_0|du\leq M\int_{t_0}^{t} M|x_0||u-t_0|du$$

$$=M^2\frac{|x_0|(t-t_0)^2}{2!}.$$

Proceeding in the same process, we have by induction

$$|x_{n+1}(t)-x_n(t)|\leq M^{n+1}\frac{|x_0|(t-t_0)^{n+1}}{(n+1)!}, \tag{3.14}$$

which is the $(n+2)$th term of a convergent series $\sum_{n=1}^{\infty} M^{n+1}\frac{|x_0|(t-t_0)^{n+1}}{(n+1)!}$.
whose sum is $|x_0|e^{M(t-t_0)}$
since the nth term of the convergent series tends to 0 as $n\to\infty$, and the right-hand side of Equation (3.14) tends to zero as $n\to\infty$.

Hence, $(x_n(t))$ is a Cauchy sequence in R^n for each $t \in I$.

Since R^n is complete, $x_n(t) \to x(t)$ in the norm $|\ |$. Since the norm is convergence and uniform, $x_n \to x$ uniformly on I.

Now consider

$$x_{n+1}(t) = x_0 + \int_{t_0}^{t} A(u)x_n(u)du. \tag{3.15}$$

Taking the limit $n \to \infty$ on both sides of Equation (3.15), we have $x_{n+1}(t) \to x(t)$.

We obtain

$$x(t) = x_0 + \lim_{n \to \infty} \int_{t_0}^{t} A(u)x_n(u)du. \tag{3.16}$$

Since $x_n(t) \to x(t)$ uniformly on I, we can take the limit under the integral of Equation (3.16). We get

$$x(t) = x_0 + \int_{t_0}^{t} A(u)x(u)du,$$

which proves that $x(t)$ is the solution of the integral equation. Since the solution of the above integral equation and the IVP Equation (3.10) are equivalent, $x(t)$ gives the solution of the IVP.

For uniqueness of the solution, we proceed as follows:

If $x(t)$ is not unique, let $y(t)$ be another solution of Equation (3.10).

We get

$$x(t) = x_0 + \int_{t_0}^{t} A(u)x(u)du,$$

$$y(t) = x_0 + \int_{t_0}^{t} A(u)y(u)du.$$

$$\therefore x(t) - y(t) = \int_{t_0}^{t} A(u)x(u)du - \int_{t_0}^{t} A(u)y(u)du$$

$$= \int_{t_0}^{t} A(u)[x(u) - x(u)]du.$$

We have, $|x(t) - y(t)| \le M \int_{t_0}^{t} |x(u) - x(u)| du.$

For any $\epsilon > 0$, we get

$$|x(t) - y(t)| < \epsilon + M\int_{t_0}^{t} |x(t) - x(t)| du.$$

Let us take $v(t) = |x(t) - y(t)|$,

$$v(t) < \epsilon + M\int_{t_0}^{t} v(u) du,\ t \in I,$$

$r(t) = \epsilon + M\int_{t_0}^{t} v(u) du$ then $r(t_0) = \epsilon$ and $v(t) < r(t)$.
We get $r'(t) = Mv(t) < Mr(t)$, so that we have $r'(t) - Mr(t) < 0$.
Integrating, we get

$$r(t)e^{-\int_{t_0}^{t} Mdt} = \left[r(t)e^{-Mt}\right]_{t_0}^{t}$$

$$\text{or,}\quad r(t)e^{-Mt} - r(t_0)e^{-Mt_0} < 0$$

from which we get $r(t) < r(t_0)e^{M(t-t_0)}$.

Hence, $v(t) < r(t) < \epsilon e^{M(t-t_0)}$, since $r(t_0) = \epsilon$.

Since the inequality is true for each ϵ, we get $v(t) < 0$, which implies

$$|x(t) - y(t)| = 0,$$

which gives $x(t) = y(t)$ on I.

This proves that the solution is unique. Hence, the theorem is complete.

3.5.2 Solution Set

Definition 3.5.1. (Solution vector) A solution vector (or solution) of the linear system $x'(t) = A(t)x(t) + f(t)$ defined in Equation (3.2) on the interval I is an $n \times 1$ matrix (or vector) of the form $x(t) = (x_1(t), x_2(t), \cdots, x_n(t))$, where all $x_i(t)$ are differentiable functions that satisfy $x'(t) = A(t)x(t) + f(t)$ on I.

Now, consider the homogeneous linear system

$$x'(t) = A(t)x(t),\ t \in I \tag{3.17}$$

for which $A(t)=a_{ij}(t)$ is continuous for all $1\leq i\leq n$ and $1\leq j\leq n$. It has n linearly independent solutions on I.

Let $\{\phi_i(t)\}_{i=1}^{m}=\left\{\begin{pmatrix}\phi_{1i}\\ \phi_{2i}\\ \vdots\\ \phi_{mi}\end{pmatrix}\right\}_{i=1}^{m}$ be a set of m solutions of Equation (3.17). We define linear dependence and independence of the set of vectors $\{\phi_i(t)\}_{i=1}^{m}$ in the same way as we define linear dependence and independence of sets of functions. If the set $\{\phi_i(t)\}_{i=1}^{m}$ is *linearly dependent* on an interval I, this means that there is a set of constants $\{c_i\}_{i=1}^{m}$, not all zero, such that $\sum_{i=1}^{m}c_i\phi_i=0$; otherwise, the set is *linearly independent.*

Example 3.5.1. $x_1'=x_1+2x_2-5e^{2t}$ and $x_2'=3x_1+2x_2+3e^{2t}$. We can write the system in matrix form as

$$\begin{pmatrix}\mathrm{x}_1\\ \mathrm{x}_2\end{pmatrix}'=\begin{pmatrix}1 & 2\\ 3 & 2\end{pmatrix}\begin{pmatrix}\mathrm{x}_1\\ \mathrm{x}_2\end{pmatrix}+\begin{pmatrix}-5\mathrm{e}^{2\mathrm{t}}\\ 3\mathrm{e}^{2\mathrm{t}}\end{pmatrix}\quad\Rightarrow\quad \mathrm{x}'=\begin{pmatrix}1 & 2\\ 3 & 2\end{pmatrix}\mathrm{x}+\begin{pmatrix}-5\mathrm{e}^{2\mathrm{t}}\\ 3\mathrm{e}^{2\mathrm{t}}\end{pmatrix}.$$

We will show that $x(t)=\begin{pmatrix}-e^{2t}\\ 2e^{2t}\end{pmatrix}$ is a solution.

By taking the derivative of $x(t)$, we obtain $x'(t)=\begin{pmatrix}-2e^{2t}\\ 4e^{2t}\end{pmatrix}$.

Next, substituting $x(t)$ into the system we obtain

$$x'(t)=\begin{pmatrix}1 & 2\\ 3 & 2\end{pmatrix}\begin{pmatrix}-e^{2t}\\ 2e^{2t}\end{pmatrix}+\begin{pmatrix}-5e^{2t}\\ 3e^{2t}\end{pmatrix}=\begin{pmatrix}-e^{2t}+4e^{2t}\\ -3e^{2t}+4e^{2t}\end{pmatrix}=\begin{pmatrix}-2e^{2t}\\ 4e^{2t}\end{pmatrix}=x'(t).$$

This shows that $x(t)$ is a solution.

Definition 3.5.2. (Fundamental set of solutions)

A set $\{\phi_i(t)\}_{i=1}^{n} = \left\{ \begin{pmatrix} \phi_{1i} \\ \phi_{2i} \\ \vdots \\ \phi_{ni} \end{pmatrix} \right\}_{i=1}^{n}$ of n linearly independent solution vectors of Equation (3.17) on an interval I is called a fundamental set of solutions of Equation (3.17) on I.

In any general solution, we can take the solution as a vector function $\phi = (\phi_1, \phi_2, \ldots, \phi_n)$. We can determine if a set of vectors is linearly independent or linearly dependent by computing the Wronskian.

Definition 3.5.3. The set $\{\phi_i(t)\}_{i=1}^{n}$ is linearly independent if and only if the Wronskian $W(\phi_1, \phi_2, \ldots \phi_n) = \det \begin{pmatrix} \phi_{11} & \phi_{12} & \cdots & \phi_{1n} \\ \phi_{21} & \phi_{22} & \cdots & \phi_{2n} \\ \vdots & \vdots & \vdots & \vdots \\ \phi_{n1} & \phi_{n2} & \cdots & \phi_{nn} \end{pmatrix} \neq 0$, where

$$\phi_1(t) = \begin{pmatrix} \phi_{11}(t) \\ \phi_{21}(t) \\ \vdots \\ \phi_{n1}(t) \end{pmatrix}, \quad \phi_2(t) = \begin{pmatrix} \phi_{12}(t) \\ \phi_{22}(t) \\ \vdots \\ \phi_{n2}(t) \end{pmatrix}, \quad \ldots, \quad \phi_n(t) = \begin{pmatrix} \phi_{1n}(t) \\ \phi_{2n}(t) \\ \vdots \\ \phi_{nn}(t) \end{pmatrix}.$$

The following theorem implies that a fundamental set of solutions cannot contain more than n vectors because the solutions could not be linearly independent.

Theorem 3.5.2. Any $n+1$ nontrivial solutions of Equation (3.17) are linearly dependent.

Theorem 3.5.3. The vector functions $\phi = (\phi_1, \phi_2, \ldots, \phi_n)$ are linearly dependent on I, then the Wronskian $W(\phi_1, \phi_2, \ldots, \phi_n)(t) = 0$ for every $t \in I$.

Proof If $\phi_1, \phi_2, \ldots, \phi_n$ are linearly dependent on I, there exist n scalars $c_1, c_2, \ldots, c_n$, not all zero, such that

$$c_1\phi_1(t) + c_2\phi_2(t) + \cdots + c_n\phi_n(t) = 0 \tag{3.18}$$

for all $t \in I$.

The above single vector equation is equivalent to the following n equations in n unknown scalars $c_1, c_2, \ldots, c_n$, for any arbitrary $t_0 \in I$, i.e.,

$$
\begin{gathered}
c_1\phi_{11}(t_0)+c_2\phi_{12}(t_0)+\cdots+c_n\phi_{1n}(t_0)=0 \\
c_1\phi_{21}(t_0)+c_2\phi_{22}(t_0)+\cdots+c_n\phi_{2n}(t_0)=0 \\
\ldots\ \ldots\ \ldots\ \ldots\ \ldots\ \ldots\ \ldots\ \ldots\ \ldots\ \ldots\ \ldots \\
c_1\phi_{n1}(t_0)+c_2\phi_{n2}(t_0)+\cdots+c_n\phi_{nn}(t_0)=0.
\end{gathered}
$$

Since the scalars $c_1, c_2, \ldots, c_n$ are not all zero, the above homogeneous system has a non-trivial solution. Hence, from the theory of algebraic equations, the determinant of the coefficients of the above system of equations is zero. So we get

$$
\begin{vmatrix}
\phi_{11}(t_0) & \phi_{12}(t_0) & \ldots & \phi_{1n}(t_0) \\
\phi_{21}(t_0) & \phi_{22}(t_0) & \ldots & \phi_{2n}(t_0) \\
\ldots & \ldots & \ldots & \ldots \\
\phi_{n1}(t_0) & \phi_{n2}(t_0) & \ldots & \phi_{nn}(t_0)
\end{vmatrix} = 0,
$$

which implies $W(\phi_1, \phi_2, \ldots \phi_n)(t_0)=0$ for every $t_0 \in I$.

Hence, we have $W(\phi_1, \phi_2, \ldots \phi_n)(t_0)=0$ for all $t_0 \in I$.

Theorem 3.5.4. The vector functions $\phi = (\phi_1, \phi_2, \ldots, \phi_n)$ are the solutions of the system Equation (3.17), and the Wronskian $W(\phi_1, \phi_2, \ldots, \phi_n)(t_0)=0$ for some $t_0 \in I$, then $(\phi_1, \phi_2, \ldots, \phi_n)$ are linearly dependent on I.

Proof The system of equations

$$
\begin{gathered}
c_1\phi_{11}(t_0)+c_2\phi_{12}(t_0)+\cdots+c_n\phi_{1n}(t_0)=0 \\
c_1\phi_{21}(t_0)+c_2\phi_{22}(t_0)+\cdots+c_n\phi_{2n}(t_0)=0 \\
\ldots\ \ldots\ \ldots\ \ldots\ \ldots\ \ldots\ \ldots\ \ldots\ \ldots\ \ldots\ \ldots \\
c_1\phi_{n1}(t_0)+c_2\phi_{n2}(t_0)+\cdots+c_n\phi_{nn}(t_0)=0
\end{gathered}
$$

Is in the n-unknown scalars $c_1, c_2, \ldots, c_n$, not all zero. In the system, the determinant of coefficients is $W(\phi_1, \phi_2, \ldots, \phi_n)(t_0)$.

By assumption, let $W(\phi_1, \phi_2, \ldots, \phi_n)(t_0) = 0.$

So, the system has non-trivial solution $c_1, c_2, \ldots, c_n$, i.e., there exist constants $c_1, c_2, \ldots, c_n$ satisfying the above equations.

If $\phi_i = (\phi_{1i}, \phi_{2i}, \ldots, \phi_{ni})$, $i = 1, 2, \ldots, n$, then using the components, the above system can be written in an equivalent form as a vector equation

$$c_1\phi_1(t_0) + c_2\phi_2(t_0) + \cdots + c_n\phi_n(t_0) = 0. \tag{3.19}$$

Now, consider the vector functions ψ defined as

$$\psi(t) = c_1\phi_1(t) + c_2\phi_2(t) + \cdots + c_n\phi_n(t),\ t \in I.$$

Since $\phi_1, \phi_2, \ldots, \phi_n$ are the solutions of the systems, their linear combination ϕ is also a solution of the system. Using Equation (3.19), we have

$$\psi(t_0) = c_1\phi_1(t_0) + c_2\phi_2(t_0) + \cdots + c_n\phi_n(t_0) = 0.$$

Since $\phi_1, \phi_2, \ldots, \phi_n$ are the solutions of the system, their linear combination ψ is also a solution of the system. That is, $c_1\phi_1(t) + c_2\phi_2(t) + \cdots + c_n\phi_n(t)$ for all $t \in I$, where $c_1, c_2, \ldots, c_n$ are not all zero. Thus, using the definition, $\phi_1, \phi_2, \ldots, \phi_n$ are linearly dependent.

Theorem 3.5.5. The vector functions $\phi_1, \phi_2, \ldots, \phi_n$ are the solutions of the homogeneous linear vector differential equation $x' = A(t)x$ on I.

Then, the n-solutions are linearly independent on I if and only if the Wronskian $W(\phi_1, \phi_2, \ldots, \phi_n)(t) \neq 0$ for all $t \in I$.

Proof By Theorems 3.5.3 and 3.5.4, the solutions are linearly dependent on $[a,b]$ if and only if the Wronskian $W(\phi_1, \phi_2, \ldots, \phi_n)(t) = 0$ for all $t \in I$. Therefore, the solutions are linearly independent on I if and only if the Wronskian $W(\phi_1, \phi_2, \ldots, \phi_n)(t) \neq 0$ for some $t \in [a,b]$. We know that $W(\phi_1, \phi_2, \ldots, \phi_n)(t) \neq 0$ for some $t_0 \in I$ if and only if $W(\phi_1, \phi_2, \ldots, \phi_n)(t) \neq 0$ for all $t \in I$.

Theorem 3.5.6. Let $\phi = (\phi_1, \phi_2, \ldots, \phi_n)$ be a set of n linearly independent solutions of Equation (3.17). Then, every solution of Equation (3.17) is a linear combination of these solutions. Hence, a general solution of Equation (3.17) is

$$c_1\phi_1(t)+c_2\phi_2(t)+\cdots+c_n\phi_n(t).$$

Theorem 3.5.7. Let $(\phi_1,\phi_2,\ldots,\phi_n)$ be a fundamental set of solutions of the homogeneous linear vector differential Equation (3.17), and let ψ be any arbitrary solution on the interval I. Then, there exist unique scalars $c_1,c_2,\ldots,c_n$ such that

$$\psi(t)=c_1\phi_1(t)+c_2\phi_2(t)+\cdots+c_n\phi_n(t),\quad t\in I, \tag{3.20}$$

that is, ψ can be expressed as suitable linear combinations of the fundamental set of solutions.

Proof Let t_0 be an interior point of I, and let a solution

$$\psi(t_0)=\alpha_0,\ \psi'(t_0)=\alpha_1,\ldots,\ \psi^{(n-1)}(t_0)=\alpha_{n-1}.$$

The, we get the following system of linear equation from Equation (3.20)

$$\begin{aligned}&k_1\phi_1(t_0)+k_2\phi_2(t_0)+\cdots+k_n\phi_n(t_0)=\alpha_0\\&k_1\phi_{1'}(t_0)+k_2\phi_{2'}(t_0)+\cdots+k_n\phi_{n'}(t_0)=\alpha_1\\&\cdots\ \cdots\ \cdots\ \cdots\ \cdots\ \cdots\ \cdots\ \cdots\ \cdots\ \cdots\ \cdots\\&k_1\phi_1^{n-1}(t_0)+k_2\phi_2^{n-1}(t_0)+\cdots+k_n\phi_n^{n-1}(t_0)=\alpha_{n-1}.\end{aligned} \tag{3.21}$$

The determinant of the above equations in $k_1,k_2,\ldots,k_n$ is $W(\phi_1(t_0),\phi_2(t_1),\ldots,\phi_n(t_n)$. Since $\phi_1,\phi_2,\ldots,\phi_n$ are linearly independent solutions, therefore,

$$W(\phi_1(t_0),\phi_2(t_1),\ldots,\phi_n(t_n)\neq 0.$$

Hence, the system of Equation (3.21) has a unique solution for $k_1,k_2,\ldots,k_n$,

$$i.e.,\ k_1=c_1,\ k_2=c_2,\ldots,\ k_n=c_n$$

considering the function $f(x)=c_1\phi_1(t)+c_2\phi_2(t)+\cdots+c_n\phi_n(t)$

$$\therefore\quad f(t_0)=c_1\phi_1(t_0)+c_2\phi_2(t_0)+\cdots+c_n\phi_n(t_0)=\alpha_0$$

and, similarly, $f'(t_0)=\alpha_0, f''(t_0)=\alpha_1,\ldots, f^{n-1}(t_0)=\alpha_{n-1}$

$$\psi=\psi(t)\equiv f(t),$$

i.e., $\psi(t)=c_1\phi_1(t)+c_2\phi_2(t)+\cdots+c_n\phi_n(t)$ for all $t\in I$.

Example 3.5.2. Verify that $\phi_1(t)=\begin{pmatrix}-2e^{2t}\\ e^{2t}\end{pmatrix}$ and $\phi_2(t)=\begin{pmatrix}-e^{5t}\\ 2e^{5t}\end{pmatrix}$ are linearly independent solutions of $x'(t)=\begin{pmatrix}1 & -2\\ 2 & 6\end{pmatrix}x(t)$. Also verify, that the linear combination of ϕ_1 and ϕ_2 is a solution of the given system.

Solution:

$$\phi_1'(t)=\begin{pmatrix}-4e^{2t}\\ 2e^{2t}\end{pmatrix}\quad\text{and}\quad\begin{pmatrix}1 & -2\\ 2 & 6\end{pmatrix}\phi_1(t)=\begin{pmatrix}1 & -2\\ 2 & 6\end{pmatrix}\begin{pmatrix}-2e^{2t}\\ e^{2t}\end{pmatrix}$$

$$=\begin{pmatrix}-2e^{2t}-2e^{2t}\\ -4e^{2t}+6e^{2t}\end{pmatrix}=\begin{pmatrix}-4e^{2t}\\ 2e^{2t}\end{pmatrix}=\phi_1(t).$$

So, $\phi_1(t)$ is a solution of the given system. Similarly, we can see that ϕ_2 is a solution of the given system. Therefore, we calculate $W(\phi_1,\phi_2)=\begin{vmatrix}-2e^{2t} & -e^{5t}\\ e^{2t} & 2e^{5t}\end{vmatrix}=-3e^{7t}$. The vector-valued functions are linearly independent because $W(\phi_1,\phi_2)=-3e^{7t}\neq 0$ for all values of t.

Let $\psi(t)=c_1\begin{pmatrix}-2e^{2t}\\ e^{2t}\end{pmatrix}+c_2\begin{pmatrix}-e^{5t}\\ 2e^{5t}\end{pmatrix}=\begin{pmatrix}-2c_1e^{2t}-c_2e^{5t}\\ c_1e^{2t}+2c_2e^{5t}\end{pmatrix}$,

where c_1,c_2 are arbitrary constants. Then, $\psi'(t)=\begin{pmatrix}-4c_1 2e^{2t}-5c_2e^{5t}\\ 2c_1e^{2t}+10c_2e^{5t}\end{pmatrix}$.

Now, $\begin{pmatrix} 1 & -2 \\ 2 & 6 \end{pmatrix}\psi(t)=\begin{pmatrix} 1 & -2 \\ 2 & 6 \end{pmatrix}\begin{pmatrix} -2c_1e^{2t}-c_2e^{5t} \\ c_1e^{2t}+2c_2e^{5t} \end{pmatrix}$

$$=\begin{pmatrix} -2c_1 2e^{2t}-c_2e^{5t}-2c_1e^{2t}-4c_2e^{5t} \\ -4c_1 2e^{2t}-2c_2e^{5t}+6c_1e^{2t}+12c_2e^{5t} \end{pmatrix}$$

$$=\begin{pmatrix} -4c_1 2e^{2t}-5c_2e^{5t} \\ 2c_1e^{2t}+10c_2e^{5t} \end{pmatrix}=\phi'(t).$$

So, the linear combination of ϕ_1 and ϕ_2 is also a solution of the system.

3.5.3 Fundamental Matrix

Definition 3.5.4. (Fundamental matrix)

Let $\{\phi_i(t)\}_{i=1}^{n}=\left\{\begin{pmatrix} \phi_{1i} \\ \phi_{2i} \\ \vdots \\ \phi_{ni} \end{pmatrix}\right\}_{i=1}^{n}$ be a fundamental set of solutions for Equation (3.17).

Then, $\phi=(\phi_1, \phi_2, \ldots, \phi_n)=\begin{pmatrix} \phi_{11} & \phi_{12} & \cdots & \phi_{1n} \\ \phi_{21} & \phi_{22} & \cdots & \phi_{2n} \\ \vdots & \vdots & \vdots & \vdots \\ \phi_{n1} & \phi_{n2} & \cdots & \phi_{nn} \end{pmatrix}$, is called a fundamental matrix of the system Equation (3.17).

Thus, a general solution of the system Equation (3.17) can be written as

$x(t)=\phi(t)C$, where $C=\begin{pmatrix} c_1 \\ c_2 \\ \vdots \\ c_n \end{pmatrix}$ is a constant vector.

Theorem 3.5.8. If $\phi=(\phi_1, \phi_2, \ldots, \phi_n)$ is a fundamental matrix for Equation (3.17), then $\phi'=A\phi$.

Proof If $\phi = (\phi_1, \phi_2, \ldots, \phi_n)$, then

$$\begin{aligned} \phi' &= (\phi_1', \phi_2', \ldots, \phi_n') \\ &= (A\phi_1, A\phi_2, \ldots, A\phi_n) \\ &= A(\phi_1, \phi_2, \ldots, \phi_n) \\ &= A\phi \end{aligned}$$

Definition 3.5.5. The matrix ϕ whose columns $\phi_i(t)$ are the solution of the system Equation (3.17) is called a solution of matrix, since it satisfies the matrix Equation (3.17).

Theorem 3.5.9. Let $A(t)$ be an $n \times n$ continuous matrix on I, and let ϕ satisfy $x' = A(t)x$. Then, $\det(\phi)$ satisfies the first-order differential equation

$$(\det\phi)' = (\operatorname{tr} A)\det\phi,$$

which is equivalent to

$$\det\phi(t) = \det\phi(\tau)\exp(\textstyle\int_\tau^t \operatorname{tr}\phi(s)ds.$$

Theorem 3.5.10 I. If ϕ is a fundamental matrix of the system

$$x'(t) = A(t)x,$$

and C is a constant non-singular matrix, then ϕC is a fundamental matrix.

II. Every fundamental matrix of the system is of this type for some non-singular matrix.

Proof I. Since ϕ is a fundamental matrix of the system, then ϕ is non-singular, given C is a constant non-singular matrix, therefore, the product ϕC is non-singular.

By Theorem 3.5.8, hence, ϕC is a fundamental matrix.

II. Let ϕ_1 and ϕ_2 be two fundamental matrices.

Suppose

$$\phi_2 = \phi_1 \theta. \tag{3.22}$$

We have to show that θ is a constant non-singular matrix.
Now, by matrix differentiation,

$$\phi_2' = \phi_1' \theta + \phi_1 \theta'. \tag{3.23}$$

Since ϕ_1 and ϕ_2 are the solution of the equation $x'(t) = A(t)x$

$$\therefore \quad \phi_{1'}(t) = A(t)\phi_1 \quad and \quad \phi_{2'}(t) = A(t)\phi_2. \tag{3.24}$$

By Equations (3.24), (3.23), and (3.22), we get

$$\phi_{2'} = A\phi_2 = \phi_{1'}\theta + \phi_1\theta' = A\phi_1\theta + \phi_1\theta' = A\phi_2 + \phi_1\theta'$$
$$\therefore A\phi_2 = A\phi_2 + \phi_1\theta'. \tag{3.25}$$

This implies that $\phi_1\theta' = 0$.
Since ϕ_1 is non-singular, then $\theta' = 0$ or $\theta = C$, is a constant matrix.
Thus, $\phi_2 = \phi_1 C$. Hence, $C = \phi_1^{-1}\phi_2$. Since both ϕ_1 and ϕ_2 are non singular. Therefore, C is a constant non singular matrix.

Example 3.5.3. Show that $\phi = \begin{pmatrix} e^{-2t} & -3e^{5t} \\ 2e^{-2t} & e^{5t} \end{pmatrix}$ is a fundamental matrix for the system $x'(t) = \begin{pmatrix} 4 & -3 \\ -2 & -1 \end{pmatrix} x(t)$. Use this fundamental matrix to find a general solution of $x'(t) = \begin{pmatrix} 4 & -3 \\ -2 & -1 \end{pmatrix} x(t)$.

Solution: Because

$$\begin{pmatrix} e^{-2t} \\ 2e^{-2t} \end{pmatrix}' = \begin{pmatrix} -2e^{-2t} \\ -4e^{-2t} \end{pmatrix} = \begin{pmatrix} 4 & -3 \\ -2 & -1 \end{pmatrix} \begin{pmatrix} e^{-2t} \\ 2e^{-2t} \end{pmatrix}$$
$$\begin{pmatrix} -3e^{5t} \\ e^{5t} \end{pmatrix}' = \begin{pmatrix} -15e^{-2t} \\ 5e^{5t} \end{pmatrix} = \begin{pmatrix} 4 & -3 \\ -2 & -1 \end{pmatrix} \begin{pmatrix} -3e^{5t} \\ e^{5t} \end{pmatrix}.$$

Both $x_1 = \begin{pmatrix} e^{-2t} \\ 2e^{-2t} \end{pmatrix}$ and $x_2 = \begin{pmatrix} -3e^{5t} \\ e^{5t} \end{pmatrix}$ are solutions of the system $x'(t) = \begin{pmatrix} 4 & -3 \\ -2 & -1 \end{pmatrix} x(t)$. Alternatively, we show that $x'(t) = \begin{pmatrix} 4 & -3 \\ -2 & -1 \end{pmatrix} x(t)$ is the same.

A general solution is given by

$$x(t) = \phi(t)C = \begin{pmatrix} e^{-2t} & -3e^{5t} \\ 2e^{-2t} & e^{5t} \end{pmatrix}\begin{pmatrix} c_1 \\ c_2 \end{pmatrix} = \begin{pmatrix} c_1 e^{-2t} - 3c_2 e^{5t} \\ 2c_1 e^{-2t} + c_2 e^{5t} \end{pmatrix} = c_1 \begin{pmatrix} e^{-2t} \\ 2e^{-2t} \end{pmatrix} + c_2 \begin{pmatrix} -3e^{5t} \\ e^{5t} \end{pmatrix}.$$

Example 3.5.4. The functions e^x, $\cos x$, $\sin x$ are the solutions of $y''' - y'' y' - y = 0$ over any interval.

Here, Wronskian is

$$W(x) = \begin{vmatrix} e^x & \cos x & \sin x \\ e^x & -\sin x & \cos x \\ e^x & -\cos x & -\sin x \end{vmatrix} = 2e^x,$$

which is not zero for any x. Thus, the above solutions are linearly independent over any interval.

Example 3.5.5. Consider the following functions over $[-2,2]$ where

$$f_1(x) = \begin{cases} 1 + x^3, & for \quad x \le 0, \\ 1, & for \quad x \ge 0. \end{cases}$$

$$f_2(x) = \begin{cases} 1, & for \quad x \le 0, \\ 1 + x^2, & for \quad x \ge 0 \end{cases}$$

$$f_3(x) = 3 + x^3, \quad for \quad each \quad x.$$

It may be seen that these functions and their derivatives up to the second order are all continuous over $[-2,2]$.

For $x \le 0$, the determinant of Wronskian is $W(x) = \begin{vmatrix} 1+x^3 & 1 & 3+x^3 \\ 3x^2 & 0 & 3x^2 \\ 6x & 0 & 6x \end{vmatrix} = 0,$

and $x \ge 0$. Its value is $W(x) = \begin{vmatrix} 1 & 1+x^3 & 3+x^3 \\ 0 & 3x^2 & 3x^2 \\ 0 & 6x & 6x \end{vmatrix} = 0.$

So, the Wronskian is identically zero over the interval $[-2,2]$.

Now, we will show that these functions are linearly independent over $[-2,2]$.

$$\text{Let } c_1 f_1(x) + c_2 f_2(x) + c_3 f_3(x) = 0. \qquad \text{(i)}$$

Here, $x \in [-2,2]$, and putting $x = -1$, we get $c_1[1+(-1)^3 + c_2.1 + c_3$ $[3+(-1)^3] = 0,$

$$\text{or, } c_2 + 2c_3 = 0. \qquad \text{(ii)}$$

$$\text{Putting } x = 0, \text{ we get } c_1 + c_2 + 3c_3 = 0, \qquad \text{(iii)}$$

$$\text{and putting } x = 1, \text{ we get, } c_1 + 2c_2 + 4c_3 = 0. \qquad \text{(iv)}$$

From Equations (ii–iv), we get $c_1 = c_2 = c_3 = 0$ showing that $f_1(x), f_2(x), f_3(x)$ are linearly independent.

Example 3.5.6. Find the fundamental matrix of the system

$$x = \begin{bmatrix} x_1 \\ x_2 \\ x_3 \end{bmatrix}, \quad A = \begin{bmatrix} -t & 0 & 0 \\ 1 & -t & 0 \\ 0 & 0 & t \end{bmatrix}.$$

The system $x' = Ax$ is equivalent to the set of differential equations

$$x_1' = Ax = -tx_1 + tx_2,\ x_2' = Ax = -tx_2,\ x_3' = Ax = tx_3.$$

We get $\log x_2 = -\dfrac{t^2}{2}$ or, $x_2 = e^{-\frac{t^2}{2}}$, $x_3 = e^{\frac{t^2}{2}}$.

$\therefore \quad x_1' = Ax = -tx_1 + tx_2 = -tx_1 + te^{-\frac{t^2}{2}}$. Hence, $x_1 = \frac{t^2}{2}e^{-\frac{t^2}{2}}$.

Hence, the solution matrix is

$$\phi(t) = \begin{bmatrix} \frac{t^2}{2}e^{-\frac{t^2}{2}} & 0 & 0 \\ 0 & e^{-\frac{t^2}{2}} & 0 \\ 0 & 0 & e^{\frac{t^2}{2}} \end{bmatrix}.$$

The column vectors of $\phi(t)$ are

$$\phi_1(t) = \begin{bmatrix} \frac{t^2}{2}e^{-\frac{t^2}{2}} \\ 0 \\ 0 \end{bmatrix}, \phi_2(t) = \begin{bmatrix} 0 \\ e^{-\frac{t^2}{2}} \\ 0 \end{bmatrix}, \phi_3(t) = \begin{bmatrix} 0 \\ 0 \\ e^{\frac{t^2}{2}} \end{bmatrix}.$$

Since $W(\phi_1, \phi_2, \phi_3)(t) = \frac{t^2}{2}e^{-\frac{t^2}{2}} \neq 0$ and ϕ_1, ϕ_2, ϕ_3 are linearly independent, hence, the solution matrix ϕ is a fundamental matrix.

Example 3.5.7. Show whether the matrix $\phi(t) = \begin{bmatrix} e^t & 1 & 0 \\ 1 & e^{-t} & 0 \\ 0 & 0 & 1 \end{bmatrix}$ is a fundamental matrix for the system $x'(t) = A(t)x$, for $t \in I$.

Solution: We know that $\phi(t)$ is a fundamental matrix if and only if $\det\phi(t) \neq 0$.

Here, the matrix, $\det\phi(t) = 1 - 1 = 0$.

Since $\det\phi(t) = 0$, therefore the given matrix $\phi(t) = \begin{bmatrix} e^t & 1 & 0 \\ 1 & e^{-t} & 0 \\ 0 & 0 & 1 \end{bmatrix}$ is not a fundamental matrix of the system.

3.6 Solution of Non-Homogeneous Linear Systems

Assume that $A(t)$ is an $n \times n$ matrix that is real and continuous on I. The system

$$x'(t) = A(t)x + f(t) \quad for \quad t \in I \tag{3.26}$$

is a non-homogeneous linear system of order n, and $f(t)$ is the continuous vector on I, which is not identically zero.

In the case of a non-homogeneous linear equation, we shall first explain how the solution Equation (3.26) is closely related to the solution of the corresponding homogeneous system

$$x'(t) = A(t)x \quad for \quad t \in I. \tag{3.27}$$

To obtain the solution of Equation (3.26), it becomes necessary to resort to the method of variation of parameters. Let $\phi(t)$ be a fundamental matrix for the system Equation (3.27). Let $\phi(t)$ be a solution of Equation (3.26) such that for some $t_0 \in I$, $\phi(t_0) = 0$.

Now assume that $\varphi(t)$ (t) is given by

$$\varphi(t) = \phi(t)u(t),\ t \in I \tag{3.28}$$

where $u(t)$ is an unknown vector function mapping $I \to R^n$ such that $u(t)$ is differentiable, and $u(t_0) = 0$.

The solution φ is determined by finding $u(t)$ in terms of known quantities $\phi(t)$ and $f(t)$. Substituting Equation (3.28) in Equation (3.26),

$$\varphi'(t) = \phi'(t)u(t) + \phi(t)u'(t) = A(t)\phi(t)u(t) + \phi(t)u'(t).$$

Also, we have seen that $\varphi'(t) = A(t)\phi(t)u(t) + f(t)$. Equating these two equations, we get

$$f(t) = \phi(t)u'(t).$$

Here, $\phi(t)$ is fundamental matrix and also non-singular on I.

$$\therefore\quad u'(t)=\phi^{-1}(t)f(t)$$
$$or,\ u(t)=0+\int_{t_0}^{t}\phi^{-1}(s)f(s)ds,\ t\in I$$

Putting this value of $u(t)$ in Equation (3.28), we get

$$\varphi(t)=\phi(t)u(t)=\phi(t)\int_{t_0}^{t}\phi^{-1}(s)f(s)ds. \tag{3.29}$$

It can be easily verified that Equation (3.29) is a solution of Equation (3.26).

Theorem 3.6.1. Let $\phi(t)$ be a fundamental matrix for the system $x'(t)=A(t)x$ for $t\in I$. Then, φ defined in Equation (3.29) is a solution of IVP.

$$x'(t)=A(t)x+f(t)\qquad \text{for}\quad x(t_0)=0. \tag{3.30}$$

Proof Let us assume $x_k(t)$ is a solution of IVP.

$$\therefore\quad x'(t)=A(t)x,\quad x(t_0)=x_0;\quad t,t_0\in I. \tag{3.31}$$

Then, $B(t)=x_k(t)+\varphi(t)$ is a solution of Equation (3.26).
Therefore,

$$f'(t)=x_k'(t)+\varphi'(t)=A(t)x_k(t)+A(t)\varphi(t)+f(t)$$
$$=A(t)\left[x_k(t)+\varphi(t)\right]+f(t)=A(t)B(t)+f(t).$$

We have $B(t_0)=x_k(t_0)+\varphi(t_0)=x_0$.

Therefore, $B(t)=x_k(t)+\varphi(t)=x_k(t)+\phi(t)\int_{t_0}^{t}\phi^{-1}(s)f(s)ds$ [from Equation (3.29)] is the solution of $x'(t)=A(t)x+f(t)$ for $x(t_0)=0$.

Since $\phi(t)$ is a fundamental matrix for the system, $x_k(t)$ may be written as

$$x_k(t)=\phi(t)c,$$

where c is a constant vector.

Since $x_k(t_0) = x_0$, $\therefore$ $x_k(t) = \phi(t)c = x_0$
gives us $c = \phi^{-1}(t)x_0$. Here, $x_k(t) = \phi(t)c = \phi(t)\phi^{-1}(t)x_0$.

Hence, $B(t) = \phi(t)\phi^{-1}(t_0)x_0 + \int_{t_0}^{t}\phi(t)\phi^{-1}(s)f(s)ds$.

Observe that $x_k(t)$ is a solution of Equation (3.31) passing through the initial point (t_0, x_0).

Denoting $x_k(t) = x_k(t, t_0, x_0) = \phi(t)\phi^{-1}(t_0)x_0$.

Hence, $x_k(t, s, f(s)) = \phi(t)\phi^{-1}(s)f(s)$.

These two relations show that $B(t) = B(t, t_0, x_0)$ of Equation (3.26) passing through (t_0, x_0) can be written as $B(t, t_0, x_0) = x_k(t, t_0, x_0) + \int_{t_0}^{t} x_k$ $(t, s, f(s))ds$.

This relation is called *variation of parameter* (constant) formula.

Example 3.6.1. Solve

$$x_{1'} = -3x_1 + 9x_2;$$
$$x_{2'} = -2x_1 + 3x_2 + \tan 3t$$
$$x_1(0) = 1,\ x_2(0) = -1.\ -\pi/6 < t < \pi/6.$$

Solution: In matrix form, the system can be written as $\mathrm{x}' = \mathrm{Ax} + \mathrm{f}(\mathrm{t})$, where $x = \begin{pmatrix} x_1 \\ x_2 \end{pmatrix}$, $A = \begin{pmatrix} -3 & 9 \\ -2 & 3 \end{pmatrix}$, and $f(t) = \begin{pmatrix} 0 \\ \tan 3t \end{pmatrix}$.

A has eigenvalues $\lambda_{1,2} = \pm 3i$ with corresponding eigenvectors $v_{1,2} = \begin{pmatrix} 3 \\ 2 \end{pmatrix} \pm \begin{pmatrix} 3 \\ 0 \end{pmatrix} i$.

Then, two linearly independent solutions of the corresponding homogeneous system, $x' = Ax$, are

$$x_1(t) = \begin{pmatrix} 3 \\ 2 \end{pmatrix}\cos 3t - \begin{pmatrix} -3 \\ 0 \end{pmatrix}\sin 3t = \begin{pmatrix} 3\cos 3t + 3\sin 3t \\ 2\cos 3t \end{pmatrix}$$

$$\textit{and}\quad x_2(t) = \begin{pmatrix} 3 \\ 2 \end{pmatrix}\sin 3t + \begin{pmatrix} -3 \\ 0 \end{pmatrix}\cos 3t = \begin{pmatrix} -3\cos 3t + 3\sin 3t \\ 2\sin 3t \end{pmatrix}.$$

So a fundamental matrix for the corresponding homogeneous system is

$$\phi(t)=\begin{pmatrix} 3\cos 3t+3\sin 3t & -3\cos 3t+3\sin 3t \\ 2\cos 3t & 2\sin 3t \end{pmatrix}$$

with inverse
$$\phi^{-1}(t)=\begin{pmatrix} \frac{1}{2}\sin 3t & \frac{1}{2}(\cos 3t-\sin 3t) \\ -\frac{1}{3}\cos 3t & \frac{1}{2}(\cos 3t+\sin 3t) \end{pmatrix}.$$

We now compute:

$$\varphi(t)=\phi(t)\int_{t_0}^{t}\phi^{-1}(s)f(s)ds$$
$$=\begin{pmatrix} -\cos 3t\ln\sec 3t+\tan 3t \\ \frac{1}{3}[-1-\cos 3t\ln\sec 3t+\tan 3t+\ln(\sec 3t+\tan 3t)]\sin 3t \end{pmatrix},$$

and a general solution is given by

$$x_1(t)=c_1(3\cos 3t+3\sin 3t)+c_2(-3\cos 3t+3\sin 3t)-\cos 3t\ln\sec 3t$$
$$+\tan 3t \quad \textit{and}$$

$$x_2(t)=2c_1\cos 3t+2c_2\sin 3t+\frac{1}{3}\big[-1-\cos 3t\ln\sec 3t+\tan 3t$$
$$+\ln(\sec 3t+\tan 3t)\big]\sin 3t.$$

Applying the initial conditions gives us $3c_1-3c_2=1;\ \ 2c_1=-1$ so $c_1=-1/2$ and $c_2=-5/6$, so the solution to the initial value problem is

$$x_1(t)=-\frac{1}{2}(3\cos 3t+3\sin 3t)-\frac{5}{6}(-3\cos 3t+3\sin 3t)-\cos 3t\ln\sec 3t$$
$$+\tan 3t \quad \text{and}$$

$$x_2(t)=-\cos 3t+\frac{5}{3}\sin 3t+\frac{1}{3}\big[-1-\cos 3t\ln\sec 3t+\tan 3t$$
$$+\ln(\sec 3t+\tan 3t)\big]\sin 3t.$$

3.7 Linear System with Constant Coefficients

In this section, we obtain the solution of the linear homogeneous system

$$x'(t) = A(t)x, \quad t \in I \tag{3.32}$$

where A is a constant matrix, and I is an interval of R.

Here, we shall find the solution in terms of distinct eigenvalues of A and also in case of eigenvalues that are repeated. When Equation (3.32) is a scalar equation, the solution is $x(t) = ce^{tA}$, where c is an arbitrary constant. Now, we shall establish the following theorem.

Theorem 3.7.1

(i) The general solution of Equation (3.32) is $x(t) = ce^{tA}$, where c is an arbitrary constant.
(ii) If (i) satisfies the initial condition $x(t_0) = x_0$, its solution is $x(t) = e^{(t-t_0)A}x_0$.
(iii) The fundamental matrix $\phi(t) = e^{tA}$.

Proof Let $x(t)$ be a solution of Equation (3.32). Then, we have

$$\begin{aligned} &x'(t) = Ace^{tA} = Ax(t) \\ &or,\ x'(t) - Ax(t) = 0,\ t \in I. \end{aligned} \tag{3.33}$$

Let us take vector $u(t) = e^{-tA}x(t)$. Then, differentiating, we get,

$$\begin{aligned} &u(t) = -Ae^{-tA}x(t) + e^{-tA}x'^{(t)}. \\ &\therefore\ u(t) = e^{-tA}\left[-Ax(t) + x'^{(t)}\right]. \end{aligned} \tag{3.34}$$

Using Equations (3.33) and (3.34) we get,

$$u(t) = e^{-tA}.0 = 0.$$

Therefore, $u(t) = c$ is a constant vector for $t \in I$.

$$\because\ u(t) = c, \text{ therefore, } u(t) = c = e^{-tA}x(t), \text{ i.e. } c = e^{-tA}x(t).$$

Hence, we get

$$x(t) = ce^{tA}. \tag{3.35}$$

(ii) We assume that $x(t_0) = x_0$. Then, $x(t_0) = x_0 = ce^{t_0 A}$. This implies that $c = e^{-t_0 A}\ x(t_0)$.

Using the value of c in Equation (3.35), we get

$$x(t) = ce^{tA} = e^{-t_0 A}\quad x(t_0)e^{tA} = e^{(t-t_0)A}x_0.$$

(iii) The fundamental matrix $\phi(t)$ satisfies the differential Equation (3.32).

Let $\det\phi(t) \neq 0$.

Here, we shall show that $\phi(t)$ has another property, i.e., e^{tA} is a fundamental matrix of the system.

Let $\phi(t) = e^{tA}$. Then, we have

$$\phi(t) = \lim_{\Delta t \to 0} \frac{e^{(t+\Delta t)A} - e^{tA}}{\Delta t} = \lim_{\Delta t \to 0} Ae^{tA} \quad \frac{e^{\Delta tA} - 1}{\Delta tA},$$

$$\text{since} \lim_{\Delta t \to 0} \frac{e^{(\Delta t)A} - 1}{\Delta tA} = 1. \quad \therefore \quad \phi(t) = Ae^{tA} = A\phi(t).$$

Thus, e^{tA} satisfies the given differential equation.

Further, we let $\phi(0) = B$, and the statement gives

$$\det\phi(t) = \det\phi(0) \quad \exp\left[(\text{tr}A)(t - t_0)\right] \neq 0.$$

Hence, $\det\phi(t) \neq 0$

so that $\phi(t)$ is a fundamental matrix of the system.

Theorem 3.7.2. Consider the vector

$$x'(t) = Ax(t),\ t \in I, \tag{3.36}$$

where A is $n \times n$ constant matrix.

(i) If $\lambda_1, \lambda_2, \ldots, \lambda_n$ are real distinct eigenvalues of A, and $c_1, c_2, \ldots, c_n$ are corresponding eigenvectors of A, then $c_1 e^{\lambda_1 t}, c_2 e^{\lambda_2 t}, \ldots, c_n e^{\lambda_n t}$ form a linearly independent set of solutions of Equation (3.32), and the general solution is

$$x(t) = \alpha_1 c_1 e^{\lambda_1 t} + \alpha_2 c_2 e^{\lambda_2 t} + \ldots + \alpha_n c_n e^{\lambda_n t}$$

where $\alpha_1, \alpha_2, \ldots, \alpha_n$ are arbitrary constants.
(ii) If $\lambda_1, \lambda_2, \ldots, \lambda_n$ are real distinct eigenvalues of A and $n_1, n_2, \ldots, n_n$ such that $n_1 + n_2 + \cdots + n_n = n$ with corresponding eigenvectors, then the solution is

$$x(t) = e^{At} \sum_{i=1}^{n} \alpha_i$$

where $\alpha_1, \alpha_2, \ldots, \alpha_n$ are arbitrary constants.

$$= \sum_{i=1}^{n} e^{\lambda_i t} \left[\sum_{p=1}^{n-1} \frac{t^p}{p!} (A - \lambda_p E)^p \right] \alpha_i,$$

where $x(0) = \alpha = (\alpha_1, \alpha_2, \ldots, \alpha_n)$.

Example 3.7.1. Determine e^{At} and a fundamental matrix for the system

$$x'(t) = Ax(t) \text{ where } A = \begin{pmatrix} -1 & 2 & 3 \\ 0 & -2 & 1 \\ 0 & 3 & 0 \end{pmatrix}.$$

Solution: Using the method of eigenvalues and eigenvectors for finding the fundamental matrix $\phi(t)$ and e^{At}.

The eigenvalues are $\lambda = (\lambda_1, \lambda_2, \lambda_3)$, and I is the identity matrix of order 3. Then,

$$A - \lambda I = \begin{pmatrix} -1 & 2 & 3 \\ 0 & -2 & 1 \\ 0 & 3 & 0 \end{pmatrix} - \lambda \begin{pmatrix} 1 & 0 & 0 \\ 0 & 1 & 0 \\ 0 & 0 & 1 \end{pmatrix} = \begin{pmatrix} -1-\lambda & 2 & 3 \\ 0 & -2-\lambda & 1 \\ 0 & 3 & -\lambda \end{pmatrix}.$$

The characteristic equation is

$$\det|A - \lambda E| = 0$$

$$\therefore \quad \det|A - \lambda E| = \begin{vmatrix} -1-\lambda & 2 & 3 \\ 0 & -2-\lambda & 1 \\ 0 & 3 & -\lambda \end{vmatrix} = 0.$$

We get $(1+\lambda)(\lambda-1)(\lambda+3)=0$. The eigenvalues are $\lambda_1 = -1, \lambda_2 = 1, \lambda_3 = -3$.

For the fundamental matrix, we find eigen vectors for $\lambda_1 = -1$. Let the eigen vectors $[x_1, x_2, x_3]$ proceed to

$$\begin{bmatrix} -1+1 & 2 & 3 \\ 0 & -2+1 & 1 \\ 0 & 3 & 1 \end{bmatrix} \begin{bmatrix} x_{1,} \\ x_{2,} \\ x_{3,} \end{bmatrix} = 0 \; or, \begin{bmatrix} 0 & 2 & 3 \\ 0 & -1 & 1 \\ 0 & 3 & 1 \end{bmatrix} \begin{bmatrix} x_1 \\ x_2 \\ x_3 \end{bmatrix} = 0,$$

which will give $2x_2 + 3x_3 = 0$

$-x_2 + x_3 = 0$

$3x_2 + x_3 = 0.$

$\Rightarrow$ Here, $x_2 = x_3 = 0$, and x_1 is arbitrary.

Let $x_1 = 1$. Hence, for $x_1 = 1$:

For $\lambda_1 = -1$, the eigen vector is $\begin{bmatrix} 1 \\ 0 \\ 0 \end{bmatrix}$.

For $\lambda_2 = 1$, let the eigen vectors $[x_1, x_2, x_3]$ proceed to

$$\begin{bmatrix} -1-1 & 2 & 3 \\ 0 & -2-1 & 1 \\ 0 & 3 & -1 \end{bmatrix} \begin{bmatrix} x_1 \\ x_2 \\ x_3 \end{bmatrix} = 0$$

$$\text{or,} \begin{bmatrix} -2 & 2 & 3 \\ 0 & -3 & 1 \\ 0 & 3 & -1 \end{bmatrix} \begin{bmatrix} \mathrm{X}_1 \\ \mathrm{X}_2 \\ \mathrm{X}_3 \end{bmatrix} = 0,$$

which will give $-2x_1 + 2x_2 + 3x_3 = 0$

$-3x_2 + x_3 = 0$

$3x_2 - x_3 = 0.$

Here, we get $x_2 = 1, x_3 = 3$ from the last two equations, and also, we can find $x_1 = \frac{11}{2}$ from the first equation.

For $\lambda_2 = 1$, the eigen vector is $\begin{bmatrix} \frac{11}{2} \\ 1 \\ 3 \end{bmatrix}$.

Again, for $\lambda_3 = -3$, let the eigen vectors $[x_1, x_2, x_3]$ proceed to

$$\begin{bmatrix} -1+3 & 2 & 3 \\ 0 & -2+3 & 1 \\ 0 & 3 & 3 \end{bmatrix} \begin{bmatrix} x_1 \\ x_2 \\ x_3 \end{bmatrix} = 0 \; or, \begin{bmatrix} 2 & 2 & 3 \\ 0 & 1 & 1 \\ 0 & 3 & 3 \end{bmatrix} \begin{bmatrix} x_1 \\ x_2 \\ x_3 \end{bmatrix} = 0,$$

which will give $2x_1 + 2x_2 + 3x_3 = 0$

$x_2 + x_3 = 0$

$3x_2 + 3x_3 = 0.$

Now, we choose $x_2 = 1$ so that $x_3 = -1$, and hence, $x_1 = \frac{1}{2}$.

For $\lambda_3 = -3$, the eigen vector is $\begin{bmatrix} \frac{1}{2} \\ 1 \\ -1 \end{bmatrix}$.

Since the eigenvectors are linearly independent, the fundamental matrix is given by

$$\phi_1(t) = e^{-t} \begin{bmatrix} 1 \\ 0 \\ 0 \end{bmatrix}, \quad \phi_2(t) = e^{t} \begin{bmatrix} \frac{11}{2} \\ 1 \\ 3 \end{bmatrix} \quad \text{and} \quad \phi_3(t) = e^{-3t} \begin{bmatrix} \frac{1}{2} \\ 1 \\ -1 \end{bmatrix},$$

which can be written as

$$\phi(t) = \begin{pmatrix} e^{-t} & \frac{11e^{t}}{2} & \frac{1e^{-3t}}{2} \\ 0 & e^{t} & e^{-3t} \\ 0 & 3e^{t} & -e^{-3t} \end{pmatrix}.$$

We know that $W(\phi_1, \phi_2, \phi_3) \neq 0$.
The general solution of the differential equation is

$$x(t) = \alpha_1 e^{-t}\begin{bmatrix} 1 \\ 0 \\ 0 \end{bmatrix} + \alpha_2 e^{t}\begin{bmatrix} \frac{11}{2} \\ 1 \\ 3 \end{bmatrix} + \alpha_3 e^{-3t}\begin{bmatrix} \frac{1}{2} \\ 1 \\ -1 \end{bmatrix},$$

where $\alpha_1, \alpha_2, \alpha_3$ are arbitrary constants.

Since ϕ_1, ϕ_2, ϕ_3 are the basis of the solution, e^{At} will correspond to the solution of the initial value problem $x'(t) = Ax(t)$, $x_1(0) = e_1$, $x_2(0) = e_2$, $x_3(0) = e_3$. For the solution $x_1(t)$, we can express e_1 as a linear combination of the eigenvectors:

$$\begin{bmatrix} 1 \\ 0 \\ 0 \end{bmatrix} = \alpha_1 \begin{bmatrix} 1 \\ 0 \\ 0 \end{bmatrix} + \alpha_2 \begin{bmatrix} \frac{11}{2} \\ 1 \\ 3 \end{bmatrix} + \alpha_3 \begin{bmatrix} \frac{1}{2} \\ 1 \\ -1 \end{bmatrix} \Rightarrow \alpha_2 = \alpha_3 = 0.$$

Hence, $x_1(t) = e^{t}\begin{bmatrix} 1 \\ 0 \\ 0 \end{bmatrix}$. Again, $\begin{bmatrix} 0 \\ 1 \\ 0 \end{bmatrix} = \alpha_1 \begin{bmatrix} 1 \\ 0 \\ 0 \end{bmatrix} + \alpha_2 \begin{bmatrix} \frac{11}{2} \\ 1 \\ 3 \end{bmatrix} + \alpha_3 \begin{bmatrix} \frac{1}{2} \\ 1 \\ -1 \end{bmatrix},$

which will give the equations $\alpha_1 + \frac{11}{2}\alpha_2 + \frac{1}{2}\alpha_3 = 0$

$0 + \alpha_2 + \alpha_3 = 1$

$0 + 3\alpha_2 - \alpha_3 = 0$

$\Rightarrow \quad \alpha_1 = -\frac{14}{8}, \quad \alpha_2 = \frac{1}{4}, \quad \text{and} \quad \alpha_3 = \frac{3}{4}.$

$$\text{Hence, } x_2(t) = -\frac{14}{81}e^{-t}\begin{bmatrix}1\\0\\0\end{bmatrix} + \frac{1}{4}e^{t}\begin{bmatrix}\frac{11}{2}\\1\\3\end{bmatrix} + \frac{3}{4}e^{-3t}\begin{bmatrix}\frac{1}{2}\\1\\-1\end{bmatrix}$$

$$= \begin{bmatrix}-\frac{14}{81}e^{-t} + \frac{11}{8}e^{t} + \frac{3}{8}e^{-3t}\\ \frac{1}{4}e^{t} + \frac{3}{4}e^{-3t}\\ \frac{3}{4}e^{t} - \frac{3}{4}e^{-3t}\end{bmatrix}.$$

For $x_3(t)$, we have

$$\begin{bmatrix}0\\0\\1\end{bmatrix} = \alpha_1\begin{bmatrix}1\\0\\0\end{bmatrix} + \alpha_2\begin{bmatrix}\frac{11}{2}\\1\\3\end{bmatrix} + \alpha_3\begin{bmatrix}\frac{1}{2}\\1\\-1\end{bmatrix},$$

which gives $\alpha_1 + \frac{11}{2}\alpha_2 + \frac{1}{2}\alpha_3 = 0$

$0 + \alpha_2 + \alpha_3 = 0$

$0 + 3\alpha_2 - \alpha_3 = 1$

$\Rightarrow \quad \alpha_1 = -\frac{10}{8}, \quad \alpha_2 = \frac{1}{4}, \quad \text{and} \quad \alpha_3 = -1$

$$\text{Hence, } x_3(t) = -\frac{10}{8}e^{-t}\begin{bmatrix}1\\0\\0\end{bmatrix} + \frac{1}{4}e^{t}\begin{bmatrix}\frac{11}{2}\\1\\3\end{bmatrix} + \frac{-1}{4}e^{-3t}\begin{bmatrix}\frac{1}{2}\\1\\-1\end{bmatrix}$$

$$= \begin{bmatrix}-\frac{10}{81}e^{-t} + \frac{11}{8}e^{t} - \frac{1}{8}e^{-3t}\\ \frac{1}{4}e^{t} - \frac{1}{4}e^{-3t}\\ \frac{3}{4}e^{t} + \frac{1}{4}e^{-3t}\end{bmatrix}.$$

Hence, $e^{At} = (x_1, x_2, x_3)$ is determined.

Theorem 3.7.3. If $\lambda_1, \lambda_2, \ldots, \lambda_n$ are the characteristic roots of the non-singular matrix C, $r_1, r_2, \ldots, r_n$ are the characteristic roots of $e^{\omega B}$, and B is a matrix, then

$$det\phi(t) = \lambda_1, \lambda_2, \ldots, \lambda_n = \exp(r_1 + r_2 + \cdots + r_n) = \omega.$$

Proof If ϕ is any other fundamental matrix, then there exists a no singular matrix D such that

$$\phi(t) = \phi(t)D. \tag{3.37}$$

$$\text{We know } \phi(t+\omega) = \phi(t)e^{\omega B}. \tag{3.38}$$

But from Equation (3.37), we get $\phi(t+\omega) = \phi(t+\omega)D$.

Since $\phi(t+\omega)D$ is a fundamental matrix, using Equation (3.37), we get $\phi(t+\omega)D = \phi(t)De^{\omega B}D^{-1}$.

Here, $e^{\omega B}$ and $De^{\omega B}D^{-1}$ are similar. Hence, they have same characteristic roots. Thus, the set of fundamental matrices determine a set of characteristic roots of $C = e^{\omega B}$.

When $t = 0$, we have from Equation (3.38)

$$\phi(\omega) = \phi(0)e^{\omega B} \Rightarrow \phi^{-1}(0)(\omega) = e^{\omega B}.$$

Hence, $\lambda_1, \lambda_2, \ldots, \lambda_n$ are the characteristic roots of $\phi^{-1}(0)\phi(\omega)$ or $\phi(\omega)$ if $\phi^{-1}(0) = E$, where E is an identity matrix.

We know that $\det\phi(t) = \det\phi(\tau)\exp\left(\int_\tau^t \operatorname{tr} A(s)ds\right)$.

Here, we have $\det\phi(\omega) = \det\phi(0)\exp\left(\int_0^\omega \operatorname{tr} A(s)ds\right)$,

so that $\det\phi(\omega) = \lambda_1\lambda_2\ldots\lambda_n = 1 \quad \exp(r_1 + r_2 + \ldots + r_n)\omega$.

[∵ Any complex $n \times n$ matrix, like $\phi(\omega), A(s)$, is similar to a diagonal matrix.]

The diagonal elements are the characteristic roots of the matrix so that the value of the $\det\phi(t)$ is the product of the roots, and tr $A(s)$ is the sum of the characteristic roots.

3.8 Exercise

1. Determine a fundamental matrix and e^{At} of the system $x'^{(t)} = Ax(t), \mathrm{x} \in \mathrm{I}$ where

$$A = \begin{bmatrix} 3 & 1 & -1 \\ 1 & 3 & -1 \\ 3 & 3 & -1 \end{bmatrix}.$$

2. Find the general solution of the system $x'(t) = Ax(t), t \in I$

where $A = \begin{bmatrix} 1 & -3 & 9 \\ 0 & -5 & 18 \\ 0 & -3 & 10 \end{bmatrix}$ and $x(0) = (a, b, c)$, and find e^{At}.

3. Find the determinant of a fundamental matrix for which $\phi(0) = E$ for the system $x'(t) = Ax(t)$ where

(i) $A = \begin{bmatrix} 0 & 1 & 0 \\ 0 & 0 & 1 \\ 2 & -5 & 4 \end{bmatrix}$ and (ii) $A = \begin{bmatrix} -2 & 2 & 0 \\ 0 & -3 & 1 \\ 0 & 0 & -3 \end{bmatrix}$.

4. Find e^{At} and the fundamental matrix for which $\phi(0) = E$ for the system $x'(t) = Ax(t)$ where

$$\text{(i) } A = \begin{bmatrix} 1 & 0 & 0 \\ 2 & 0 & 2 \\ 3 & -3 & 5 \end{bmatrix}, \text{ (ii) } A = \begin{bmatrix} 1 & -1 & -1 \\ 1 & 3 & 1 \\ -3 & 1 & -1 \end{bmatrix}, \text{ and}$$

$$\text{(iii) } A = \begin{bmatrix} 1 & 1 & 1 \\ 0 & -2 & 3 \\ 0 & 1 & 0 \end{bmatrix}.$$

5. Determine a fundamental matrix and e^{At} of the system

$$x'(t) = Ax(t), \quad t \in I \text{ where } A = \begin{bmatrix} 3 & 1 & -1 \\ 1 & 3 & -1 \\ 3 & 3 & -1 \end{bmatrix}.$$

6. Write the system of first-order equations in matrix form.

$$\begin{aligned} x' &= x + 2y \\ y' &= 4x + 3y \end{aligned} \quad \begin{pmatrix} x \\ y \end{pmatrix}' = \begin{pmatrix} 1 & 2 \\ 4 & 3 \end{pmatrix} \begin{pmatrix} x \\ y \end{pmatrix}.$$

7. Put the following initial value problems into fundamental form:

(i) $\dfrac{d^2x}{dt} - 2\dfrac{dx}{dt} - 3x = 0; \quad x(0) = 4, \quad \dfrac{dx(0)}{dt} = 5.$

(iii) $e^t \dfrac{d^2x}{dt} - 2e^{2t} \dfrac{dx}{dt} - 3e^t = 2; \quad x(0) = 0, \quad \dfrac{dx(0)}{dt} = 0.$

(iii) $4x''' + tx'' - x = 0; \quad x(-1) = 2, \quad x'(-1) = 1, \quad x''(-1) = 205.$

8. Use the given eigenvalues and corresponding (linearly independent) eigenvectors of the matrix A to find a general solution of $x' = Ax$.

(i) $A = \begin{pmatrix} 5 & -1 \\ 3 & 1 \end{pmatrix}; \ \lambda_1 = 2, \ v_1 = \begin{pmatrix} 1 \\ 3 \end{pmatrix}; \ \lambda_2 = 4, \ v_1 = \begin{pmatrix} 1 \\ 1 \end{pmatrix}.$

(ii) $A = \begin{pmatrix} -4 & 0 \\ -1 & 2 \end{pmatrix}; \ \lambda_1 = -4, \ v_1 = \begin{pmatrix} 6 \\ 1 \end{pmatrix}; \ \lambda_2 = 2, \ v_1 = \begin{pmatrix} 0 \\ 1 \end{pmatrix}.$

(iii) $A = \begin{pmatrix} 0 & -6 \\ 6 & 0 \end{pmatrix}$; $\lambda_{1,2} = \pm 6i$, $v_{1,2} = \begin{pmatrix} \pm i \\ i \end{pmatrix}$.

9. Find the fundamental matrix for the system $x'(t) = Ax(t)$ where

$$A = \begin{bmatrix} 7 & -1 & 6 \\ -10 & 4 & -12 \\ -2 & 1 & -1 \end{bmatrix}.$$

10. Find the eigenvalues and the fundamental matrix for the linear system $\dfrac{dy}{dx} = \begin{pmatrix} 3 & 1 \\ 4 & 3 \end{pmatrix} y$.

4

Systems of Linear Differential Equations-II

Abstract

In this chapter, we continue our discussion of linear differential equations, focusing on second- and higher-order cases. We introduce the method of variation of parameters to find particular solutions to non-homogeneous linear differential equations—especially when the method of undetermined coefficients is not applicable.

The Wronskian is used to determine whether a set of solutions is linearly independent. Additionally, we explore the general solution to an nth-order non-homogeneous equation using the annihilator method in special cases.

Keywords: Wronskian, variation of parameters, linear independence, particular solution, annihilator method

4.1 Introduction

An ordinary differential equation is a differential equation in which the dependent variable and its derivative do not appear in products and they occur only in first degree.

A linear differential equation of order n is an equation of the form:

$$a_0(x)\frac{d^n y}{dx^n}+a_1(x)\frac{d^{n-1}y}{dx^{n-1}}+\cdots\ldots+a_{n-1}(x)\frac{dy}{dx}+a_n(x)y=b(x) \tag{4.1}$$

where $a_0(x)\neq 0$ and $a_0,a_1,a_2,\ldots a_n,b$ are continuous real functions defined on a real interval I.

$b(x)$ is called the *non homogeneous* term.

Points $x\in I$ such that $a_0(x)=0$ are called singular points of Equation (4.1). Here, we consider $a_0(x)\neq 0$ on I.

Dividing Equation (4.1) by $a_0(x)$, we obtain the equation of the same form, with $a_0(x)$ replaced by constant 1. Thus, we consider a linear differential equation of order n of the form:

$$\frac{d^n y}{dx^n}+a_1(x)\frac{d^{n-1}y}{dx^{n-1}}+\cdots\ldots+a_{n-1}(x)\frac{dy}{dx}+a_n(x)y=b(x) \quad (4.2)$$

It will be convenient to denote the differential expression on the left-hand side of Equation (4.1) by $L(y)$,

i.e., $L(y)=\frac{d^n y}{dx^n}+a_1(x)\frac{d^{n-1}y}{dx^{n-1}}+\cdots\ldots+a_{n-1}(x)\frac{dy}{dx}+a_n(x)y$.

Therefore, Equation (4.3) becomes

$$L(y)=\mathrm{b}(\mathrm{x}).$$

If $b(x)=0$ *for all x in I*, the corresponding equation

$$L(y)=0$$

is called a *homogeneous equation*,
whereas if $b(x)\neq 0$ for some x *in* I,

$$L(y)=\mathrm{b}(\mathrm{x}) \quad (4.3)$$

is called a *non homogeneous equation*.

L is called differential operator which operates on functions which have n derivative on I, and transforms such a function ϕ into a function $L(\phi)$ whose value at x is given by

$$L(\phi)(x)=\phi^{(n)}(x)+a_1(x)\phi^{(n-1)}(x)+\cdots\ldots+a_{n-1}(x)\phi^{(1)}(x)$$
$$+a_n(x)\phi(x).$$

Thus, $L(\phi)=\phi^{(n)}+a_1\phi^{(n-1)}+\cdots\ldots+a_{n-1}\phi^{(1)}+a_n\phi$.

A solution of $L(y) = \text{b}(\text{x})$ is therefore a function ϕ having an n derivative on I such that

$$L(\phi) = \text{b}.$$

Theoritically, if b is continuous on I, it is possible to find all solutions of Equation (4.3).

In this chapter, we first consider a simple case of a second-order equation, $n = 2$. All solutions of the homogeneous equation can be found by a simple device, which reduces the problem to the algebraic one of locating roots of a polynomial. The solution of a non-homogeneous equation can be generated by using the solutions of the corresponding homogeneous equation together with an integration involving the function b.

Secondly, we show how the method which work for second order case can be extended to solve the order equation.

Examples are:

1. Legendre's equation:

$$(1-x^2)\frac{d^2y}{dx^2} - 2x\frac{dy}{dx} + n(n+1)y = 0 \tag{4.4}$$

2. Bessel's equation:

$$x^2\frac{d^2y}{dx^2} + x\frac{dy}{dx} + (x^2 - n^2)y = 0 \tag{4.5}$$

3. $x^3\dfrac{d^3y}{dx^3} + 5x^2\dfrac{d^2y}{dx^2} + 6x\dfrac{dy}{dx} + 7y = 4sinx$ (4.6)

4. $\dfrac{d^2y}{dx^2} + w^2y = 4sinx = sinpt$ (4.7)

We here noted in the Equation (4.2):

If $b(x) = 0$, then it is called a homogeneous linear differential equation.
If $b(x) \neq 0$, then it is called a non-homogeneous linear differential equation

If $a_1, a_2, \ldots\ldots a_n$ are constants, then the equation is called linear differential equation with constant coefficients.

Here, Equations (4.4) and (4.5) are homogeneous, whereas Equations (4.6) and (4.7) are non-homogeneous. Equation (4.6) is a Cauchy equation, whereas Equation (4.7) is an equation with constant coefficients.

Lastly we studied nth order homogeneous and non-homogeneous differential equation and method for solving them.

4.2 Linearly Dependent and Independent Functions

A set of n functions $\phi_1, \phi_2, \ldots\ldots \phi_n$ defined on an interval I are said to be *linearly independent* on I if they are not linearly dependent.

Thus, a set of n functions $\phi_1, \phi_2, \ldots\ldots \phi_n$ defined on an interval I is said to be *linearly independent* on I if there exist constants $c_1, c_2, \ldots n$ such that

$$c_1\phi_1 + c_2\phi_2 + \ldots\ldots\ldots\ldots\ldots + c_n\phi_n = 0 \tag{4.8}$$

for all $x \in I$, which implies that

$$c_1 = c_2 =, \ldots - c_n = 0.$$

In particular, two functions, ϕ_1, ϕ_2, are linearly dependent on I if there exist two constants c_1, c_2, not both zero, such that $c_1\phi_1 + c_2\phi_2 = 0$ for all $x\ in\ I$, which implies that $c_1 = c_2 = 0$.

Note. For the case $n = 2$, we see that the functions ϕ_1, ϕ_2 are linearly dependent if and only if one of them is a constant multiple of the other on the interval I,

i.e., if $c_1\phi_1 + c_2\phi_2 = 0$ with $c_1 \neq 0$, then

$$\phi_1 = -\frac{c_2}{c_1}\phi_2 \tag{4.9}$$

for all $x \in I$.

Equation (4.10) can be written as $\phi_1 = c\phi_2$.

The following may be observed:

(i) Any set of functions containing a function, which is identically zero on I, is linearly dependent on I.
(ii) Any set of functions containing a linearly dependent set is linearly dependent.
(iii) A subset of a linearly independent set is linearly independent.
(iv) If a set of functions is a linearly dependent set on I, then it is also linearly dependent on any interval contained in I.
(v) If a set of functions is a linearly independent set on I, then it is also linearly independent on any interval contained in I.

There is a simple test, which enables us to find whether two solutions ϕ_1, ϕ_2 of $L(y)=0$ *are linearly independent* or not.
For this,

$$W(\phi_1,\phi_2)=\begin{bmatrix} \phi_1 & \phi_2 \\ \phi_1' & \phi_2' \end{bmatrix}=\phi_1\phi_2'-\phi_2\phi_1'$$

is called the *Wronskian of* ϕ_1,ϕ_2, and its value at x *is* $W(\phi_1,\phi_2)(x)$.

Definition 4.2.1. Let ϕ_1, ϕ_2 be two solutions of $L(y)=0$ such that $W(\phi_1,\phi_2) \neq= 0 \quad \forall x \in I$, then $\phi = c_1\phi_1 + c_2\phi_2 . \, x \in I$ is called the *general solution* of $L(y)=0$, and ϕ_1 and ϕ_2 are called a *fundamental set of solutions* of $L(y)=0$.

Theorem 4.2.1. Let two solutions ϕ_1, ϕ_2 of $L(y)=0$ be linearly independent on an interval I. Every solution ϕ of $L(y)=0$ can be written as

$$\phi = c_1\phi_1 + c_2\phi_2, \quad \text{where } c_1 \text{ and } c_2 \text{ are constants.}$$

Proof Let x_0 be a point in I, since ϕ_1, ϕ_2 of $L(y)=0$ are linearly independent. We know that $W(\phi_1,\phi_2)(x_0) \neq 0$.
Let $\phi_1(x_0)=\gamma, \quad \phi_2=\delta$

$$c_1\phi_1(x_0)+c_2\phi_2(x_0)=\gamma$$
$$c_1\phi_1'(x_0)+c_2\phi_2'(x_0)=\delta$$

since the determinant of the coefficients c_1, c_2 is just $W(\phi_1,\phi_2)(x_0) \neq 0$.

Here, c_1, c_2 are a unique pair of constants satisfying the equation. Choose c_1, c_2 as such constants.

Then, the functions $\psi = c_1\phi_1(x) + c_2\phi_2(x)$ is such that $\psi(x_0) = \phi(x_0)$ and $\psi'(x_0) = \phi'(x_0)$ and $L(\psi) = 0$.

By the uniqueness theorem, it follows that $\psi = \phi$ on I, i.e., $\phi = c_1\phi_1(x) + c_2\phi_2(x)$.

Theorem 4.2.2. Two solutions ϕ_1, ϕ_2 of $L(y) = 0$ *are linearly independent* on an interval I, if and only if

$$W(\phi_1, \phi_2)(x) \neq 0 \text{ for all } x \textit{ in } I.$$

Proof First, we suppose $W(\phi_1, \phi_2)(x) \neq 0$ for all $x \textit{ in } I$ and let c_1, c_2 be constants such that

$$c_1, \phi_1(x) + c_2, \phi_2(x) = 0 \tag{4.10}$$

for all $x \textit{ in } I$. Then, also

$$c_1, \phi_1'(x) + c_2, \phi_2'(x) = 0 \tag{4.11}$$

for all $x \textit{ in } I$.

For fixed x, Equations (4.10) and (4.11) are linear homogeneous equations satisfied by c_1, c_2.

The determinant of the coefficients is just $W(\phi_1, \phi_2)(x)$, which is not zero. Therefore, $c_{1,} = c_{2,} = 0$ is the the only solution of Equations (4.10) and (4.11).

This proves that ϕ_1, ϕ_2 are linearly independent.

Conversely, we suppose that ϕ_1, ϕ_2 are linearly independent on I, and also, suppose $x_0 \textit{ in } I$ such that

$$W(\phi_1, \phi_2)(x_0) = 0.$$

This implies that $c_1, \phi_1(x) + c_2, \phi_2(x) = 0$

$$c_1, \phi_1'(x) + c_2, \phi_2'(x) = 0 \tag{4.12}$$

has solution c_1, c_2 where at least one of these numbers is not zero. Let c_1, c_2 be such a solution, and consider the function $\varphi = c_1, \phi_1 + c_2, \phi_2$.

Now, $L(\varphi)=0$.
From Equation (4.12), we have

$$\varphi(x_0)=0 \text{ and } \varphi'(x_0)=0.$$

By the uniqueness theorem, we write $\varphi(x)=0$ for all $x\ in\ I$.
Hence, $c_1,\phi_1(x)+c_2,\phi_2(x)=0$ for all $x\ in\ I$.
But it contradicts assumption that ϕ_1,ϕ_2 are linearly independent on I.
Thus, our supposition must be false.
We have consequently proved that $W(\phi_1,\phi_2)(x)\neq 0$ for all $x\ in\ I$.

Theorem 4.2.3. Two solutions ϕ_1,ϕ_2 of $L(y)=0$ *are linearly dependent* on an interval I, if and only if $W(\phi_1,\phi_2)(x)$ is identically zero for all $x\ in\ I$.

Proof Suppose that ϕ_1,ϕ_2 are linearly dependent on I. *Then,* $\phi_2(x)=k\phi_1(x)$, for all $x\ in\ I$ and for some $k\in R$, and their Wronskian

$$W(\phi_1,\phi_2)(x)=\phi_1(x)\phi_2'(x)-\phi_2(x)\phi_1'(x)=\phi_1(x)\phi_2'(x)-k\phi_1(x)\phi_1'(x)$$
$$=\phi_1(x)[\phi_2'(x)-k\phi_1'(x)]=0$$

[Differentiating the above equation, we get $\phi_2'(x)=k\phi_1'(x)$].

$$\therefore W(\phi_1,\phi_2)(x)=0.$$

Conversely, we suppose that

$$W(\phi_1,\phi_2)(x)\equiv 0 \tag{4.13}$$

on I.

If either of $\phi_1\ or\ \phi_2$ is identically zero on I, then they are linearly dependent on I, so that we can take

$$\phi_1(x_0)\neq 0$$

for some $x_0\ on\ I$.

By continuity of ϕ_1 (ϕ_1 being a solution, is differentiable on I), there exists a subinterval I' of I containing x_0 such that

$$\phi_1(x)\neq 0 \tag{4.14}$$

for all $x \in I'$.

By Equations (4.13) and (4.14), we have

$$\frac{d}{dx}\left(\frac{\phi_2(x)}{\phi_1(x)}\right) = \frac{\phi_1(x)\phi_2'(x) - \phi_2(x)\phi_1'(x)}{[\phi_1(x)]^2} = 0.$$

on I'.

Hence, we get $\phi_2(x) = K\phi_1(x)$ for some K, and all $x \in I'$
by the uniqueness theorem for all $x \in I$ showing thatthey are linearly dependent on I.

Example 4.2.1. Show that two functions $y_1 = 1 + x$ and $y_2 = e^x$ are independent solutions of the differential equations $x\frac{d^2y}{d^2x} - (1+x)\frac{dy}{dx} + y = 0$.

Solution: Here, $W(y_1, y_2)(x) = \begin{bmatrix} y_1 & y_2 \\ y_1' & y_2' \end{bmatrix} = y_1y_2' - y_2y_1' = (1+x)e^x - e^x.1 =$ xe^x, which is never zero for all x. Therefore, it follows that they are independent.

Example 4.2.2. Test the functions

$$y_1 = 2x^2 \text{ and } y_2 = x\lceil x \rceil$$

on I contained in $\{1 \geq x \geq 0\}$ *and* $-1 < x < 1$ for a linear dependent or independent function.

Solution: These are linearly dependent because for
$x \geq 0$, $y_2 = x^2$
so that $y_1(x) = 2x^2 = 2y_2$, for all $x \in I$.
For $I = \{-1 < x < 1\}$,
the functions are linearly independent on I because

$$y_1(x) = \begin{cases} -2y_2(x), x < 0 \\ 2y_2(x), x > 0 \end{cases}. \tag{4.15}$$

Here, we cannot find p for which

$$y_1(x) = p\, y_2(x) \quad , x \in I.$$

We come to the conclusion that $y_1(x)$, $y_2(x)$ are linearly dependent on $0 \le x < 1$ but linearly independent on $-1 < x < 1$.

Also, $w(y_1, y_2) = \begin{vmatrix} 2x^2 & x^2 \\ 4x & 2x \end{vmatrix} = 0$ and $w(y_1, y_2) = \begin{vmatrix} 2x^2 & -x^2 \\ 4x & -2x \end{vmatrix} = 0$ [from Equation (4.16)].

Here, $w(y_1, y_2)$ is zero for $(-\infty, \infty)$.

Hence, we conclude that functions that are linearly dependent on an interval may not remain linearly dependent on a larger interval.

Theorem 4.2.4. Let y_1 and y_2 be two solutions of

$$y'' + a_1(x)y' + a_2(x)y = 0 \text{ on } a \le x \le b \text{ iff}$$
$$W(y_1, y_2)(x) \ne 0 \text{ for all } x \text{ on } a \le x \le b.$$

Proof Let $W(y_1, y_2)(x) \ne 0$ for all x on $a \le x \le b$.

Let c_1, c_2 be constants such that

$$c_1 y_1(x) + c_2 y_2(x) = 0 \text{ for all } x \text{ } on \text{ } a \le x \le b. \tag{4.16}$$

Then, also

$$c_1 y_1'(x) + c_2 y_2'(x) = 0 \text{ for all } x \text{ } on \text{ } a \le x \le b. \tag{4.17}$$

For a fixed x, Equations (4.17) and (4.18) are linear homogeneous equations satisfied by c_1, c_2.

$$\therefore W(y_1, y_2)(x) \ne 0.$$

Here, there is only a solution such that $c_1, = c_2 = 0$.

Imply that y_1 *and* y_2 are linearly independent.

Conversely,

let y_1, y_2 be linearly independent on $a \le x \le b$

Let there be a point x_0 in $a \le x \le b$ such that

$$W(y_1, y_2)(x_0) = 0.$$

This implies that the system of two equations

$$c_1 y_1(x_0) + c_2 y_2(x_0) = 0$$
$$c_1 y_1'(x_0) + c_2 y_2'(x_0) = 0$$

has a solution c_1, c_2 where at least one of these is not zero.

Let c_1, *and* c_2 be such a solution and consider the function

$$\psi = c_1 y_1 + c_2 y_2.$$

Here, $L(\psi) = 0$.

From Equation (4.17), $\psi(x_0) = 0$ *and* $\psi'(x_0) = 0$.

$$\therefore \quad \psi(x) = 0 \text{ for all } x \text{ on } a \le x \le b.$$

Thus, $c_1 y_1(x) + c_2 y_2(x) = 0$ for all x *on* $a \le x \le b$.

It results in a contradiction,

i.e., y_1, y_2 are linearly independent on $a \le x \le b$.

Hence, there is a point x_0 in $a \le x \le b$ such that

$$W(y_1, y_2)(x_0) = 0 \text{ is false.}$$

Hence, $W(y_1, y_2)(x_0) \neq 0$ for all x i.e., *on* $a \le x \le b$.

Example 4.2.3. Show that the Wronskian $w((y_1, y_2)(x)$ of any two independent solutions y_1, y_2 of the differential equation $y'' + a_1(x)y' + a_2(x)y = 0$ on I satisfy

$$w((y_1, y_2)(x) = w(y_1, y_2)(x_0)e^{\int_{x0}^{x} a_1(z)dz}.$$

Solution: Since y_1, y_2 is the solution of the differential equation $y'' + a_1(x)y' + a_2(x)y = 0$,

then

$$y_1'' + a_1(x)y_1' + a_2(x)y_1 = 0$$

$$y_2'' + a_1(x)y_2' + a_2(x)y_2 = 0.$$

From these equations, we get

$$\left[y_1y_2' - y_2y_1'\right]' + a_1\left[y_1y_2' - y_2y_1'\right] = 0$$

or, $w' + a_1 w = 0$, which is a linear differential equation.

The solution is $w = w(x_0)e^{\int_{x_0}^{x} a_1(z)dz}$.

Example 4.2.4. Show that $e^x \sin t$ and $e^x \cos x$ form a set of two linearly independent solutions of $y'' - 2y' + 2y = 0$. What is the general solution? Find the solution $y(x)$ satisfying the initial conditions $y(0) = 2; y'(0) = -3$.

Solution: Let $\phi_1 = e^x \sin x$ and $\phi_2 = e^x \cos x$. Now, $\phi_1' = e^x \sin x + e^x \cos x = e^x(\sin x + \cos x)$ and $\phi_1'' = e^x(\sin x + \cos x) + e^x(\cos x - \sin x)$.

Now, $\phi_1''(x) - 2\phi_1'(x) + 2\phi_1(x) = 2e^x \cos x - 2^x(\sin x + \cos x) + 2e^x \sin x = 0$ showing that $\phi_1(x)$ is a solution of the given equation.

Similarly, we can show that $\phi_2(x)$ is a solution of the given equation.

Now, the Wronskian $W(x) = \begin{pmatrix} e^x \sin x & e^x \cos x \\ e^x(\sin x + \cos x) & e^x(\cos x + \sin x) \end{pmatrix} = e^{2x} \neq 0$.

Hence, $\phi_1 = e^x \sin x$ and $\phi_2 = e^x \cos x$ are linearly independent solutions of the given equation.

The general solution is given by $\phi(x) c_1\phi_1(x) + c_2\phi_2(x) = e^x(c_1 \sin x + c_2 \cos x)$ where c_1 and c_2 are arbitrary constants.

Putting $x = 0$ in the general solution and using the given result $y(0) = 2$, we get $c_2 = 2$.

Putting $x = 0$ in the derivative of general solution and using the given result $y'(0) = -3$, we get $c_2 + c_1 = -3 \Rightarrow c_1 = -5$.

Hence, the solution of the given equation satisfying the initial value is $y = e^x(2\cos x - 5\sin t)$.

4.3 The Second-Order Homogeneous Equation

We are concerned with the second-order homogeneous linear equations with constant coefficients:

$$L(y) = y'' + a_1 y' + a_2 y = 0 \tag{4.18}$$

where a_1, a_2 are constants. We know that the equation

$$y' + ay = 0$$

has the solution $y = e^{-ax}$ where a can be taken as the root of the equation $r + a = 0$.

Let us try it for Equation (4.18). We find $L(e^{rx}) = (r^2 + a_1 r + a_2)e^{rx}$,
and e^{rx} will be a solution of $L(y) = 0$;
i.e., $L(e^{rx}) = 0$ if r satisfied $r^2 + a_1 r + a_2 = 0$.
Let $p(r) = r^2 + a_1 r + a_2$
and call p the *characteristic polynomial of* L.

Note. $p(r)$ can be obtained from $L(y)$ by replacing $y^{(k)}$ by r^k, where we use the conventions that the zero-th derivative of $y, y^{(0)}$, is y itself, and that $r^0 = 1$.

From the fundamental theorem of Algebra, we know that the polynomial p always has two complex roots r_1, r_2 (which may be real).

Case 1. If $r_1 \neq r_2$, we see that $e^{r_1 x}$ *and* $e^{r_2 x}$ are two distinct solutions of $L(y) = 0$.

Case 2. If possible, let $r_1 = r_2$, then

$$L(e^{rx}) = p(r)e^{rx} \tag{4.19}$$

for all r *and* x

We know if r_1 is the repeated root of p, then

$$p(r_1) = 0 \text{ and } p'(r_1) = 0.$$

This suggests that differentiating Equation (4.19) with respect to x,

$$\therefore \frac{\partial}{\partial r} L(e^{rx}) = L\left(\frac{\partial}{\partial r} e^{rx}\right) = L(xe^{rx}).$$

Hence, $L(xe^{rx}) = [p'(r) + xp(r)]e^{rx}$ $[\because \frac{\partial}{\partial r} L(e^{rx}) = \frac{\partial}{\partial r}[p(r)e^{rx}] = p'(r)$ $e^{rx} + p(r)xe^{rx}]$.

Now, putting $r = r_1$ in the above equation, we have

$$L\left(xe^{r_1 x}\right) = 0.$$

It shows that $xe^{r_1 x}$ is another solution in case $r_1 = r_2$.

Case 3. If the equation has complex roots $a + ib; a - ib$, then solutions are

$$\phi_1(x) = e^{r_1 x} \cos bx \; and \; \phi_2(x) = e^{r_2 x} sinbx.$$

Before getting the general solution, let us first note the following fact: If $c(x)$ is a complex valued solution of $L(y) = 0$, *then*

$$L(c(x)) = L[Rec(x) + iImc(x)] = L[Rec(x)] + iL[Imc(x)].$$

Since $L(c(x)) = 0$, this implies that

$$L[Rec(x)] = 0 \text{ and } L[Imc(x)] = 0.$$

This shows that if $L(y) = 0$ has a complex valued function $c(x)$ as a solution, then its real and imaginary parts are also solutions of $L(y) = 0$. *Now, we can write the following two solutions:*

$$e^{r_1 x} = e^{(a+ib)x} = e^{ax}[cosbx + isibbx]$$
$$e^{r_2 x} = e^{(a-ib)x} = e^{ax}[cosbx - isibbx].$$

Hence, the real part $e^{ax} cosbx = y_1(x)$ and
the imaginary part $e^{ax} sinbx = y_2(x)$ are the solutions of $L(y) = 0$.
Also, $\dfrac{y_1(x)}{y_2(x)} = \dfrac{e^{ax} cosbx}{e^{ax} sinbx} = cotbx \neq \text{constant}.$
So, the two solutions are linearly independent.
Hence, the general solution is

$$y(x) = e^{ax}[c_1 cosbx + c_2 sinbx].$$

4.3.1 Initial Value Problems for Second-Order Equations

Every solution of the equation

$$L(y) = y'' + a_1 y' + a_2 y = 0$$

is a linear combination of the solutions Equations (4.21) or (4.22), which will depend on showing that the initial value problems for this equation have unique solutions.

An *initial value problem* for $L(y) = 0$ is a problem of finding a solution ϕ satisfying

$$\Phi(x_0) = \alpha \quad \phi'(x_0) = \beta \tag{4.23}$$

where x_0 is some real number, and α and β are two given constants.

This problem is denoted

$$L(y) = 0, \quad y(x_0) = \alpha \quad y'(x_0) = \beta. \tag{4.24}$$

Theorem 4.3.3. For any real x_0, and α, β are two given constants, there exists a solution ϕ of initial value problem

$$L(y) = 0, \quad y(x_0) = \alpha \quad y'(x_0) = \beta. \tag{4.25}$$

on $-\infty < x < \infty$.

Proof We show that for unique constants c_1, c_2 such that $\phi = c_1\phi_1 + c_2\phi_2$ satisfies

$$\Phi(x_0) = \alpha \; \phi'(x_0) = \beta \tag{4.26}$$

where ϕ_1, ϕ_2 are the solutions given by

$$\begin{aligned} &\phi_1(x) = e^{r_1 x} \; and \; \phi_2(x) = e^{r_2 x} \\ &\text{or, } \phi_1(x) = e^{r_1 x} \; and \; \phi_2(x) = xe^{r_1 x}, \end{aligned} \tag{4.27}$$

we have

$$c_1\phi_1(x_0)+c_2\phi_2(x_0)=\alpha$$
$$\text{and } \phi'(x_0)=c_1\phi_1'(x_0)+c_2\phi_2'(x_0)=\beta \tag{4.28}$$

These equations will have a unique solution c_1, c_2 if the determinant

$$\Delta=\begin{vmatrix}\phi_1(x_0) & \phi_2(x_0)\\ \phi_1'(x_0) & \phi_2'(x_0)\end{vmatrix}=\phi_1(x_0)\phi_2'(x_0)-\phi_2(x_0)\phi_1'(x_0)\neq 0.$$

In case $r_1 \neq r_2$,

$$\phi_1(x)=e^{r_1x} \textit{ and } \phi_2(x)=e^{r_2x}.$$

$$\therefore \Delta=\phi_1(x_0)\phi_2'(x_0)-\phi_2(x_0)\phi_1'(x_0)=r_2e^{r_1x_0}e^{r_2x_0}-r_1e^{r_1x_0}e^{r_2x_0}$$
$$=(r_2-r_1)e^{(r_1+r_2)x_0},$$

which is not zero because $e^{(r_1+r_2)x_0}\neq 0.$

Hence, the determinant condition is satisfied in either case. Thus, if c_1, c_2 are unique constants satisfying Equation (4.25), the function

$$\phi=c_1\phi_1+c_2\phi_2$$

will be the desired solution satisfying

$$\Phi(x_0)=\alpha \quad \phi'(x_0)=\beta.$$

4.4 Non-Homogeneous Equation of Second-Order: Method of Variation of Parameters

We now return to the problem of finding all solutions of the equation

$$L(y)=y''+a_1y'+a_2y=b(x) \tag{4.29}$$

where b is some continuous function on the interval I.

The general solution of the above equation (non-homogeneous) in terms of the corresponding homogeneous equation is

$$y'' + a_1 y' + a_2 y = 0. \tag{4.30}$$

Suppose we know that ψ_p is a particular solution of this equation and that ψ is any other solution. Then,

$$L(\psi - \psi_p) = l(\psi) - L(\psi_p) = b - b = 0 \textit{ on } I.$$

This shows that $\psi - \psi_p$ is a solution of the homogeneous equation $L(y) = 0$.

If ϕ_1, ϕ_2 are the linearly independent solutions of $L(y) = 0$, *there are unique constants* c_1, c_2 such that

$$\psi - \psi_p = c_1\phi_1 + c_2\phi_2.$$

We may write in other words, i.e., every solution of $L(y) = b(x)$ can be written as

$$\psi = \psi_p + c_1\phi_1 + c_2\phi_2.$$

We see that the problem of finding all solutions of

$$L(y) = b(x)$$

reduces to finding a particular solution ψ_p from $L(y) = b(x)$ and two linearly independent solutions ϕ_1, ϕ_2 of $L(y) = 0$.

Note that if

$$L(\psi_p) = b \text{ and } L(\phi_1) = L(\phi_2) = 0,$$

and c_1, c_2 are any constants, then

$$\psi = \psi_p + c_1\phi_1 + c_2\phi_2$$

satisfies $L(y)=b(x)$.

Find the particular solution of

$$L(y)=b(x).$$

Every solution of $L(y)=0$ is of the form $L(y)=c_1\phi_1+c_2\phi_2$ where c_1,c_2 are any constants and two linearly independent solutions ϕ_1,ϕ_2 of $L(y)$.

Such a function $c_1\phi_1+c_2\phi_2$ cannot be a solution of $L(y)=b(x)$ unless $b(x)=0$ on I.

Suppose we allow c_1,c_2 to become functions of u_1,u_2 (not necessarily constants) on I, and then ask whether there is a solution of $L(y)=b(x)$ of the form $u_1\mathbf{f}_1+u_2\phi_2$ on I.

This procedure is known as the *variation of parameters*. We argue in reverse. Suppose we have a solution of $L(y)=b(x)$ of the form $u_1\mathbf{f}_1+u_2\phi_2$ where u_1,u_2 are functions. Then

$$\begin{aligned}&(u_1\phi_1+u_2\phi_2)''+a_1(u_1\phi_1+u_2\phi_2)'+a_2(u_1\phi_1+u_2\phi_2)=b(x)\\&=u_1L(\phi_1)+u_2L(\phi_2)+\left(u_1''\phi_1+u_2''\phi_2\right)+2\left(u_1'\phi_1''+u_2'\phi_2''\right)+a_1\left(u_1'\phi_1+u_2'\phi_2\right)\\&=\left(u_1''\phi_1+u_2''\phi_2\right)+2\left(u_1'\phi_1''+u_2'\phi_2''\right)+a_1\left(u_1'\phi_1+u_2'\phi_2\right)=b.\end{aligned}$$

We observe that if

$$u_1'\phi_1+u_2'\phi_2=0, \tag{4.31}$$

then

$$0=\left(u_1'\phi_1+u_2'\phi_2\right)'=\left(u_1'\phi_1''+u_2'\phi_2''\right)+\left(u_1''\phi_1+u_2''\phi_2\right),$$

and we must have

$$u_1'\phi_1'+u_2'\phi_2'=b. \tag{4.32}$$

Looking at this reasoning in reverse, we see that if we can find two functions u_1,u_2 satisfying Equations (4.28) and (4.29), then, indeed, $u_1\phi_1+u_2\phi_2$ will satisfy $L(y)=b(x)$.

Equations (4.28) and (4.29) are two linear equations for u_1',u_2' with a determinant, which is just the Wronskian $W(\phi_1,\phi_2)$, since we assumed ϕ_1,ϕ_2 to be a linear independent, and this determinant is never zero on

I, and there exist unique solutions u_1', u_2' Indeed, a little calculation shows that

$$u_1' = \frac{-\phi_2 b}{W(\phi_1, \phi_2)} \text{ and } u_2' = \frac{\phi_1 b}{W(\phi_1, \phi_2)},$$

i.e., if x_0 is in I, we may take for u_1, u_2

$$u_1(x) = -\int_{x_0}^{x} \frac{\phi_2(t) b(t)}{W(\phi_1, \phi_2)(t)} dt \quad u_2(x) = \int_{x_0}^{x} \frac{\phi_1(t) b(t)}{W(\phi_1, \phi_2)(t)} dt.$$

The solution $\psi_p(\mathrm{x}) = u_1\phi_1(x) + u_2\phi_2(x)$ takes the form

$$\psi_p(x) = \int_{x_0}^{x} \frac{[\phi_2(x)\phi_1(t) - \phi(x)_1 \phi_2(t)] b(t)}{W(\phi_1, \phi_2)(t)} dt. \tag{4.33}$$

Remark 4.4.1. The particular solution $\psi_p(x)$ used in the above general solution of Equation (4.27) may indeed be taken to be any particular solution because if ψ is any other particular solution of Equation (4.27), then

$$\psi + \phi(\mathrm{x}) = \phi(\mathrm{x}) + (\psi - y_p) + y_p.$$

Since $\psi - y_p$ is a solution of Equation (4.29),
hence $\phi(x) + (\psi - y_p)$
is again a solution of Equation (4.30) having two constants, and it may be taken as a general solution of Equation (4.30).

Remark 4.4.2. The general solution $\phi(\mathrm{x})$ of the homogeneous Equation (4.30) is often called the *complementary function* for the differential Equation (4.29), and $y_p(x)$ is called its *particular integral*.

Here, for finding the *particular integral* of obtaining the general solution of the linear non-homogeneous differential Equation (4.27), the procedure is known as *method of variation of parameters*.

We summarize our above results by the following theorem:

Theorem 4.4.1. Let $b(x)$ be continuous on an interval I. Every solution ψ of $L(y)=b(x)\, on\, I$ can be written as

$$\psi=\psi_p+c_1\phi_1+c_2\phi_2$$

where ψ_p is a particular solution, ϕ_1,ϕ_2 are two linearly independent solutions of $L(y)=0\, and\, c_1,c_2$ are constants. A particular solution ψ_p is given by Equation (4.30). Conversely, every such ψ is a solution of

$$L(y)=b(x).$$

For example, let us solve $L(y)=b(x)$. In this case,

$$p(r)=r^2+a_1r+a_2$$

has two distinct roots r_1,r_2 We get

$$\phi_1(x)=e^{r_1.x}\quad \phi_2(x)=e^{r_2.x},$$

$then\, W(\phi_1,\phi_2)(x)=(r_2-r_1)e^{(r_1+r_2)x}$

$$\phi_1(t)\phi_2(x)-\phi_1(x)\phi_2(t)=e^{r_1t}\, e^{r_2x}-e^{r_2t}e^{r_1x}.$$

Thus, every solution ψ of $L(y)=b(x)$ in this case has the form

$$\psi(x)=c_1e^{r_1.x}+c_2e^{r_2.x}+\frac{1}{(r_1.-r_2)}\int_{x_0}^{x}[e^{r_1.(x-t)}-e^{r_2.(x-t)}]b(t)dt$$

where x_0 is a real number, and c_1,c_2 are constants.

Alternative Working Method: This method is quite general and applies to equations of the following form:

$$y''+a_1y'+a_2y=b(x) \tag{4.34}$$

where a_1, a_2, b are some functions of x on the interval I.

It gives $P.I. = \int_{x_0}^{x} \frac{[\phi_2(x)\phi_1(t) - \phi(x)_1 \phi_2(t)]b(t)}{W(\phi_1,\phi_2)(t)} dt$

where ϕ_1, ϕ_2 are the solutions of $y'' + a_1 y' + a_2 y = 0$ and $W(\phi_1,\phi_2) = \begin{vmatrix} \phi_1(x_0) & \phi_2(x_0) \\ \phi_1'(x_0) & \phi_2'(x_0) \end{vmatrix}$.

Example 4.4.1. Solve $y'' - y' - 2y = e^{-x}$.

Solution: The characteristic polynomial

$$r^2 - r - 2 = (r+1)(r-2),$$

and therefore, two linearly independent solutions ϕ_1, ϕ_2 of the homogeneous equation are

$$\phi_1 = e^{-x},\ \phi_2 = e^{2x}.$$

A particular solution ψ_p of the non-homogeneous equation is of the form

$$\psi_p = u_1(x)e^{-x} + u_2(x)e^{2x}$$

where u_1', u_2' satisfy the equations

$$u_1'\phi_1 + u_2'\phi_2 = 0 \tag{i}$$

$$u_1'\phi_1' + u_2'\phi_2' = b \tag{ii}$$

i.e., $u_1'e^{-x} + u_2'e^{2x} = 0$

and $-u_1'e^{-x} + 2u_2'e^{2x} = e^{-x}$,

Solving, u_1', u_2', we get,

$$u_1'(x) = -\frac{1}{3} \text{ and } u_2'(x) = \frac{1}{3}e^{-3x}$$

and $u_1(x) = -\frac{x}{3}$ $u_2(x) = -\frac{1}{9}e^{-3x}$.

Thus, $\psi_p = u_1(x)e^{-x} + u_2(x)e^{2x} = -\frac{x}{3}e^{-x} = -\frac{1}{9}e^{-3x}e^{2x} = -\frac{x}{3}e^{-x} - \frac{1}{9}e^{-x}$.

Since $\frac{1}{9}e^{-x}$ is the solution of the homogeneous equation, we take $-\frac{x}{3}e^{-x}$ as a particular solution of the non-homogeneous equation.

The general solution ψ of the non-homogeneous equation is

$$\psi = \psi_p + c_1\phi_1 + c_2\phi_2 = -\frac{x}{3}e^{-x} + c_1e^{-x} + c_2e^{2x}$$

where c_1, c_2 are constants.

4.5 Higher-Order Homogeneous Linear Differential Equations with Constant Coefficients

Everything we have done for second-order differential equations can be extended over to the case of order n. Let us consider the equation

$$L(y) = y^{(n)} + a_1y^{(n-1)} + 2y^{(n-1)} + \ldots + a_ny \tag{4.35}$$

where $a_1, a_2, \ldots\ldots\ldots a_n$ are constants. We try to solve $L(y) = 0$ as before by trying an exponential e^{rx}.

We have

$$L(y) = p(r)e^{rx} \tag{4.36}$$

where $p(r) = r^n + a_1r^{n-1} + a_2r^{n-2} +, \ldots\ldots\ldots + a_n$.

$p(r)$ is a *characteristic polynomial* of L.

If r_1 is a root of p then

$$L(e^{r_1x}) = 0,$$

and we have a solution e^{r_1x}.

If r_1 is a root of multiplicity m_1 of p, *then*

$$p(r_1)=0,\quad p'(r_1)=0,\ldots\ldots p^{(m_1-1)}(r_1)=0$$

If we differentiate Equation (4.31) k times with respect to r_1 we obtain[1]

$$\frac{\partial^k}{\partial r^k}L(e^{rx})=L\left(\frac{\partial^k}{\partial r^k}e^{rx}\right)=L(x^k e^{rx})$$
$$=\left[p^{(k)}(r)+kp^{(k-1)}(r)x+\frac{k(k-1)}{2!}p^{(k-2)}(r)x^2+\cdots.+p(r)x^k\right]e^{rx}.$$

Thus, for $k=0,1,\ldots m_1-1$, we see that $x^k e^{r_1 x}$ is a solution of $L(y)=0$. By repeating this process, we get the following result.
Note. If f,g are two functions having k derivatives, then

$$(fg)^{(k)}\ f^{(k)}g+kf^{(k-1)}g'+\frac{k(k-1)}{2!}f^{(k-2)}g''+\ldots fg^{(k)}$$

Theorem 4.5.1. Let $r_1,\ldots r_s$ be the distinct roots of the characteristic polynomial p, and suppose r_i has multiplicity m_i (thus $m_1+m_2+\ldots+m_s=n$). Then, the n *functions*

$$e^{r_1 x}, xe^{r_1 x},\ldots x^{m_1-1}e^{r_1 x}$$
$$e^{r_2 x}, xe^{r_2 x},\ldots x^{m_2-1}e^{r_2 x}$$
$$\cdots\cdots\cdots\cdots\cdots\cdots$$
$$e^{r_s x}, xe^{r_s x},\ldots x^{m_s-1}e^{r_s x}$$

are solutions of $L(y)=0$.

Note 1. The n functions $\phi_1,\phi_2,\ldots\phi_n$ on an interval I are said to be linearly dependent on I if there are constants $c_1,c_2,\ldots c_n$, not all zero, such that

$$c_1\phi_1+c_2\phi_2\ldots.+c_n\phi_n=0 \text{ for all } x \text{ in } I.$$

The n functions $\phi_1,\phi_2,\ldots\phi_n$ on an interval I are said to be linearly independent on I if they are not linearly dependent.

Note 2. The n solutions of $L(y)=0$ given above in the theorem are linearly independent on any interval I.

If $\phi_1,\phi_2,\ldots\phi_m$. *are any n* solutions of $L(y)=0$ on an interval I, and $c_1,c_2,\ldots.c_m$ are any m constants, then

$$\phi = c_1\phi_1 + c_2\phi_2 \ldots . + c_m\phi_m$$

is also a solution since

$$L(\phi) = c_1L(\phi_1) + c_2L(\phi_2) \ldots . + c_mL(\phi_m) = 0.$$

As an example, consider the equation

$$y'' - 3y' + 2y = 0.$$

The characteristic polynomial is

$$p(r) = r^2 - 3r + 2,$$

and its roots are 1,1,−2. Thus, three linearly independent solutions are

$$e^x,\ xe^x \text{ and } e^{-2x},$$

and any solution ϕ has the form

$$\phi(x) = (c_1 + c_2x)e^x + c_3e^{-2x}$$

where c_1,c_2,c_3 are any constants.

4.5.1 Initial Value Problems for nth-Order Equations

An *initial value problem* for L(y)=0 is a problem of finding a solution ϕ, which has prescribed values for it, and its first $n-1$ derivatives, at some point x_0 (the initial point).

If $\alpha_1,\alpha_2,\ldots\alpha_n$ are given constants, and x_0 is some real number, the problem of finding a solution ϕ of $L(y)=0$ satisfying

$$\Phi(x_0)=\alpha_1,\ \Phi'(x_0)=\alpha_2,\ldots\ldots\ldots\Phi^{(n-1)}(x_0)=\alpha_n$$

is denoted by

$$L(y)=0\ y(x_0)=\alpha_1,\ y'(x_0)=\alpha_2,\ldots\ldots\ldots y^{(n-1)}(x_0)=\alpha_n.$$

There is only one solution to such an initial value problem, and the demonstration of this will depend on an estimate for the rate of growth of a solution ϕ of $L(y)=0$, *together with its derivatives* $\Phi',\ldots\ \Phi^{(n-1)}$. We define

$$[\![\phi(x)]\!]=[|\Phi'(x)|^2+\cdots.+|\Phi^{(n-1)}(x)|^2]^{1/2}.$$

Theorem 4.5.2. Let ϕ be any solution of

$$L(y)=y^{(n)}+a_1y^{(n-1)}+2y^{(n-1)}+\cdots+a_ny=0$$

on an interval I containing a point x_0. Then, for all x *in* I

$$[\Phi(x_0)]e^{-k|x-x_0|}\le[\Phi(x)]\le[\Phi(x_0)]e^{k|x-x_0|}.$$

Theorem 4.5.3. Let $\alpha_1,\alpha_2,\ldots\alpha_n$ be given constants, and x_0 is some real number on any interval I containing x_0. There exists at most one solution ϕ of $L(y)=0$ satisfying

$$\Phi(x_0)=\alpha_1,\Phi'(x_0)=\alpha_2,\ldots\Phi^{(n-1)}(x_0)=\alpha_n.$$

Theorem 4.5.4. If $\phi_1,\phi_2,\ldots\phi_n$ are n solutions of $L(y)=0$ on any interval I containing x_0, they are linearly independent there, if and only if

$$W(\phi_1,\phi_2,\ldots\phi_n)\neq 0\ \textit{for all x in I}.$$

Theorem 4.5.5. If $\phi_1, \phi_2, \ldots \phi_n$ are linearly independent solutions of $L(y)=0$ on any interval I, if $W(\phi_1, \phi_2, \ldots \phi_n) \neq 0$ *for all x in I*, and if $c_1, c_2, \ldots . c_n$ are any n constants, then

$$\phi = c_1\phi_1 + c_2\phi_2 \ldots . + c_m\phi_n \tag{4.37}$$

is a solution, and every solution may be represented in this form.

Theorem 4.5.6. Let $\phi_1, \phi_2, \ldots \phi_n$ be n solutions of $L(y)=0$ on any interval I containing x_0, then

$$W(\phi_1, \phi_2, \ldots \phi_n)(\mathrm{x}) = e^{-\alpha_1(x-x_0)} W(\phi_1, \phi_2, \ldots \phi_n)(x_0). \tag{4.38}$$

Example 4.5.1. Let us consider a homogeneous equation of order 3, which has a root r_1 with multiplicity 3. Its characteristic polynomial is

$$p(r) = (r - r_1)^3 = r^3 - 3r_1r^2 + 3r_1^2 r - r_1^3.$$
$$\therefore L(y) = y''' - 3r_1 y'' + 3r_1^2 y' - r_1^3 y,$$

and we get $\alpha_1 = -3r_1$

$$\Phi_1(x) = e^{r_1 x}, \Phi_2(x) = xe^{r_1 x}\ \ \Phi_2(x) = x^2 e^{r_1 x}.$$

We get

$$W(\phi_1, \phi_2, \ldots \phi_n) = \begin{vmatrix} e^{r_1 x} & xe^{r_1 x} & x^2 e^{r_1 x} \\ r_1 e^{r_1 x} & (1 + r_1 x)e^{r_1 x} & (2x + r_1 x^2)e^{r_1 x} \\ r_1^2 e^{r_1 x} & (2r_1 + r_1^2 x)e^{r_1 x} & (2x + r_1^2 x^2)e^{r_1 x} \end{vmatrix}$$

Using Equation (4.34), with $x_0 = 0$, we obtain

$$W(\phi_1, \phi_2, \ldots \phi_n) = \begin{vmatrix} 1 & 0 & 0 \\ r_1 & 1 & 0 \\ r_1^2 & 2r_1 & 2 \end{vmatrix} = 2.$$

Hence, $W(\phi_1, \phi_2, \ldots \phi_n)(x) = 2e^{3r_1 x}$.

4.5.2 The n-th Order Non-Homogeneous Equation

Let b be a continuous function on interval I, and consider the equation

$$L(y) \equiv y^{(n)} + a_1 y^{(n-1)} + a_2 y^{(n-2)} + \cdots + a_n y = b(x)$$

where $a_1, a_2, \ldots a_n$ are constants. If ψ_p is a particular silution of $L(y) = b(x)$, and ψ is any other solution, then

$$L(\psi - \psi_p) = L(\psi) - L(\psi_p) = b - b = 0.$$

Then, $\psi - \psi_p$ is a solution of the homogeneous equation $L(y) = 0$. It implies that any solution ψ of $L(y) = b(x)$ can be written as

$$\psi = \psi_p + c_1\phi_1 + c_2\phi_2 + \cdots .. + c_n\phi_n$$

where ψ_p is a particular silution of $L(y) = b(x)$, and the functions $\phi_1, \phi_2 \ldots, \phi_n$ are n linearly independent solutions of $L(y) = 0$, and $c_1, c_2, \ldots c_n$ are constants.

To find a particular solution ψ_p, we proceed, i.e., we use the variation of parameter method. We try to find n functions $u_1, u_2, \ldots\ldots u_n$, so that

$$\psi_p = u_1\phi_1 + u_2\phi_2 + \ldots\ldots + u_n\phi_n$$

is a solution. We see before

$$u_1'\phi_1 + u_2'\phi_2 + \ldots\ldots + u_n'\phi_n = 0,$$

then $\psi_p' = u_1\phi_1' + u_2\phi_2' + \ldots\ldots + u_n\phi_n'$,
and if $u_1'\phi_1' + u_2'\phi_2' + \ldots\ldots + u_n'\phi_n' = 0$,
we have $\psi_p'' = u_1\phi_1'' + \ldots\ldots + u_n\phi_n''$.

Thus, if $u_1', u_2', \ldots\ldots u_n'$ satisfy

$$\begin{aligned}
&u_1'\phi_1 + u_2'\phi_2 + \ldots\ldots + u_n'\phi_n = 0\\
&u_1'\phi_1' + u_2'\phi_2' + \ldots\ldots + u_n'\phi_n' = 0\\
&\ldots\ldots\ldots\ldots\ldots\ldots\ldots\ldots\\
&\ldots\ldots\ldots\ldots\ldots\ldots\ldots\ldots\\
&u_1'\phi_1^{(n-2)} + \ldots + u_n'\phi_n^{(n-2)} = 0\\
&u_1'\phi_1^{(n-1)} + \ldots + u_n'\phi_n^{(n-1)} = b,
\end{aligned} \tag{4.39}$$

we see that

$$\begin{aligned}
&\psi_p = u_1\phi_1 + u_2\phi_2 + \ldots\ldots + u_n\phi_n\\
&\psi_p' = u_1\phi_1' + u_2\phi_2' + \ldots\ldots + u_n\phi_n'\\
&\ldots\ldots\ldots\ldots\ldots\ldots\ldots\ldots\\
&\psi_p^{(n-1)} = u_1\phi_1^{(n-1)} + u_2\phi_2^{(n-1)} + \ldots\ldots + u_n\phi_n^{(n-1)}\\
&\psi_p^{(n)} = u_1\phi_1^{(n)} + u_2\phi_2^{(n)} + \ldots\ldots + u_n\phi_n^{(n)} + \mathrm{b}.
\end{aligned} \tag{4.40}$$

Hence,

$$L(\psi_p) = u_1 L(\phi_1) + u_2 L(\phi_2) + \ldots\ldots + u_n L(\phi_n) + b = b,$$

and ψ_p is a solution of $L(y) = b(x)$.

The whole problem is now reduced to solving the linear system Equation (4.38) for $u_1', u_2', \ldots u_n'$. The determinant of the coefficients is just $W(\phi_1, \phi_2 \ldots, \phi_n)$, which is never zero when $\phi_1, \phi_2 \ldots, \phi_n$ are linearly independent solutions of $L(y) = 0$.

Therefore, there are unique functions $u_1', u_2', \ldots u_n'$ satisfying Equation (4.37). The solutions are given by

$$u_k'(x) = \frac{W_k(x) b(x)}{W(\phi_1, \phi_2 \ldots, \phi_n)(x)}, (\mathrm{k} = 0, 1, \ldots \mathrm{n})$$

where W_k is the determinant obtained from $W(\phi_1,\ldots,\phi_n)$ by replacing the k-th column (i.e., $\phi_k,\phi_k',..\phi_k^{(n-1)}$ by 0,0,…0,1.

If x_0 be any point in I, we may take u_k given by

$$u_k = \int_{x_0}^{x} \frac{W_k(t)b(t)}{W(\phi_1,\phi_2\ldots,\phi_n)(t)}\,\mathrm{dt}\,(\mathrm{k}=0,1,\ldots\mathrm{n}).$$

The particular solution ψ_p now takes the form

$$\psi_p(x) = \sum_{k=1}^{n} \phi_k(x) \int_{x_0}^{x} \frac{W_k(t)b(t)}{W(\phi_1,\phi_2\ldots,\phi_n)(t)}\,\mathrm{dt}. \tag{4.41}$$

Theorem 4.5.7. Let $b(x)$ be continuous on an interval I. Every solution ψ of $L(y)=b(x)\, on\, I$ can be written as

$$\psi = \psi_p + c_1\phi_1 + \ldots\ldots + c_n\phi_n$$

where ψ_p is a particular solution, $\phi_1,\phi_2\ldots,\phi_n$ are n linearly independent solutions of $L(y)=0, and$

$c_1,c_2,\ldots c_n$ are constants. A particular solution ψ_p is given by Equation (4.40).

Note. It can be shown that the particular solution ψ_p in Equation (4.40) satisfies

$$\psi_p(x_0) = \psi_p'(x_0) = \psi_p^{(n-1)}(x_0) = 0. \tag{4.42}$$

Example 4.5.2. Solve

$$y''' + y'' + y' + y = 1, \tag{i}$$

which satisfies

$$\psi(0) = 0,\ \psi'(0) = 1,\ and\ \psi''(0) = 0. \tag{ii}$$

Solution: The homogeneous equation is

$$y''' + y'' + y' + y = 0. \qquad \text{(iii)}$$

Its characteristic polynomial corresponding to it is

$$p(r) = r^3 + r^2 + r + 1.$$

The roots of p are $i, -i,$ and -1. Since we are interested in a solution satisfying real initial conditions, we take the independent solutions of Equation (iii):

$$\phi_1(x) = cosx \quad , \phi_1(x) = sinx \quad , \phi_1(x) = e^{-x}.$$

To get a particular solution of Equation (i) of the form

$$u_1\phi_1 + u_2\phi_2 + u_3\phi_3,$$

we must solve the following equations: u_1', u_2', u_3'

$$\begin{aligned} u_1'\phi_1 + u_2'\phi_2 + u_3'\phi_3 &= 0 \\ u_1'\phi_1' + u_2'\phi_2' + u_3'\phi_3' &= 0 \\ u_1'\phi_1'' + u_2'\phi_2'' + u_3'\phi_3'' &= 1, \end{aligned}$$

which, in this case, reduce to

$$\begin{aligned} &u_1' cosx + u_2' sinx + u_3' e^{-x} = 0 \\ &-sinx u_1' + cosx u_2' - e^{-x} u_3' = 0 \\ &-cosx u_1' - sinx u_2' + e^{-x} u_3' = 1. \end{aligned} \qquad \text{(iv)}$$

The determinant of these coefficients is

$$W(\phi_1, \phi_2, \phi_3)(x) = \begin{vmatrix} cosx & sinx & e^{-x} \\ -sinx & cosx & -e^{-x} \\ -cosx & -sinx & e^{-x} \end{vmatrix}.$$

Using Equation (4.35), we have

$$W(\phi_1,\phi_2,\phi_3)(x)=e^{-x}W(\phi_1,\phi_2,\phi_3)(0)$$

since $\alpha_1=1$ in this case. Now,

$$W(\phi_1,\phi_2,\phi_3)(0)=\begin{vmatrix} 1 & 0 & 1 \\ 0 & 1 & -1 \\ -1 & 0 & 1 \end{vmatrix}=2$$

$$(\phi_1,\phi_2,\phi_3)(x)=2e^{-x}$$

Solving Equation (4.41) for u_1, we find that

$$u_1'(x)=\frac{1}{2}e^{x}\begin{vmatrix} 0 & sinx & e^{-x} \\ 0 & cosx & -e^{-x} \\ 1 & -sinx & e^{-x} \end{vmatrix}=-\frac{1}{2}(cosx+sinx) \quad \text{(v)}$$

Similarly,

$$u_2'(x)=\frac{1}{2}(cosx-sinx) \quad \text{(vi)}$$

$$and\ u_3'(x)=\frac{1}{2}e^{x}. \quad \text{(vii)}$$

Integrating all the above equations, we get

$$u_1(x)=\frac{1}{2}(cosx-sinx)$$

$$u_2(x)=\frac{1}{2}(sinx+cosx)$$

$$u_3(x)=\frac{1}{2}e^{x}.$$

The particular solution of Equation (i) is

$$u_1\phi_1 + u_2\phi_2 + u_3\phi_3 = \frac{1}{2}(cosx - sinx)cosx + \frac{1}{2}(sinx + cosx)sinx + \frac{1}{2}e^x e^{-x}$$
$$= 1.$$

The general solution ψ of Equation (i) is

$$\psi(x) = 1 + c_1 cosx + c_2 sinx + c_3 e^{-x}$$

where c_1, c_2, c_3 are constants.

We choose these constants so that the conditions Equation (ii) are valid. Now, we get

$$c_1 + c_3 {}_{.} = -1, c_2 - = 1, c_1 - c_3 = 0,$$

which implies $c_1 = -\frac{1}{2}, c_2 = \frac{1}{2}, c_3 = -\frac{1}{2}$.

Therefore, the solution of the given equation is

$$\psi(x) = 1 + c_1 cosx + c_2 sinx + c_3 e^{-x} = 1 + \frac{1}{2}(sinx - cosx - e^{-x}).$$

The solution corresponding to Equation (4.36) *with* $x_0 = 0$ *is to be* $\psi_p(x) = 1 - \frac{1}{2}(sinx + cosx + e^{-x}) = 1$, and this satisfies

$$\psi_p(0) = 0 \quad \psi_p{}'(0) = 0 \quad \psi_p{}''(0) = 0.$$

4.5.3 A Special Method for Solving the Non-Homogeneous Equation

We now give a method, which is often faster, in solving the non-homogeneous equation

$$L(y) = b(x)$$

when b is a solution of some homogeneous equation

$$M(y)=0$$

with constant coefficients. Thus, $b(x)$ must be a sum of terms of type $P(x)e^{ax}$, where P is a polynomial and a is a constant.

Suppose $L\, and\, M$ have constant coefficients, and have orders $n\, and\, m$, respectively.

If ψ is a solution of $L(y)=b(x) and\ M(b)=0$, then clearly

$$M(L(\psi))=M(b)=0.$$

This shows that ψ is a solution of a homogeneous equation

$$M(L(y))=0$$

with constant coefficients of order $m+n$ linearly independent solutions of

$$M(L(y))=0.$$

Not every linear combination will be a solution of $L(y)=b(x)$, however.

Thus, to find out what conditions must be satisfied by the constants, we substitute back into

$$L(y)=b(x).$$

This always leads to a determination of a set of coefficients.

4.6 Examples

Example 4.6.1. Solve $y''-3y'+2y=x^2$.
Solution: Since x^2 is a solution of $M(y)=y'''=0$,

we see that every solution ψ of $\mathrm{L}(\mathrm{y})=x^2$ is a solution of

$$M(L(y))=y^5-3y^4+2y^3=0.$$

The characteristic polynomial of this equation is

$p(r) = r^3(r^2 - 3r + 2)$, just the product of the characteristic polynomials for L *and* M. The roots are 0,0,0,1,2, and hence,

$$\psi(x) = c_0 + c_1x + c_2x^2 + c_3e^x + c_4e^{2x}.$$

Here, $c_3e^x + c_4e^{2x}$ is just a solution of $L(y) = 0$, since we are interested only in a particular solution ψ_p *of* $L(y) = x^2$

We can assume ψ_p has the form

$$\psi_p(x) = c_0 + c_1x + c_2x^2.$$

The problem is to determine the constants c_0, c_1, c_2 so that

$$L(\psi_p) = x^2.$$

We have $\psi_p'(x) = c_1 + 2c_2x$; $\psi_p''(x) = 2c_2$

and $L(\psi_p) = \psi_p'' - 3\psi_p' + 2\psi_p$

$$= 2c_2 - 3(c_1 + 2c_2x) + 2(c_0 + c_1x + c_2x^2)$$

$$= (2c_2 - 3c_1 + 2c_0) + (-6c_2 + 2c_1)x + 2c_2x^2,$$

i.e., $(2c_2 - 3c_1 + 2c_0) + (-6c_2 + 2c_1)x + 2c_2x^2 \equiv x^2$,

which implies $2c_2 - 3c_1 + 2c_0 = 0$

$-6c_2 + 2c_1 = 0$

$2c_2 = 1.$

Solving these, we get $c_0 = \frac{7}{4}$, $c_1 = \frac{3}{2}$ and $c_2 = \frac{1}{2}$.

Therefore, $\psi_p(x) = \frac{7}{4} + \frac{3}{2}x + \frac{1}{2}x^2 = \frac{1}{4}(7 + 6x + 2x^2)$ is a particular solution of $L(y) = x^2$

We call this method the *annihilator method* because to solve $L(y) = b(x)$, we find an M, which makes $M(b) = 0$, i.e., *annihilates b.*

Example 4.6.2. Solve

$$L(y) = Ae^{ax} \tag{i}$$

where L has characteristic polynomial p, *and* A, a are constants.

Suppose a is not a root of p.

The operator M given by

$$M(y) = y' - ay,$$

with characteristic polynomial $r - a$, annihilates Ae^{ax}.

The characteristic polynomial of

ML *is* $(r-a)p(r)$, and a is a simple root (multiplicity 1) of this.

Thus, any solution ψ of Equation (i) has the form

$$\psi = \mathrm{B}e^{ax} + \phi$$

where $L(\phi) = 0$, and B is a constant.

Placing ψ back into Equation (i), we obtain

$$L(\psi) = BL(e^{ax}) + L(\phi) = Bp(a)e^{ax} = Ae^{ax}.$$

Since $p(a) \neq 0$,
we get $B = A / p(a)$.

Now, if a is not a root of the characteristic polynomial of L, there is a solution *of* Equation (i) of the form

$$\psi(\mathrm{x}) = \frac{A}{p(a)} e^{ax}.$$

Now, Equation (i) of Example 4.5.2. is:

$$y''' + y'' + y' + y = 1.$$

The right hand is of the form Ae^{ax}
for this $A = 1, a = 0$.

The characteristic polynomial is

$p(r) = r^3 + r^2 + r + 1$ with points $i, -i, -1$.

Therefore, a solution of this equation is given by

$$\psi(x)\frac{A}{p(a)}e^{ax}=\frac{1}{p(0)}e^{0x}=1.$$

4.7 Exercise

1. Show that the functions $y_1(x)=x$ and $y_2(x)=sinx$ and $y_3(x)=e^x$ are linearly independent on I.
2. Solve the equation

$$y''-\frac{2}{x}y'+\frac{2y}{x^2}=xsinx,\quad x\in[1,\infty].$$

3. Find all the solutions of

(i) $y^{(4)}-4y^{(3)}+6y''-4y'+y=e^x$
(ii) $y''+2i\,y'+y=0$
(iii) $y''-4y'+5y=0.$

4. Show that the differential equation

$$x^3y'''-6xy'+12y=0$$

has three linearly independent solutions of the form $y=x^r$, and compute the Wronskian of the solution.

5. Compute the solutions $y'''+y''+y'+y=1$ with the initial conditions

$$y(0)=0, y'(o)=1, y''(0)=0.$$

6. Find the particular solution of the following equations:
(i) $y''+4y=tan2x$
(ii) $y^{(4)}+8y^{(3)}+24y''+32y'+16y=\frac{120}{x^2}e^{-2x}.$

7. (a) Show that the functions ϕ_1, ϕ_2 defined by

$$\phi_1(x) = x^2 \cdots, \cdots \phi_1(x) = x|x|$$

are linearly independent for $-\infty < x < \infty$.
(b) Compute the Wronskian of these functions.

8. Suppose the functions ϕ_1, ϕ_2 are linearly independent solutions of equation

$$y'' + \alpha_1 y' + \alpha_2 y = 0,$$

and let $W(\phi_1, \phi_2)$ be abbreviated to W. Show that W is a constant if and only if $\alpha_1 = 0$.

9. Find all solutions of the following equations:

(i) $y'' + 9y = sin3x$
(ii) $4y'' - y = e^x$
(iii) $4y'' + 5y' - 6y = x.$

10. Find all solutions of the following equations:

(i) $y''' - 8y = 0$
(ii) $y^{(4)} + 16y = 0$
(iii) $y^{(100)} + 100y = 0$
(iv) $y''' - 3iy'' - 3y' + iy = 0.$

11. (a) Compute the Wronskian of four independent solutions of the equation

$$y^{(4)} + 16y = 0.$$

(b) Compute the solution ϕ of this equation, which satisfies

$$\phi(0) = 1 \phi'(0) = 0, \phi''(0) = 0 \; and \; \phi'''(0) = 0.$$

12. Find four linearly independent solutions of the equation

$$y^{(4)} + \lambda y = 0$$

in case (a)λ=0, (b) λ>0, (c) λ<0.

13. Consider the equation

$$y^{(5)} - y^{(4)} - y' + y = 0.$$

(a) Compute five linearly independent solutions.
(b) Compute the Wronskian of the solutions found in (a) using Theorem 4.5.6.
(c) Find the solution ϕ satisfying

$$\phi(0) = 1\ \phi'(0) = \phi''(0) = \phi'''(0) = \phi^{(4)}(0) = 0.$$

14. Find all solutions of the following equations:

(i) $y''' - y' = x$
(ii) $y^{(4)} + 16y = cosx$
(iii) $y^{(4)} - y = cosx.$

15. Let

$$L(y) = y^{(n)} + a_1 y^{(n-1)} + \ldots + a_n y$$

and p be the characteristic polynomial

$$\mathrm{p(r)} = r^n + a_1 r^{n-1} + \ldots + a_n.$$

If $A\ and\ \alpha$ are constants, and $p(r) \neq 0$, show that there is a solution of $L(y) = Ae^{ax}$ of the form Be^{ax}, where B is a constant. What is B?

16. Using the annihilator method, find a particular solution of following equations:

    (i) $y'' - y' - 2y = x^2 + cosx$
    (ii) $y''' = x^2 + e^{-x} sinx$
    (iii) $y''' + 3y'' + 3y' + y = x^2 e^{-x}$.

17. Consider the constant coefficient operator L with characteristic polynomial p. Consider the equation $L(y) = e^{ax}$ where a is a root of p with multiplicity k. Show by annihilator method that a solution is given by

$$\psi(x) = \frac{x^k e^{ax}}{p^{(k)}(a)}$$

5
Adjoint Equation

Abstract
The aim of this chapter is to find solutions of second-order linear differential equations using integrating factors. To determine the integrating factor, we construct another differential equation called the adjoint equation, whose solution serves as the integrating factor for the original equation. We also study self-adjoint equations and Green's formula. Adjoint equations are powerful tools used in various fields, particularly in optimization and sensitivity analysis.

***Keywords*:** Lagrange identity, adjoint equation, self-adjoint equation, Green's formula, Bilinear concomitant

5.1 Introduction

This chapter's objective is to use an integrating factor to solve linear differential equations. We will construct another differential equation known as the *adjoint equation*, in order to determine the necessary integrating factor. An integrating factor of the given differential equation is its solution. In this chapter, we also define the self-adjoint equation. It is a special case where the adjoint equation is identical to the original equation. We shall present a simple method of finding all solutions of a linear differential equation of order two when one of its solution is known.

Adjoint equations are used to efficiently calculate the sensitivity of a specific output of the primal equation with respect to changes in the input parameters. This is particularly useful in optimization problems where we need to know how changes in design parameters affect the outcome. We discussed the properties of adjoint and self adjoint differential equations.

5.2 Adjoint Equation

In the theory of first-order differential equations, the idea of an integrating factor is useful. Higher-order equations might make advantage of it.

Dipankar De. Ordinary Differential Equations and Special Functions, (181–202) © 2026 Scrivener Publishing LLC

Now let the equation

$$L(y) \equiv p_0(x)\frac{d^n y}{dx^n} + p_1(x)\frac{d^{n-1}y}{dx^{n-1}} + \ldots\ldots. + p_{n-1}(x)\frac{dy}{dx} + p_n(x)y = 0. \tag{5.1}$$

Suppose there exists a function $v(x)$ such that vL(y) is an exact differential.

We have the result:

$$WU^{(r)} = \frac{d}{dx}\left[U^{(r-1)}W - U^{(r-2)}W' + \ldots\ldots + (-1)^{r-1}UW^{(r-1)}\right] + (-1)^r UW^{(r)}. \tag{5.2}$$

Therefore [applying this differentiation on Equation (5.1) (multiplied by v)], step by step, we get the following equation:

$$\begin{aligned} vL(y) &= \frac{d}{dx}\left[y^{(n-1)}(p_0v) - U^{(n-2)}(p_0v)' + \ldots\ldots + (-1)^{n-1}U(p_0v)^{(n-1)}\right] \\ &+ \frac{d}{dx}\left[y^{(n-2)}(p_1v) - U^{(n-3)}(p_1v)' + \ldots\ldots + (-1)^{n-2}U(p_1v)^{(n-2)}\right] + \ldots.. \\ &+ \frac{d}{dx}\left[y'(p_{n-2}v) - y(p_{n-2}v)'\right] + \frac{d}{dx}\left[yp_{n-1}v\right] \\ &+ \left\{y\left[(-1)^n\frac{d^n}{dx^n}(p_0v) + (-1)^{n-1}\frac{d^{n-1}}{dx^{n-1}}(p_1v) + \ldots - \frac{d}{dx}(p_{n-1}v) + p_nv\right]\right\} = \\ &= \frac{d}{dx}\left[y^{(n-1)}(p_0v) - \ldots + \frac{d}{dx}y(p_{n-1}v)\right] + y\bar{L}(v) \end{aligned} \tag{5.3}$$

where

$$\bar{L}(v) = (-1)^n\frac{d^n}{dx^n}(p_0v) + (-1)^{n-1}\frac{d^{n-1}}{dx^{n-1}}(p_1v) + \ldots - \frac{d}{dx}(p_{n-1}v) + p_nv. \tag{5.4}$$

The differential expression $\bar{L}(v)$ is called to be *adjoint* to $L(y)$, and

$$\bar{L}(v) = 0 \tag{5.5}$$

is said to be the *adjoint equation* corresponding to $L(y)=0$.

Relation Equation (5.3) can be expressed as

$$vL(y)-y\bar{L}(v)=\frac{d}{dx}P(y,v) \tag{5.6}$$

where $P(y,v)=y^{(n-1)}(p_0v)-\ldots+\frac{d}{dx}y(p_{n-1}v)$.

The relation Equation (5.6) is known as *Lagrange identity.*

$P(y,v)$ is linear and homogeneous in $y,y',\ldots.y^{(n-1)}$ and $v,v',\ldots.v^{(n-1)}$, which is known as *bilinear concomitant.*

Note. From Equation (5.3), we observe that v will be an integrating factor for $L(y)$ iff the function v satisfies the adjoint equation $\bar{L}(v)$. If v is a solution of equation $\bar{L}(v)$, then from Equations (5.5) and (5.6), we get

$$\frac{d}{dx}P(y,v)=vL(y)=0.$$

Integrating, we get

$$P(y,v)=c \tag{5.7}$$

where c is a constant. That is, equation $L(y)=0$ reduces to the linear Equation (5.7) of order $(n-1)$.

5.2.1 Lagrange Identity

Theorem5.2.1. If $L_n=a_0(x)\frac{d^n}{dx^n}+a_1(x)\frac{d^{n-1}}{dx^{n-1}}+\ldots\ldots+a_{n-1}(x)\frac{d}{dx}+a_n(x)$ and $a_k(x)$ have continuous $(n-k)th$ order derivatives for $k=0,1,2\ldots n$ on $a\le x\le b$.

Let u *and* v be any two functions of x in $[a,b]$, then

$$vL_n(u)-u\overline{L_n}(v)=\frac{d}{dx}P(u,v)$$

where $P(u,v)=u^{(n-1)}(p_0v)-\ldots+\frac{d}{dx}u(p_{n-1}v)$

$$=\sum_{k=1}^{n}\left[\sum_{j=1}^{k}-(1)^{j-1}u^{(k-j)}(a_{n-k}v)^{j-1}\right]$$

where $\overline{L}_n(v)$ is the adjoint operator of $L_n(v)$.

Proof. Since $v\mathrm{L}(\mathrm{u}) - u\overline{L}(v) = \frac{d}{dx}P(u,v)$

where $P(u,v) = \sum_{k=1}^{n}\left[\sum_{j=1}^{k} -(1)^{j-1} u^{(k-j)}(a_{n-k}v)^{j-1}\right]$,
first, we will prove that

$$vu^k = (-1)^k v^k u + \frac{d}{dx}[u^{(k-1)}(v) - u^{(k-2)}(v)' + \ldots\ldots + (-1)^{k-1} u(v)^{(k-1)}$$

$$for\ k = 0,1,2,\ldots n. \tag{5.8}$$

To prove Equation (5.8),

$$\frac{d}{dx}\left(u^{(k-1)}v\right) = u^{k-1}v' + u^k v = u^k v + u^{k-1}v'$$

$$-\frac{d}{dx}\left(u^{k-2}\,v'\right) = -u^{k-1}v' - u^{k-2}v''$$

$$\ldots\ldots\ldots\ldots\ldots\ldots\ldots\ldots\ldots\ldots\ldots$$

$$(-1)^{k-1}u(v)^{(k-1)} = (-1)^{k-1}u'(v)^{(k-1)} + (-1)^{k-1}uv^k.$$

Adding the above equations, we get

$$\frac{d}{dx}\left[u^{(k-1)}v - u^{k-2}\,v' + \ldots\ldots. + (-1)^{k-1}u(v)^{(k-1)}\right] = u^k v + (-1)^{k-1}uv^k$$

$$\Rightarrow \frac{d}{dx}\left[u^{(k-1)}v - u^{k-2}\,v' + \ldots\ldots. + (-1)^{k-1}u(v)^{(k-1)}\right] - (-1)^{k-1}uv^k = u^k v$$

$$\Rightarrow vu^k = (-1)^k uv^k + \frac{d}{dx}\left[u^{(k-1)}v - u^{k-2}\,v' + \ldots\ldots. + (-1)^{k-1}u(v)^{(k-1)}\right].$$

Equation (5.8) is proved.

Now, $vL_n(\mathrm{u}) = va_0(x)\frac{d^n u}{dx^n} + va_1(x)\frac{d^{n-1}u}{dx^{n-1}} + \ldots\ldots + va_{n-1}(x)\frac{du}{dx} + va_n(x)u.$

$$\Rightarrow vL_n(\mathrm{u}) = (va_0)u^n + (va_1)u^{n-1} + \ldots\ldots\ldots\ldots + (va_{n-1})u' + (va_n)u. \tag{5.9}$$

Applying Equation (5.8) in each term of Equation (5.9), we get

$$(va_0)u^n = (-1)^n (va_0)^n u + \frac{d}{dx}[u^{(n-1)}(va_0) - u^{(n-2)}(va_0)' + \ldots\ldots + (-1)^{n-1} u(va_0)^{(n-1)}$$

$$(va_1)u^{n-1} = (-1)^{n-1}(va_1)^{n-1} u + \frac{d}{dx}[u^{(n-2)}(va_1) - u^{(n-3)}(va_1)' + \ldots\ldots + (-1)^{n-2} u(va_1)^{(n-2)}$$

....

$$(va_{n-1})u' = (-1)^1 (va_{n-1})' u + \frac{d}{dx}\left[u^{n-1}(va_{n-1})\right].$$

Using all the above terms in Equation (5.9), we get

$$vL_n(\mathrm{u}) = (-1)^n (va_0)^n u + \frac{d}{dx}[u^{(n-1)}(va_0) - u^{(n-2)}(va_0)' + \ldots\ldots + (-1)^{n-1} u(va_0)^{(n-1)} + (-1)^{n-1}(va_1)^{n-1} u + \frac{d}{dx}[u^{(n-2)}(va_1) - u^{(n-3)}(va_1)' + \ldots\ldots + (-1)^{n-2} u(va_1)^{(n-2)} + +(-1)^1 (va_{n-1})' u + \frac{d}{dx}\left[u^{n-1}(va_{n-1})\right] + (va_n)u$$

$$vL_n(\mathrm{u}) = u\left[(-1)^n (va_0)^n + (-1)^{n-1}(va_1)^{n-1} + \ldots + (-1)^1 (va_{n-1})' + (va_n)\right] + \frac{d}{dx}\left[\sum_{j=1}^{n}(-1)^{j-1} u^{(n-j)}(a_0 v)^{j-1}\right] + \frac{d}{dx}\left[\sum_{j=1}^{n-1}(-1)^{j-1} u^{(n-1-j)}(a_1 v)^{j-1}\right] + \ldots + \frac{d}{dx}\left[\sum_{j=1}^{1}(-1)^{j-1} u^{1-j}(a_{n-1} v)^{j-1}\right]$$

$$\Rightarrow vL_n(\mathrm{u}) = \mathrm{u}\sum_{\mathrm{k}=0}^{\mathrm{n}}(-1)^{n-k}(va_k)^{n-k} + \sum_{k=1}^{n}\left[\frac{d}{dx}\sum_{j=1}^{k}(-1)^{j-1} u^{(k-j)}(a_{n-k} v)^{j-1}\right]. \tag{5.10}$$

We have

$$L_n(v) = a_0(x)\frac{d^n v}{dx^n} + a_1(x)\frac{d^{n-1}v}{dx^{n-1}} + \ldots\ldots + a_{n-1}(x)\frac{dv}{dx} + a_n(x)v.$$

$$\therefore \bar{L}_n(v) = (-1)^n \frac{d^n}{dx^n}(a_0 v) + (-1)^{n-1}\frac{d^{n-1}}{dx^{n-1}}(a_1 v) + \ldots\ldots + a_n v$$

$$= \sum_{k=0}^{n}(-1)^{n-k}(a_k v).$$

Therefore, Equation (5.10) becomes

$$vL_n(\mathrm{u}) = \mathrm{u}\sum_{k=0}^{n}(-1)^{n-k}(va_k)^{n-k} + \sum_{k=1}^{n}\left[\frac{d}{dx}\sum_{j=1}^{k}(-1)^{j-1}u^{(k-j)}(a_{n-k}v)^{j-1}\right]$$

$$= u\,\bar{L}_n(v) + \frac{d}{dx}\sum_{k=1}^{n}\sum_{j=1}^{k}(-1)^{j-1}u^{(k-j)}(a_{n-k}v)^{j-1}$$

or $vL_n(\mathrm{u}) - u\,\bar{L}_n(v) = \frac{d}{dx}\sum_{k=1}^{n}\sum_{j=1}^{k}(-1)^{j-1}u^{(k-j)}(a_{n-k}v)^{j-1}$

or $vL_n(\mathrm{u}) - u\,\bar{L}_n(v) = \frac{d}{dx}P(u,v)$

where $P(u,v) = \sum_{k=1}^{n}\left[\sum_{j=1}^{k} -(1)^{j-1}u^{(k-j)}(a_{n-k}v)^{j-1}\right]$.

Hence, it is proved.

5.2.2 Self-Adjoint Equation

If the adjoint equation of any linear homogeneous differential equation is identical with the equation itself, then the given equation is called a *self-adjoint equation.*

That is, let $\bar{L}(v) = 0$ be adjoint to $L(y) = 0$, and if $\bar{L}(v) = L(y)$, then $L(y) = 0$ is called a self-adjoint equation.

Theorem 5.2.2. Prove that if $\bar{L}(v)$ is adjoint to $L(y)$, then $L(y)$ is adjoint to $\bar{L}(v)$.

Proof. If $L(y)$ is not adjoint to $\bar{L}(v)$, let $L_1(y)$ be adjoint to $\bar{L}(v)$. Then, there exists a function $P_1(y,v)$ such that

$$vL_1(y)-y\bar{L}(v)=\frac{d}{dx}P_1(y,v). \tag{5.11}$$

Also, we know that

$$vL(y),-y\bar{L}(v)=\frac{d}{dx}P(y,v). \tag{5.12}$$

[$\because \bar{L}(v)$ is adjoint to $L(y)$].

Hence, we get

$$v\left[L_1(y)-L(y)\right]=\frac{d}{dx}[P_1(y,v)-P(y,v)]. \tag{5.13}$$

Now, $[P_1(y,v)-P(y,v)]$ is linear in $v,v',v'',\ldots v^{(n-1)}$, *but* $v\left[L_1(y)-L(y)\right]$ does not contain $v^{(n-1)}$.

Hence, the coefficient of $v^{(n-1)}$ on the right-hand side of Equation (5.13) must be zero.

Similarly, the coefficient of $v^{(n-2)},\ldots v,v',v''$ on the right-hand side of Equation (5.13) must be zero separately.

It implies that $\left[L_1(y)-L(y)\right]=0$, i.e., $L_1(y)$ is equal to $L(y)$.

Thus, $L(y)$ is adjoint to $\bar{L}(v)$.

5.2.3 Adjoint Equations for Linear Equations of Order One

Consider a linear differential equation of order one

$$L(y)\equiv a_0(x)y'+a_1(x)y=0 \tag{5.14}$$

where $a_0(x)\neq 0$, $a_1(x)$ is a continuous function on interval I, *and* $a_0'(x)$ exists and is continuous on I. The integrating factor

$$p(x)=\frac{1}{a_0(x)}e^{\int\frac{a_1(x)}{a_0(x)}dx}.$$

Multiplying the integrating factor in Equation (5.11), we have

$$p(x)[a_0(x)y' + a_1(x)y] = [r(x)y]'$$

where r is to be determined.

Comparing both sides of the above equation:

$$a_0(x)p(x) = r(x) \text{ and } a_1(x)p(x) = r'(x)$$

or,
$$r'(x) = \left[a_0(x)p(x)\right]' = a_1(x)p(x)$$

or,
$$\left[a_0(x)p(x)\right]' - a_1(x)p(x) = 0. \tag{5.15}$$

Thus, integrating factor p of Equation (5.14) is the solution of Equation (5.15),

i.e.,
$$\left[a_0(x)p(x)\right]' = a_1(x)p(x).$$

Now,
$$a_0(x)p(x) = r(x).$$

Then,
$$a_1(x)p(x) = r'(x)$$

and,
$$\frac{r'(x)}{r(x)} = \frac{a_1(x)}{a_0(x)}$$

$$\therefore r(x) = e^{\int \frac{a_1(x)}{a_0(x)}dx}$$

$$\because a_0(x)p(x) = r(x) = e^{\int \frac{a_1(x)}{a_0(x)}dx}.$$

Hence, $p(x)=\frac{1}{a_0(x)}e^{\int\frac{a_1(x)}{a_0(x)}dx}$,

which is the integrating factor of Equation (5.14).

This shows that every solution of Equation (5.15) is an integrating factor of Equation (5.14).

The equation defined by

$$M_1(p)=[a_0(x)p(x)]'-a_1(x)p(x)=0$$

is called the *adjoint equation* of $L(y)=0$,
and the equation $L(y)=0$ is called the *self-adjoint*, iff $L_1(y)=M_1(y)$.

Theorem 5.2.3. The equation $L_1(y)=0$ is self-adjoint, iff $a_0'(x)=2a_1(x)$.

Proof. The equation $L_1(y)=0$ is self-adjoint iff

$$M_1(y)=L_1(y)$$
$$\textit{iff}\quad a_0(x)y'+a_1(x)y=[a_0(x)y(x)]'-a_1(x)y(x),$$
$$\text{i.e.,}\quad a_1(x)=a_0'(x)-a_1(x).$$
$$\therefore a_0'(x)=2a_1(x).$$

5.2.4 Adjoint Equations for Linear Equations of Order Two

We consider a second-order homogeneous equation

$$L_2(y)=a_0(x)y''+a_1(x)y'+a_2(x)y=0 \tag{5.16}$$

where $a_0(x)\neq 0$; a_1,a_2 are continuous functions on an interval I, and a_0'',a_1' exist and are continuous on I.

Multiplying the integrating factor p in Equation (5.16),

$$p\left[a_0(x)y''+a_1(x)y'+a_2(x)y\right]=[l(x)y'+m(x)y]' \tag{5.17}$$

where l, m are to be determined and are continuous functions defined on I.

Comparing the coefficients of $y'' y', y$, we get

$$a_0(x)p = l(x)$$
$$a_1(x)p = l'(x) + m(x)$$
$$a_2(x)p = m'(x).$$

From the above equations, we get

$$[a_1(x)p]' = l''(x) + m'(x) = [a_0(x)p]'' + a_2(x)p$$

or,
$$[a_0(x)p]'' - [a_1(x)p]' + a_2(x)p = 0. \tag{5.18}$$

Thus, integrating factor p of Equation (5.16) is the solution of Equation (5.18).

This shows that every solution of Equation (5.15) is an integrating factor of Equation (5.13).

The equation defined by

$$M_2(p) = [a_0(x)p]'' - [a_1(x)p]' + a_2(x)p = 0,$$

i.e.,
$$\frac{d^2}{dx^2}[a_0(x)p] - \frac{d}{dx}[a_1(x)p] + a_2(x)p = 0$$

or,
$$a_0(x)\frac{d^2y}{dx^2} + [2a_0'(x) - a_1(x)]\frac{dy}{dx} + [a_0''(x) - a_1'(x) + a_2(x)]y = 0$$

[changing variable p by y]

is called an *adjoint equation* of Equation (5.16), and Equation (5.16) is called *self-adjoint* iff $L_2(y) = M_2(y)$.

Theorem 5.2.4. The equation $L_2(y) = 0$ is self adjoint, if and only if $a_0'(x) = a_1(x)$.

Proof. The equation $L_2(y)=0$ is self adjoint iff

$$L_2(y)=M_2(y),$$

i.e., $a_0(x)y''+a_1(x)y'+a_2(x)y=[a_0(x)y]''-[a_1(x)y]'+a_2(x)y$
$[[a_0(x)y]''-[a_1(x)y]'=\{[a_0(x)y'+a_0'(x)y]'-[a_1(x)y]'\}=$
$a_0(x)y''+a_0'(x)y'+a_0''(x)y+a_0'(x)y'-a_1'(x)y-a_1(x)y'=$
$a_0(x)y''+[2a_0'(x)-a_1(x)]y'+[a_0''(x)-a_1'(x)]y\}$
or $a_0(x)y''+a_1(x)y'+a_2(x)y=a_0(x)y''+[2a_0'(x)-a_1(x)]y'+$
$[a_0''(x)-a_1'(x)]y+a_2(x)y=a_0(x)y''+\left[2a_0'(x)-a_1(x)\right]y'+$
$\left[a_0''(x)-a_1'(x)+a_2(x)\right]y.$

Comparing the coefficients of y', we get

$$a_1(x)=2a_0'(x)-a_1(x)$$
$$\Rightarrow \quad a_0'(x)=a_1(x).$$

Alternative proof: We consider a second-order homogeneous equation

$$L_2(y)=a_0(x)y''+a_1(x)y'+a_2(x)y=0.$$

Its adjoint equation is

$$a_0(x)\frac{d^2y}{dx^2}+\left[2a_0'(x)-a_1(x)\right]\frac{dy}{dx}+\left[a_0''(x)-a_1'(x)+a_2(x)\right]y=0. \tag{5.19}$$

Since they are equivalent, then

$$a_1(x)=2a_0'(x)-a_1(x) \text{ and } a_0''(x)-a_1'(x)=0.$$

From these two equations, it can be

$$\frac{d}{dx}\left[a_0(x)\right]=a_1(x). \text{ This is a necessary part of proof}$$

Conversely, Equation (5.19) is self-adjoint if $a_1(x)=2a_0'(x)-a_1(x)\Rightarrow a_0'(x)=a_1(x)$ and $a_0''(x)-a_1'(x)+a_2(x)=a_2(x)$.

It implies that $a_0(x)y''+a_1(x)y'+a_2(x)y=0$, which is Equation (5.19).

Remark 5.2.1. If the equation $L_2(y)=0$ is self-adjoint, then $a_0'(x)=a_1(x)$.

It can be expressed as

$$\begin{aligned}L_2(y)&=a_0(x)y''+a_1(x)y'+a_2(x)y\\&=a_0(x)y''+a_0'(x)y'+a_2(x)y\\&=[a_0(x)y']'+a_2(x)y=0.\end{aligned}$$

Thus, every self-adjoint equation of order two can be expressed as

$$[r(x)y']'+p(x)y=0$$

where $r(x)=a_0(x)$ *and* $p(x)=a_2(x)$.

That is, a linear differential equation of order two can be expressed as:

$$L_2(y)=[r(x)y']'+p(x)y=0,$$

which is necessarily self-adjoint, where $r(x)\neq 0, p$ are continuous functions on I, and $r'(x)$ exists and is continuous.

Theorem 5.2.5. Every homogeneous linear differential equation of order two can be reduced to a self-adjoint form.

Proof. Let the homogeneous linear equation of order two:

$$a_0(x)y''+a_1(x)y'+a_2(x)y=0 \tag{5.16}$$

where $a_0(x) \neq 0$ and a_1, a_2 are continuous functions on an interval I, and a_0'', a_1' exist and are continuous on I.

Multiplying the integrating factor p on Equation (5.16), $p(x)\left[a_0(x)y'' + a_1(x)y' + a_2(x)y\right]$ becomes self-adjoint, i.e., the equation

$$p(x)y'' + \frac{a_1(x)}{a_0(x)}p(x)y' + \frac{a_2(x)}{a_0(x)}p(x)y = 0 \tag{5.20}$$

is self-adjoint.

Now, applying the condition of self-adjoint, we get

$$p'(x) = \frac{a_1(x)}{a_0(x)}p(x)$$

or,

$$\frac{p'(x)}{p(x)} = \frac{a_1(x)}{a_0(x)}.$$

Integrating, we get

$$p(x) = e^{\int \frac{a_1(x)}{a_0(x)}dx}.$$

Now, Equation (5.20) becomes

$$e^{\int \frac{a_1(x)}{a_0(x)}dx}y'' + \frac{a_1(x)}{a_0(x)}e^{\int \frac{a_1(x)}{a_0(x)}dx}y' + \frac{a_2(x)}{a_0(x)}e^{\int \frac{a_1(x)}{a_0(x)}dx}y = 0$$

or,

$$\left[e^{\int \frac{a_1(x)}{a_0(x)}dx}y'\right]' + \frac{a_2(x)}{a_0(x)}e^{\int \frac{a_1(x)}{a_0(x)}dx}y = 0,$$

which is the self-adjoint form

$$\left[r(x)y'\right]' + p(x)y = 0$$

where, $$r(x) = e^{\int \frac{a_1(x)}{a_0(x)}dx} \quad and\ p(x) = \frac{a_2(x)}{a_0(x)} e^{\int \frac{a_1(x)}{a_0(x)}dx}.$$

Note. A linear differential equation of an order higher than two may not be reduced to self-adjoint form.

Theorem 5.2.6. If $a_0(x)y'' + a_1(x)y' + a_2(x)y = 0$ is self adjoint, then it can be written as $\frac{d}{dx}\left[a_0(x)y'\right] + a_2(x)y = 0$

Proof.

$$\text{Since } a_0(x)y'' + a_1(x)y' + a_2(x)y = 0 \tag{i}$$

is self adjoint, we use the condition $\frac{d}{dx}\left[a_0(x)\right] = a_0'(x) = a_1(x)$

Equation (i) can be written as

$$a_0(x)y'' + a_0'(x)y' + a_2(x)y = 0$$

or, $$\frac{d}{dx}[a_0(x)y'] + a_2(x)y = 0.$$

5.3 Green's Formula

Definition 5.3.1. The identity $\int_{x_1}^{x_2}(vL_n u - u\bar{L}_n v)dx = \left[H(u,v)\right]_{x_1}^{x_2}$,
where $H(u,v) = u\left[a_1 v - (a_0 v)'\right] + u'(a_0 v)$
is called *Green's formula.*

Theorem 5.3.1. Let a_n, the coefficients in

$$L_n = a_0(x)\frac{d^n}{dx^n} + a_1(x)\frac{d^{n-1}}{dx^{n-1}} + \ldots\ldots + a_{n-1}(x)\frac{d}{dx} + a_n(x)$$

have continuous $(n-k)th$ –order derivatives for $k=0,1,2\ldots n$ on $a\le x\le b$.

Let u *and* v be any two functions of x. Then, for any two points x_1 and x_2 in $[a,b]$

$$\int_{x_1}^{x_2}(vL_n u - u\bar{L}_n v)dx = \left[P(u,v)\right]_{x=x_1} - \left[P(u,v)\right]_{x=x_2}$$

Proof. We know the Lagrange's identity,

$$vL_n(u) - u\bar{L}_n(v) = \frac{d}{dx}P(u,v).$$

Integrating the above with limits x_1 to x_2, we have

$$\int_{x_1}^{x_2}(vL_n(u) - u\bar{L}_n(v))dx = \left[P(u,v)\right]_{x_1}^{x_2}$$

$$\therefore \quad \int_{x_1}^{x_2}(vL_n(u) - u\bar{L}_n(v))dx = \left[P(u,v)\right]_{x=x_2} - \left[P(u,v)\right]_{x=x_1}.$$

Theorem 5.3.2. If the homogeneous linear equation of order two

$$a_0(x)y'' + a_1(x)y' + a_2(x)y = 0$$

is self-adjoint, then it can be written as $\frac{d}{dx}\left[a_0(x)\frac{dy}{dx}\right] + a_2(x)y = 0.$

Proof. Since the homogeneous linear equation of order two

$$a_0(x)y'' + a_1(x)y' + a_2(x)y = 0$$

is self-adjoint, then $\frac{d}{dx}\left[a_0(x)\right] = a_1(x)$ on $[a,b]$ is satisfied.

The homogeneous linear equation of order two can be written as

$$a_0(x)y'' + \frac{d}{dx}\left[a_0(x)\right]y' + a_2(x)y = 0$$

or, $$\frac{d}{dx}\left[a_0(x)y'\right] + a_2(x)y = 0.$$

5.4 Examples

Example 5.4.1. Reduce the following equation to self-adjoint form:

$$x^2y'' + xy' + y = 0 \quad (x > 0).$$

Solution: Here, $a_0(x) = x^2, a_1(x) = x, a_2(x) = 1$

$$r(x) = e^{\int \frac{a_1(x)}{a_0(x)}dx} = e^{\int \frac{x}{x^2}dx} = e^{\log x} = x, and$$

$$p(x) = \frac{a_2(x)}{a_0(x)} e^{\int \frac{a_1(x)}{a_0(x)}dx} = \frac{1}{x^2} e^{\log x} = \frac{1}{x^2} x = \frac{1}{x}.$$

Hence, the self-adjoint equation is

$$[r(x)y']' + p(x)y = 0$$

or,

$$\frac{d}{dx}[xy'] + \frac{1}{x}y = 0$$

or,

$$y' + xy'' + \frac{1}{x}y = 0$$

or,

$$x^2y' + xy' + y = 0.$$

Example 5.4.2. Find the adjoint equation of

$$x^2y'' + (2x^3 + 1)y' + y = 0.$$

Solution: Here, $a_0(x)=x^2,,a_1(x)=2x^3+1,a_2(x)=1$.
Its adjoint equation is in the form:

$$\left[a_0(x)y\right]''-\left[a_1(x)y\right]'+a_2(x)y=0.$$

Putting these values,

$$\left[x^2y\right]''-\left[\left(2x^3+1\right)y\right]'+y=0$$

or, $$\left(2xy+x^2y'\right)'-\left[6x^2y+\left(2x^3y'+y'\right)\right]+y=0$$

or, $$2y+2xy'+2xy'+x^2y''-6x^2y-2x^3y'-y'+y=0$$

or, $$x^2y''+\left(4x-2x^3-1\right)y'+3y\left(1-2x^2\right)=0.$$

Example 5.4.3. Find the adjoint equation of

$$t^2x''+7tx'+8x=0 \quad \text{where} \quad x'=\frac{dx}{dt}.$$

Solution: Here, $a_0(x)=t^2,a_1(x)=7t,a_2(x)=8$.
Its adjoint equation is in the form:

$$\left[a_0(x)y\right]''-\left[a_1(x)y\right]'+a_2(x)y=0.$$

Putting these values,

$$[t^2x]'' - [7tx]' + 8x = 0$$

or, $$[2tx + t^2x']' - (7x + 7tx') + 8x = 0$$

or, $$2x + 2tx' + 2tx' + t^2x'' - 7x - 7tx' + 8x = 0$$

or, $$t^2x'' - 3tx' + 3x = 0,$$

Example 5.4.4. Find the self-adjoint equation of

(i) $sinxy'' + cosxy' + 2x = 0$ where $y' = \frac{dy}{dx}$.

(ii) $\frac{x+1}{x}y'' - \frac{1}{x^2}y' + \frac{1}{x^3}y = 0.$

Solution:

(i) Here, $a_0(x) = sinx, a_1(x) = cosx, a_2(x) = 2$

$$r(x) = e^{\int \frac{a_1(x)}{a_0(x)}dx} = e^{\int \frac{cosx}{sinx}dx} = e^{-logx} = \frac{1}{x}, and$$

here, $a_0'(x) = cosx = a_1(x)$.

∴ The given equation is self-adjoint.

(ii) Here, $a_0(x) = \frac{x+1}{x}, a_1(x) = -\frac{1}{x^2}, a_2(x) = \frac{1}{x^3}$.

Here, $a_0'(x) = \left(\frac{x+1}{x}\right)' = -\frac{1}{x^2} = a_1(x)$.

∴ The given equation is self-adjoint.

Example 5.4.5. Show that

$$\left(1-x^2\right)y''-2xy'+n(n+1)=0.$$

Solution:

Here, $a_0(x)=\left(1-x^2\right),, a_1(x)=-2x, a_2(x)=n(n+1)$.
Here, $a_0{}'(x)=\left(1-x^2\right)'=-2x=a_1(x)$.
∴ The given equation is self-adjoint.

Example 5.4.6. Transform $\left(x^4+x^2\right)y''+2x^3y'+3y=0$

into an equivalent self - adjoint equation

Solution. Here, $a_0(x)=\left(x^4+x^2\right), a_1(x)=2x^3, a_2(x)=3$.

Here, $$\frac{1}{a_0(x)}e^{\int\frac{a_1(x)}{a_0(x)}dx}=\frac{1}{x^4+x^2}e^{\int\frac{2x^3}{x^4+x^2}dx}=\frac{1}{x^4+x^2}e^{\int\frac{2x}{1+x^2}dx}$$
$$=\frac{1}{x^4+x^2}e^{\log\left(1+x^2\right)}=\frac{1+x^2}{x^4+x^2}=\frac{1}{x^2}.$$

Multiplying the given equation by $\frac{1}{x^2}$, we get

$$\frac{1}{x^2}\left[\left(x^4+x^2\right)y''+2x^3y'+3y\right]=0$$

or,
$$\left(x^2+1\right)y''+2xy'+3\frac{1}{x^2}y=0.$$

Here, $a_0(x)=x^2+1, a_1(x)=2x, a_2(x)=\frac{3}{x^2}$.
Also, $\left[a_0(x)\right]'=\left(x^2+1\right)'=2x=a_1(x)$.

Therefore, the equation is self-adjoint, and its adjoint equation is in the form:

or,
$$\frac{d}{dx}\left[\left(x^2+1\right)y'\right]+3\frac{1}{x^2}y=0,$$

which is the required equivalent self-adjoint equation.

Example 5.4.7. Show that the equation

$$t^2\frac{d^2x}{dt^2}-2t\frac{dx}{dt}+2x=0$$

is not self-adjoint. How to make it self-adjoint.

Solution: The equation

$$t^2\frac{d^2x}{dt^2}-2t\frac{dx}{dt}+2x=0.$$

Compare with

$$a_0(t)y''+a_1(t)y'+a_2(t)y=0.$$

Here, $a_0(t)=t^2,\ a_1(t)=-2t,\ a_2(t)=2$
since $\left[a_0(t)\right]=-2t\neq a_1(t)$.
Therefore, it is not self-adjoint.
Let us find the factor

$$\frac{1}{a_0(x)}e^{\int\frac{a_1(t)}{a_0(t)}dt}=\frac{1}{t^2}e^{\int\frac{-2t}{t^2}dx}=\frac{1}{t^2}e^{-logt^2}=\frac{1}{t^2}.\frac{1}{t^2}=\frac{1}{t^4}.$$

Multiplying the given equation by $\frac{1}{t^3}$ on any interval $a \le t \le b$, which does not include $t = 0$, we get

$$\frac{1}{t^4}.t^2\frac{d^2x}{dt^2} - 2t\frac{1}{t^4}.\frac{dx}{dt} + 2x.\frac{1}{t^4} = 0$$

or,
$$\frac{1}{t^2}\frac{d^2x}{dt^2} - 2\frac{1}{t^3}.\frac{dx}{dt} + 2x.\frac{1}{t^4} = 0$$

or,
$$\frac{d}{dt}\left[\frac{1}{t^2}\frac{dx}{dt}\right] + \frac{2x}{t^4} = 0.$$

This equation is self-adjoint.

5.5 Exercise

1. Reduce the following equation to self-adjoint form:

   (i) $xy'' - y' + x^2y = 0$
   (ii) $y'' - 3y' + 2y = 0$
   (iii) $(1 - x^2)y'' - 2xy' + (n^2 + n)y = 0$
   (iv) $x^2y'' + xy' + (x^2 - n^2)y = 0$

2. Show that the equation

$$t^3\frac{d^2x}{dt^2} + 3t^2\frac{dx}{dt} + x = 0 \text{ is self-adjoint.}$$

3. Transform $f(t)\frac{d^2x}{dt^2}+g(t)x=0$ into an equivalence self-adjoint equation.
4. Find the adjoint equation of the following equations:

   (i) $t^2\frac{d^2x}{dt^2}+(2t^3+7t)\frac{dx}{dt}+(8t^2+8)x=0$

   (ii) $2\frac{d^2x}{dt^2}+3t\frac{dx}{dt}+3x=0$.

5. Transform $\frac{d^2y}{dx^2}-tanx\frac{dy}{dx}+y=0$ into self adjoint equation.
6. What do you understand by 'Self adjoint equation'?

   Specify the condition and transform the following differential equation into the equivalent self adjoint equation:

$$p(t)\frac{d^2y}{dt^2}+q(t)\frac{dy}{dt}+r(t)x=0$$

6
Boundary Value Problem

Abstract

In this chapter, we discuss boundary value problems (BVPs), which involve differential equations with solutions that must satisfy specified conditions at two or more points within the domain. These problems require that the solution fulfill constraints at multiple locations. We also study Green's functions, which are used to solve non-homogeneous linear differential equations—especially those with specific boundary or initial conditions. Green's functions transform differential equations into integral equations, simplifying the solution process.

***Keywords*:** Singular boundary value problem, regular boundary value problem, Green's function, Volterra integral equation

6.1 Introduction

We discuss about boundary value problem in this chapter. Green's functions and Strum–Liouville boundary value issues are also discussed. Here, we review what a boundary value problem is.

We are aware that there are n independent constants in the general solution of an equation of $n-$th order. These conditions are referred to as initial conditions if they apply to the solution's behavior at a single point in time.

Consider a linear differential equation of order n of the form

$$L(y)=a_0(x)\frac{d^n y}{dx^n}+a_1(x)\frac{d^{n-1}y}{dx^{n-1}}+\cdots\ldots+a_{n-1}(x)\frac{dy}{dx}+a_n(x)y=b(x) \tag{6.1}$$

where $a_0(x)\neq 0,\ a_1,a_2,\ldots.a_n$ and b are real or complex-valued continuous functios defined on an interval $I=[c,d]$.

Dipankar De. Ordinary Differential Equations and Special Functions, (203–216) © 2026 Scrivener Publishing LLC

A boundary condition is a condition imposed on the solutions of Equation (6.1) at two or more points on the interval I. The points of I where the conditions are imposed are known as *boundary points*, and the value of $y, y', \ldots y^{(n-1)}$ at the boundary points are known as *boundary values.*

A differential equation with some boundary conditions or with conditions imposed at more than one point is known as *boundary value problem.*

Such a problem requires the solution of a differential equation in a region R subject to the various conditions on the boundary R. Practical applications give rise to many such problems. We shall discuss two point linear boundary value problems of the following differential equations:

(i) $\dfrac{d^n y}{dx^n} + \alpha(x)\dfrac{dy}{dx} + \beta(x)y = \gamma(x) with\ y(x) = a, y(x_0) = a\ , y(x_n) = b$

(ii) $\dfrac{\partial u}{\partial x} = c^2 \dfrac{\partial^2 u}{\partial x^2}\ (x > 0, t > 0)$ [heat conduction]

subject to the initial condition $u(x,0) = f(x)$ and the boundary condition $u(0,t) = 0$.

(It can be solved using Fourier transformation.)

We define some important forms of boundary conditions for Equation (6.1).

Let the boundary conditions of the form

$$\mathrm{y(c) = y(d),\ y'\ (c) = y'(d), \ldots\ldots} y^{(n-1}\mathrm{(c)} = y^{(n-1)}\ \mathrm{(d)}$$

be called *periodic boundary conditions* stated at $x = c\ and\ x = d.$

For the other forms of boundary conditions, we have, for any scalars $\alpha_k^i, \beta_k^j, (k = 1,2,..n\ and\ i = 0,1,..\ n-1)$, the quantities denoted and defined by

$$v_k(y) = \alpha_k^0\ y(c) + \alpha_k^1 y'(c) + \cdots \ldots\ldots + \alpha_k^{n-1} y^{(n-1)}(c) + \beta_k^0 y(d) +$$
$$\beta_k^1 y'(d) + \cdots + \beta_k^{n-1} y^{(n-1)}(d),$$

which are called *linear forms* as stated at the values $y(c), y'(c), \ldots y^{(n-1)}(c);$ $y(d), y'(d), \ldots\ldots y^{(n-1)}(d).$

If any linear form $v_k(y) \neq 0$, then $v_k, v_k, \ldots\ldots v_k$.*are called nonhomogen linear forms*, and if $v_k(y) = 0$, then $v_k, v_k, \ldots\ldots v_k$.*are called homogeneous linear forms.*

If $v_k(y)$ $(\mathrm{k = 1,2,..n,})$ are linearly independent linear forms, then the problems

$L(y) = 0$ satisfying the boundary conditions

$$v_k(y) = 0\,, k = 1, 2, \ldots n \tag{6.2}$$

is called *a linear homogeneous boundary value problem,* and the conditions Equation (6.2) are called *linear homogeneous boundary conditions.*

Example 6.1.1.

(i) The differential equation

$$y'' + y' + y = 0, (c \le x \le d),$$

with boundary conditions

$$y(c) = 0 = y(d),$$

constitutes a linear homogeneous BVP
since we define linear homogeneous boundary conditions by

$$v_1[y(c), y'(c), y(d), y'(d)] = y(c) = 0$$

$$\text{and, } v_2[y(c), y'(c), y(d), y'(d)] = y(d) = 0.$$

(ii) The differential equation

$$y'' + e^x y' + 2y = \sin 2\pi x, (c \le x \le d),$$

with periodic boundary conditions $y(c) = y(d), y'(c) = y'(d)$,
constitutes a linear non-homogeneous BVP equation
since we define linear homogeneous boundary conditions by

$$v_1[y(c), y'(c), y(d), y'(d)] = y(c) = y(d) = 0$$

$$\text{and, } v_2[y(c), y'(c), y(d), y'(d)] = y'(c) = y'(d) = 0.$$

Example 6.1.2.

(i) $y'' + y = 0, y(0) = 1\, and\; y'\left(\frac{\pi}{2}\right) = 2$

Solution: $y = cosx + 2sinx.$
It is a solution of the boundary value problem.

(ii) The differential equation $\ddot{x} = -g$ (equation of projectile),

with an initial condition at time $t = t_0,\; x = x_0$ and $velocity, \dot{x} = v_0$ and $\ddot{x} = -g,\; x(t_1) = x_1$ and $x(t_2) = x_2$ leads to a boundary condition.
It can be easily solved with initial and boundary conditions.

Note. It must be observed that a solution may not always exist, and if it exists, then it may not be unique.

Example 6.1.3. Consider the boundary value problem

$$\frac{d^2y}{dx^2} + y = 0\; with\; conditions$$
$$y(0) = 0, y(l) = y_1.$$

Solution: The general solution of the equation is

$$y = A\; cosx + B\; sinx$$

where $A\, and\, B$ are constants.

Now, $y(0) = A = 0$ and $y(l) = Acosl + Bsinl = y_1$

$\Rightarrow A = 0\; and\; B = \frac{y_1}{sinl}.$

$\therefore$ The solution: $y = \frac{y_1}{sinl} sinx.$

If $l = n\pi,\; y_1 = 0,\; \Rightarrow$A=0.

$\therefore$ The solution: $y = Bsinx.$

Here, we have infinite number of solutions to the problem.

If $l = n\pi$ and $y_1 \neq 0$, let $y_1 = 1$. Then, there is no solution to the boundary value problem. For, if a solution exists, then the equality $1 = Bsinn\pi = 0$ is satisfied, but it is not possible.

Note. A linear BVP is called *a regular boundary value problem* if

(i) $c\; and\; d$ are finite and
(ii) $a_0(x) \neq 0\; on\, [c,d].$

A regular BVP, which is not regular, is called a *singular boundary value problem.*

6.2 Green's Function

Consider the linear homogeneous differential equation

$$L(y) + \phi(x) = 0 \tag{6.3}$$

where, $$L(y) = \left[\frac{d}{dx}\left(p\frac{d}{dx}\right) + q\right]y = py'' + p'y' + qy, \tag{6.4}$$

together with the homogeneous boundary conditions of the form

$$\alpha y + \beta \frac{dy}{dx} = 0. \tag{6.5}$$

Now, let us find a function $G(x,t)$, which, for a given number t, is given by

$$G_1(x) \text{ when } x < t$$

and by, $$G_2(x) \text{ when } x > t$$

and which has the following properties:

I. G_1 *and* G_2 satisfy the equation $L(G) = 0$ in their defined intervals,
   i.e., $L(G_1) = 0$ when $x < t$
   $L(G_2) = 0$ when $x > t$.
II. G_1 *and* G_2 satisfy the boundary conditions at the end points $x = a$ *and* $x = b$, respectively.

III. $G(x,t)$ is continuous at $x = t$, i.e., $G_1(t) = G_2(t)$.

IV. The derivative of G is continuous at every point within the range of x except at $x = t$, i.e.,

$$G_2'(t) - G_1'(t) = -\frac{1}{p(t)}.$$

$G(x,t)$ as defined above is called the *Green's function.*

If $G(x,t)$ exists, then the solution of the given problem can be transformed to the integral equation#

$$y(x) = \int_a^b G(x,t)\phi(t)\,dt. \tag{6.6}$$

Let $y = y_1(x)$ *and* $y = y_2(x)$ be the non-trival solutions of the equation $L(y) = 0$, which satisfy the homogeneous conditions at $x = a$ *and* $x = b$, respectively.

The above conditions *I and II* are satisfied if we write

$$G = \begin{cases} C_1 y_1(x) & , when\ x < t, \\ C_2 y_2(x) & , when\ x > t \end{cases} \tag{6.7}$$

Imposing condition III on Equation (6.7), we get

$$C_2 y_2(t) - C_1 y_1(t) = 0. \tag{6.8}$$

Imposing condition IV on Equation (6.7), we get

$$C_2 y_2'(t) - C_1 y_1'(t) = -\frac{1}{p(t)}. \tag{6.9}$$

#The general type of linear integral equation is of the form

$$y(x) = F(x) + \lambda \int_a^b \kappa(x,t) y(t) dt;$$

where $F(x)$ *and* $\kappa(x,t)$ are known functions, while $y(x)$ is to be determined. The function $\kappa(x,t)$ is called the *kernel* of the integral equation and (i) if a,b are constants, the equation is known as *Fredholm integral equation;* (ii) if a is a constant, while b is a variable, it is called a *Volterra integral equation.*

Equations (6.8) and (6.9) give a unique solution, if

$$\begin{vmatrix} y_1(t) & y_2(t) \\ y_1'(t) & y_2'(t) \end{vmatrix} = y_1(t)y_2'(t) - y_2(t)y_1'(t) \neq 0.$$

By Abel's formula*, we find that

$$y_1(t)y_2'(t) - y_2(t)y_1'(t) = \frac{c}{p(t)} \tag{6.10}$$

multiplying Equation (6.9) by $y_2(t)$ and subtracting Equation (6.8) multiplied by $y_1(t)$.

Using Equation (6.10), we get

$$C_1 = -\frac{1}{C}y_2(t) \text{ and } C_2 = -\frac{1}{C}y_1(t).$$

Putting C_1 *and* C_2, Equation (6.7) becomes

$$G = \begin{cases} -\dfrac{1}{C}y_2(t)y_1(x) \text{ , } when\ x < t, \\ -\dfrac{1}{C}y_1(t)y_2(x) \text{ , } when\ x > t \end{cases} \tag{6.11}$$

Conversely, it can be shown that the integral equation

$$y(x) = \int_a^b G(x,t)\phi(t)\,dt$$

where $G(x,t)$ is defined by Equation (6.11) and satisfies the differential equation $L(y) + \phi(x) = 0$ with its given boundary conditions.

---

*The conditions that $y_1(x)$ *and* $y_2(x)$ satisfy the equation $L(y) = 0$ are $(py_1')' + qy_1 = 0$ and $(py_2')' + qy_2 = 0$,

which gives $[p(y_1y_2' - y_2y_1')]' \Rightarrow y_1y_2' - y_2y_1' = \dfrac{C}{p}$.

6.3 Examples

Example 6.3.1. Transform the differential equation

$$y'' + y = x, y(0) = 1, y'(1) = 0$$

to a Fredholm integral equation, finding the corresponding Green's function.

Solution: Given the equation

$$L(y) + \phi(x) = 0. \quad \text{(i)}$$

Here, $L(y) = y''$, $\phi(x) = y - x\,[\because L(y) = py'' + p'y' + qy]$.
Conditions: $y(0) = 0,\ y'(1) = 0$.
The associated equation:

$$L(y) = y'' = 0 \quad \text{(ii)}$$

Its solution:

$$y = C_1x + C_2. \quad \text{(iii)}$$

Let y_1 be a particular solution of Equation (ii) satisfying the condition $y(0) = 0, \Rightarrow C_2 = 0, C_1 = 1.$
We get from Equation (iii):

$$y_1(x) = x. \quad \text{(iv)}$$

Let y_2 be another solution of Equation (ii) satisfying the condition $y'(1) = 0 \Rightarrow C_2 = 0, C_1 = 1.$
From Equation (iii), $y'(1) = C_1 = 0$. Then, $y_2(x) = C_2$.
Taking $C_2 = 1$,
the particular solution is $y_2(x) = 1$.
From Abel's formula $y_1y_2' - y_2y_1' = \dfrac{C}{p}$,
we can find the constant C.
Since $L(y) = py'' + p'y' + qy = y''$
$\Rightarrow p = 1.$
Thus, $C = p(y_1y_2' - y_2y_1') = y_1y_2' - y_2y_1' = x.0 - 1.1 = -1.$

∴ The Green's function is

$$G = \begin{cases} -\frac{1}{C} y_2(t) y_1(x) \text{ , when } x < t, \\ -\frac{1}{C} y_1(t) y_2(x) \text{ , when } x > t \end{cases} \quad \text{(v)}$$

$$= \begin{cases} y_2(t) y_1(x) \text{ , } x < t, \\ y_1(t) y_2(x) \text{ , } x > t \end{cases} = \begin{cases} 1.x = x \text{ ,} x < t \\ t.1 = t, x > t \end{cases}$$

Hence, Equation (i) is equivalent to the integral equation

$$y(x) = \int_a^b G(x,t)\phi(t)\, dt = \int_0^1 G(x,t)(y-t)\, dt \quad [\because \phi(t) = y - t]$$

$$= \int_0^1 G(x,t) y dt - [\int_0^x xtdt + \int_x^1 t.tdt]$$

$$= \int_0^1 G(x,t) y dt - [x.\left(\frac{t^2}{2}\right)_0^x + \left(\frac{t^3}{3}\right)_x^1]$$

$$= \int_0^1 G(x,t) y dt - \frac{1}{6}(x^3 + 2)$$

where $G(x,t)$ is given by Equation (v).

Example 6.3.2. Find the Green's function for the boundary value problem

$$\frac{d^2y}{dx^2} + \mu^2 x = 0; \; y(0) = y(1) = 0.$$

Solution: We observe that the solution

$$y_1 = sin\mu x$$

$[y = Acos\mu x + Bsin\mu x;$ for $y(0) = 0 \Rightarrow \mathrm{A} = 0 \colon y = Bsin\mu x. \, taking \; B = 1]$

satisfies the boundary conditions $y(0) = 0$, and the solution $y_2 = \sin\mu(x - 1)$ satifies the second condition $y(1) = 0$, and both these solutions are linearly independent.

The constant C is found from $y_1 y_2' - y_2 y_1' = \frac{C}{p}$.

Since $L(y) = py'' + p'y' + qy = y''$, $\therefore p = 1$.

$\therefore C = p(y_1 y_2' - y_2 y_1') = \mu sin\mu x cos\mu(x-1) - \mu sin\mu(x-1)cos\mu = \mu sin\mu$.

Hence, the Green's function is given by

$$G = \begin{cases} -\frac{1}{C} y_2(t) y_1(x) \ , \ x < t, \\ -\frac{1}{C} y_1(t) y_2(x) \ , \ x > t \end{cases} = \begin{cases} \frac{sin\mu x sin\mu(t-1)}{\mu sin\mu}, x < t, \\ \frac{sin\mu t sin\mu(x-1)}{\mu sin\mu}, x > t \end{cases}$$

Example 6.3.3. Find the Green's function for the boundary value problem

$$\frac{d^2 y}{dx^2} - y = 0 \ ; \ y(0) = y(1) = 0.$$

Solution: We observe that the solution is

$$y = Ae^x + Be^{-x}.$$

With boundary condition, $y_1 = e^x - e^{-x} = 2sinhx$ and $y_2 = e^{x-1} - e^{1-x} =$ 2sinh $(x-1)$.

Since $L(y) = py'' + p'y' + qy = y''$. $\therefore p = 1$.

$\therefore C = p(y_1 y_2' - y_2 y_1') = (e^x - e^{-x})(e^{x-1} + e^{1-x}) - (e^{x-1} - e^{1-x})(e^x + e^{-x}) =$

$$2(e - \frac{1}{e}) = 2\frac{e^2 - 1}{e} = 2sinh1$$

$[y_2' = e^{x-1} + e^{1-x}$ and $y_1' = e^x + e^{-x}]$.

Hence, the Green's function is given by

$$G = \begin{cases} -\frac{1}{C} y_2(t) y_1(x) \ , \ x < t, \\ -\frac{1}{C} y_1(t) y_2(x) \ , \ x > t \end{cases} = \begin{cases} \frac{sinhx sinh(t-1)}{sinh1}, x < t \\ \frac{sinht sinh(x-1)}{sinh1}, x > t \end{cases}$$

Example 6.3.4. Construct Green's function for the boundary value problem, and find the general solution of

$$\frac{d^2y}{dx^2}-3\frac{dy}{dx}+2y=x;\ (-\infty<x<\infty).$$

Solution: It is found that the functions with the boundary conditions $y_1(x)=e^x$, $y_2(x)=e^{2x}$ form the basis for the solutions of the corresponding homogeneous equation.

$$L(y)=y''-3y'+2y=0$$

Therefore, $W(t)=\begin{vmatrix} e^t & e^{2t} \\ e^t & 2e^{2t} \end{vmatrix}=e^{3t}.$

Since $a_0(t)=1$, Green's function $G(x,t)$ is given by

$$G(x,t)=\frac{1}{a_0(t)W(t)}\begin{vmatrix} e^t & e^{2t} \\ e^x & e^{2x} \end{vmatrix}=\frac{1}{e^{3t}}(e^{2x+t}-e^{x+2t})=e^{2x-2t}-e^{x-t}.$$

Hence, a particular solution $y_p(x)$ is given by

$$y_p(x)=\int_0^x G(x,t)b(t)dt=\int_0^x[e^{2x-2t}-e^{x-t}]tdt=e^x+\frac{1}{4}e^{2x}+\frac{x}{2}+\frac{3}{4},$$

and the general solution y is given by

$$y(x)=c_1e^x+c_2e^{2x}+\frac{x}{2}+\frac{3}{4}$$

where c_1,c_2 are arbitrary constants.

Example 6.3.5. Using the Green's function, solve the BVP

$$y''-y=x,\ y(0)=y(1)=0.$$

Solution: It is found that the functions $y_1(x)=e^x$, $y_2(x)=e^{-x}$ form the basis for the solutions of the corresponding homogeneous equation.

$$L(y) = y'' - y = 0,$$

and hence, the general solution is given by

$$y(x) = Ae^{x} + Be^{-x}.$$

The boundary conditions are satisfied if and only if $A = B = 0$, i.e., $y(x) = 0$ is the only solution of $y'' - y = 0$.

Hence, Green's function $G(x,t)$ exists and is given by

$$G = \begin{cases} -\frac{1}{C} y_2(t) y_1(x) \ , if \ 0 \le t \le x, \\ -\frac{1}{C} y_1(t) y_2(x) \ , \ if \ x \le t \le 1 \end{cases}$$

$$= \begin{cases} \frac{sinht sinh(x-1)}{sinh1}, if \ 0 \le t \le x \\ \frac{sinhx sinh(t-1)}{sinh1}, if \ x \le t \le 1 \end{cases}$$

[all values c, p, y_1', y_2' are calculated in Example (6.3.3.)].

Hence, a particular solution $y_p(x)$ of the given BVP is

$$y_p(x) = \int_0^1 G(x,t) b(t) dt.$$

Thus, we have

$$y_p(x) = \int_0^1 G(x,t) b(t) dt = \int_0^x \frac{t.sinht sinh(x-1)}{sinh1} dt + \int_x^1 \frac{t.sinhx sinh(t-1)}{sinh1} dt$$

$$= \frac{sinh(x-1)}{sinh1} \int_0^x t.sinht dt + \frac{sinhx}{sinh1} \int_x^1 t.sinh(t-1) dt +$$

$$= \frac{1}{sinh1} \{ sinh(x-1)[xcosh - sinhx] + sinhx[1 - xcosh(x-1)] \}$$

$$= \frac{sinhx}{sinh1} - x.$$

Hence, the general solution of the given problem is

$$y(x) = Ae^x + Be^{-x} + \frac{sinhx}{sinh1} - x.$$

6.4 Exercise

1. Transform the problem $\frac{d^2y}{dx^2} + xy = 1; y(0) = y(1) = 0$ *to* an integral equation, finding the corresponding Green's function.
2. Transform the problem $\frac{d^2y}{dx^2} + y = x; y(0) = 1, y'(1) = 0$ *to* an integral equation using Green's function.
3. Construct an integral equation corresponding to the boundary value problem

$$\frac{d^2u}{dx^2} + e^u u = x; u(0) = 0, u(1) = 1.$$

4. Solve the boundary value problem $y'' + 4y = 3, y'(0) = 0$, $y(\frac{\pi}{2}) = 0.$
5. Solve the boundary value problem $x^2y'' - 3xy' + 3y = 24x^5$, $y(1) = 0, y(2) = 0.$
6. Find the Green's function for the boundary value problem

$$\frac{d^2y}{dx^2} - y = 0 \text{ when } y(0) = y(1) = 0$$

7. Construct Green's function for Boundary value problem

$$y^{(4)} = 0 \text{ when } y(0) = y'(0) = 0, y(1) = y'(1) = 0$$

8. Solve the BVP

$$\frac{d^2y}{dx^2} + y = x \text{ when } y(0) = y\left(\frac{\pi}{2}\right) = 0$$

7
Strum Liouville Problem

Abstract
The Strum–Liouville problem is a boundary value problem for a second-order differential equation involving the determination of eigenvalues and corresponding eigenfunctions that satisfy both the differential equation and specific boundary conditions. This chapter discusses the orthogonality properties of eigenfunctions, the formation of orthonormal sets, and the Gram–Schmidt process for orthonormalization.

Keywords: Eigenfunctions, eigenvalues, Strum–Liouville operator, completeness

7.1 Introduction

The French mathematicians Joseph Liouville (1809–1882) and Charles-Francois Strum (1803–1855) worked separately on the problem of heat conduction through a metal bar in the 1830s, developing methods for resolving a broad range of PDEs, the most basic of which takes the form

$$[r(x)y']' + [\lambda p(x) + q(x)]y = 0$$

where λ is an eigenvalue, or parameter, that constrains the equation so that y fulfills the boundary values at the end points of the interval over which the variable x runs, and y is a physical quantity (or the quantum mechanical wave function). The equation will have a family of solutions, known as eigen functions, that correspond to the eigenvalue solutions if the functions p, q, r meet the necessary requirements. This chapter discusses the following properties: realism of eigenvalues, orthogonality of eigen functions, and orthogonality of the Strum–Liouville problem and also study on Gram-Schmidt process of orthonomarlisation.

7.2 Strum–Liouville Equation

A differential equation of the form:

$$[r(x)y']' + [\lambda p(x) + q(x)]y = 0 \tag{7.1}$$

Dipankar De. Ordinary Differential Equations and Special Functions, (217–234) © 2026 Scrivener Publishing LLC

is known as the *Strum–Liouville equation.*

We assume that the functions p, q, r in Equation (7.1) are continuous in interval $a \le x \le b$ and $p(x) > 0$. Here, λ is a parameter independent of x.

Equation (7.1) is considered on some interval $a \le x \le b$ satisfying boundary conditions at the two ends at a *and* b, so that

$$\alpha_1 y(a) + \alpha_2 y'(a) = 0, \ \beta_1 y(b) + \beta_2 y'(b) = 0, \tag{7.2}$$

with the real constants α_1, α_2, not both zero, and β_1, β_2, not both zero. The conditions Equation (7.2) at the end points are called *boundary conditions.*

A differential equation, together with the boundary conditions, is called a *boundary value problem.* Equation (7.1), together with the boundary condition Equation (7.2), is called a *Strum–Liouville problem.*

Clearly, $y = 0$ is a solution of the problem for any value of the parameter λ, which is a trivial solution and, as such, is of no practical utility. Any other solution of Equation (7.1) satisfying Equation (7.2) is called an *eigen function* of the problem, and the corresponding value of λ is called an *eigenvalue* of the problem.

Special Case: Taking $r = p = 1 \; and \; q = 0 \; in$ Equation (7.1), then

$$y'' + \lambda y = 0 \tag{7.3}$$

and also $\alpha_1 = \beta_1 = 1 \; and \; \alpha_2 = \beta_2 = 0$, then the boundary conditions Equation (ii) become

$$y(a) = 0 \, , y(b) = 0. \tag{7.4}$$

Thus, Equations (7.3) and (7.4) constitute the *simplest form of Strum–Liouville problem.*

Note. Legendre's equation

$$(1 - x^2)y'' - 2xy' + n(n+1) = 0 \tag{i}$$

can be written as $[((1 - x^2)y']' + \lambda y = 0$ [here, λ=n(n+1)].

Bessel's equation $t^2 \dfrac{d^2 y}{dx^2} + t \dfrac{dy}{dx} + (t^2 - n^2)y = 0$ can be transformed by $t = kx$

so that $\dfrac{dy}{dt} = \dfrac{dy}{dx} \cdot \dfrac{dx}{dt} = y' \dfrac{1}{k}$ and $\dfrac{dy^2}{dt^2} = \dfrac{y''}{k^2}$ takes the form

$$x^2 y'' + xy' + (k^2 x^2 - n^2)y = 0$$

or, $$(xy'' + y') + (k^2 x - \frac{n^2}{x})y = 0$$

or, $$(xy')' + (\text{»}x - \frac{n^2}{x})\text{y} = 0 \, [\text{here, } \text{»} = k^2] \qquad \text{(ii)}$$

Both Legendre Equation (i) and Bessel equations Equation (ii) are of the form

$$[r(x)y']' + [\lambda p(x) + q(x)]y = 0, \qquad (7.5)$$

which is known as the *Strum – Liouville equation.* Similarly, Laguerre's, Hermite's equation can be reduced to Equation (7.5).

Definition 7.2.1. Two functions $y_m(x)$ *and* $y_n(x)$, defined on some interval $a \le x \le b$, are said to be *orthogonal* on this interval with respect to the weight function $p(x) > 0$, if

$$\int_a^b p(x)\, y_m(x) y_n(x) dx = 0 \quad \textit{for } m \ne n.$$

Definition 7.2.2. The *norm* of y_m, denoted by $\| y_m \|$, is defined to be the non-negative square root of

$$\int_a^b p(x) \,\lceil y_m(x) \rceil dx.$$

Thus, $$\| y_m \| = \sqrt{\int_a^b p(x) \,\lceil y_m(x) \rceil dx}.$$

Definition 7.2.3. The functions that are orthogonal on interval $a \le x \le b$ and have a norm equal to 1 are called *orthonormal* on the interval.

7.3 Orthogonality of Eigen Functions

Theorem 7.3.1. If (a) the functions p,q,r,r' in the Strum–Liouville equation (i) are continuous in $a \le x \le b$,
(ii) $y_m(x), y_n(x)$ are two eigen functions of the Strum–Liouville problem corresponding to eigenvalues $\lambda_m(x) and\ \lambda_n(x)$, respectively,
then, $\lambda_m(x) and\ \lambda_n(x)$ $(m \ne n)$ are orthogonal on that interval with respect to the weight function $p(x)$.

Proof. Since $y_m(x), y_n(x)$ are satisfied Equation (i)

$$[r(x)y_m']' + [\lambda_m p(x) + q(x)]y_m = 0$$

$$[r(x)y_n']' + [\lambda_n p(x) + q(x)]y_n = 0.$$

Multiplying the first equation by y_n and the second by y_m and subtracting, we get

$$\begin{aligned}(\lambda_m - \lambda_n)py_m y_n &= y_m(ry_n')' - y_n(ry_m')' \\ &= \frac{d}{dx}[(ry_n')y_m - (ry_m')y_n].\end{aligned}$$

Now, integrating both sides with respect to from $a\ to\ b$, we obtain

$$\begin{aligned}(\lambda_m - \lambda_n)\int_a^b py_m y_n dx &= [(ry_n')y_m - (ry_m')y_n]_a^b \\ &= r(b)[y_n'(b)y_m(b) - y_m'(b)y_n(b)] \\ &\quad - r(a)[y_n'(a)y_m(a) - y_m'(a)y_n(a)].\end{aligned} \tag{A}$$

The right hand side of Equation (A) will vanish if the boundary conditions are of one of the following forms:

I. $y(a) = y(b) = 0,$
II. $y'(a) = y'(b) = 0,$ and
III. $\alpha_1 y(a) + \alpha_2 y'(a) = 0, \beta_1 y(b) + \beta_2 y'(b) = 0$ where either $\alpha_1\ or$ α_2 is not zero, and either $\beta_1\ or\ \beta_2$ is not zero.

Thus for each case we get from Equation (A) $\int_a^b p y_m y_n \, dx = 0 \quad (m \neq n)$, which shows that the eigen functions y_m *and* y_n are orthogonal on $a \leq x \leq b$ with respect to the weight function $p(x) = 0$.

(The third form of the boundary conditions, in fact, contains the first two forms as special cases.)

Corollary 7.3.1. *Orthogonality of Legendre polynomials* has already been established. But it follows at once from the above theorem.

We have already seen that Legendre's equation is a Strum–Liouville equation

$$[(1-x^2)y']' + \lambda y = 0, \text{ with } r(x) = 1 - x^2, p(x) = 1 \, and \, q(x) = 0;$$

$$= 0; \text{here, } \lambda = n\ (n+1).$$

Since $y(-1) = y(1) = 0$ *and for* $n = 0,1,2,\ldots;$ $\lambda = 0,1.2,2.3,\ldots$ the Legendre polynomials are the solutions of the problem, i.e., these are the eigen functions. Thus, it follows, by the above theorem, that they are orthogonal on $-1 \leq x \leq 1$.

Corollary 7.3.2. *Orthogonality of Bessel functions* has been established. But it can easily be seen to follow from the above theorem.

Now, we transformed the Bessel's equation

$$X^2 \frac{d^2 J_n}{dx^2} + X \frac{dJ_n}{dx} + (X^2 - n^2) J_n(x) = 0$$

into $[xJ_n'(kx)]' + \left(k^2 x - \frac{n^2}{x}\right) J_n(kx) = 0$, which is a Strum–Liouville equation with $r(x) = x$, $p(x) = x$, $q(x) = -\frac{n^2}{x}$ *and* $\lambda = k^2$, since $r(0) = 0$. It follows from the above theorem that these solutions of $J_n(kx)$, which are zero at $x = 0$, form an orthogonal set on $-1 \leq x \leq 1$ with weight function $p(x) = x$.

Corollary 7.3.3. Orthogonality of Hermite polynomial $H_n(x)$: *Hermite polynomials* are orthogonal over $(-\infty, \infty)$ with respect to the weight function e^{-x^2}, i.e.,

$$\int_{-\infty}^{\infty} e^{-x^2} H_m(x) H_n(x) dx = 0, for \ \ m \neq n.$$

7.4 Orthonormal Set of Functions

Definition 7.4.1. Let $\{f_n(x)\}$, where $n = 1,2,3,\ldots$ is a collection of functions defined on some interval $a \le x \le b$. Then, the set $\{f_n(x)\}$ is said to be *orthonormal on* $a \le x \le b$ *where each function is orthogonal to every other function on* $a \le x \le b$, and each function has a norm of unity within domain, *essentially acting like a set of perpendicular vectors with unit length in a function space. This is often used in areas like Fourier analysis and signal processing to represent complex functions as a combination of simpler, orthogonal components.*

Thus, the set $\{f_n(x)\}$ is *orthonormal* on $a \le x \le b$

if $\int_a^b f_n(x) f_n(x) dx = \begin{cases} 0, \text{ when } m \neq n \\ 1, \text{ when } m = n \end{cases}.$

Definition 7.4.2. Let $\{f_n(x)\}$, where $n = 1,2,3,\ldots$ is a set of functions defined on some interval $a \le x \le b$. Then, the set $\{f_n(x)\}$ is said to be *orthonormal* with respect to a weight function $p(x) > 0$ *if they are orthogonal* with respect to weight function $p(x)$*on* $a \le x \le b$, and each has a norm of *1*. Thus, the set $\{f_n(x)\}$ is *orthonormal with respect to weight function* $q(x)$

if $\int_a^b q(x) f_m(x) f_n(x) dx = \delta_n^m = \begin{cases} 0, \text{when } m \neq n \\ 1, \text{when } m = n \end{cases}.$

7.5 Gram–Schmidt Process of Orthonormalization

The Gram–Schmidt process or procedure is a sequence of operations that allow us to transform a set of linear independent functions (vectors) into a set of orthonormal functions (vectors) that span the same space spanned by the original set.

Let $\{f_n(x)\}$, where $n = 1,2,3,\ldots$ is a set of linearly independent functions for each of which norm $||f_n(x)||$ exists and is non zero. Then, we wish to obtain an orthonormal set $\{\varphi_n(x)\}$, where $n = 1,2,3,\ldots$ such that

$$\int_a^b \varphi_m(x) \varphi_n(x) dx = \begin{cases} 0, \text{ when } m \neq n \\ 1, \text{ when } m = n \end{cases}. \tag{7.6}$$

We select $f_1(x)$ and obtain

$$\varphi_1(x) = \frac{f_1(x)}{\| f_1(x) \|}, \tag{7.7}$$

and
we next choose $f_2(x)$, and let

$$F_2(x) = f_2 + c\varphi_1 \tag{7.8}$$

where c is chosen, such that F_2 may be orthogonal to φ_1,
i.e., $\int_a^b F_2\varphi_1 dx = 0$ or $\int_a^b (f_2 + c\varphi_1)\varphi_1 dx = 0$

or,
$$\int_a^b f_2\varphi_1 dx + c\int_a^b \varphi_1^2 dx = 0$$

or,
$$\int_a^b f_2\varphi_1 dx + c = 0$$

using Equation (7.6) $\therefore\ c = -\int_a^b f_2\varphi_1 dx.$

Now, Equation (7.8) gives

$$F_2(x) = f_2 - \varphi_1 \int_a^b f_2\varphi_1 dx. \tag{7.9}$$

Now, we take

$$\varphi_2 = \frac{F_2}{\| F_2 \|} \tag{7.10}$$

and choose f_3,

$$F_3(x) = f_3 + c_1\varphi_1 + c_2\varphi_2 \tag{7.11}$$

where c_1, c_2 are chosen so that F_2 may be orthogonal to φ_1 *and* φ_2,

i.e., $\int_a^b (f_3 + c_1\varphi_1 + c_2\varphi_2)\varphi_1 dx = 0$ and $\int_a^b (f_3 + c_1\varphi_1 + c_2\varphi_2)\varphi_2 dx = 0$

or,
$$\int_a^b f_3\varphi_1 dx + c_1 = 0 \text{ and } \int_a^b f_3\varphi_2 dx + c_2 = 0.$$

$$\therefore c_1 = -\int_a^b f_3\varphi_1 dx \; and \; c_2 = -\int_a^b f_3\varphi_2 dx. \tag{7.12}$$

Using Equation (7.11), Equation (7.12) gives

$$F_3(x) = f_3 - \varphi_1 \int_a^b f_3\varphi_1 dx - \varphi_2 \int_a^b f_3\varphi_2 dx. \tag{7.13}$$

So, we take

$$\varphi_3 = \frac{F_3}{\|F_3\|}. \tag{7.14}$$

Continuing this process, we get
the *nth* normalized function φ_n, which is given by

$$\varphi_n = \frac{F_n}{\|F_n\|} \tag{7.15}$$

where

$$F_n(x) = f_n - \varphi_1 \int_a^b f_n\varphi_1 dx - \varphi_2 \int_a^b f_n\varphi_2 dx - \cdots\cdots - \varphi_{n-1} \int_a^b f_n\varphi_{n-1} dx. \tag{7.16}$$

This process will fail iff at the same stage

$$F_r = 0 \ for \ 1 \le r \le n.$$

But $F_r = 0 \Rightarrow f_r - \varphi_1 \int_a^b f_r\varphi_1 dx - \varphi_2 \int_a^b f_r\varphi_2 dx - \cdots .. - \varphi_{r-1} \int_a^b f_r\varphi_{r-1} dx = 0,$

$$\text{i.e., } f_r = \varphi_1 \int_a^b f_r\varphi_1 dx + \varphi_2 \int_a^b f_r\varphi_2 dx + \cdots .. + \varphi_{r-1} \int_a^b f_r\varphi_{r-1} dx$$

implies that f_r is a linear combination of $\varphi_1, \varphi_2, \ldots \varphi_{r-1}$ ($\because \int_a^b f_r\varphi_{r-1} dx$, $r = 1, 2, \ldots$ *are constants*),

i.e., f_r is a linear combination of $f_1, f_2, \ldots . f_{r-1}$.

$\Rightarrow f_1, f_2, \ldots . f_{r-1}$ is a linearly dependent set for $1 \le r \le n$ (since the norm of all functions is not zero).

We have arrived at a contradiction because $\{f_n(x)\}$ is a linearly independent set, and by hypothesis, it follows that its every subset is linearly independent.

Hence, $F_r \ne 0 \ for \ 1 \le r \le n$, and we would always get a set $\{\varphi_n(x)\}$, which would be orthonormal.

Example 7.5.1. (a) Show that the functions $f_1(x) = 1, f_2(x) = x$ are orthogonal on the interval $(-1,1)$, and determine the constants $A \ and \ B$ so that the function $f_3(x) = 1 + Ax + Bx^2$ is orthogonal to both $f_1 \ and \ f_2$ on the interval $(-1,1)$.

(b) Show that the functions $f_1(x) = 4, f_2(x) = x^2$ are orthogonal on the interval $(-2,2)$, and determine the constants $A \ and \ B$ so that the function $f_3(x) = 1 + Ax + Bx^2$ is orthogonal to both $f_1 \ and \ f_2$.

7.6 Reality of Eigenvalues

To prove that all eigenvalues of the Strum–Liouville problem are real, we can use the property of self-adjointness of the Strum–Liouville operator, which allows us to show that if λ is an eigenvalue, then its complex conjugate must also be an eigenvalue leading to the conclusion that λ must be a real number.

Theorem 7.6.1. Prove that all eigenvalues of Strum–Liouville problem are real.

Proof. Consider the following Strum–Liouville problem:

$$[r(x)y']' + [\lambda p(x) + q(x)]y = 0 \quad (7.17)$$

$$a_1 y(a) + ay'(a) = 0 \quad (7.18)$$

$$b_1 y(b) + b_2 y'(b) = 0 \quad (7.19)$$

where $p, q, r, and\, r'$ are real valued and continuous, and $p(x) \geq 0\; on\; a \leq x \leq b.$ Let a_1, a_2 in Equation (7.18) be given constants, not both zero, and similarly b_1, b_2 in Equation (7.19).

Let $y(x)$ be an eigen function corresponding to an eigenvalue $\lambda = \alpha + i\beta\; where\; \alpha, \beta$ are real given constants. This eigen function $y(x)$ satisfies Equations (7.17)–(7.19) and may be a complex-valued function.

Taking the complex conjugates of all terms in Equations (7.17)–(7.19), we get

$$[r(x)\bar{y}']' + [\bar{\lambda}\, p(x) + q(x)]\bar{y} = 0 \quad (7.20)$$

$$a_1 \bar{y}(a) + a_2 \bar{y}'(a) = 0 \quad (7.21)$$

$$b_1 \bar{y}(b) + b_2 \bar{y}'(b) = 0. \quad (7.22)$$

The above Equations (7.20)–(7.22) show that $\bar{y}(x)$ is the eigen function corresponding to the eigenvalue $\bar{\lambda} = \alpha - i\beta$. Multiplying Equation (7.17) by $\bar{y}$ and Equation (7.20) by y and subtracting, we get

$$(ry')'\bar{y} - (r\bar{y}')'y + (\lambda - \bar{\lambda})py\bar{y} = 0$$

or,

$$(\lambda - \bar{\lambda})py\bar{y} = (r\bar{y}')'y - (ry')'\bar{y}.$$

$$\therefore \quad (\lambda - \bar{\lambda})py\bar{y} = \frac{d}{dx}[(r\bar{y}')y - (ry')\bar{y}]. \quad (7.23)$$

Integrating both sides of Equation (7.23) with respect to x from *a to b*, we get

$$(\lambda-\bar{\lambda})\int_a^b py\bar{y}\,dx=[(r\bar{y}')y-(ry')\bar{y}]_a^b$$

or,

$$(\lambda-\bar{\lambda})\int_a^b py\bar{y}\,dx=r(b)[\bar{y}'(b)y(b)-y'(b)\bar{y}(b)]-r(a)[\bar{y}'(a)y(a)-y'(a)\bar{y}(a)]. \tag{7.24}$$

We verify various cases depending on whether $r(x)$ vanishes or not at *a or b*, respectively.

(i) Let $r(a)=r(b)=0,$ then

$$(\lambda-\bar{\lambda})\int_a^b py\bar{y}\,dx=0. \tag{7.25}$$

(ii) Let $r(b)=0,$ *but* $r(a)\neq 0.$ Then Equation (7.24) reduces to

$$(\lambda-\bar{\lambda})\int_a^b py\bar{y}\,dx=-r(a)[\bar{y}'(a)y(a)-y'(a)\bar{y}(a)]. \tag{7.26}$$

Consider the relations Equations (7.18) and (7.21). Let $a_2\neq 0.$ Multiplying Equation (7.21) by $y(a)$ *and* Equation (7.18) by $\bar{y}(a)$ and subtracting, we get

$$a_2[\bar{y}'(a)y(a)-y'(a)\bar{y}(a)]=0.$$

Since, $a_2\neq 0,$

$$\bar{y}'(a)y(a)-y'(a)\bar{y}(a)=0. \tag{7.27}$$

By Equation (7.26), with this relation, we get Equation (7.25).

(iii) Let $r(a)=0, but\ r(b)\neq 0.$ Then Equation (7.24) reduces to

$$(\lambda-\bar{\lambda})\int_a^b py\bar{y}\,dx = r(b)[\bar{y}'(b)y(b)-y'(b)\bar{y}(b)]. \tag{7.28}$$

Consider the relations Equations (7.19) and (7.22). Let $b_2\neq 0.$

Multiplying Equation (7.22) by $y(b)\ and$ Equation (7.19) by $\bar{y}(b)$ and subtracting, we get

$$b_2[\bar{y}'(b)y(b)-y'(b)\bar{y}(b)]=0$$

$b_2\neq 0.$ So,

$$\bar{y}'(b)y(b)-y'(b)\bar{y}(b)=0. \tag{7.29}$$

Using Equations (7.28), Equation (7.28) reduces to Equation (7.25).

If $b_2=0,$ then assume $b_1\neq 0.$ Now, multiplying Equation (7.19) by $y\bar{y}'(b)$ and Equation (7.22) by $y'(b)$ and then subtracting, we get

$$b_1[\bar{y}'(b)y(b)-y'(b)\bar{y}(b)]=0.$$

Since $b_1\neq 0,$ therefore, $\bar{y}'(b)y(b)-y'(b)\bar{y}(b)=0.$

Hence, as before, Equation (7.28) reduces to Equation (7.25).

(iv) Let $r(a)\neq 0, but\ r(b)\neq 0.$ Then, assuming $a_2\neq 0$ and $b_2\neq 0,$ then proceeding as cases II and III, relations Equations (7.27) and (7.29) can be proved.

Then, using Equations (7.27) and (7.29), (iv) reduces to Equation (7.25)

(v) Let $r(a)=r(b),$ and proceed as in case (iv) to show that (iv) reduces to Equation (7.25).

From all cases, we see that in all situations, we get Equation (7.25).

Now, $\lambda=\alpha+i\beta,$ so $\bar{\lambda}=\alpha-i\beta.$

$$\therefore \lambda-\bar{\lambda}=2i\beta.$$

Again, $y\bar{y}=|y|^2,$ where $|y|$ stands for the modulus of $y.$ Then Equation (7.23) reduces to

$$2i\beta\int_a^b p\overline{|y|^2}dx=0\ or\ \beta\int_a^b p\overline{|y|^2}dx=0. \tag{7.30}$$

Since $\int_a^b p\overline{|y|^2}dx$ has a positive value in the given interval $a \leq x \leq b$, Equation (7.28) reduces to $\beta = 0$.

Therefore, $\lambda = \alpha + i\beta = \alpha$, which indicates that λ is real.

Since λ is an arbitrary eigenvalue, it follows that eigenvalues of the Strum–Liouville problem are all real.

7.7 Examples

Example 7.7.1. For the Strum–Liouville problem

$$y'' + \lambda y = 0,\ y(0) = 0, y(l) = 0.$$

Find the eigen functions, and show that they are orthogonal.

Solution. Let $\lambda = -\rho^2$. The general solution is $y = c_1 e^{\rho x} + c_2 e^{-\rho x}$.
For the given boundary conditions,
$c_1 + c_2 = 0$ and $c_1 e^{\rho l} + c_2 e^{-\rho l} = 0 \rightarrow c_1(e^{\rho l} - e^{-\rho l}) = 0$.
$\therefore c_1 = c_2 = 0,\ \ y = 0$, which is not an eigen function.
For $\lambda = \rho^2,\ \ y'' + \rho^2 y = 0$,
The solution is $y = a\cos\rho x + b\sin\rho x$.
The first boundary condition gives $y(0) = 0 = a$,
and the boundary condition $y(l) = 0 = b\sin\rho l$ implies that $\rho l = n\pi$.

$$\therefore \rho = 0, \pm\frac{\pi}{l}, \pm 2\pi/l.$$

Thus, the eigenvalues are $\lambda = 0, \frac{\pi^2}{l^2}, 4\frac{\pi^2}{l^2}$, and $b = 1$, and the corresponding eigen functions are

$$y_n(x) = \sin\left(\frac{n\pi x}{l}\right), \quad n = 0,1,2.$$

It follows that the said eigen functions are orthogonal on the interval $0 \leq x \leq l$.

Example 7.7.2. For the Strum–Liouville problem

$$y'' + \lambda y = 0,\ y(0) = 0, y'(l) = 0.$$

Find the eigen functions, and show that they are orthogonal.

Solution. Let $\lambda = -\rho^2$. The general solution is $y = c_1 e^{\rho x} + c_2 e^{-\rho x}$ *and* $y' = c_1 \rho e^{\rho x} - c_2 \rho e^{-\rho x}$.

For the given boundary conditions,

$c_1 + c_2 = 0$ and $c_1 \rho e^{\rho l} - c_2 \rho e^{-\rho l} = 0 \rightarrow c_1 \rho (e^{\rho l} + e^{-\rho l}) = 0$.

$\therefore c_1 = c_2 = 0$ *and also* $y(x) = 0$. Since $(x) \neq 0$, there is no eigen function corresponding to λ.

Let $\lambda = \rho^2$. Therefore, the solution is $y = A\cos\rho x + B\sin\rho x$

and $y' = -A\rho\sin\rho x + B\rho\cos\rho x$

using the boundary conditions $A = 0$

and $0 = B\rho\cos\rho l \rightarrow \cos\rho l = 0$.

$\therefore \rho l = (2n-1)\pi/2$ for n=1,2,3,...

Hence, $\rho = (2n-1)\pi/2l$.

The solution is $y = A\cos\rho x + B\sin\rho x = B\sin(2n-1)\dfrac{\pi}{2l}x$

and $\lambda = \rho^2 = [(2n-1)\pi/2l]^2 = \dfrac{(2n-1)^2\pi^2}{4l^2}$, n=1,2,3,...

So, the required eigen functions $y_n(x)$ with the corresponding eigenvalues λ_n are

$$y_n(x) = B_n \sin(2n-1)\frac{\pi}{2l}x \text{ and } \lambda_n = \frac{(2n-1)^2\pi^2}{4l^2}.$$

Example 7.7.3. Find the eigenvalues and eigen functions of

$$[xy'(x)]' + \left(\frac{\lambda}{x}\right)y(x) = 0,\ y'(1) = y'(e^{2\pi}) = 0.$$

Solution. The equation is $xy'' + y' + \left(\dfrac{\lambda}{x}\right)y = 0$

or,
$$x^2y'' + xy' + \lambda y = 0$$

or, $$(x^2D^2 + xD + \lambda)y = 0 \text{ where } D = \frac{d}{dx}. \tag{i}$$

This is a homogeneous differential equation. To solve it,

let, $$x = e^z, \textit{ so that } z = logx. \tag{ii}$$

If we let

$$D_1 = \frac{d}{dz} = \frac{dx}{dz}\frac{d}{dx} = xD, D_1^2 = \frac{d}{dz}\left(\frac{d}{dz}\right) = \frac{d}{dz}(xD) = \frac{dx}{dz}\frac{d}{dx}(xD)$$
$$= x(D + xD^2), x^2D^2 = D_1^2 - D_1 = D_1(D_1 - 1), \tag{iii}$$

using Equation (iii), Equation (i) becomes $[D_1(D_1 - 1) + D_1 + \lambda]y = 0$

$$[D_1^2 + \lambda]y = 0 \tag{iv}$$

with boundary conditions

$$y'(1) = y'(e^{2\pi}) = 0. \tag{v}$$

Case I. λ=0. The solution of Equation (iv) $D_1^2 y = 0$ is

$$y = Az + B \;\; or, \;\; y = Alogx + B. \tag{vi}$$

$$\therefore y' = \frac{A}{x}. \tag{vii}$$

Using Equation (v), $A = 0$ *and* $0 = \frac{A}{2\pi}$.
These give $A = 0$, while B may be taken as arbitrary.

Let $B = 1$. Equation (vi) gives $y(x) = 1$ as an eigen function corresponding to the eigenvalue λ=0.

Case II. Let $\lambda = -\rho^2$ where $\neq 0$. Then, Equation (iv) gives

$$[D_1^2 - \rho^2]y = 0$$

$$y = Ae^{\rho z} + Be^{-\rho z} = A(e^z)^\rho + B(e^z)^{-\rho}$$

$$y = A(e^z)^\rho + B(e^z)^{-\rho} = A(x)^\rho + B(x)^{-\rho} \quad \text{(viii)}$$

From Equation (viii),

$$y'(x) = A\rho(x)^{\rho-1} - B\rho(x)^{-\rho-1}. \quad \text{(ix)}$$

$$\therefore 0 = A\rho - B\rho \text{ and } 0 = A\rho(e^{2\pi})^{\rho-1} - B\rho(e^{-2\pi})^{\rho+1}.$$

Therefore, $A - B = 0$ and $Ae^{2\rho\pi} - Be^{-2\rho\pi} = 0$

$$\Rightarrow \rho = 0;\ A = B.$$

$\therefore y(x) = 0$, which is not an eigen function.
Case III. Let $\lambda = \rho^2$ where $\neq 0$. Then, Equation (iv) gives

$$[D_1^2 + \rho^2]y = 0.$$

The solution is $y = Acos\rho z + Bsin\rho z$

or,
$$y = Acos(\rho logx) + Bsin(\rho logx). \quad \text{(x)}$$

$$\therefore y'(x) = -A\frac{\rho}{x}sin(\rho logx) + B\frac{\rho}{x}cos(\rho logx). \quad \text{(xi)}$$

Using Equation (v) Equation (xi) reduces to $log1 = 0$ *and* $loge^{2\pi} = 2\pi$.

$0 = B\rho$ and $0 = -\left(\frac{A\rho}{e^{2\pi}}\right)sin2\pi\rho + \left(\frac{B\rho}{e^{2\pi}}\right)cos2\pi\rho$ (Using boundary condition).

$$\therefore B = 0 \ and \ Asin2\pi\rho = 0. \tag{xii}$$

Consider

$$Asin2\pi\rho = 0. \tag{xiii}$$

If $A = 0$, *then with* $B = 0$, Equation (x) reduces to $y(x) = 0$, which is not an eigen function. So, we take $A \neq 0$ for existence of eigen functions.

Since $A \neq 0$, Equation (xiii) $sin2\pi\rho = 0 \Rightarrow 2\pi\rho = \pi, \ n = 1,2,3,\ldots$

Thus,

$$\rho = n/2, \ \text{n=1,2,3,...} \tag{xiv}$$

$n = 0$ is not being considered because $n = 0 \Rightarrow \rho = 0 \Rightarrow \lambda = 0$, which is already being considered in case I.

Using $B = 0$ and Equation (xiv), Equation (x) becomes

$$y(x) = Acos\left[\left(\frac{n}{2}\right)logx\right] \text{ with } \lambda = \rho^2 = \frac{n^2}{4}, \ n = 1,2,3,\ldots$$

So, the required eigen functions are

$$y_n(x) = cos\left[\left(\frac{n}{2}\right)logx\right], \ n = 1,2,3,\ldots \text{ and } \lambda_n = \frac{n^2}{4}, \ n = 1,2,3,\ldots$$

and $y(x) = 1 \ with \ \lambda = 0$.

7.8 Exercise

1. For the Strum–Liouville problem

$$y'' + \lambda y = 0, \ y'(0) = 0, y'(l) = 0,$$

find the eigen functions and eigenvalues.

2. For the Strum–Liouville problem

$$y'' + \lambda y = 0,\ y(0) = 0, y(\pi) = 0,$$

find the eigen functions and eigenvalues.

3. Find the eigen functions and eigenvalues of the Strum–Liouville problem

$$y'' + \lambda y = 0,\ y(0) = 0, y'(\pi/2) = 0.$$

4. Find all eigen functions and eigenvalues of the Strum–Liouville problem

$$y'' + \lambda^2 y = 0,\ y'(0) = 0, y'(l) = 0,\ \ 0 \le x \le l.$$

5. Find the eigen values and eigen functions of the Strum-Liouville problem

(i) $y''' + \mu y = 0, y'(0) = 0, y'(l) = 0$
(ii) $y''' + \mu y = 0, y'(0) = 0, y(\pi) = 0$

Part II

SPECIAL FUNCTIONS

8

Preliminaries II

Abstract
The series solution of differential equations involves finding solutions in the form of infinite series, particularly power series, when direct analytic solutions are not readily available. In this chapter, we discuss different types of infinite series, their properties, convergence tests, infinite products, and some theorems related to functions of complex variables. These concepts are used throughout the second part of the book as ready references.

Keywords: Power series, infinite integrals, uniform convergence, infinite products

8.1 Introduction

Most of the special functions, as we shall see, are the series solutions of differential equations. Since we shall be constantly dealing with infinite series, improper integrals, and series solutions, it is essential to recall some definitions and results concerning them.

Consider the linear differential equations $\frac{d^2y}{dx^2}-8\frac{dy}{dx}+15y=0$, its solution is given by $y=c_1e^{3x}+c_2e^{5x}=c_1\sum_{n=0}^{\infty}\frac{(3x)^n}{n!}+c_2\sum_{n=0}^{\infty}\frac{(5x)^n}{n!}$, which is a series solution. Some differential equation with variable coefficient cannot be solved by the usual method, and we need to apply this series solution method in terms of infinite convergent series. This chapter discusses infinite series, infinite integrals, uniform series with their convergent test as ready reference for next chapters.

8.2 Infinite Series

An ordered set of real numbers $u_1,u_2,u_3,\ldots u_n$ is called a sequence and is denoted by (u_n). If the number of terms is unlimited, then the sequence is said to be an infinite sequence.

Dipankar De. Ordinary Differential Equations and Special Functions, (237–252) © 2026 Scrivener Publishing LLC

If $u_1, u_2, u_3, \ldots u_n$ is an infinite sequence of real numbers, then $u_1 + u_2 + u_3, \ldots + \infty$ is called an infinite series. An infinite series is denoted by $\sum u_n$ and the sum of the first n terms is denoted by s_n.

Clearly, s_n is a function of n, and as n increases indefinitely, three possibilities arise:

8.2.1 Convergent and Divergent Series

(i) If s_n tends to a finite limit as $n \to \infty$, the series $\sum u_n$ is said to be *convergent.*

(ii) If s_n tends to $\pm\infty$ as $n \to \infty$, the series $\sum u_n$ is said to be *divergent.*

(iii) If s_n tends to a unique limit as $n \to \infty$, then the series $\sum u_n$ is said to be *oscillatory* or non-convergent.

(iv) The series $\sum u_n$ is said to be *absolutely convergent* if the series $\sum |u_n|$ is convergent. We know that an absolutely convergent series is itself convergent.

8.2.2 Alternating Series

A series in which the terms are alternately positive or negative is called an *alternating series.*

Leibnitz's series: An alternating series $u_1 - u_2 + u_3 - u_4 + \cdots$ converges if (i) each term is numerically less than its preceding term, and (ii) $\lim_{n\to\infty} u_n = 0$.

If $\lim_{n\to\infty} u_n \neq 0$, then the given series is oscillatory.

Example 8.2.1. Discuss the convergence of the series

(i) $1 - \frac{1}{\sqrt{2}} + \frac{1}{\sqrt{3}} - \frac{1}{\sqrt{4}} + \ldots$ (ii) $\frac{5}{2} - \frac{7}{4} + \frac{9}{6} - \frac{11}{8} + \ldots$

Solution: (i) The term of the given series is alternately positive and negative, each term numerically less than its preceding term, since $u_n - u_{n-1} = \frac{1}{\sqrt{n}} - \frac{1}{\sqrt{n-1}} < 0.$

Also, $\lim_{n\to\infty} u_n = \lim_{n\to\infty} \frac{1}{\sqrt{n}} = 0.$

Hence, by Leibnitz rule, the given series is convergent.

(ii) The term of the given series is alternately positive and negative, each term numerically less than its preceding term, since $u_n - u_{n-1} = \frac{2n+3}{2n} - \frac{2n+1}{2n-2} = \frac{(2n+3)(n-1) - n(2n+1)}{2n(n-1)} = \frac{-3}{2n(n-1)} < 0 \; if \; n > 1.$

Here, $u_n < u_{n-1}$ for $n > 1$ and $\lim_{n\to\infty} u_n = \lim_{n\to\infty} \frac{2n+3}{2n} = \lim_{n\to\infty} \frac{2+\frac{3}{n}}{2} = 1 \neq 0$.
Hence, by Leibnitz rule, the given series is not convergent; it is oscillatory.

Example 8.2.2. For what values of x does the following series converge?

$$x - \frac{x^2}{2} + \frac{x^3}{3} - \frac{x^4}{4} + \frac{x^5}{5} - \cdots \ldots \infty$$

Solution: Here, $u_n = (-1)^{n-1} \frac{x^n}{n}$ *and* $u_{n+1} = (-1)^n \frac{x^{n+1}}{n+1}$

$$\frac{u_{n+1}}{u_n} = -\frac{n}{n+1} x$$

$$\lim_{n\to\infty} \left| \frac{u_{n+1}}{u_n} \right| = \lim_{n\to\infty} \frac{n}{n+1} |x| = \lim_{n\to\infty} \frac{1}{1+\frac{1}{n}} |x| = |x|.$$

By ratio test, the given series converges for $|x| < 1$ and diverges for $|x| > 1$.

8.2.3 Uniform Convergence

Let

$$u_1(x) + u_2(x) + \cdots + \cdots \infty = \sum_{n=1}^{\infty} u_n(x). \tag{8.1}$$

be an infinite series of functions each of which is defined in the interval (a,b). Let $s_n(x)$ be the sum of its first n terms, i.e., $s_n(x) = u_1(x) + u_2(x) + \cdots + u_n(x)$.

At some point $x = x_1$, *if* $\lim_{n\to\infty} s_n(x_1) = s(x_1)$,

Then, the series 1.1 is said to converge to the sum $s(x_1)$ at that point. This means that at $x = x_1$, given a positive number ϵ we can find a number N such that $|s(x_1) - s_n(x_1)| < \epsilon$ *for* $n > N$.

Definition 8.2.1. The series $\sum u_n(x)$ is said to be *uniformly convergent* in the interval (a,b) if for a given $\epsilon > 0$, a number N can be found independent of x such that for every x in the interval (a,b)

$$|s(x) - s_n(x)| < \epsilon \text{ for all } n > N.$$

In complex analysis, the most important series with variable terms is power series. A power series has the form

$$\sum_{n=0}^{\infty} a_n z^n = a_0 + a_1 z + a_2 z^2 + \cdots \ldots$$

where z is a complex variable, and $a_0, a_1, \ldots$ are constants (complex numbers). With every power series, there is an associated circle called the *circle of convergence.* In the interior of the circle, the series converges, and in the exterior, it diverges. If a_n *and* z are real, the circle of convergence reduces to an interval.

8.2.4 Convergent Test

To determine the absolute convergence of a series of complex terms, we apply the Cauchy's root test, D'Alembert's ratio test, comparison test, or Raabe's test, as we do in the case of real series of positive terms. In the real case, $|a_n|$ stands for the numerical value of a_n, while in a complex case, it denotes the modulus of the complex number a_n.

For a non-absolutely convergent complex series, we apply the Dedekind's test, which states that the series $\sum a_n u_n$ is convergent if

(i) the series $\sum a_n$ has bounded partial sums.
(ii) the series $\sum (u_n - u_{n+1})$ is absolutely convergent. and
(iii) $u_n \to 0$ *as* $n \to \infty$.

Weierstrass M-Test: A series $\sum u_n(x)$ is uniformly convergent in an interval (a,b), if there exists a convergent series $\sum M_n$ of positive constants such that $|u_n(x)| < M_n$ for all values of x in (a,b).

Since $\sum M_n$ is convergent, therefore, for a given $\epsilon > 0$, we can find a number N, such that $|s - s_n| < \epsilon$ for every $n > N$,

where $s = M_1 + M_2 + \cdots . + M_n + M_{n+1} + \cdots$ *and* $s_n = M_1 + M_2 + \cdots . + M_n$

$$\Rightarrow |M_{n+1} + M_{n+2} \cdots| < \epsilon \text{ for every } n > N.$$

Since $|u_n(x)| < M_n$

$$|u_{n+1}(x) + u_{n+2}(x) + \cdots| \le |u_{n+1}(x)| + |u_{n+2}(x)| + \cdots . \le M_{n+1} + M_{n+2} + \cdots < \epsilon \text{ for every } n > N,$$

i.e., $|s - s_n| < \epsilon$ for every $n > N$, where $s(x)$ is the sum of the series $\sum u_n(x)$.

Since N does not depend on x, the series $\sum u_n(x)$ converges uniformly in (a,b).

8.2.5 Properties of Uniformly Convergent Series

I. If the series $\sum u_n(x)$ converged uniformly to the sum $s(x)$ in the interval (a,b), and each of the functions $u_n(x)$ is continuous in the interval, then the sum $s(x)$ is also continuous in (a,b).

II. If the series $\sum u_n(x)$ converged uniformly in the interval (a,b), and each of the functions $u_n(x)$ is continuous in the interval, then the series can be integrated term by term,

i.e., $\int_a^b [u_1(x)+u_2(x)+\cdots]dx = \int_a^b u_1(x)dx + \int_a^b u_2(x)dx + \cdots$.

III. If the series $\sum u_n(x)$ is convergent, having continuous derivatives of its terms, and the series $\sum u_n(x)$ converges uniformly, then the series can be differentiated term by term,

$$\frac{d}{dx}[u_1(x)+u_2(x)+\cdots] = u_1'(x)+u_2'(x)+\cdots.$$

Example 8.2.3. Examine the following series for uniform convergence:

(i) $\sum \frac{cosnx}{n^p}$ (ii) $\sum_{n=1}^{\infty} \frac{\sin(nx+x^2)}{n(n+2)}$

Solution: $\left|\frac{cosnx}{n^p}\right| = \frac{|cosnx|}{n^p} \le \frac{1}{n^p} = M_n\,(let)$.

But $\sum M_n$ is a p-series with $p>1$, so it is convergent.

By Weierstrass M-Test, the given series is uniformly convergent for all real values of x with $p>1$.

(ii) Here, $\left|\frac{\sin(nx+x^2)}{n(n+2)}\right| = \frac{|\sin(nx+x^2)|}{n(n+2)} \le \frac{1}{n^2} = M_n\,(let)$.

$\sum M_n = \sum_{n=1}^{\infty} \frac{1}{n^2}$ is convergent;

therefore, By Weierstrass M-Test, the given series is uniformly convergent for all real values of x.

Example 8.2.4. Prove that $\int_0^1\left(\sum \frac{x^n}{n^2}\right)dx = \sum_{n=1}^{\infty} \frac{1}{n^2(n+1)}$.

Solution: $|x^n| \le 1 \; for \; 0 \le x \le 1$

$$\left|\frac{x^n}{n^2}\right| \le \frac{1}{n^2} = M_n(let) \text{ and } for\; 0 \le x \le 1.$$

Hence, $\sum M_n = \sum_{n=1}^{\infty} \frac{1}{n^2}$ is convergent.

By M-Test, the series $\sum \frac{x^n}{n^2}$ is uniformly convergent in $0 \le x \le 1$ and also, $\frac{x^n}{n^2}$ is continuous in $0 \le x \le 1$.

The series $\sum \frac{x^n}{n^2}$ can be integrated term by term in $0 \le x \le 1$.

$$\int_0^1\left(\sum \frac{x^n}{n^2}\right)dx = \sum\left(\int_0^1 \frac{x^n}{n^2}dx\right) = \sum\left(\frac{1}{n^2}\int_0^1 x^n dx\right) = \sum \frac{1}{n^2}\frac{1}{(n+1)}.$$

8.3 Infinite Integrals

The definite integrals, in which either the interval of integration is infinite, or the integrand has discontinuous at finite number of points in the interval of integration, are called *infinite integrals.*

(i) Let $f(x)$ be bounded and integrable in the arbitrary interval (a,b). We define the integral $\int_a^{\infty} f(x)dx$ as

$$\lim_{b\to\infty}\int_a^b f(x)dx$$

when this limit exists. The integral $\int_a^{\infty} f(x)dx$ is referred to as an infinite integral, and it *converges* in this case. We similarly define the integral $\int_{-\infty}^{b} f(x)dx$ as

$$\lim_{a\to-\infty}\int_a^b f(x)dx \text{ when this limit exists.}$$

If the infinite integrals $\int_{-\infty}^{b} f(x)dx$ and $\int_a^{\infty} f(x)dx$ are both convergent, we say that the infinite integrals $\int_{-\infty}^{\infty} f(x)dx$ exists and is equal to their sum.

(ii) The integral $\int_a^\infty f(x)dx$ is said to be *absolutely convergent* if $f(x)$ is bounded and integrable in the arbitrary interval (a,b) and $\int_a^\infty |f(x)|dx$ is convergent.

The test of convergence for the infinite integral $\int_a^\infty f(x)dx$ is known as the μ*-test and is given below*

Let $f(x)$ be bounded and integrable in the arbitrary interval (a,b) where $a > 0$. If there is a number $\mu > 1$ such that

$\lim_{x\to\infty}\left(x^\mu f(x)\right)$ exists, then $\int_a^\infty f(x)dx$ is *absolutely convergent.*

If this limit exists and is equal to zero for $\mu \le 1$, then $\int_a^\infty f(x)dx$ *diverges.*

Convergence of the infinite integrals $\int_{-\infty}^b f(x)dx$ *and* $\int_{-\infty}^\infty f(x)dx$ can be determined by changing them to the form $\int_b^\infty f(x)dx$.

(iii) We discuss the case of infinite discontinuities of the integrand in the interval of integration. Let us suppose first that a is the only point of such discontinuity of $f(x)$ in the interval (a,b), and $f(x)$ is bounded and integrable in the arbitrary interval $(a+\xi, b), a < a+\xi < b$

Then, if the integral $\int_{a+\xi}^b f(x)dx$ has a limit $\xi \to 0$, we define the integral $\int_a^b f(x)dx$ as

$$\lim_{\xi\to 0}\int_{a+\xi}^b f(x)dx.$$

Similarly, if point b is the only point of discontinuity of f(x) in (a,b), and $f(x)$ is bounded and integrable in the arbitrary interval (a,b-ξ), $a < b-\xi < b$ and we define the integral infinite as

$$\lim_{\xi\to 0}\int_a^{b-\xi} f(x)dx \text{ when this limit exists.}$$

In case a and b are both points of infinite discontinuity of f(x), we define the integral $\int_a^b f(x)dx$ as the sum of the integrals $\int_a^c f(x)dx$ and $\int_c^b f(x)dx$ when these exist, c being an arbitrary point between a and b.

Finally, if there are finite numbers of points of infinite discontinuity of f(x) in the interval (a,b), say $x_1, x_2, \cdots .x_n$, we define the integral $\int_a^b f(x)dx$ as

$\int_a^{x_1} f(x)dx + \int_{x_1}^{x_2} f(x)dx + \cdots ..... + \int_{x_n}^b f(x)dx$ when these integrals exist by definitions given. Infinite integrals of this type are sometimes referred to as *improper integrals.*

The infinite integrals $\int_a^b f(x)dx$, where a is a point of infinite discontinuity of (x), are said to be absolutely convergent, if f(x) is bounded and integrable in the arbitrary interval $(a+\xi, b)$, *and* $\int_a^b |f(x)|dx$ converges.

The test of convergence of the improper integral $\int_a^b f(x)dx$, $f(x)$ having an infinite discontinuity only at x=0, also known as $\mu-$*test, may be stated* as below:

If there is a number μ lying between 0 and 1 such that $(x-a)^{\mu} f(x)$ is bounded when $a \le x \le b$,

then $\int_a^b f(x)dx$ converges absolutely. If there is a number $\mu \ge 1$ such that $(x-a)^{\mu} f(x)$ has a positive lower bound or a negative upper bound when $a \le x \le b$, then the same integral $\int_a^b f(x)dx$ diverges to ∞ in the first case, and to $-\infty$ in the second case.

Example 8.3.1. Discuss the convergence of the integrals

(i) $\int_o^{\infty} \frac{x^3}{(a^2+x^2)^2}dx$ (ii) $\int_0^1 x^{n-1} logx\, dx$.

Solution: (i) The form of the integral $\int_o^{\infty} f(x)dx$.

Now, apply $\mu-test$.

Here, we take $\mu=1$, then

$$\lim_{x\to\infty}\left(x^{\mu} f(x)\right) = \lim_{x\to\infty}\left(x.\frac{x^3}{(a^2+x^2)^2}\right) = 1.$$

Hence, the integral is divergent.

(ii) When $n>1$, the integrand is finite in the interval $0<x<1$

since $\lim_{x\to 1}\left(x^{n-1} logx\right) = \lim_{x\to 1} logx = 0$ *and* $\lim_{x\to 0}\left(x^{n-1} logx\right) = \lim_{x\to 0} x^{n-1} logx = 0$.

Thus, when $n>1$, the integral $\int_0^1 x^{n-1} logx\, dx$ is an ordinary integral, and hence, it is convergent.

When $n=1$,

$$\int_0^1 logx\, dx = \lim_{x\to 0}\int_x^1 logx\, dx = \lim_{x\to 0}\left[xlogx - x\right]_x^1 = -1.$$

Therefore, the integral again converges.

When $0<n<1$, the point $x=0$ is a point of infinite discontinuity.

Since $\lim_{x\to 0}\left(x^{\mu} f(x)\right) = 0$, *if* $\mu > 1-n$ the integral converges, for it is possible to get $0<\mu<1$.

Finally, when $n \leq 0$

$$\lim_{x\to 0}\left(x.x^{n-1}\left|logx\right|\right) = \lim_{x\to 0} x^n \left|logx\right| = \infty.$$

Thus, we conclude that the given integral converges when $n > 0$ and diverges when $n \leq 0$.

8.4 Infinite Products

Let us suppose that a sequence $\{u_n\}$ is defined for all positive integral values of k, and

$$P_n = \Pi_{k=1}^{n}(1+u_k) = (1+u_1)(1+u_2)\cdots(1+u_n).$$

If the limit of P_n as $n \to \infty$ is equal to $P(\neq 0)$, then the infinite product $\Pi_{k=1}^{n}(1+u_k)$ *converges to P.*

If one or a finite number of factors are zero, and if the infinite product with the zero factors are deleted, converges to a value $P \neq 0$, we say that the infinite product **converges to zero.**

If the infinite product is not convergent, it is said to be divergent. If $\lim_{n\to\infty} P_n \to 0$, the product is said to *diverge to zero.*

(i) If $\Pi_{k=1}^{n}(1+u_k)$ converges, then the condition $\lim_{n\to\infty} P_n = 0$ is necessary but not sufficient.
(ii) If $u_n \neq -1$, $\Pi_{k=1}^{n}(1+u_k)$ *and* $\sum_{n=1}^{\infty} log(1+u_n)$ converges or diverges together.

The product $\Pi_{k=1}^{n}(1+u_k)$ with zero factors deleted is said to be absolutely convergent iff the series $\sum_{n=1}^{\infty} log(1+u_n)$ is absolutely convergent. Absolute convergence of $\sum_{n=1}^{\infty} log(1+u_n)$ implies the absolute convergence of $\sum_{n=1}^{\infty} u_n$

Definition 8.4.1. Orthogonal Polynomials
If $\{\phi_n(x)\}$ is a simple set of polynomials and if $P(x)$ is a polynomial of degree m, there exist constants C_k such that

$$P(x) = \sum_{k=1}^{\infty} C_k \phi_k(x). \tag{8.2}$$

Let $\{\phi_n(x)\}$ be a simple set of real polynomials. The set is said to be orthogonal over the interval $a < x < b$ with respect to a weight function $w(x) > 0$, *such that*

$$\int_a^b w(x)\phi_n(x)\phi_m(x)dx = 0 \quad ,m \neq n. \tag{8.3}$$

Since we have taken $w(x) > 0$ and $\phi_n(x)$ as real, Equation (8.2) implies that

$$\int_a^b w(x)\phi^2{}_n(x)dx \neq 0. \tag{8.4}$$

If $\{\phi_n(x)\}$ is an orthogonal set of polynomials over the interval (a,b) with respect to the weight function $w(x) > 0$, then an application of the property Equation (8.2) yields

$$\int_a^b w(x)\phi^k{}_n(x)dx = 0, k = 1,2,\ldots,(n-1). \tag{8.5}$$

8.5 Some Theorems on Functions of Complex Variables

A function $f(z)$, which is single valued and possesses a unique derivative with respect to z at all points of a region R, is called an *analytical function* of z in that region. An analytic function is also called a *regular function or a holomorphic function.*

A function that is analytic everywhere in the complex plane is known as an *entire function.*

Some important theorems concerning functions of a complex variable are as follows:

(i) If $f(z)$ is a function of z, which is analytic at all points on and inside a closed contour C, then

$$\int_C f(z)dz = 0.$$

This result is known as *Cauchy's integral theorem.*

(ii) If $f(z)$ is an analytic within and on a closed contour C, and if a is any point within C, then

$$f(a)=\frac{1}{2\pi i}\int_C \frac{f(z)}{(z-a)}dz.$$

This result is known as *Cauchy's Integral formula.*

(iii) If $f(z)$ is an analytic within a given region R, it possesses derivatives of all orders in the region. At a point $z=a$ within R, we have

$$f^{(n)}(a)=\frac{n!}{2\pi i}\int_C \frac{f(z)}{(z-a)^{n+1}}dz$$

where n is a positive integer, $f^{(n)}(a)$ is the nth derivative of $f(z)$ at $z=a$, and C is any closed contour in R surrounding a.

(iv) If $f(z)$ is an analytic within and on the circle $C:|z-a|=r$ then

$$\left|f^{(n)}(a)\right|\le\frac{Mn!}{r^n}, \tag{8.6}$$

where M is the maximum value of $|f(z)|$ on C. This result is known as *Cauchy's inequality.*

(v) If $f(z)$ is an analytic and bounded for all z in the entire complex plane, then $f(z)$ is a constant.

Taking $n=1$ and replacing a *by* z, Equation (8.6) gives $|f'(z)|\le\frac{M}{r}$.

As $r\to\infty$, it gives $f'(z)=0$, i.e., $f(z)$ is constant for all z.

This is called *Liouville's theorem.*

(vi) If $f(z)$ is an analytic function in a circle C of radius R about the point $z=a$, then at any point z within the circle, $f(z)=f(a)+(z-a)f'(a)+\cdots+\frac{(z-a)^n}{n!}f^n(a)+\cdots$.

This is *Taylor's theorem.*

Maclaurin's theorem is a special case of this when $a=0$,

i.e., $f(z)=f(0)+zf'(0)+\cdots+\frac{z^n}{n!}f^n(0)+\cdots$.

(vii) If $f(z)$ is analytic in the ring-shaped region R bounded by two concentric circles C *and* C_1 *of radii* r *and* $r_1\ (r > r_1)$ and with the center a, then for all z in R

$$f(z)=a_0+a_1(z-a)+a_2(z-a)^2+\cdots.+a_{-1}(z-a)^{-1}+a_{-2}(z-a)^{-2}+\cdots \tag{8.7}$$

where $a_n = \dfrac{1}{2\pi i}\oint \dfrac{f(t)}{(t-a)^{n+1}}dt$. This result is known as *Laurent's theorem*

(viii) A *zero* of an analytic function $f(z)$ is that value of z for which $f(z)=0$.

If $f(z)$ is analytic in the neighborhood of a point $z=a$, then by Taylor's theorem,

$f(z)=a_0+a_1(z-a)+a_2(z-a)^2+\cdots.+a_n(z-a)^n+\cdots$, where $a_n=\dfrac{f^n(a)}{n!}$.

If $a_0=a_1=a_2\ldots\ldots=a_{m-1}=0$ *and* $a_m\neq 0$, then $f(z)$ is said to have a zero of order m at $z=a$.

When $m=1$, the zero is said to be *simple.* In the neighborhood of zero $(z=a)$ of order m,

$$\begin{aligned} f(z) &= a_m(z-a)^m + a_{m+1}(z-a)^{m+1}+\cdots.+\infty \\ &= (z-a)^m\phi(z) \text{ where } \phi(z)=a_m+a_{m+1}(z-a)+\cdots. \end{aligned}$$

This $\phi(z)$ is analytic and non-zero in the neighborhood of $z=a$.

A *singular point* of a function is the point at which the function ceases to be analytic.

(a) *Isolated singularity:* If $z=a$ is a singularity of $f(z)$ such that $f(z)$ is analytic at each point in its neighborhood (i.e., there exists a circle with a center a, which has no other singularity), then $z=a$ is called an *isolated singularity.*

In such case, $f(z)$ can be expanded in a Laurent's series around $z=a$

$$f(z)=a_0+a_1(z-a)+a_2(z-a)^2+\cdots.+a_{-1}(z-a)^{-1}+a_{-2}(z-a)^{-2}+\cdots.$$

For example, $f(z) = \cot\left(\frac{\pi}{z}\right)$ is not analytic, but $f(z) = \tan\left(\frac{\pi}{z}\right) = 0$, i.e., $\frac{\pi}{z} = 4\pi \Rightarrow z = \frac{1}{4}$ implying $z = \frac{1}{n}$.

Thus, $z = 1, \frac{1}{2}, \frac{1}{4}, \ldots$ are all isolated singularities as there are no other singularty in their neighborhood.

(b) Removable singularity. If all the negative powers of $(z-a)$ in Equation (8.7) are zero, then

$$f(z) = \sum_{n=0}^{\infty} a_n (z-a)^n.$$

Here, the singularity can be removed by defining $f(z)$ *at* $z = a$ in such a way that it becomes analytic at $z = a$. Such singularity is called a *removable singularity.* If they are infinite in number, then $z = a$ is called an essential singularity of $f(z)$.

If all the negative powers of $(z-a)$ in Equation (8.7) after the nth are missing, then the singularity at $z = a$ is called a pole of order n.

(ix) The coefficient of $(z-a)^{-1}$ in the expansion of $f(z)$ arround an isolated singularity is called the residue of $f(z)$ at that point. Thus, Laurent's series expansion is around $z = a$, i.e.,

$$f(z) = a_0 + a_1(z-a) + a_2(z-a)^2 + \cdots . + a_{-1}(z-a)^{-1} + a_{-2}(z-a)^{-2} + \cdots .$$

The residue of $f(z)$ at $z = a$ is a_{-1}.

Residue Theorem 8.5.1. If $f(z)$ is analytic in a closed curve C except at a finite number of singular points within C, then $\oint_C f(z)dz = 2\pi i \times ($sum of the residues at the singular points within $C)$.

This is called Cauchy's theorem of residues.

In the case of a simple pole at a, the residue of $f(z)$ at the pole is $\lim_{z \to a} (z-a) f(z)$.

If $f(z)$ has multiple poles of order $n(>1)$ at the point $z=a$, we may write

$$f(z)=\frac{\phi(z)}{(z-a)^n}.$$

The residue at pole $z=a$ is then given by

$$\frac{\phi^{n-1}(z)}{(n-1)!}=\frac{1}{2\pi i}\int_C\frac{\phi(z)}{(z-a)^n}dz.$$

A function that is analytic in a region except for a finite number of poles is said to be *meromorphic* in the region.

(x) If $f(z)$ is analytic in a closed contour C and (z) is meromorphic inside C and has no zeros on C, then $\frac{1}{2\pi i}\oint_C f(z)\frac{\phi'(z)}{\phi(z)}dz=\sum r_1 f(a_1)-\sum s_1 f(b_1)$ where $a_1,a_2,\ldots$ are the zeros inside C of orders $r_1,r_2,\ldots$ respectively, and $b_1,b_2,\ldots$ are the poles of orders $s_1,s_2,\ldots$, respectively.

(xi) If $f(z)$ is meromorphic inside a closed contour C, and has no zeros on C, then

$$\frac{1}{2\pi i}\oint_C\frac{f'(z)}{f(z)}dz=N_z-N_p$$

where N_z is the number of zeros, and N_p is the number of poles of $f(z)$ within C.

8.6 Exercise

1. Examine the convergence of the series

$$\text{(i) }\sum_{n=1}^{\infty}\frac{(-1)^{n-1}.n}{2n-1}\quad\text{(ii) }\sum_{n=2}^{\infty}\frac{(-1)^{n-1}.x^n}{n(n-1)},0<x<1.$$

2. Show that the following series converges uniformly:

$$\text{(i)} \sum \frac{1}{n^3 + n^4 x^2} \quad \text{(ii)} \sum_{n=1}^{\infty} \frac{1}{n^p + n^q x^2}.$$

3. Test the uniform convergence of the series

$$\text{(i)} \frac{sinx}{1^2} + \frac{sin2x}{2^2} + \frac{sin3x}{3^2} + \frac{sin4x}{4^2} + \cdots \ldots \infty.$$

$$\text{(ii)} \ 1 + \frac{1}{2^2} - \frac{1}{3^2} - \frac{1}{4^2} + \frac{1}{5^2} + \frac{1}{6^2} - \frac{1}{7^2} - \frac{1}{8^2} \cdots \ldots \infty.$$

4. For what values of x is the following series convergent?

$$x - \frac{x^2}{\sqrt{2}} + \frac{x^3}{\sqrt{3}} - \frac{x^4}{\sqrt{4}} + \cdots \ldots \infty$$

5. Discuss the convergence of the integrals

$$\text{(i)} \int_0^{\frac{\pi}{2}} \frac{sinx}{x^{n+1}} dx \text{ and (ii)} \int_0^1 \frac{1}{\sqrt{\{x(1-x)\}}} dx.$$

9

Series Solution of Differential Equations

Abstract
The series solution of certain differential equations gives rise to special functions such as Bessel functions, Legendre polynomials, Laguerre polynomials, and Hermite polynomials. The Strum–Liouville problem, based on the orthogonality of functions, is also included in this chapter. We discuss the Frobenius method for solving second-order differential equations using series solutions. Both series solutions around an ordinary point and a regular singular point are covered.

Keywords: Ordinary point, singular point, regular singular point, Frobenius method, hypergeometric equation

9.1 Introduction

Many differential equations arising from physical problems do not permit a general solution in terms of known functions, namely, polynomials, rational functions, exponentials, trigonometric functions, hyperbolic functions, logarithms, etc. Such equations can be solved by numerical methods, but in many cases, it is easier to find a solution in the form of an infinite convergent series, arranged according to powers of the independent variable. The solution in the form of infinite series is called *integration in series.* In this chapter we discuss some methods of getting solution in the form of infinite series of ordinary differential equations of second order which are linear. This chapter discusses two methods – power series method, Frobenius method and more examples are given using these methods.

Power series method is applicable when we solve the equations at an ordinary point. We shall illustrate this method by using Hermite equations and Legendre equation. The Forbenius method gives the solutions of differential equations at regular singular points. In this method we obtain an equation called indicial equation. Depending upon the roots of the indicial equation, we obtain the general solution. This method is the best illustrated with the Bessel equation of order n given by

Dipankar De. Ordinary Differential Equations and Special Functions, (253–308) © 2026 Scrivener Publishing LLC

$$x^2 y'' + xy' + (x^2 - n^2) y = 0$$

where n is a positive constant.

9.2 Power Series

An infinite series of the form

$$\sum_{n=0}^{\infty} C_n (x - x_0)^n = C_0 + C_1(x - x_0) + C_2(x - x_0)^2 + \cdots \ \ldots \quad (9.1)$$

is called a series in $(x - x_0)$.

In particular, a power series in x is an infinite series

$$\sum_{n=0}^{\infty} C_n x^n = C_0 + C_1 x + C_2 x^2 + \ldots \quad (9.2)$$

For example, the exponential function

$$e^x = \sum_{n=0}^{\infty} \frac{x^n}{n!} = 1 + x + \frac{x^2}{2!} + \frac{x^3}{3!} + \cdots\ldots\ldots$$

The power series Equation (9.1) converges absolutely for $|x| < R$ where

$$R = \lim_{n \to \infty} \left| \frac{C_n}{C_{n+1}} \right| \text{ provided the limit exists.} \quad (9.3)$$

R is said to be the *radius of convergence* of the power series Equation (9.1). The interval $(-R, R)$ is said to be the *interval of convergence.*

Since $R = \infty$ for the power series Equation (9.2), the interval of convergence of the power series Equation (9.2) is $(-\infty, \infty)$, i.e., the real line.

A function $f(x)$, defined on an interval containing the point $x = x_0$, is called analytic at x_0 if the Taylor's series

$$\sum_{n=0}^{\infty} \frac{f^{(n)}(x_0)}{n!} (x - x_0)^n \quad (9.4)$$

exists and converges to $f(x)$ for all x in the interval of convergence of Equation (9.4).

Hence, we find that all polynomial functions are analytic everywhere, i.e., $cosx, sinx, e^x$. A rational function is analytic except the values at which the denominator is zero.

Example 9.2.1. The function

$$f(z) = (1-z)^s$$

is analytic for any real s *with* z in the interval $-1 < z < 1$. Its derivatives are

$$\begin{aligned}
&f'(z) = -s(1-z)^{s-1}\\
&f''(z) = (-)^2 s(s-1)(1-z)^{s-2}\\
&\therefore =\\
&f^k(z) = (-1)^k s(s-1)\ldots(s-k+1)(1-z)^{s-k}
\end{aligned}$$

from which we can construct the Taylor series

$$\begin{aligned}
(1-z)^s &= \sum_{k=0}^{\infty} \frac{[f^{(k)}(k)]_{z=0}}{k!} z^k\\
&= \sum_{k=0}^{\infty} \frac{(-1)^k s(s-1)\ldots.(s-k+1)}{k!} z^k.
\end{aligned}$$

With this definition, we can write

$$\begin{aligned}
&(1-z)^s\\
&= \sum_{k=0}^{\infty} \frac{.(-s)_k}{k!} z^k.
\end{aligned}$$

Using the Pochhammer symbols $(a)_n$,
where $(a)_n = 1; (a)_n = a(a+1)(a+2)\ldots..(a+n-1), n = 1,2,3,\ldots.,$
we get $(-s)_k = (-1)^k (s-k+1)_k$.
Putting $s = -1$, Equation (9.4) *reduces to* the geometric series

$$\frac{1}{1-z} = \sum_{k=0}^{\infty} z^k,$$

which converges for $|z| < 1$.

Example 9.2.2. Construct the Taylor series for the function

$$f(x) = \log(1+x) \text{ about } x = 0.$$

The *nth* derivative is

$$f^{(n)}(x) = \frac{(-1)^{n-1}}{(1+x)^n}(n-1)!$$

Clearly, $f(0)=0$

$$f^{(n)}(0) = (-1)^{n-1}(n-1)! \qquad n \geq 1.$$

Thus, $\log(1+x) = \sum_{n=1}^{\infty} \frac{(-1)^{n-1}}{n} x^n, |x|<1.$

Remark 9.2.1.
A point $x = x_0$ is called an *ordinary point* of the differential equation

$$y'' + P(x)y' + Q(x)y = 0 \tag{9.5}$$

if both $P(x)$ *and* $Q(x)$ are analytic at $x = x_0$.

If the point $x = x_0$ is not an ordinary point of the differential Equation (9.5), then it is called the *singular point* of the differential equation.

There are two types of singular points as follows:

(i) regular singular point and (ii) irregular singular point.

A singular point $x = x_0$ of the differential Equation (9.5) is called the *regular singular point* if both

$(x - x_0)P(x)$ and $(x - x_0)Q(x)$ are analytic at $x = x_0$.

If this singular point is not regular, it is called an *irregular singular point.*

9.3 Power Series Solution Near the Ordinary Point $x = x_0$

Let the given equation be

$$y'' + P(x)y' + Q(x)y = 0. \tag{9.6}$$

If $x = x_0$ is an ordinary point of Equation (9.6), then it has two non-trivial linearly independent power series solution of the form

$$\sum_{n=0}^{\infty} C_n (x - x_0)^n, \tag{9.7}$$

and these power series converges in some interval of convergence $|x - x_0| < R$ where R is the radius of convergence of Equation (9.7) about x_0. In order to get the coefficients $C_n's$ in Equation (9.6), we take

$$y = \sum_{n=0}^{\infty} C_n (x - x_0)^n. \tag{9.8}$$

Differentiating twice in succession, Equation (9.7) gives

$$y' = \sum_{n=0}^{\infty} nC_n (x - x_0)^{n-1}$$
$$y'' = \sum_{n=0}^{\infty} n(n-1)C_n (x - x_0)^{n-2}. \tag{9.9}$$

Putting the above values of y, y', y'' in Equation (9.6), we get an equation of the form

$$A_0 + A_1(x - x_0) + A_2(x - x_0)^2 + \cdots + A_n(x - x_0)^n + \cdots .. = 0 \tag{9.10}$$

where the coefficients $A_0, A_1, A_2 \ldots$ etc., are now some functions of coefficients of $C_0, C_1, C_2 \ldots$etc.

Since Equation (9.10) is an identity, all the coefficients of $A_0, A_1, A_2 \ldots$ of Equation (9.10) must be zero, i.e.,

$$A_0, = 0, A_1 = 0, A_2 = 0, \ldots . A_n = 0 \ldots$$

Substituting the values of A's we obtain the coefficient of Equation (9.8) in terms of $C_0 \; and \; C_1$. Substituting these values in Equation (9.8), we obtain the required series solution of Equation (9.6) in powers of $(x - x_0)$.

Example 9.3.1. Determine whether $x = 0$ is an ordinary or regular singular point of the differential equation $2x^2 \frac{d^2y}{dx^2} + 7x(x+1)\frac{dy}{dx} - 3y = 0.$

Solution: The differential equation is

$$2x^2 \frac{d^2y}{dx^2} + 7(x+1)\frac{dy}{dx} - 3y = 0$$

or, $$\frac{d^2y}{dx^2}+\frac{7(x+1)}{2x}\frac{dy}{dx}-\frac{3}{2x^2}y=0. \quad \text{(i)}$$

Comparing Equation (i) with the standard equation

$$y''+P(x)y'+Q(x)y=0,$$

here, $$P(x)=\frac{7(x+1)}{2x} \text{ and } Q(x)=-\frac{3}{2x^2}. \quad \text{(ii)}$$

Since both $P(x)\,and\,Q(x)$ are undefined at $x=0$, both $P(x)\,and\,Q(x)$ are not analytic at $x=0$.

Therefore, $x=0$ is not an ordinary point; it is a singular point.

Second, $(x-0)P(x)=\frac{7(x+1)}{2}$ and $(x-0)^2Q(x)=-\frac{3}{2}$ showing that both $(x-0)P(x)\,and\,(x-0)^2Q(x)$ are analytic at $x=0$.

Therefore, $x=0$ is a regular singular point.

Example 9.3.2. Show that $x=0$ is an ordinary point of

$$(x^2-1)y''+xy'-y=0,$$

but $x=1$ is a regular singular point.

Solution: The differential equation is

$$(x^2-1)y''+xy'-y=0. \quad \text{(i)}$$

Dividing by (x^2-1), the given differential equation becomes

$$\frac{d^2y}{dx^2}+\frac{x}{(x^2-1)}\frac{dy}{dx}-\frac{1}{(x^2-1)}y=0$$

or, $$\frac{d^2y}{dx^2}+\frac{x}{(x-1)(x+1)}\frac{dy}{dx}-\frac{1}{(x-1)(x+1)}y=0.$$

Comparing with $y''+P(x)y'+Q(x)y=0$, we have

$$P(x)=\frac{x}{(x-1)(x+1)} \text{ and } Q(x)=-\frac{1}{(x-1)(x+1)} \tag{ii}$$

Since both $P(x),Q(x)$ are analytic at $x=0$, so $x=0$ is an ordinary point of the given Equation (i).

Here, both $P(x),Q(x)$ are undefined at $x=1$; it is not an analytic at $x=1$. Thus, it is not an ordinary point, so $x=1$ is a singular point of the given Equation (i).

Also, $(x-1)P(x)=\frac{x}{(x-1)(x+1)}=\frac{x}{(x+1)}$ and $(x-1)^2Q(x)=-\frac{x-1}{(x+1)}$ showing that both $(x-1)P(x)$ and $(x-1)^2Q(x)$ are analytic at $x=1$.

Therefore, $x=1$ is a regular singular point.

Example 9.3.3. Show that $x=0$ *and* $x=-1$ are singular points of

$$x^2(x+1)^2y''+(x^2-1)y'+2y=0$$

where the first point is irregular, and the other point is a regular singular point.

Solution: The differential equation is

$$x^2(x+1)^2y''+(x^2-1)y'+2y=0$$

or,
$$y''+\frac{(x^2-1)}{x^2(x+1)^2}y'+\frac{2}{x^2(x+1)^2}2y=0$$

or,
$$y''+\frac{(x-1)}{x^2(x+1)}y'+\frac{2}{x^2(x+1)^2}2y=0.$$

Here, $P(x)=\frac{(x-1)}{x^2(x+1)}$ and $Q(x)=\frac{2}{x^2(x+1)^2}$.

Here, *at* $x=0,\ x=-1, P(x), Q(x)$ are not defined, and so, it is not analytic at this point.

So, $x=0$ is not a ordinary point; it is a singular point.

Now, $(x-0)P(x)=\dfrac{x(x-1)}{x^2(x+1)}=\dfrac{x-1}{x(x+1)}$ and $(x-0)^2Q(x)=\dfrac{2x^2}{x^2(x+1)^2}=\dfrac{2}{(x+1)^2}$;

$(x-0)P(x)$ is not defined at $x=0$.

So, it is an irregular singular point at $x=0$.

For the point $x=-1$,

$P(x)=\dfrac{(x-1)}{x^2(x+1)}$ and $Q(x)=\dfrac{2}{x^2(x+1)^2}$ are not defined. So, it is not an analytic at $x=-1$.

Therefore, $x=-1$ is a singular point.

Now, $(x+1)P(x)=\dfrac{(x+1)(x-1)}{x^2(x+1)}=\dfrac{x-1}{x^2}$ and $(x+1)^2Q(x)=\dfrac{2(x+1)^2}{x^2(x+1)^2}=\dfrac{2}{x^2}$.

Both $(x+1)P(x)$ and $(x+1)^2Q(x)$ are defined at $x=-1$. So, it is a regular singular point at $x=-1$.

So, $x=-1$ is an ordinary point.

Example 9.3.4. Discuss the singularity of the differential equation

$x^2y''+xy'+(x^2-n^2)y=0$ at $x=0$ *and* $\boldsymbol{x}=\infty$ [$x=0$, regular singular; $\boldsymbol{x}=\infty$, irregular singular].

Solution: By solving, we have $x=0$ *is* a regular singular point.

Here is a discussion about singularity at $x=\infty$:

Let $x=\dfrac{1}{t},\ t=\dfrac{1}{x}\ \therefore \dfrac{dt}{dx}=-\dfrac{1}{x^2}$.

Now, $y'=\dfrac{dy}{dx}=\dfrac{dy}{dt}\dfrac{dt}{dx}=\dfrac{dy}{dt}\left(-\dfrac{1}{x^2}\right)=-t^2\dfrac{dy}{dt}$

$$y''=\frac{d}{dx}(y')=\frac{d}{dt}\frac{dt}{dx}\left(\frac{dy}{dx}\right)=\frac{d}{dt}\left(-t^2\frac{dy}{dt}\right)\left(-\frac{1}{x^2}\right)$$
$$=\frac{1}{x^2}\left(2t\frac{dy}{dt}+t^2\frac{d^2y}{dt^2}\right)=t^4\frac{d^2y}{dt^2}+2t^3\frac{dy}{dt}.$$

The given equation becomes

$$\frac{1}{t^2}(t^4\frac{d^2y}{dt^2}+2t^3\frac{dy}{dt})+\frac{1}{t}(-t^2\frac{dy}{dt})+(\frac{1}{t^2}-n^2)y=0$$

or,
$$t^2\frac{d^2y}{dt^2}+t\frac{dy}{dt}+\frac{1-n^2t^2}{t^2}y=0$$

or,
$$\frac{d^2y}{dt^2}+\frac{1}{t}\frac{dy}{dt}+\left(\frac{1-n^2t^2}{t^4}\right)y=0.$$

Compare it with $\frac{d^2y}{dt^2}+P(t)\frac{dy}{dt}+Q(t)y=0,$

$$\therefore P(t)=\frac{1}{t}Q(t)=\frac{1-n^2t^2}{t^4}.$$

Here, $(t\text{-}0)P(t)=1$, $(t\text{-}0)^2Q(t)=\frac{1-n^2t^2}{t^2}$.

Since $(t\text{-}0)^2Q(t)$ is not analytic at $t=0$, so $t=0$, i.e., $x=\infty$ is irregular of the given differential equation.

9.4 Series Solution About Regular Singular Point $x=0$: Frobenius Method

We shall use the Frobenius method for finding a series solution about $x=0$, *if it is a regular point.*

However, if we wish to find the series solution about the regular singular point $x=a$, then we shift the origin to a point $x=a$, and then, we proceed as before.

Now, we discuss the *Frobenius method.*

Consider the differential equation of the form:

$$y_2 + \frac{F(x)}{x} y_1 + \frac{G(x)}{x^2} y = 0 \tag{9.11}$$

where $y_2 = \frac{d^2 y}{dx^2}$ and $y_1 = \frac{dy}{dx}$ and the functions $F(x)\, and\, G(x)$ are analytic at $x = 0$.

The following method for solving Equation (9.10) is called the *Frobenius method.*

We assume a trail solution

$$y = x^{r\sum_{m=0}^{\infty} C_m x^m} = x^r (c_0 + c_1 x + c_2 x^2 + c_3 x^3 + \cdots \ldots\ldots), \text{ where } c_0 \neq 0. \tag{9.12}$$

Differentiating Equation (9.11) with respect to, term by term, we have

$$\begin{aligned} y_1 &= rc_0 x^{r-1} + (r+1)c_1 x^r + (r+2)c_2 x^{r+1} + \cdots \ldots \ldots \\ &= x^{r-1}[rc_0 + (r+1)c_1 x + (r+2)c_2 x^2 + \cdots \ldots \ldots] \\ y_2 &= r(r-1)c_0 x^{r-2} + (r+1)rc_1 x^{r-1} + (r+2)(r+1)c_2 x^r + \cdots \ldots \ldots \\ &= x^{r-2}[r(r-1)c_0 + (r+1)rc_1 x + (r+2)(r+1)c_2 x^2 + \cdots \ldots \ldots]. \end{aligned}$$

Since $F(x)\, and\, G(x)$ are analytic at $x = 0$, we can write

$$F(x) = a_0 + a_1 x + a_2 x^2 + a_3 x^3 + \cdots \ldots \ldots \text{ and } G(x) = b_0 + b_1 x + b_2 x^2 + b_3 x^3 + \cdots \ldots$$

Putting the value of $y_1, y_2, F(x)$ and $G(x)$ in Equation (9.10) and multiplying both sides by x^2, we get

$$\begin{aligned} &x^r[r(r-1)c_0 + (r+1)rc_1 x + (r+2)(r+1)c_2 x^2 + \cdots \ldots \ldots] \\ &\quad + x(a_0 + a_1 x + a_2 x^2 + a_3 x^3 + \cdots \ldots \ldots) \\ &\quad x^{r-1}[rc_0 + (r+1)c_1 x + (r+2)c_2 x^2 + \cdots \ldots \ldots] \\ &\quad + (b_0 + b_1 x + b_2 x^2 + b_3 x^3 + \cdots)\, x^r (c_0 + c_1 x + c_2 x^2 + \cdots) = 0 \end{aligned}$$

$$x^r\{[r(r-1)c_0+(r+1)rc_1x+(r+2)(r+1)c_2x^2+\cdots\ \ldots\ \ldots]$$
$$+(a_0+a_1x+a_2x^2+a_3x^3+\cdots\ \ldots\ \ldots)$$
$$[rc_0+(r+1)c_1x+(r+2)c_2x^2+\cdots\ \ldots\ \ldots]$$

or, $$+(b_0+b_1x+b_2x^2+b_3x^3+\cdots)(c_0+c_1x+c_2x^2+c_3x^3+\cdots\ \ldots\ \ldots)\}=0.$$

Equating this identity, i.e., we equate all the coefficients of powers of x.

This will give us a system of equations involving the unknown coefficient C_m.

The smallest power of the above identity is x^r, and equating its coefficient to zero, we get

$$r(r-1)c_0+a_0rc_0+b_0c_0=0.$$

Since $c_0\neq 0$, we get

$$r(r-1)+a_0r+b_0=0 \text{ or, } r^2+(a_0-1)r+b_0=0 \tag{9.13}$$

This quadratic equation is called an indicial equation of Equation (9.11).

This method will give rise to a fundamental system of solutions. One of these solutions will always be of the form Equation (9.11), but the other solutions will be three different possibilities corresponding to the following cases:

Case I: *The indicial equation has distinct roots, but roots do not differ by an integer.*

Let r_1 and r_2 be the roots of Equation (9.12). Now, we substitute $r=r_1$ in the abovementioned system of solutions Equation (9.11) and obtained the solutions:

$$u(x)=x^{r_1}(c_0+c_1x+c_2x^2+\cdots\ \ldots\ \ldots),$$

and for $r=r_2$,

$$v(x)=x^{r_2}(c_0+c_1x+c_2x^2+\cdots\ \ldots\ \ldots).$$

Since r_2-r_1 is not an integer, $\dfrac{u}{v}$ is never a constant.

Therefore, u *and* v are two independent solutions of Equation (9.11). Hence, the general solution will be

$$y = au + bv,$$

where a,b are arbitrary constants.

Case II. *The indicial equation has equal roots.*

The indicial Equation (9.13) has equal roots if

$$(a_0 - 1)^2 - 4b_0 = 0 \text{ and } r = \frac{1-a_0}{2}. \tag{9.14}$$

We first obtain $c_1, c_2, c_3, \ldots..$ successively from the system of equations connecting them. The first solution is

$$u(x) = x^r(c_0 + c_1 x + c_2 x^2 + \cdots \ldots) \text{ where in we write } r = \frac{1-a_0}{2}. \tag{9.15}$$

To determine another solution, we use the method of variation of parameters, i.e., we replace constant c in the solution cu by a function $w(x)$ to be determined such that

$$v(x) = u(x)w(x) \tag{9.16}$$

is the solution of Equation (9.11).

Differentiating Equation (9.15), we get

$$v_1(x) = u_1(x)w(x) + u(x)w_1(x) \text{ and}$$
$$v_2(x) = u_2(x)w(x) + 2u_1(x)w_1(x) + u(x)w_2(x). \tag{9.17}$$

Since $v(x)$ is the solution of Equation (9.11), by assumption,

$$x^2 v_2 + xF(x)v_1 + G(x)v = 0. \tag{9.18}$$

Putting the value of v, v_1, v_2 from Equations (9.16) and (9.17) in Equation (9.18), we get

$$x^2[u_2w + 2u_1w_1 + uw_2] + xF(x)[u_1w + uw_1 + G(x)uw = 0$$

or, $[x^2u_2 + xF(x)u_1 + G(x)u]w + x^2[2u_1w_1 + uw_2] + xF(x)uw_1 = 0.$

Since $u(x)$ is the solution of Equation (9.10), we have

$$x^2u_2 + xF(x)u_1 + G(x)u = 0.$$

Then, the above equation reduces to

$$x^2[2u_1w_1 + uw_2] + xF(x)uw_1 = 0.$$

Dividing by x^2u and putting the value of (x), we get

$$w_2 + \left(2\frac{u_1}{u} + \frac{F(x)}{x}\right)w_1 = 0$$

or, $$w_2 + \left(2\frac{u_1}{u} + \frac{a_0 + a_1x + a_2x^2 + \cdots}{x}\right)w_1 = 0$$

or, $$w_2 + \left(2\frac{u_1}{u} + \frac{a_0}{x} + a_1 + ..\right)w_1 = 0. \tag{9.19}$$

Now, from Equation (9.14), we obtain,

$$\frac{u_1}{u} = \frac{x^{r-1}(rc_0 + c_1(r+1)x + c_2(r+2)x^2 + \cdots}{x^r(c_0 + c_1x + c_2x^2 + c_3x^3 + \cdots)} = \frac{r}{x} + \cdots.$$

Equation (9.18) becomes $w_2 + \left[2\left(\frac{r}{x}+\cdots\right)+\frac{a_0}{x}+a_1+..\right]w_1 = 0$

or,
$$w_2 + \left[\frac{2r+a_0}{x}+a_1+\cdots\right]w_1 = 0, \tag{9.20}$$

but from Equation (9.13), we have $2r + a_0 = 1$.

Hence, Equation (9.19) reduces to

$$w_2 + \left[\frac{1}{x}+a_1\ldots\right]w_1 = 0$$

or,
$$\frac{w_2}{w_1} = -\frac{-1}{x} + (\ldots)\,[\text{Here, } w_2 = d(w_1)\ w_1 = d(w)]$$

Integrating, we get, $logw_1 = -logx + (\ldots)$

or,
$$logw_1x = -(\ldots)$$

$$w_1 = \frac{1}{x}e^{(\ldots)}.$$

Expanding the exponential function in powers of x and integrating once more, we obtain the expression of w always in the following form

$$w = logx + k_1x + k_2x^2 + \cdots\ldots..$$

Putting the value of w in Equation (9.15), we obtain the desired another independent solution $v(x)$. The general solution is

$$y = au + bv,$$

a and b being arbitrary constants.

Case III: *The roots of an indicial equation differ by an integer.*

Let the roots r_1 and r_2 differ by p, i.e., *let* $r_1 = r, r_2 = r - p$, where p is a positive integer.

Then, one solution will be given by

$$u(x) = x^{r_1}(c_0 + c_1 x + c_2 x^2 + c_3 x^3 \ldots\ldots)$$

From Equation (9.14), we can write $r_1 + r_2 = -(a_0 - 1)$.

Here, $r_1 = r, r_2 = r - p$. $\therefore 2r - p = 1 - a_0$

or,
$$2r + a_0 = 1 + p.$$

Thus, Equation (9.19) becomes $w_2 + \left[\dfrac{1+p}{x} + a_1 + \cdots\right] w_1 = 0$

or,
$$\frac{w_2}{w_1} = -\left(\frac{1+p}{x} + \cdots\right).$$

Integrating, we get $\log w_1 = -(1+p)\log x + (\ldots.)$

or,
$$w_1 = x^{-(1+p)} e^{(\ldots.)} = \frac{1}{x^{1+p}}\{1 + (\ldots.) + \cdots\}$$

$$= \frac{1}{x^{1+p}} + \frac{k_1}{x^p} + \cdots. + \frac{k_p}{x} + k_{p+1} + k_{p+2} x + \cdots.$$

Integrating again, $w = -\dfrac{1}{px^p} - k_1 \dfrac{1}{(p-1)x^{p-1}} + \cdots k_p \log x + k_{p+1} x + \cdots.$

Putting this value of w *in* Equation (9.15), we obtain another solution $v(x)$ and the general solution

$$y = au + bv,$$

where a *and* b are arbitrary constants.

9.4.1 Method of Differentiation

We remember the following results for this method

(i) Maclaurin's theorem

$$y=(y)_0+\frac{x}{1!}(y_1)_0+\frac{x^2}{2!}(y_2)_0+\frac{x^3}{3!}(y_3)_0+\cdots$$

(ii) Leibnitz theorem. If *u and v* be two functions, then

$$D^n(uv)=u_nv+\binom{n}{1}u_{n-1}v_1+\binom{n}{2}u_{n-2}v_2+\cdots..+uv_n$$

where $u_n=\frac{d^nu}{dx^n}$ and $v_n=\frac{d^nv}{dx^n}$

Example 9.4.1. Solve $(1-x^2)y''-xy'+\mathrm{m}^2\mathrm{y}=0$ when $\mathrm{x}=0$, then $\mathrm{y}=0, \frac{\mathrm{dy}}{\mathrm{dx}}=\mathrm{m}$

Solution. Given

$$(1-x^2)y''-xy'+\mathrm{m}^2\mathrm{y}=0 \qquad \text{(i)}$$

Differentiating this equation n times by Leibnitz theorem with respect to x, we have

$$(1-x^2)y^{(n+2)}+\binom{n}{1}y^{(n+1)}(-2x)+\binom{n}{2}y^{(n)}(-2)-xy^{(n+1)}$$
$$-\binom{n}{1}y^{(n)}.1+\mathrm{m}^2y^{(n)}=0$$

or,

$$(1-x^2)y^{(n+2)}-2nxy^{(n+1)}-\frac{n(n-1)}{2}y^{(n)}.2-xy^{(n+1)}-ny^{(n)}+\mathrm{m}^2y^{(n)}=0$$

or, $$(1-x^2)y^{(n+2)}-(2n+1)xy^{(n+1)}+(\mathrm{m}^2-\mathrm{n}^2)y^{(n)}=0 \qquad \text{(ii)}$$

Here, putting

$$(y)_0 = 0, (y_1)_0 = m \tag{iii}$$

and putting in Equation (i)

$$y_0^{(2)} = -m^2(y)_0 \text{ [by Equation (iii)]} \tag{iv}$$

$$\text{Putting x=0 in (ii), } (y^{(n+2)})_0 + (m^2 - n^2)(y^{(n)})_0 = 0 \tag{v}$$

Putting $n = 2,4,6,..\, in\ (v) and\ (y^{(2)})_0 = 0, we\ get$

$$(y^{(2)})_0 = (y^{(4)})_0 = (y^{(6)})_0 = 0......., \tag{vi}$$

Putting $n = 1,3,5,..\, in$ Equation (v), *we get*

$$(y^{(3)})_0 = (1^2 - m^2)(y^{(1)})_0 = (1^2 - m^2)m, by \text{ Equation } (iii) \tag{vii}$$

so on.
Now from Maclaurin's theorem

$$y = (y)_0 + \frac{x}{1!}(y_1)_0 + \frac{x^2}{2!}(y_2)_0 + \frac{x^3}{3!}(y_3)_0 + \cdots$$

or, $$y = mx + \left(1^2 - m^2\right)m\frac{x^3}{3!} + \left(3^2 - m^2\right)\left(1^2 - m^2\right)m\frac{x^5}{5!} + \cdots....$$

9.5 Examples

Example 9.5.1. Find the power solution of the equation

$$(x^2 + 1)y'' + xy' - xy = 0$$

in powers of x.

Solution: Here, the differential equation is

$$(x^2+1)y''+xy'-xy=0 \tag{i}$$

or,
$$y''+\frac{x}{(x^2+1)}y'-\frac{x}{(x^2+1)}y=0. \tag{ii}$$

$$\therefore P(x)=\frac{x}{(x^2+1)} \text{ and } Q(x)=-\frac{x}{(x^2+1)}.$$

Showing that $P(x)$, $Q(x)$ are analytic at $x=0$,
i.e., $x=0$ is an ordinary point.
Let the solution

$$y=\sum_{n=0}^{\infty} c_n x^n \tag{iii}$$

$$y'=\sum_{n=1}^{\infty} nc_n x^{n-1}$$
$$y''=\sum_{n=2}^{\infty} n(n-1)c_n x^{n-2}. \tag{iv}$$

Substituting y, y', y'' *from* Equations (*iii*) *and* (*iv*) in Equation (i), we get

$$(x^2+1)\sum_{n=2}^{\infty} n(n-1)c_n x^{n-2}+x\sum_{n=1}^{\infty} nc_n x^{n-1}-x\sum_{n=0}^{\infty} c_n x^n=0$$

or,

$$\sum_{n=2}^{\infty} n(n-1)c_n x^n+\sum_{n=2}^{\infty} n(n-1)c_n x^{n-2}+\sum_{n=1}^{\infty} nc_n x^n-\sum_{n=0}^{\infty} c_n x^{n+1}=0$$

or,
$$\sum_{n=2}^{\infty} n(n-1)c_n x^n+\sum_{n=0}^{\infty}(n+2)(n+1)c_{n+2}x^n$$
$$+\sum_{n=1}^{\infty} nc_n x^n-\sum_{n=1}^{\infty} c_{n-1}x^n=0$$

or,
$$\begin{aligned}&2c_2+6c_3x+c_1x-c_0x+\sum_{n=2}^{\infty}(n+2)(n+1)c_{n+2}x^n\\&+\sum_{n=2}^{\infty} n(n-1)c_n x^n+\sum_{n=2}^{\infty} nc_n x^n-\sum_{n=2}^{\infty} c_{n-1}x^n=0\end{aligned} \tag{v}$$

or,
$$2c_2 + (6c_3 + c_1 - c_0)x$$
$$+\sum_{n=2}^{\infty}[(n+2)(n+1)c_{n+2} + n(n-1)c_n + nc_n - c_{n-1}]x^n = 0.$$

Comparing the coefficients

$$2c_2 = 0 \text{ or } c_2 = 0, \tag{vi}$$

$$6c_3 + c_1 - c_0 = 0 \Rightarrow c_3 = \frac{1}{6}(c_0 - c_1), \tag{vii}$$

and $(n+2)(n+1)c_{n+2} + n(n-1)c_n + nc_n - c_{n-1} = 0 \;\; for\; n \geq 2$

or,
$$c_{n+2} = \frac{c_{n-1} - n^2 c_n}{(n+2)(n+1)} \quad for\; n \geq 2. \tag{viii}$$

The above relation Equation (viii) is known as a *recurrence relation*. Putting $n = 2\; in$ Equation (*viii*), we get

$$c_4 = \frac{c_1 - 4c_2}{4.3} = \frac{c_1}{12} \quad as\;\; c_2 = 0 \tag{ix}$$

$$c_5 = \frac{c_2 - 9c_3}{5.4} = -\frac{9c_3}{20} = -\frac{9.\frac{1}{6}(c_0 - c_1)}{20} = -\frac{3}{40}(c_0 - c_1). \tag{x}$$

Putting the values of $c_2, c_3, c_4, c_5, \ldots,$ etc., in Equation (iii), we get

$$y = c_0 + c_1x + c_2x^2 + c_3x^3 + c_4x^4 + c_5x^5 + \cdots$$
$$= c_0 + c_1x + 0.x^2 + \frac{1}{6}(c_0 - c_1)x^3 + \frac{c_1}{12}x^4 + -\frac{3}{40}(c_0 - c_1)x^5 + \cdots$$
$$= c_0\left(1 + \frac{1}{6}x^3 - \frac{3}{40}x^5 + \cdots\right) + c_1(x - \frac{1}{6}x^3 + \frac{1}{12}x^4 + \frac{3}{40}x^5 + \cdots),$$

which is the required solution near $x = 0$, where $c_1\; and\; c_0$ are arbitrary constants.

Example 9.5.2. Find the solution in series of

$$\frac{d^2y}{dx^2}+x\frac{dy}{dx}+x^2y=0 \text{ about } x=0.$$

Solution: The given the differential equation is

$$y''+xy'+x^2y=0. \tag{i}$$

Here, $P(x)=x$ and $Q(x)=x^2$.

At $x=0$, $P(x)$, $Q(x)$ are analytic, and $x=0$ is an ordinary point of the differential equation.

We take the solution

$$\begin{aligned} y &= \sum_{n=0}^{\infty} c_n x^n \\ &= c_0+c_1x+c_2x^2+c_3x^3+\cdots\ldots\ldots \end{aligned} \tag{ii}$$

$$\begin{aligned} y' &= \sum_{n=1}^{\infty} nc_nx^{n-1} \\ y'' &= \sum_{n=2}^{\infty} n(n-1)c_nx^{n-2}. \end{aligned} \tag{iii}$$

Substituting y, y', y'' *from* Equations (iii) and (iv) in Equation (i), we get

$$\sum_{n=2}^{\infty} n(n-1)c_nx^{n-2}+x\sum_{n=1}^{\infty} nc_nx^{n-1}+x^2\sum_{n=0}^{\infty} c_nx^n=0$$

or, $$\sum_{n=2}^{\infty} n(n-1)c_nx^{n-2}+\sum_{n=1}^{\infty} nc_nx^n+\sum_{n=0}^{\infty} c_nx^{n+2}=0$$

or, $$\sum_{n=0}^{\infty}(n+2)(n+1)c_{n+2}x^n+\sum_{n=1}^{\infty} nc_nx^n+\sum_{n=2}^{\infty} c_{n-2}x^n=0$$

or, $$2c_2+3.2c_3x+\sum_{n=2}^{\infty}(n+2)(n+1)c_{n+2}x^n+c_1x+\sum_{n=2}^{\infty} nc_nx^n + \sum_{n=2}^{\infty} c_{n-2}x^n=0$$

or, $$2c_2+3.2c_3x+c_1x+\sum_{n=2}^{\infty}[(n+2)(n+1)c_{n+2}+nc_n+c_{n-2}]x^n=0. \tag{iv}$$

Equating the coefficients, we get

$$2c_2 = 0 \Rightarrow \quad c_2 = 0 \tag{v}$$

$$6c_3 + c_1 = 0 \Rightarrow c_3 = -\frac{1}{6}c_1 \tag{vi}$$

and $(n+2)(n+1)c_{n+2} + nc_n + c_{n-2} = 0 \;\; for\; n \geq 2$

$$c_{n+2} = -\frac{nc_n + c_{n-2}}{(n+2)(n+1)} \tag{vii}$$

$$For\; n = 2,\; in \text{ Equation (vii)}: c_4 = -\frac{2c_2 + c_0}{4.3} = -\frac{c_0}{12}. \tag{viii}$$

Putting $n = 3$ *in* Equation (vii), $c_5 = -\dfrac{3c_3 + c_1}{5.4} = -\dfrac{3(-\frac{1}{6}c_1) + c_1}{5.4} = -\dfrac{c_1}{40}.$

For $n = 4$, $c_6 = -\dfrac{4c_4 + c_2}{6.5} = -\dfrac{4(-\frac{c_0}{12}) + 0}{6.5} = \dfrac{c_0}{90}.$

So, putting these values in Equation (ii),

$$y = c_0 + c_1x + c_2x^2 + c_3x^3 + \cdots$$

$$= c_0 + c_1x + 0x^2 - \frac{1}{6}c_1x^3 - \frac{c_0}{12}x^4 - \frac{c_1}{40}x^5 + \frac{c_0}{90}x^6 + \ldots\ldots\ldots$$

$$= c_0\left(1 - \frac{1}{12}x^4 + \frac{1}{90}x^6 + \ldots\right) + c_1(x - \frac{1}{6}x^3 - \frac{1}{40}x^5 - \cdots \ldots).$$

Example 9.5.3. Find the power series solution in powers of $(x-1)$ of the initial value problem:

$$xy'' + y' + 2y = 0 \quad ; y(1) = 1,\; y'(1) = 2.$$

Solution: The given differential equation is

$$xy'' + y' + 2y = 0 \tag{i}$$

or,
$$y'' + \frac{1}{x}y' + \frac{2}{x}y = 0.$$

Here, $P(x) = \frac{1}{x}$ *and* $Q(x) = \frac{2}{x}$.

$$P(x) \text{ and } Q(x) \text{ is analytic at } x = 1.$$

Hence, $x = 1$ is an ordinary point of Equation (i). To find the solution near $x = 1$, we shall find the series solution in powers of $(x-1)$.

Let

$$y = c_0 + c_1(x-1) + c_2(x-1)^2 + c_3(x-1)^3 + \cdots = \sum_{n=0}^{\infty} c_n(x-1)^n. \quad \text{(ii)}$$

Differentiating with respect to x, we get

$$y' = \sum_{n=1}^{\infty} nc_n(x-1)^{n-1}$$

$$y'' = \sum_{n=2}^{\infty} n(n-1)c_n(x-1)^{n-2}. \quad \text{(iii)}$$

Putting the value of y'' *and* y' in Equation (i), we have

$$x\sum_{n=2}^{\infty} n(n-1)c_n(x-1)^{n-2} + \sum_{n=1}^{\infty} nc_n(x-1)^{n-1} + 2\sum_{n=0}^{\infty} c_n(x-1)^n = 0$$

or,
$$[(x\text{-}1)+1]\sum_{n=2}^{\infty} n(n-1)c_n(x-1)^{n-2} + \sum_{n=1}^{\infty} nc_n(x-1)^{n-1} + 2\sum_{n=0}^{\infty} c_n(x-1)^n = 0$$

or,
$$\sum_{n=2}^{\infty} n(n-1)c_n(x-1)^{n-1} + \sum_{n=2}^{\infty} n(n-1)c_n(x-1)^{n-2} + \sum_{n=1}^{\infty} nc_n(x-1)^{n-1} + 2\sum_{n=0}^{\infty} c_n(x-1)^n = 0$$

or,
$$\sum_{n=1}^{\infty} n(n+1)c_{n+1}(x-1)^n + \sum_{n=0}^{\infty}(n+2)(n+1)c_{n+2}(x-1)^n + \sum_{n=0}^{\infty}(n+1)c_{n+1}(x-1)^n + 2\sum_{n=0}^{\infty} c_n(x-1)^n = 0$$

or, $$2c_2 + c_1 + 2c_0$$

$$+\sum_{n=1}^{\infty}[n(n+1)c_{n+1} + (n+2)(n+1)c_{n+2} + (n+1)c_{n+1} + 2c_n](x-1)^n = 0.$$

Equating the coefficients of powers of $(x-1)$, we get

$$2c_2 + c_1 + 2c_0 = 0, \text{ so that } c_2 = -\frac{1}{2}(c_1 + 2c_0) \quad \text{(iv)}$$

and $n(n+1)c_{n+1} + (n+2)(n+1)c_{n+2} + (n+1)c_{n+1} + 2c_n = 0 \quad \textit{for } n \geq 1$

or, $$(n+1)^2 c_{n+1} + (n+2)(n+1)c_{n+2} + 2c_n = 0$$

or, $$c_{n+2} = -\frac{(n+1)^2 c_{n+1} + 2c_n}{(n+2)(n+1)} \text{ for } n \geq 1. \quad \text{(v)}$$

Given $y = 1 \; and \; y' = 2 \; when \; x = 1$,
hence, putting $x = 1$ in Equations (ii) and (iii), we have

$$c_0 = 1 \text{ and } c_1 = 2. \quad \text{(vi)}$$

Using Equations (vi) and (iv) gives

$$c_2 = -\frac{1}{2}(2 + 2.1) = -2. \quad \text{(vii)}$$

Putting $n = 1, 2, 3, \ldots,$ we get

for $n = 1$, $c_3 = -\frac{2^2 c_2 + 2c_1}{3.2} = -\frac{2^2.(-2) + 2.2}{3.2} = \frac{2}{3}$;

for $n = 2, 3$, $c_4 = -\frac{3^2 c_3 + 2c_2}{4.3.} = -\frac{3^2 \frac{2}{3} + 2.(-2)}{4.3.} = -\frac{1}{6}$, $c_5 = -\frac{4^2 c_4 + 2c_3}{5.4}$

$= -\frac{4^2(-\frac{1}{6}) + 2.\frac{2}{3}}{4.5} = \frac{1}{15}$;

and so on. Putting the values of c's in Equation (ii), we get the required solution

$$y=1+2(x-1)-2(x-1)^2+\frac{2}{3}(x-1)^3-\frac{1}{6}(x-1)^4+\frac{1}{15}(x-1)^5$$
$$+\cdots\ldots\ldots$$

Example 9.5.4. Solve the series

$$9x(1-x)y''-12y'+4y=0.$$

It is case I of the Frobenius method: when roots of the indicial equation are distinct and do not differ by an integer. The complete solution is $y=c_1(y)_{m_1}+c_2(y)_{m_2}$, where m_1,m_2 are the roots.

Solution: The given equation is

$$9x(1-x)y''-12y'+4y=0 \tag{i}$$

or,
$$y''-\frac{4}{3x(1-x)}y'+\frac{4}{9x(1-x)}y=0.$$

Here, $P(x)=-\frac{4}{3x(1-x)}$ and $Q(x)=\frac{4}{9x(1-x)}$.

Since $P(x)$ and $Q(x)$ are not analytic at $x=0$, i.e., it is not an ordinary point of Equation (i).

Now, $xP(x)=-\frac{4}{3(1-x)}$ and $x^2Q(x)=\frac{4x}{9(1-x)}$ are analytic at $x=0$.

Therefore, $x=0$ is a regular singular point of Equation (i).

Let the solution of Equation (i) be

$$y=\sum_{n=0}^{\infty}c_n x^{k+n} \tag{ii}$$

Differentiating with respect to x, we get

$$y'=\sum_{n=0}^{\infty}(k+n)c_n x^{k+n-1}$$
$$y''=\sum_{n=0}^{\infty}(k+n)(k+n-1)c_n x^{k+n-2}. \tag{iii}$$

Substituting these values in Equation (i), we get

$$9x(1-x)\sum_{n=0}^{\infty}(k+n)(k+n-1)c_n x^{k+n-2} - 12\sum_{n=0}^{\infty}(k+n)c_n x^{k+n-1}$$
$$+4\sum_{n=0}^{\infty} c_n x^{k+n} = 0$$

or, $$\sum_{n=0}^{\infty} 9(k+n)(k+n-1)c_n x^{k+n-1} - \sum_{n=0}^{\infty} 9(k+n)(k+n-1)c_n x^{k+n}$$
$$-12\sum_{n=0}^{\infty}(k+n)c_n x^{k+n-1} + 4\sum_{n=0}^{\infty} c_n x^{k+n} = 0$$

or, $$\sum_{n=0}^{\infty} c_n\{9(k+n)(k+n-1) - 12(k+n)\}x^{k+n-1})$$
$$+\sum_{n=0}^{\infty} c_n\{4 - 9(k+n(k+n-1)\}x^{k+n} = 0. \qquad \text{(iv)}$$

Here,

$$9(k+n)(k+n-1) - 12(k+n) = 3(k+n)(3k+3n-7) \qquad \text{(v)}$$

and, $$4 - 9(k+n)(k+n-1) = -[9(k+n)^2 - 9(\mathrm{k}+\mathrm{n}) - 4]$$
$$= -[\{3(\mathrm{k}+\mathrm{n}) - 4\}\{3(\mathrm{k}+\mathrm{n}) + 1\}]. \qquad \text{(vi)}$$

Using Equations (v) and (vi), Equation (iv) can be written as

$$\sum_{n=0}^{\infty} c_n 3(k+n)(3k+3n-7)x^{k+n-1}$$
$$-\sum_{n=0}^{\infty} c_n\{3(\mathrm{k}+\mathrm{n}) - 4\}\{3(\mathrm{k}+\mathrm{n}) + 1)\}x^{k+n} = 0. \qquad \text{(vii)}$$

Equating to zero, the coefficient of the smallest power of x, i.e., x^{k-1} Equation (vii) gives us an indicial equation.

$$3c_0(k)(3k-7) = 0 \text{ or } k(3k-7) = 0, \text{ since } c_0 \neq 0.$$

$$\therefore\ k = 0 \ \ or \ \ k = \frac{7}{3}, \qquad \text{(viii)}$$

which are unequal and not differing by an integer.

To obtain the recurring relations, we equate to zero the coefficients of x^{k+n-1} and obtain

$$3c_n(k+n)(3k+3n-7)-3c_{n-1}(\mathrm{k}+\mathrm{n}-1)-4\}\{3(\mathrm{k}+\mathrm{n}-1)+1\}=0$$

$$c_n=\frac{3\mathrm{k}+3\mathrm{n}-2}{3(k+n)}c_{n-1}. \qquad \text{(ix)}$$

For $n=1$,

$$c_1=\frac{c_0}{3}\frac{3\mathrm{k}+1}{(k+1)}. \qquad \text{(x)}$$

For $n=2$,

$$c_2=\frac{3\mathrm{k}+4}{3(k+2)}c_1=\frac{3\mathrm{k}+4}{3(k+2)}\frac{3\mathrm{k}+1}{3(k+1)}c_0=\frac{c_0}{3^2}\frac{3\mathrm{k}+4}{(k+2)}\frac{3\mathrm{k}+1}{(k+1)} \qquad \text{(xi)}$$

and so on.

$$y=\sum_{n=0}^{\infty}c_nx^{k+n}$$
$$=x^k(c_0+c_1x+c_2x^2+c_3x^3+c_4x^4+c_5x^5+\cdots..)$$
$$=x^k\left(c_0+\frac{c_0}{3}\frac{3\mathrm{k}+1}{(k+1)}x+\frac{c_0}{3^2}\frac{3\mathrm{k}+4}{(k+2)}\frac{3\mathrm{k}+1}{(k+1)}x^2+\frac{c_0}{3^3}\frac{3\mathrm{k}+4}{(k+2)}\frac{(3\mathrm{k}+1)(3\mathrm{k}+7)}{(k+1)(k+3)}x^3++\cdots..\right).$$

Putting $k=0$, we get

$$\mathrm{y}=\left(c_0+\frac{c_0}{3}x+\frac{c_0}{3^2}\frac{4}{2}\frac{1}{1}x^2+\frac{c_0}{3^3}\frac{1.4.7}{1.2.3}x^3++\cdots..\right)$$
$$=a\left(1+\frac{1}{3}x+\frac{1}{3^2}\frac{1.\ 4}{1.\ 2}+\frac{1}{3^3}\frac{1.4.7}{1.2.3}x^3+\cdots.....\right)=au\ (\text{let}). \qquad \text{(xii)}$$

For $k=\frac{7}{3}$ (putting) and replacing c_0 *by b* in Equation (xii), we get

$$\mathrm{y}=bx^{\frac{7}{3}}\left(1+\frac{8}{10}x+\frac{8}{10}\frac{11}{13}x^2+\frac{8}{10}\frac{11}{13}\frac{14}{16}x^3+\cdots.....\right)=bv.$$

The solution is

$$y = a\left(1+\frac{1}{3}x+\frac{1}{3^2}\frac{1.4}{1.2}+\frac{1}{3^3}\frac{1.4.7}{1.2.3}x^3+\cdots.....\right)$$
$$+bx^{\frac{7}{3}}\left(1+\frac{8}{10}x+\frac{8}{10}\frac{11}{13}x^2+\frac{8}{10}\frac{11}{13}\frac{14}{16}x^3+\cdots.....\right).$$

Example 9.5.5. Solve in series the equation $x\frac{d^2y}{dx^2}+\frac{dy}{dx}+xy=0.$

This is case II of the Frobenius method: when roots of the indicial equation are equal, the complete solution is $y = c_1(y)_{m_1} + c_2(\frac{\partial y}{\partial m})_{m_1}$ where m_1, m_1 are the roots.

Solution: The given equation is

$$xy'' + y' + xy = 0 \tag{i}$$

or,

$$y'' + \frac{1}{x}y' + y = 0.$$

Here, $P(x) = \frac{1}{x}$ and $Q(x) = 1$.

Since $P(x)$ and $Q(x)$ are not analytic at $x = 0$, it is not an ordinary point of Equation (i).

Now, $xP(x) = 1$ and $x^2Q(x) = x$ are analytic at $x = 0$.

Therefore, $x = 0$ is a regular singular point of Equation (i).

Let the solution of Equation (i) be

$$y = \sum_{n=0}^{\infty} c_n x^{k+n}. \tag{ii}$$

Differentiating with respect to x, we get

$$y' = \sum_{n=0}^{\infty}(k+n)c_n x^{k+n-1} \tag{iii}$$
$$y'' = \sum_{n=0}^{\infty}(k+n)(k+n-1)c_n x^{k+n-2}.$$

Substituting these values in Equation (i), we get

$$x\sum_{n=0}^{\infty}(k+n)(k+n-1)c_n x^{k+n-2} + \sum_{n=0}^{\infty}(k+n)c_n x^{k+n-1} + \mathrm{x}\sum_{n=0}^{\infty} c_n x^{k+n} = 0$$

$$\sum_{n=0}^{\infty}(k+n)(k+n-1)c_n x^{k+n-1} + \sum_{n=0}^{\infty}(k+n)c_n x^{k+n-1} + \sum_{n=0}^{\infty} c_n x^{k+n+1} = 0$$

$$\sum_{n=0}^{\infty} c_n[(k+n)(k+n-1)+(\mathrm{k+n})]x^{k+n-1} + \sum_{n=0}^{\infty} c_n x^{k+n+1} = 0 \quad \text{(iv)}$$

$$\sum_{n=0}^{\infty} c_n[(k+n)(k+n-1)+(\mathrm{k+n})]x^{k+n-1} + \sum_{n=2}^{\infty} c_{n-2} x^{k+n-1} = 0 \quad \text{(v)}$$

The lowest power of x is x^{k-1}.
Its coefficient equated to zero gives

$$c_0[k(k-1)+\mathrm{k}] = 0.$$

Since, $$c_0 \neq 0, \therefore k^2 = 0 \Rightarrow k, k = 0 \quad \text{(vi)}$$

For, $$n = 1\, in\, (v),\, c_1[(k+1)k+(k+1)] = 0 \Rightarrow c_1 = 0. \quad \text{(vii)}$$

To obtain the recurring relations, we equate to zero the coefficients of x^{k+n-1} and obtain

$$c_n[(k+n)(k+n-1)+(k+n)] + c_{n-2} = 0$$

or, $$c_n = -\frac{1}{(k+n)(k+n-1)+(k+n)}c_{n-2},\ for\ n \geq 2. \quad \text{(viii)}$$

$$For\ n = 2,\ c_2 = -\frac{1}{(k+2)(k+1)+(k+2)}c_0 = -\frac{1}{(k+2)^2}c_0.$$

$$For\ n = 4,\ c_4 = -\frac{1}{(k+4)(k+3)+(k+4)}c_2 = \frac{1}{(k+2)^2(k+4)^2}c_{0.}$$

$$For\ n = 6,\ c_6 = -\frac{1}{(k+2)^2(k+4)^2(k+6)^2}c_0, \text{ and so on.}$$

$$\text{For } n=3,\ c_3=-\frac{1}{(k+3)(k+2)+(k+3)}c_1=-\frac{1}{(k+3)^2}c_1=-\frac{1}{(k+3)^2}0=0$$

$$c_3=c_5=c_7=\cdots\ldots=0.$$

The first solution is

$$y=\sum_{n=0}^{\infty}c_nx^{k+n}=x^k(c_0+c_1x+c_2x^2+c_3x^3+c_4x^4+c_5x^5+\cdots)$$

$$=c_0x^k(1++c_2x^2+c_4x^4+c_5x^5+c_5x^5+\cdots)+(c_1x+c_3x^3+c_5x^5+c_5x^5+\cdots)$$

$$=c_0x^k\left(1-\frac{1}{(k+2)^2}x^2+\frac{1}{(k+2)^2(k+4)^2}x^4-\frac{1}{(k+2)^2(k+4)^2(k+6)^2}x^6+\ldots\right)+0. \tag{ix}$$

For $k=0$, the first solution is

$$y_1=c_0\left(1-\frac{1}{2^2}x^2+\frac{1}{4^2}\frac{1}{2^2}x^4-\frac{1}{6^2}\frac{1}{4^2}\frac{1}{2^2}x^6+\cdots.\right). \tag{x}$$

This gives only one solution instead of two. To get the second solution, differentiate Equation (viii) partially with respect to k:

$$[y=c_0x^k\left(1-\frac{1}{(k+2)^2}x^2+\frac{1}{(k+2)^2(k+4)^2}x^4-\frac{1}{(k+2)^2(k+4)^2(k+6)^2}x^6+\cdots\right)$$

$$=c_0x^kf(k)\ logy=logc_0+klogx+logf(k),$$

$$\therefore\frac{1}{y}\frac{\partial y}{\partial k}=logx+\frac{f'(k)}{f(k)}\Rightarrow\frac{\partial y}{\partial k}=ylogx+c_0x^kf(k)\frac{f'(k)}{f(k)}=ylogx+c_0x^kf'(k)\ ]$$

$$\frac{\partial y}{\partial k}=ylogx+c_0x^k[\frac{x^2}{(k+2)^2}\frac{2}{k+2}-\frac{1}{(k+2)^2(k+4)^2}x^4\left(\frac{2}{k+2)}+\frac{2}{k+4)}\right)+\cdots.$$

The second solution is

$$y_2=(\frac{\partial y}{\partial k})_{k=0}$$

$$=\left[ylogx+c_0x^k\left\{\frac{x^2}{(k+2)^2}\frac{2}{k+2}-\frac{1}{(k+2)^2(k+4)^2}x^4\left(\frac{2}{k+2)}+\frac{2}{k+4)}\right)+\cdots\right\}\right]_{k=0}.$$

$$=y_1logx+c_0\left\{\frac{x^2}{2^2}-\frac{1}{2^2.4^2}x^4\left(1+\frac{1}{2}\right)+\frac{1}{2^2.4^2..6^2}x^6\left(1+\frac{1}{2}+\frac{1}{3}\right)\ldots\right\} \tag{xi}$$

Hence, the complete solution is

$$y = a(y)_{m_1} + b(\frac{\partial y}{\partial m})_{m_1}$$

$$= ay_1 + \text{b}\left\{ y_1 logx + c_0 \left\{ \frac{x^2}{2^2} - \frac{1}{2^2.4^2} x^4 \left(1+\frac{1}{2}\right) + \frac{1}{2^2.4^2..6^2} x^6 \left(1+\frac{1}{2}+\frac{1}{3}\right)\ldots \right\}\right\}$$

$$= (a + blogx)y_1 + bc_0 \left\{ \frac{x^2}{2^2} - \frac{1}{2^2.4^2} x^4 \left(1+\frac{1}{2}\right) + \frac{1}{2^2.4^2..6^2} x^6 \left(1+\frac{1}{2}+\frac{1}{3}\right)\ldots \right\}$$

$$= (a + blogx)c_0 \left(1 - \frac{1}{2^2} x^2 + \frac{1}{4^2}\frac{1}{2^2} x^4 - \frac{1}{6^2}\frac{1}{4^2}\frac{1}{2^2} x^6 + \cdots.\right)$$

$$+ bc_0 \left\{ \frac{x^2}{2^2} - \frac{1}{2^2.4^2} x^4 \left(1+\frac{1}{2}\right) + \frac{1}{2^2.4^2..6^2} x^6 \left(1+\frac{1}{2}+\frac{1}{3}\right)\ldots \right\}$$

$$= (C_1 + C_2 logx) \left(1 - \frac{1}{2^2} x^2 + \frac{1}{4^2}\frac{1}{2^2} x^4 - \frac{1}{6^2}\frac{1}{4^2}\frac{1}{2^2} x^6 + \cdots.\right)$$

$$+ C_2 \left\{ \frac{x^2}{2^2} - \frac{1}{2^2.4^2} x^4 \left(1+\frac{1}{2}\right) + \frac{1}{2^2.4^2..6^2} x^6 \left(1+\frac{1}{2}+\frac{1}{3}\right)\ldots \right\} \text{ where } C_1 = ac_0, C_2 = bc_0.$$

Example 9.5.6. Solve in series the equation $x(1-x)\frac{d^2y}{dx^2} - (1+3x)y' - y = 0$ [let m_1 and m_2 be the roots such that $m_1 < m_2$. If some of the coefficients of the y series become infinite when $m = m_1$, we modify the form of y by replaceing a_0 by $b_0(m - m_1)$. Then, the complete solution is

$$y = C_1(y)_{m_2} + C_2(\frac{\partial y}{\partial m})_{m_1}].$$

Solution: The given differential equation is

$$x(1-x)y'' - (1+3x)\frac{dy}{dx} - y = 0. \tag{i}$$

Here, $x = 0$ is a regular singular point.
Substituting $y = x^k(c_0 + c_1x + c_2x^2 + c_3x^3 + \cdots\ldots)$

$$y' = c_0kx^{k-1} + c_1(k+1)x^k + c_2(k+2)x^{k+1} + c_3(k+3)x^{k+2} + \cdots\ldots$$

$$y'' = c_0k(k-1)x^{k-2} + c_1(k+1)kx^{k-1} + c_2(k+2)(k+1)x^k$$
$$+ c_3(k+3)(k+2)x^{k+1} + \ldots.$$

The given differential equation becomes

$$x(1-x)y''-(1+3x)\frac{dy}{dx}-y=0$$

$$x(1-x)\sum_{n=0}^{\infty}(k+n)(k+n-1)c_nx^{k+n-2}-(1+3x)\sum_{n=0}^{\infty}(k+n)c_nx^{k+n-1}$$
$$-\sum_{n=0}^{\infty}c_nx^{k+n}=0$$

or, $$\sum_{n=0}^{\infty}(k+n)(k+n-1)c_nx^{k+n-1}-\sum_{n=0}^{\infty}(k+n)(k+n-1)c_nx^{k+n}$$
$$-\sum_{n=0}^{\infty}(k+n)c_nx^{k+n-1}--3\sum_{n=0}^{\infty}(k+n)c_nx^{k+n}-\sum_{n=0}^{\infty}c_nx^{k+n}=0$$

or, $$\sum_{n=0}^{\infty}c_n((k+n)(k+n-1)-(k+n))x^{k+n-1}$$
$$-\sum_{n=0}^{\infty}c_n[(k+n)(k+n-1)+3(k+n)+1]x^{k+n}=0.$$

Equating to zero the lowest power of x, we get

$$c_0((k)(k-1)-(k))=0, \text{ since } c_0\neq 0.$$

$\therefore k^2-2k=0$, which is an indicial equation $\Rightarrow k=0,2$
and $c_1[(k+1)(k+1-1)-(k+1)]-c_0[(k)(k-1)+3(k)+1]=0$

or, $$c_1(k+1)(k-1)-c_0(k+1)^2=0.$$

$$\therefore c_1=\frac{k+1}{k-1}c_0$$

$n\geq 2, c_n[(k+n)(k+n-1)-(k+n)]-c_{n-1}[(k+n-1)(k+n-2)+3(k+n-1)$
$+1]=0$
or

$$c_n=\frac{(k+n)^2}{(k+n)(k+n-2)}c_{n-1}=\frac{k+n}{k+n-2}c_{n-1} \text{ for } n\geq 2$$

$$\therefore c_2 = \frac{k+2}{k} c_1 = \frac{k+2}{k}\frac{k+1}{k-1} c_0,\ c_3 = \frac{k+3}{k+1} c_2 = \frac{k+3}{k+1}\frac{k+2}{k}\frac{k+1}{k-1} c_0.$$

The solution is $y = \sum_{n=0}^{\infty} c_n x^{k+n}$

$$y = c_0 x^k \left(1 + \frac{k+1}{k-1}x + \frac{k+2}{k}\frac{k+1}{k-1}x^2 + \frac{k+3}{k+1}\frac{k+2}{k}\frac{k+1}{k-1}x^3 + \cdots\cdots\cdots\right). \tag{ii}$$

Putting $k = 2$ in Equation (ii), the first solution is $y_1 = c_0 x^2\left(1 + \frac{3}{1}x + \frac{3.4}{2.1}x^2 + \frac{3.4.5.}{1.2.3}x^3 + \cdots\ldots\right)$ (if we put $m = 0$ in Equation (ii), the coefficients become infinite).

To avoid this difficulty, put $c_0 = b_0(k-0)$ so that

$$y = b_0 x^k \left(k + \frac{k(k+1)}{k-1}x + \frac{(k+2)\,(k+1)}{k-1}x^2 + \frac{(k+3)\,(k+2)\,(k+1)}{k+1 \quad\quad k-1}x^3 + \cdots\ldots\ \ldots\ldots\right)$$

$$\frac{\partial y}{\partial k} = b_0 x^k \log x\left(k + \frac{k(k+1)}{k-1}x + \frac{(k+2)\,(k+1)}{k-1}x^2 + \frac{(k+3)\,(k+2)\,(k+1)}{k+1 \quad\quad k-1}x^3 + \cdots\ldots\ \ldots\ldots\right)$$
$$+ b_0 x^k\left(1 + \frac{k^2-2k-1}{(k-1)^2}x + \frac{k^2-k-5}{(k-1)^2}x^2 + \frac{k^2-2k-11}{(k-1)^2}x^3 + \cdots\right).$$

The second solution is $y_2 = \left(\frac{\partial y}{\partial k}\right)_{m=0}$

$$= b_0 \log x[-1.2x^2 - 2.3x^3 - 3.4x^4 - \cdots]$$
$$+ b_0[1 - x - 5x^2 - 11x^3 - \cdots].$$

Hence, the complete solution is $y = c_1 y_1 + c_2 y_2$

$$y = \frac{1}{2}c_1 c_0[1.2x^2 + 2.3x^3 + 3.4x^4 + \cdots] - b_0 c_2 \log x[1.2x^2 + 2.3x^3 + 3.4x^4 + ..]$$
$$- b_0 c_2[-1 + x + 5x^2 + 11x^3 + \cdots]$$
$$= (C_1 + C_2 \log x)[1.2x^2 + 2.3x^3 + 3.4x^4 + \cdots] - C_2[-1 + x + 5x^2 + 11x^3 + \cdots].$$

Example 9.5.7. Solve (Bessel equation)
(a) $x^2 y'' + xy' + (x^2 - n^2)y = 0$ in series taking $2n$ as a non-integral.

Solution: The given differential equation is

$$x^2y'' + xy' + (x^2 - n^2)y = 0 \tag{i}$$

or,

$$y'' + \frac{1}{x}y' + \frac{(x^2 - n^2)}{x^2}y = 0.$$

Comparing with $y'' + P(x)y' + Q(x)y = 0$, we get

Here, $P(x) = \frac{1}{x}$ *and* $Q(x) = \frac{(x^2 - n^2)}{x^2}$ are not analytic at $x = 0$.

$xP(x) = 1$ *and* $x^2Q(x) = (x^2 - n^2)$ are analytic at $x = 0$. So, $x = 0$ is a regular singular point of Equation (i).

Let the series solution of Equation (i) be of the form

$$y = \textstyle\sum_{m=0}^{\infty} c_m x^{k+m} \quad \text{where } c_0 \neq 0. \tag{ii}$$

Differentiating with. respect to x, we get

$$y' = \sum_{m=0}^{\infty}(k+m)c_m x^{k+m-1}$$

$$y'' = \textstyle\sum_{m=0}^{\infty}(k+m)(k+m-1)c_n x^{k+m-2}. \tag{iii}$$

Substituting these values in Equation (i), we get

$$x^2\textstyle\sum_{m=0}^{\infty}(k+m)(k+m-1)c_m x^{k+m-2} + x\sum_{m=0}^{\infty}(k+m)c_m x^{k+m-1}$$
$$+(x^2 - n^2)\textstyle\sum_{m=0}^{\infty} c_m x^{k+m} = 0$$

or,

$$\textstyle\sum_{m=0}^{\infty}(k+m)(k+m-1)c_m x^{k+m} + \sum_{m=0}^{\infty}(k+m)c_m x^{k+m}$$
$$-n^2\textstyle\sum_{m=0}^{\infty} c_m x^{k+m} + \sum_{m=0}^{\infty} c_m x^{k+m+2} = 0$$

or, $\sum_{m=0}^{\infty} c_m[(k+m)(k+m-1) + (k+m) - n^2]x^{k+m} + \sum_{m=0}^{\infty} c_m x^{k+m+2} = 0$

or, $$\sum_{m=0}^{\infty} c_m[(k+m)^2 - n^2]x^{k+m} + \sum_{m=0}^{\infty} c_m x^{k+m+2} = 0$$

or, $$\sum_{m=0}^{\infty} c_m[(k+m+n)(k+m-n)]x^{k+m} + \sum_{m=0}^{\infty} c_m x^{k+m+2} = 0,$$
which is an identity (iv)

Equating to zero the coefficient of the lowest power of x, i.e., x^k in Equation (iv) gives the indicial equation

$$c_0[(k+n)(k-n)] = 0$$

Since $$c_0 \neq 0 \Rightarrow k = n, -n \quad \text{(v)}$$

Since $2n$ is a non-integral, the roots are unequal and not differing by an integer. To obtain the recurrence relation, we equate to zero the coefficient of x^{k+m} and obtain

$$c_m[(k+m+n)(k+m-n)] + c_{m-2} = 0$$

$$\therefore c_m = -\frac{1}{(k+m+n)(k+m-n)} c_{m-2}. \quad \text{(vi)}$$

[If the recurrence relation connects c_m *and* c_{m-2}, then we, in general, determine c_1 by equating to zero the coefficient of the next higher power (than already used for getting the indicial equation). On the other hand, the recurrence relation connects c_m *and* c_{m-1}. This step may be omitted.]

Equating the coefficient of x^{k+1} in Equation (iv), we get

$$c_1[(k+1+n)(k+1-n)] = 0 \text{ gives } c_1 = 0 \text{ for both } k = n, -n.$$

From Equation (vi), we get

$$c_1 = c_2 = c_5 \ldots = 0. \quad \text{(vii)}$$

From Equation (vi), for $m = 2$,

$$c_2 = -\frac{1}{(k+2+n)(k+2-n)}c_0. \tag{viii}$$

For $m = 4$,

$$c_4 = -\frac{1}{(k+4+n)(k+4-n)}c_2 = \frac{1}{(k+4+n)(k+4-n)}\frac{1}{(k+2+n)(k+2-n)}c_0, \tag{ix}$$

and so on.

Putting these values of c's in Equation (ii),

$$y = \sum_{m=0}^{\infty} c_m x^{k+m} = x^k(c_0 + c_1 x + c_2 x^2 + c_3 x^3 + \cdots..)$$

$$= x^k c_0 \left(1 - \frac{1}{(k+2+n)(k+2-n)}x^2 + \frac{1}{(k+4+n)(k+4-n)}\frac{1}{(k+2+n)(k+2-n)}x^4 - \cdots\ldots\ldots..\right). \tag{x}$$

Putting $k = n$ and replacing c_0 by a in Equation (x), we get

$$y = x^n a\left(1 - \frac{1}{4(n+1)}x^2 + \frac{1}{4.8(n+1)(n+2)}x^4 - \cdots\ldots\ldots..\right) = au\ldots.$$

Next, $k = -n$, and replacing c_0 by b in Equation (x), we get

$$y = x^{-n} b\left(1 - \frac{1}{4(1-n)}x^2 + \frac{1}{4.8(1-n)(2-n)}x^4 - \cdots\ldots\ldots..\right) = bv.$$

The required general solution is given by

$y = au + bv$ where a,b are arbitrary constants.

Example 9.5.8. Show that $x=0$ is a regular singular point of $(2x+x^3)y''-y'-6xy=0$, and find its solution about $x=0$.

Solution: The given equation is

$$(2x+x^3)y''-y'-6xy=0 \tag{i}$$

or,
$$y''-\frac{1}{(2x+x^3)}y'-\frac{6}{(2+x^2)}y=0.$$

Comparing it with the equation

$$y''+P(x)y'+Q(x)y=0,$$

$\therefore P(x)=-\dfrac{1}{(2x+x^3)}$ *and* $Q(x)=-\dfrac{6}{(2+x^2)}$ are not analytic at $x=0$.

Now, $xP(x)=-\dfrac{1}{(2+x^2)}$ *and* $x^2Q(x)=-\dfrac{6x^2}{(2+x^2)}$ are analytic at $x=0$.

Hence, $x=0$ is a regular singular point of Equation (i).

Let the series solution of Equation (i) be of the form

$$y=\sum_{n=0}^{\infty}c_nx^{k+n} \text{ where } c_0\neq 0. \tag{ii}$$

Differentiating with respect to x, we get

$$y'=\sum_{n=0}^{\infty}(k+n)c_nx^{k+n-1}$$
$$y''=\sum_{n=0}^{\infty}(k+n)(k+n-1)c_nx^{k+n-2}. \tag{iii}$$

Substituting these values in Equation (i), we get

$$(2x+x^3)\sum_{n=0}^{\infty}(k+n)(k+n-1)c_nx^{k+n-2}-\sum_{n=0}^{\infty}(k+n)c_nx^{k+n-1}$$
$$-6x\sum_{n=0}^{\infty}c_nx^{k+n}=0$$

or, $\sum_{n=0}^{\infty} 2(k+n)(k+n-1)c_n x^{k+n-1} + \sum_{n=0}^{\infty}(k+n)(k+n-1)c_n x^{k+n+1}$
$$-\sum_{n=0}^{\infty}(k+n)c_n x^{k+n-1} - 6\sum_{n=0}^{\infty} c_n x^{k+n+1} = 0$$

or,

$$\sum_{n=0}^{\infty} c_n[(2(k+n)(k+n-1)) - (k+n)]x^{k+n-1} + \sum_{n=0}^{\infty} c_n[(k+n)(k+n-1)-6]x^{k+n+1} = 0$$

or,

$$\sum_{n=0}^{\infty} c_n[(k+n)(2k+2n-3)]x^{k+n-1} + \sum_{n=0}^{\infty} c_n[(k+n-3)(k+n+2)]x^{k+n+1} = 0. \quad \text{(iv)}$$

Equating the coefficient of the smallest power of x, i.e., x^{k-1}, the above Equation (iv) gives the indicial equation $c_0 k(2k-3) = 0$.

Since $c_0 \neq 0$, $k(2k-3) = 0 \Rightarrow k = 0 \; and \; k = \frac{3}{2}$.

The coefficient of x^k in the first term of Equation (iv): $c_1(k+1)(2k-1) = 0$ $\Rightarrow c_1 = 0$ (for both $k = 0 \; and \; k = \frac{3}{2}$).

For $n \geq 2$, we get from Equation (iv)

$$c_n[(k+n)(2k+2n-3)] + c_{n-2}(k+n-5)(k+n) = 0$$

or,
$$c_n = -\frac{k+n-5}{2k+2n-3}c_{n-2}. \quad \text{(v)}$$

For $n = 3,5,7 \; in$ Equation (v), we get

$$c_1 = c_3 = c_5 = 0, \text{ and so on.} \quad \text{(vi)}$$

For $n = 2,4,6,8$ *in* Equation (v), we get

$$c_2 = -\frac{k-3}{2k+1}c_0$$

$$c_4 = -\frac{k-1}{2k+5}c_2 = \frac{k-1}{2k+5}\frac{k-3}{2k+1}c_0, \text{ and so on.} \tag{vii}$$

$$\therefore y = c_0 x^k \left(1 - \frac{k-3}{2k+1}x^2 + \frac{k-1}{2k+5}\frac{k-3}{2k+1}x^4 - \cdots\ldots\right). \tag{viii}$$

Putting k=0 and replacing c_0 by a in Equation (viii), we get

$$y = a\left(1 + 3x^2 + \frac{3}{8}x^4 - \cdots\ldots\right) = au.$$

Putting $k = \frac{3}{2}$ and replacing c_0 by b in Equation (viii), we get

$$y = bx^{\frac{3}{2}}\left(1 + \frac{3}{8}x^2 - \frac{3}{128}x^4 - \cdots\ldots\right) = bv.$$

Hence, the required solution is

$$y = au + bv.$$

Example 9.5.9. Solve the *hypergeometric equation* $x(1-x)y'' + \{\gamma - (\alpha + \beta + 1)x\}y' - \alpha\beta\gamma = 0$ near $x = 0$, where γ is not an integer.

Solution: The given equation is

$$x(1-x)y'' + \{\gamma - (\alpha + \beta + 1)x\}y' - \alpha\beta\gamma = 0 \tag{i}$$

or,
$$y'' + \frac{\{\gamma - (\alpha + \beta + 1)x\}}{x(1-x)}y' - \frac{\alpha\beta}{x(1-x)}\gamma = 0.$$

Comparing the above equation with

$$y'' + P(x)y' + Q(x)y = 0,$$

here, $P(x) = \frac{\{\gamma - (\alpha+\beta+1)x\}}{x(1-x)}$ and $Q(x) = -\frac{\alpha\beta\gamma}{x(1-x)}$

$$xP(x) = \frac{\{\gamma - (\alpha+\beta+1)x\}}{(1-x)} \text{ and } x^2Q(x) = -\frac{\alpha\beta\gamma x}{(1-x)} \text{ are analytic at } x = 0.$$

So, $x = 0$ is a regular singular point of Equation (i).
Let the series solution of Equation (i) be

$$y = \sum_{n=0}^{\infty} c_n x^{k+n} \quad \text{where } c_0 \neq 0. \tag{ii}$$

Differentiating with respect to x, we get

$$y' = \sum_{n=0}^{\infty} (k+n)c_n x^{k+n-1}$$

$$y'' = \sum_{n=0}^{\infty} (k+n)(k+n-1)c_n x^{k+n-2}. \tag{iii}$$

Substituting these values in Equation (i), we get

$$x(1-x)\sum_{n=0}^{\infty}(k+n)(k+n-1)c_n x^{k+n-2} + \{\gamma - (\alpha+\beta+1)x\}$$
$$\sum_{n=0}^{\infty}(k+n)c_n x^{k+n-1} - \alpha\beta\sum_{n=0}^{\infty} c_n x^{k+n} = 0$$

or, $$\sum_{n=0}^{\infty}(k+n)(k+n-1)c_n x^{k+n-1} - \sum_{n=0}^{\infty}(k+n)(k+n-1)c_n x^{k+n}$$
$$+\gamma\sum_{n=0}^{\infty}(k+n)c_n x^{k+n-1} - (\alpha+\beta+1)\sum_{n=0}^{\infty}(k+n)c_n x^{k+n}$$
$$-\alpha\beta\sum_{n=0}^{\infty} c_n x^{k+n} = 0$$

or, $$\sum_{n=0}^{\infty} c_n[(k+n)(k+n-1) + \gamma(k+n)]x^{k+n-1}$$
$$-\sum_{n=0}^{\infty} c_n[(k+n)(k+n-1) + (\alpha+\beta+1)(k+n) + \alpha\beta]x^{k+n} = 0$$

or,
$$\sum_{n=0}^{\infty} c_n(k+n)(k+n+\gamma-1)x^{k+n-1}$$
$$-\sum_{n=0}^{\infty} c_n(k+n+\alpha)(k+n+\beta)]x^{k+n}=0. \qquad \text{(iv)}$$

Equating the coefficient of the lowest power of x, i.e., x^{k-1}, we get

$$c_0 k(k+\gamma-1)=0$$

Since $c_0 \neq 0$, $k(k+\gamma-1)=0$, which is an indicial equation.
The roots of the indicial equation are

$$k=0\,, k=1-\gamma,$$

which are unequal and do not differ by an integer.
To obtain the recurrence relation, we equate to zero the coefficient of x^{k+n-1}

$$c_n(k+n)(k+n+\gamma-1)-c_{n-1}(k+n-1+\alpha)(k+n-1+\beta)=0$$
$$c_n=\frac{(k+n-1+\alpha)(k+n-1+\beta)}{(k+n)(k+n+\gamma-1)}c_{n-1}. \qquad \text{(v)}$$

Case I: When $k=0$
Putting $n=1,2,3,\ldots$ successively in Equation (v),

$$c_1=\frac{(1-1+\alpha)(1-1+\beta)}{(1)(1+\gamma-1)}c_0=\frac{\alpha\beta}{\gamma}c_0$$
$$c_2=\frac{(\alpha+1)(\beta+1)}{2(\gamma+1)}c_1=\frac{(\alpha+1)(\beta+1)}{2(\gamma+1)}\frac{\alpha\beta}{\gamma}c_0=\frac{\alpha(\alpha+1)\beta(\beta+1)}{1.2\gamma(\gamma+1)}c_0$$
$$c_3=\frac{(\alpha+2)(\beta+2)}{3(\gamma+2)}c_2=\frac{\alpha(\alpha+1)(\alpha+2)\beta(\beta+1)(\beta+2)}{1.2.3\gamma(\gamma+1)(\gamma+2)}c_0, \text{ and so on.}$$

Putting these values of c's in Equation (ii) for $=0$ and replacing c_0 by a, we get

$$y=a\left[1+\frac{\alpha\beta}{\gamma}x+\frac{\alpha(\alpha+1)\beta(\beta+1)}{1.2\gamma(\gamma+1)}x^2+\frac{\alpha(\alpha+1)(\alpha+2)\beta(\beta+1)(\beta+2)}{1.2.3\gamma(\gamma+1)(\gamma+2)}x^3+\cdots\cdots\right]$$
(vi)

If we take $a=1$ in Equation (vi), the right-hand side of Equation (vi) is called a hypergeometric series and is represented by ${}_2F_1(\alpha,\beta,\gamma,x)$. ${}_2F_1(\alpha,\beta,\gamma,x)$ is the solution of Equation (i).

Case II: When $k=1-\gamma$, c_n in Equation (v) reduces to

$$\begin{aligned} c_n &= \frac{(1-\gamma+n-1+\alpha)(1-\gamma+n-1+\beta)}{(1-\gamma+n)(1-\gamma+n+\gamma-1)}c_{n-1} \\ &= \frac{(-\gamma+n+\alpha)(-\gamma+n+\beta)}{n(1-\gamma+n)}c_{n-1} \\ &= \frac{(\alpha'+n-1)(\beta'+n-1)}{n(\gamma'+n-1)}c_{n-1} \end{aligned} \quad \text{(vii)}$$

where $$\gamma'=2-\gamma, \alpha'=\alpha-\gamma+1, \beta'=\beta-\gamma+1. \quad \text{(viii)}$$

Putting $n=1,2,3,\ldots$ successively in Equation (v) gives

$$c_1=\frac{\alpha'\beta'}{1.\gamma'}c_0, c_2=\frac{(\alpha'+1)(\beta'+1)}{1.2.(\gamma'+1)}\frac{\alpha'\beta'}{1.\gamma'}c_0=\frac{\alpha'(\alpha'+1)\beta'(\beta'+1)}{1.2.\gamma'(\gamma'+1)}c_0$$

$$c_3=\frac{\alpha'(\alpha'+1)(\alpha'+2)\beta'(\beta'+1)(\beta'+2)}{1.2.3\gamma'(\gamma'+1)(\gamma'+2)}c_0, \text{ and so on.}$$

Putting $k=1-\gamma$, the values of c's, and replacing c_0 by b in Equation (ii), we get

$$y=bx^{1-\gamma}\left[1+\frac{\alpha'\beta'}{1.\gamma'}x+\frac{\alpha'(\alpha'+1)\beta'(\beta'+1)}{1.2.\gamma'(\gamma'+1)}x^2+\frac{\alpha'(\alpha'+1)(\alpha'+2)\beta'(\beta'+1)(\beta'+2)}{1.2.3\gamma'(\gamma'+1)(\gamma'+2)}x^3+\cdots.\right] \quad \text{(ix)}$$

If we take $b=1$ in Equation (ix), the series right-hand side of Equation (ix) would be

$$\begin{aligned} y&=x^{1-\gamma}\left[1+\frac{\alpha'\beta'}{1.\gamma'}x+\frac{\alpha'(\alpha'+1)\beta'(\beta'+1)}{1.2.\gamma'(\gamma'+1)}x^2+\frac{\alpha'(\alpha'+1)(\alpha'+2)\beta'(\beta'+1)(\beta'+2)}{1.2.3\gamma'(\gamma'+1)(\gamma'+2)}x^3+\cdots.\right] \\ &=x^{1-\gamma}\,{}_2F_1(\alpha',\beta',\gamma',x), \end{aligned}$$

i.e., $y = x^{1-\gamma}\, {}_2F_1(\alpha',\beta',\gamma',x) = x^{1-\gamma}\, {}_2F_1(\alpha-\gamma+1,\beta-\gamma+1,2-\gamma,x)$ is another independent solution of Equation (i).

Hence, the general solution is

$$y = a\, {}_2F_1(\alpha,\beta,\gamma,x) + bx^{1-\gamma}\, {}_2F_1(\alpha-\gamma+1,\beta-\gamma+1,2-\gamma,x).$$

Example 9.5.10. Find the series solution of

$$4xy'' + 2y' + y = 0.$$

Solution: The given equation is

$$4xy'' + 2y' + y = 0 \tag{i}$$

or,
$$y'' + \frac{1}{2x}2y' + \frac{1}{4x}y = 0.$$

Comparing it with the equation

$$y'' + P(x)y' + Q(x)y = 0,$$

$\therefore P(x) = \frac{1}{2x}$ *and* $Q(x) = \frac{1}{4x}$ are not analytic at $x = 0$.

Now, $xP(x) = \frac{1}{2}$ *and* $x^2Q(x) = \frac{1}{4}$ are analytic at $x = 0$.

Hence, $x = 0$ is a regular singular point of Equation (i).

Let the series solution of Equation (i) be of the form

$$y = \sum_{n=0}^{\infty} c_n x^{k+n} \quad \text{where } c_0 \neq 0. \tag{ii}$$

Differentiating with respect to x, we get

$$y' = \sum_{n=0}^{\infty}(k+n)c_n x^{k+n-1}$$
$$y'' = \sum_{n=0}^{\infty}(k+n)(k+n-1)c_n x^{k+n-2}. \tag{iii}$$

Substituting these values in Equation (i), we get

$$4x\sum_{n=0}^{\infty}(k+n)(k+n-1)c_n x^{k+n-2}+2\sum_{n=0}^{\infty}(k+n)c_n x^{k+n-1}+\sum_{n=0}^{\infty}c_n x^{k+n}=0$$

or,

$$4\sum_{n=0}^{\infty}(k+n)(k+n-1)c_n x^{k+n-1}+2\sum_{n=0}^{\infty}(k+n)c_n x^{k+n-1}+\sum_{n=0}^{\infty}c_n x^{k+n}=0$$

or, $\sum_{n=0}^{\infty}c_n[4(k+n)(k+n-1)+2(k+n)]x^{k+n-1}+\sum_{n=0}^{\infty}c_n x^{k+n}=0.$

Equating the coefficient of the lowest *power of* x, i.e., x^{k+-1} and putting $n=0$, we get

$$c_0[4k(k-1)+2k]=0\Rightarrow k(2k-1)=0, \textit{ which is an indicial equation.}$$

The roots are $k=0,\frac{1}{2}$, which do not differ by integer.

For recurrence relation:

$$c_n[4(k+n)(k+n-1)+2(k+n)]+c_{n-1}=0$$

or, $$c_n 2(k+n)(2k+2n-1)+c_{n-1}=0$$

or, $$c_n=-\frac{1}{2(k+n)(2k+2n-1)}c_{n-1}. \tag{iv}$$

Putting $n=1,2,3,\ldots.$

$$c_1=-\frac{1}{2(k+1)(2k+1)}c_0 \tag{v}$$

$$c_2 = -\frac{1}{2(k+2)(2k+3)}c_1 = \frac{1}{2(k+2)(2k+3)}\frac{1}{2(k+1)(2k+1)}c_0$$
$$= \frac{1}{4(k+1)(k+2)(2k+1)}\frac{1}{(2k+3)}c_0$$
$$c_3 = -\frac{1}{8(k+1)(k+2)(k+3)(2k+1)}\frac{1}{(2k+3)(2k+5)}c_0.$$

Putting the values of $c_1, c_2, c_3, \ldots$ in Equation (ii) and $k=0$ and replacing c_0 by a, we get

$$y = a(1 + \frac{1}{1.2}x + \frac{1}{4.1.2.3}x^2 - \frac{1}{8.1.2.3.3.5}x^3 + \cdots\ldots..)$$
$$= a(1 - \frac{1}{1.2}x + \frac{1}{4.1.2.3}x^2 - \frac{1}{1.2.3.4.5.6}x^3 + \cdots\ldots..)$$
$$= a(1 - \frac{(\sqrt{x})^2}{2!} + \frac{(\sqrt{x})^4}{4!} - \frac{(\sqrt{x})^6}{6!} + \cdots..)$$
$$= a\cos(\sqrt{x}).$$

Taking $k = \frac{1}{2}$ and replacing c_0 by b in Equation (ii), we get

$$y = bx^{\frac{1}{2}}(1 - \frac{1}{2.3}x + \frac{1}{2.3.4.5}x^2 - \frac{1}{2.3.3.4.5.6.7}x^3 + \cdots\ldots..)$$
$$= bx^{\frac{1}{2}}(1 - \frac{1}{2.3}x + \frac{1}{2.3.4.5}x^2 - \frac{1}{2.3.3.4.5.6.7}x^3 + \cdots\ldots..)$$
$$= b(\sqrt{x} - \frac{(\sqrt{x})^3}{3!} + \frac{(\sqrt{x})^5}{5!} - \frac{(\sqrt{x})^7}{7!} + \cdots..)$$
$$= bsin(\sqrt{x}).$$

The general solution is

$$y = a\cos\left(\sqrt{x}\right) + bsin\left(\sqrt{x}\right).$$

Example 9.5.11. Using the Frobenius method, solve the differential equation

$$y'' + \frac{1}{4x}y' + \frac{1}{8x^2}y = 0.$$

Solution: The given differential equation is

$$y'' + \frac{1}{4x}y' + \frac{1}{8x^2}y = 0$$

$$8x^2y'' + 2xy' + y = 0 \tag{i}$$

$x = 0$ is a regular singular point of Equation (i).
Let the series solution of Equation (i) be of the form

$$y = \sum_{n=0}^{\infty} c_n x^{k+n} \qquad \text{where } c_0 \neq 0. \tag{ii}$$

Differentiating with respect to x, we get

$$y' = \sum_{n=0}^{\infty} (k+n)c_n x^{k+n-1}$$

$$y'' = \sum_{n=0}^{\infty} (k+n)(k+n-1)c_n x^{k+n-2}. \tag{iii}$$

Substituting these values in Equation (i), we get

$$8x^2 \sum_{n=0}^{\infty} (k+n)(k+n-1)c_n x^{k+n-2} + 2x \sum_{n=0}^{\infty} (k+n)c_n x^{k+n-1} + \sum_{n=0}^{\infty} c_n x^{k+n} = 0$$

or,

$$\sum_{n=0}^{\infty} 8(k+n)(k+n-1)c_n x^{k+n} + \sum_{n=0}^{\infty} 2(k+n)c_n x^{k+n} + \sum_{n=0}^{\infty} c_n x^{k+n} = 0.$$

Equating the coefficient of the lowest power of x, i.e., x^k, we get

$$c_0[8k(k-1)+2k+1]=0$$

or,

$$c_0[8k^2-6\text{k}+1]=0.$$

Since $c_0 \neq 0 \Rightarrow (2k-1)(4k-1)=0$, which is an indicial equation.

The roots are $k=\frac{1}{2},\frac{1}{4}$.

The recurrence relation is

$$c_n[8(k+n)(k+n-1)+2(k+n)+1]=0 \text{ for all } n \geq 1$$

$$\Rightarrow c_1=c_2=c_3=\cdots\ldots\ldots=0.$$

The general solution is

$$y=ax^{\frac{1}{2}}+bx^{\frac{1}{4}}.$$

Example 9.5.12. Solve the series solution $(1\text{-}\,x^2)y''-xy'+4\text{y}=0$.

Solution: $x=0$ is an ordinary point of the equation.

Let the series solution of Equation (i) be of the form

$$y=\sum_{n=0}^{\infty} c_m x^{k+n} \qquad \text{where } c_0 \neq 0. \tag{ii}$$

Differentiating with respect to x, we get

$$y'=\sum_{n=0}^{\infty}(k+n)c_n x^{k+n-1}$$

$$y''=\sum_{n=0}^{\infty}(k+n)(k+n-1)c_n x^{k+n-2}. \tag{iii}$$

Substituting in Equation (i),

$$(1-x^2)\sum_{n=0}^{\infty}(k+n)(k+n-1)c_n x^{k+n-2} - x\sum_{n=0}^{\infty}(k+n)c_n x^{k+n-1} + 4\sum_{n=0}^{\infty} c_m x^{k+n} = 0$$

or, $$\sum_{n=0}^{\infty}(k+n)(k+n-1)c_n x^{k+n-2} - \sum_{n=0}^{\infty}(k+n)(k+n-1)c_n x^{k+n} - \sum_{n=0}^{\infty}(k+n)c_n x^{k+n} + 4\sum_{n=0}^{\infty} c_n x^{k+n} = 0.$$

Equating the coefficient of the lowest power of x, i.e., x^{k-2}, we get

$$k(k-1)c_0 = 0.$$

Since $c_0 \neq 0$, $k(k-1)=0$, which is an indicial equation.
Its roots are $k = 0,1$.
The recurrence relation:

$$(k+n)(k+n-1)c_n - c_{n-2}[(k+n-2)(k+n-3)+(k+n-2)-4] = 0$$

or, $$(k+n)(k+n-1)c_n - c_{n-2}[(k+n-2)^2 - 4] = 0.$$

$$\therefore c_n = \frac{(k+n-4)}{(k+n-1)}c_{n-2} \quad \text{(iv)}$$

and $(k+1)kc_1 = 0$,
if $k=0$, c_1 is indeterminate.

With $k=0$, using Equation (iv), we can express $c_2, c_4, c_6, \ldots\ldots$ in terms of c_0 and $c_3, c_5, c_7, \ldots$ interms of c_1.

For $k=0$, $c_n = \dfrac{(n-4)}{(n-1)}c_{n-2}$.

$$\therefore c_2 = -2c_0,\ c_4 = 0,\ c_6 = c_8 = \cdots. = 0$$

and $c_3 = -\dfrac{1}{2}c_1, c_5 = \dfrac{1}{4}c_3 = -\dfrac{1}{4}\dfrac{1}{2}c_1 = -\dfrac{1}{8}c_1, c_7 = -\dfrac{1}{16}c_1,$

$$y = x^k(c_0 + c_1x + c_2x^2 + \cdots \ldots\ldots)$$
$$= x^0(c_0 + c_2x^2 + c_4x^4 \ldots\ldots\ldots) + x^0(c_1x + c_3x^3 + c_5x^5 + \cdots.)$$
$$= c_0(1\text{-}2x^2) + c_1(x - \frac{1}{2}x^3 - \frac{1}{8}x^5 - \frac{1}{16}x^7 - \cdots.). \tag{v}$$

Remark: The root $k = 1$,
we can easily verify that indicial equation gives another solutions of type I on Frobenius method. But this solution will be a constant multiple of one the series occurring in Equation (v).

Example 9.5.13. Solve the series of Legendre's equation

$$(1\text{-}x^2)y'' - 2xy' + n(n+1)y = 0.$$

Solution: The given equation is

$$(1\text{-}x^2)y'' - 2xy' + n(n+1)y = 0. \tag{i}$$

Here, $x = 0$ is an ordinary point of Equation (i).
Let the series solution of Equation (i) be of the form

$$y = \sum_{n=0}^{\infty} c_m x^{k+n} \quad \text{where } c_0 \neq 0. \tag{ii}$$

Differentiating with respect to x, we get

$$y' = \sum_{m=0}^{\infty} (k+m)c_m x^{k+m-1}$$
$$y'' = \sum_{m=0}^{\infty} (k+m)(k+m-1)c_m x^{k+m-2}. \tag{iii}$$

Substituting in Equation (i),

$$(1-x^2)\sum_{m=0}^{\infty}(k+m)(k+m-1)c_m x^{k+m-2} - 2x\sum_{m=0}^{\infty}(k+m)c_m x^{k+m-1}$$
$$+ n(n+1)\sum_{n=0}^{\infty} c_m x^{k+m} = 0$$

or, $$\sum_{m=0}^{\infty}(k+m)(k+m-1)c_m x^{k+m-2} - \sum_{m=0}^{\infty}(k+m)(k+m-1)c_m x^{k+m}$$
$$-2\sum_{m=0}^{\infty}(k+m)c_m x^{k+m} + n(n+1)\sum_{n=0}^{\infty} c_m x^{k+m} = 0$$

or, $$\sum_{m=0}^{\infty}(k+m)(k+m-1)c_m x^{k+m-2}$$
$$-[\sum_{m=0}^{\infty} c_m[(k+m)(k+m-1)+2(k+m)-n(n+1)]x^{k+m} = 0.$$

Equating the coefficient of the lowest power of x, i.e., x^{k-2}, we get

$$k(k-1)c_0 = 0, \tag{iv}$$

which is an indicial equation. Since $c_0 \neq 0$, its roots are $k = 0,1$.

These are unequal and differ by an integer.

To get the recurrence relation, we equate to zero the coefficient of x^{k+m-2}, and we get

$$(k+m)(k+m-1)c_m - \sum_{m=0}^{\infty} c_{m-2}[(k+m-2)(k+m-3)$$
$$+2(k+m-2)-n(n+1)] = 0[(k+m-2)(k+m-1)-n(n+1) = (k+m)^2$$
$$-3(k+m)-(n+2)(n-1) = (k+m-n-2)(k+m+n-1)]$$

or, $$c_m = \frac{(k+m-n-2)(k+m+n-1)}{(k+m)(k+m-1)} c_{m-2}. \tag{v}$$

If we take $k = 0$, Equation (v) reduces to

$$c_m = \frac{(m-n-2)(m+n-1)}{m(m-1)} c_{m-2}. \tag{vi}$$

We now get $c_2, c_4, c_6, \ldots\ldots..$ by putting $m = 2,4,6\ldots..,$

$$c_2 = -\frac{n(n+1)}{2.1}c_0,\; c_4 = \frac{(n-2)(n+3)}{4.3}.\frac{n(n+1)}{2.1}c_0 = \frac{(n-2)n(n+1)(n+3)}{1.2.3.4}c_0,$$

and the coefficient of the next lower power of x^{k+m-1}

$(k+1)kc_1 = 0 \Rightarrow c_1$ is indeterminate since $k = 0$

$$c_3 = -\frac{(n-1)(n+2)}{3.2}c_1, c_5 = -\frac{(n-3)(n+4)}{5.4}.-\frac{(n-1)(n+2)}{3.2}c_1$$
$$= \frac{(n-3)(n-1)(n+2)(n+4)}{2.3.4.5}.$$

The solution is

$$y = x^k(c_0 + c_2x^2 + c_4x^4 + \ ....) + x^k(c_1x + c_3x^3 + \cdots..)$$
$$= x^0c_0\left(1 - \frac{n(n+1)}{2.1}x^2 + \frac{(n-2)n(n+1)(n+3)}{1.2.3.4}x^4 + \ ....\right)$$
$$+ x^0c_1(x - \frac{(n-1)(n+2)}{3.2}x^3 + \frac{(n-3)(n-1)(n+2)(n+4)}{2.3.4.5}x^5.....)$$
$$= c_0\left(1 - \frac{n(n+1)}{2.1}x^2 + \frac{(n-2)n(n+1)(n+3)}{1.2.3.4}x^4 + \ ....\right)$$
$$+ c_1(x - \frac{(n-1)(n+2)}{3.2}x^3 + \frac{(n-3)(n-1)(n+2)(n+4)}{2.3.4.5}x^5.....),$$

which is the required general solution, and c_0 and c_1 are arbitray constants.

For $k = 1$, Equation (vi) shows that $c_1 = 0 \Rightarrow c_3 = c_5 = c_7 = \cdots. = 0$.

We get $c_2 = -\frac{(n-1)(n+2)}{3!}c_0, c_4 = \frac{(n-3)(n-1)(n+2)(n+4)}{5!}c_0$.

The solution is

$$y = x^k(c_0 + c_2x^2 + c_4x^4 + \ ....) + x^k(c_1x + c_3x^3 + \ \cdots..)$$
$$= c_0x\left(1 - \frac{(n-1)(n+2)}{3!}x^2 + \frac{(n-3)(n-1)(n+2)(n+4)}{5!}x^4 + \ ....\right).$$

Example 9.5.14. Find the series solution of $x^2y'' - xy' - \left(x^2 + \frac{5}{4}\right)y = 0$.

Solution: The given equation is

$$x^2y'' - xy' - \left(x^2 + \frac{5}{4}\right)y = 0 \qquad \text{(i)}$$

or, $$y'' - \frac{1}{x}y' - \frac{4x^2+5}{x^2}y = 0.$$

Here, $x = 0$ is a regular singular point of Equation (i).
Let the series solution of Equation (i) be of the form

$$y = \sum_{n=0}^{\infty} c_m x^{k+n} \qquad \text{where } c_0 \neq 0. \tag{ii}$$

Differentiating with respect to x, we get

$$y' = \sum_{m=0}^{\infty} (k+m)c_m x^{k+m-1}$$
$$y'' = \sum_{m=0}^{\infty} (k+m)(k+m-1)c_m x^{k+m-2}. \tag{iii}$$

Substituting in Equation (i),

$$x^2 \sum_{m=0}^{\infty} (k+m)(k+m-1)c_m x^{k+m-2} - x \sum_{m=0}^{\infty} (k+m)c_m x^{k+m-1} - \left(x^2 + \frac{5}{4}\right)$$
$$\sum_{n=0}^{\infty} c_m x^{k+m} = 0$$

or, $$\sum_{m=0}^{\infty} (k+m)(k+m-1)c_m x^{k+m} - \sum_{m=0}^{\infty} (k+m)c_m x^{k+m}$$
$$-\sum_{n=0}^{\infty} c_m x^{k+m+2} - \sum_{n=0}^{\infty} \frac{5}{4} c_m x^{k+m} = 0 \tag{iv}$$

or, $$\sum_{m=0}^{\infty\infty\infty} c_m[(k+m)(k+m-1) - (k+m) - \frac{5}{4}]x^{k+m} - \sum_{n=0}^{\infty} c_m x^{k+m+2} = 0.$$

The indicial equation: the coefficient of x^k is zero.

$$c_0 k(k-1) - k - \frac{5}{4} = 0, \; c_0 \neq 0.$$

$$\therefore k^2 - 2k - \frac{5}{4} = 0 \text{ or } 4k^2 - 8k - 5 = 0$$

or (2k−5)(2k+1)=0. The roots are $k = -\frac{1}{2}, \frac{5}{2}$.

For $m = 1$, $c_1\left[(k+1)k - (k+1) - \frac{5}{4}\right] = 0 \Rightarrow c_1 = 0.$

For the recurrence relation,

$$c_m\left[(k+m)(k+m-1) - (k+m) - \frac{5}{4}\right] - c_{m-2} = 0$$

$$c_m = \frac{1}{(k+m)(k+m-2) - \frac{5}{4}} c_{m-2}. \tag{v}$$

Putting $k = -\frac{1}{2}$,

$$c_2 = \frac{1}{-\frac{3}{2}.\frac{1}{2} - \frac{5}{4}} c_0 = -\frac{1}{2} c_0, c_4 = \frac{1}{\frac{73}{22} - \frac{5}{4}} c_2 = -\frac{1}{4}.\frac{1}{2} c_0 = -\frac{1}{8} c_0$$

$c_3 = c_5 = c_7 = \cdots \ldots = 0.$

The solution is $y = x^{-\frac{1}{2}}(c_0 + c_2x^2 + c_4x^4 + \ \ldots.) + x^{-\frac{1}{2}}(c_1x + c_3x^3 + \cdots..)$

$$= x^{-\frac{1}{2}} c_0(1 - \frac{1}{2}x^2 - \frac{1}{8}x^4 - \cdots\ldots..).$$

For $k = \frac{5}{2}$, all $c_1 = c_3 = c_5 = \cdots\ldots. = 0$

and $c_2 = \frac{1}{\frac{9.5}{2.2} - \frac{5}{4}} c_0 = \frac{1}{10} c_0$

$c_4 = \frac{1}{\frac{13.9}{2.2} - \frac{5}{4}} c_0 = \frac{1}{28} c_0$, and so on.

The solution is $y = c_0 x^{\frac{5}{2}}\left(1 + \frac{1}{10}x^2 + \frac{1}{28}x^4 + \ \ldots.\right).$

Example 9.5.15. Find the power series solution of $(1-x^2)\frac{d^2y}{dx^2}-2x\frac{dy}{dx}+6y=0$.

Solution. $(1-x^2)\frac{d^2y}{dx^2}-2x\frac{dy}{dx}+6y=0$. (i)

$$\text{Let } x=\frac{1}{t}, \frac{dt}{dx}=-\frac{1}{t^2}, \quad \text{(ii)}$$

$$\frac{dy}{dx}=-t^2\frac{dy}{dt} \quad \text{(iii)}$$

$$\frac{d^2y}{dx^2}=t^2(2t\frac{dy}{dx}+t^2\frac{d^2y}{dt^2} \quad \text{(iv)}$$

Using Equations (ii), (iii), (iv); (i) transform to

$$\left(1-\frac{1}{t^2}\right)\left(2t^3\frac{dy}{dt}+t^4\frac{d^2y}{dt^2}\right)-\frac{2}{t}\left(-t^2\frac{dy}{dt}\right)+6y=0$$

or, $$t^2(t^2-1)\frac{d^2y}{dt^2}+2t^3\frac{dy}{dt}+6y=0 \quad \text{(v)}$$

or, $$\frac{d^2y}{dt^2}+\frac{2t}{(t^2-1)}\frac{dy}{dt}+\frac{6}{t^2(t^2-1)}y=0$$

Comparing with, $\frac{d^2y}{dt^2}+P(t)\frac{dy}{dt}+Q(t)y=0$

$$P(t)=\frac{2t}{(t^2-1)} \textit{ and } Q(t)=\frac{6}{t^2(t^2-1)}$$

Here $t=0$ *is analytic*, so $t=0$ is regular singular point of Equation (v)

Let the series solution be

$$y=t^k(C_0+C_1t+C_2t^2+\cdots\ldots..=\sum_{m=0}^{\infty}C_mt^{k+m} \quad \text{(vi)}$$

$$\frac{dy}{dt}=\sum_{m=0}^{\infty}(k+m)c_m t^{k+m-1}$$

$$\frac{d^2y}{dt^2}=\sum_{m=0}^{\infty}(k+m)(k+m-1)c_m t^{k+m-2}$$

Putting these values in Equation (v), finally we get

$$\textstyle\sum_{m=0}^{\infty}(k+m-3)(k+m+2)c_m t^{k+m}-\sum_{m=0}^{\infty}c_m[(k+m)(k+m+1)t^{k+m+2}=0 \tag{vii}$$

The indicial equation is $C_0(k+3)(k-2)=0$
$\therefore k=3,-2.$
Next equating to zero, the coefficient of t^{k+1} in Equation (vii) we get

$$C_1(k-2)(k+3)=0\Rightarrow C_1=0\ \mathit{for}\ k=3\ \mathit{and}\ k=-2.$$

Finally, equating to zero, coefficient of t^{k+m} in Equation (iv)

$$(k+m-3)(k+m+2)c_m-(k+m-2)(k+m-1)c_{m-2}=0$$

$$c_m=\frac{(k+m-2)(k+m-1)}{(k+m-3)(k+m+2)}c_{m-2} \tag{viii}$$

Putting m = 3,5,7,….

Gives,
$$C_1=C_3=C_5=\cdots=0 \tag{ix}$$

Putting m = 2,4,6,…. In Equation (viii) we get

$$C_2=\frac{k(k+1)}{(k-1)(k+4)}C_0, \tag{x}$$

$$C_4 = \frac{(k+2)(k+3)}{(k+6)} \frac{k}{(k-1)(k+4)} C_0 \text{ [Using Equation (x)] and so on.}$$

Putting these values in Equation (vi), we have

$$y = t^k C_0 \left[1 + \frac{k(k+1)}{(k-1)(k+4)} t^2 + \frac{(k+2)(k+3)}{(k+6)} \frac{k}{(k-1)(k+4)} t^4 + \cdots .. \right] \tag{xi}$$

Putting k=3 in Equation (xi) and replacing C_0 *by a*, Equation (xi) becomes

$$y = at^3 \left[1 + \frac{3.4}{2.7} t^2 + \frac{5.6.3}{2.7.9} t^4 + \cdots .. \right] = au, say \tag{xii}$$

Putting k=-2 in Equation (xi) and replacing C_0 *by b*, Equation (xi) becomes

$$y = bt^{-2} \left[1 - \frac{t^2}{3} \right] = bv, say \tag{xiii}$$

The required solution is $y = au + bv, ie$

$$y = at^3 \left[1 + \frac{3.4}{2.7} t^2 + \frac{5.6.3}{2.7.9} t^4 + \cdots .. \right] + bt^{-2} \left[1 - \frac{t^2}{3} \right]$$
$$= \frac{a}{x^3} \left[1 + \frac{3.4}{2.7} \frac{1}{x^2} + \frac{5.6.3}{2.7.9} \frac{1}{x^4} + \cdots .. \right] + bx^2 \left[1 - \frac{1}{3x^2} \right]$$

9.6 Exercise

1. Solve $x^2 y'' + xy' + (x^2 - 4)y = 0$ in a series solution.
2. Find the solution near $x = 0$ of

$$x^2 y'' + (x + x^2) y' + (x - 9) y = 0.$$

3. Solve in series $xy''+2y'+xy=0.$

4. Solve $x\frac{d^2y}{dx^2}+(1+x)\frac{dy}{dx}+2y=0.$

5. Solve $8x^2\frac{d^2y}{dx^2}+10x\frac{dy}{dx}-(1+x)y=0.$

6. Solve $xy''+y'+my=0$ *given that* $y=1$ *when* $x=0$

7. Solve $8x^2\frac{d^2y}{dx^2}+10x\frac{dy}{dx}-(1+x)y=0$

8. Solve $3x\frac{d^2y}{dx^2}+(1-x)\frac{dy}{dx}-y=0$

9. Find the power series solution of $(1-x^2)\frac{d^2y}{dx^2}-2x\frac{dy}{dx}+12y=0$ *at* $x=\infty$

10. Find the power series solution of $(x-x^2)\frac{d^2y}{dx^2}+(1-5x)\frac{dy}{dx}-4y=0$

11. Find the power series solution of $(x-x^2)\frac{d^2y}{dx^2}+(1-x)\frac{dy}{dx}-y=0$, *near* $x=0$

10

Hypergeometric Functions

Abstract
In this chapter, we study hypergeometric functions, a class of special functions that generalize elementary functions and arise as solutions to hypergeometric differential equations. These functions are defined by power series and are widely applicable in physics and engineering. The most well-known example is the ordinary hypergeometric function, which satisfies a second-order differential equation.

We discuss the properties of hypergeometric functions, as well as various related topics including the hypergeometric equation and series, confluent hypergeometric functions, and contiguous hypergeometric functions. The chapter also introduces the generalized hypergeometric function and explores its properties.

Finally, we study notable results and theorems by Ramanujan, Saalschütz, Whipple, and Dixon.

Keywords**:** Hypergeometric series, confluent hypergeometric function, Barnes' contour integral, generalized hypergeometric function, Kummer's function

10.1 Introduction

Hypergeometric functions, initially studied as power series solutions to differential equations, have found extensive applications in various fields, including number theory, algebraic geometry and theoretical physics. Hypergeometric functions appeared in the mid-seventeenth century, since then, they have played an important role in the development of mathematical and physical theories. Most of the elementary and special functions are members of the large hypergeometric class. This functions have history over two centuries, starting with Euler and Gauss, and then contributed by Kummer, Riemann, Paul Appell and others.

The differential equation of the form

$$x(1-x)y''+\{y-(\alpha+\beta+1)x\}y'-\alpha\beta y=0$$

Dipankar De. Ordinary Differential Equations and Special Functions, (309–372) © 2026 Scrivener Publishing LLC

or, $$y'' + \frac{\{y-(\alpha+\beta+1)x\}}{x(1-x)}y' - \frac{\alpha\beta}{x(1-x)}y = 0\ldots\ldots \qquad \text{(i)}$$

is known as Gauss' equation or hypergeometric equation, where α, β, γ are parameters. The points $x = 0$, $x = 1$ and $x = \infty$ are singular points of the differential equation.

A power series of the form

$$1 + \frac{\alpha\beta}{1.\gamma}x + \frac{\alpha(\alpha+1)\beta(\beta+1)}{1.2.\gamma(\gamma+1)}x^2 + \frac{\alpha(\alpha+1)(\alpha+2)\beta(\beta+1)(\beta+2)}{1.2.3\gamma(\gamma+1)(\gamma+2)}x^3 + \ldots\ldots\ldots \qquad \text{(ii)}$$

is known as a *Hypergeometric series.* When it converges to function

$$F(\alpha,\beta,\gamma:x) = {}_2F_1(\alpha,\beta,\gamma:x) = \sum_{n=0}^{\infty}\frac{(\alpha)_n(\beta)_n x^n}{(\gamma)_n}\frac{x^n}{n!} \qquad \text{(iii)}$$

is called Hypergeometric function. The notation $F(\alpha, \beta, \gamma: x)$ is sometimes replaced by

$${}_2F_1\begin{bmatrix}\alpha,\beta; & \\ & x\\ \gamma; & \end{bmatrix} \text{or simply } F\begin{bmatrix}\alpha,\beta; & \\ & x\\ \gamma; & \end{bmatrix}$$

We study the generalized hypergeometric function. If the series $\sum_{n=0}^{\infty}\frac{(\alpha_1)_n(\alpha_2)_n\ldots(\alpha_p)_n.}{(\beta_1)_n(\beta_2)_n\ldots.(\beta_q)_n.}\frac{x^n}{n!}$ converges to a function F, we call F the *generalized hypergeometric function.* and it is in the form of ${}_pF_q(\alpha_1,\alpha_2,\ldots.\alpha_p;\beta_1,\beta_2,\ldots\beta_q;x)$,or ${}_pF_q$.

In this chapter we discuss confluent hypergeometric function, contiguous hypergeometric function, some special generalized hypergeometric functions and studied some theorems on hypergeometric functions.

10.1.1 Factorial Function

We are familiar with the factorial of a positive integer. We write it as

$$n! = n.(n-1).(n-2)\ldots 3.2.1$$

We now define its generalization $(\alpha)_n = \alpha(\alpha+1)(\alpha+2)\ldots.(\alpha+n-1)$, n≥1

I. $(\alpha)_n = \dfrac{\Gamma(\alpha+n)}{\Gamma(\alpha)}$

II. $(\alpha)_{n+1} = \alpha(\alpha+1)_n$

III. $(\alpha)_{n+1} = (\alpha+n)(\alpha)_n$

Deductions:

I. $(\alpha)_n = \alpha(\alpha+1)(\alpha+2)\ldots.(\alpha+n-1)$

$$= \frac{1.2.3\ldots.(\alpha-1)\alpha(\alpha+1)(\alpha+2)\ldots.(\alpha+n-1)}{1.2.3\ldots.(\alpha-1))}$$

$$= \frac{\Gamma(\alpha+n)}{\Gamma(\alpha)}$$

II. $(\alpha)_{n+1} = \alpha(\alpha+1)(\alpha+2)\ldots.(\alpha+n)$

$$= \alpha[(\alpha+1)(\alpha+2)\ldots.\{(\alpha+1)+n-1\}]$$

$$= \alpha(\alpha+1)_n$$

III. $(\alpha+n)(\alpha)_n = \alpha(\alpha+1)(\alpha+2)\ldots.(\alpha+n-1)(\alpha+n)$

$$= \alpha(\alpha+1)(\alpha+2)\ldots.(\alpha+n-1)(\alpha+n+1-1)$$

$$= (\alpha)_{n+1}.$$

10.1.2 Hypergeometric Series

A power series of the form

$$1+\frac{\alpha\beta}{1.\gamma}x+\frac{\alpha(\alpha+1)\beta(\beta+1)}{1.2.\gamma(\gamma+1)}x^2+\frac{\alpha(\alpha+1)(\alpha+2)\beta(\beta+1)(\beta+2)}{1.2.3\gamma(\gamma+1)(\gamma+2)}x^3+\cdots. \tag{10.1}$$

is known as a *hypergeometric series.* It includes, as particular cases, most of the familiar series like binomial, exponential, logarithmic, and trigonometric series.

For $\alpha = 1, \beta = \gamma$, *the series reduces to a simple* geometric series $1 + x + x^2 + \cdots$.., *so* the series Equation (10.1) is a generalization of the geometric series, which is called a hypergeometric series.

In terms of factorial function, we use the notation

$$(\alpha)_n = \alpha(\alpha+1)(\alpha+2)(\alpha+3)\ldots(\alpha+n-1) = \frac{\Gamma(\alpha+n)}{\Gamma(\alpha)}.$$

The general term of the series Equation (10.1) can be written as

$$u_n = \frac{\alpha(\alpha+1)(\alpha+2)\ldots(\alpha+n-1)\beta(\beta+1)(\beta+2)\ldots.(\beta+n-1)}{1.2.3\gamma(\gamma+1)(\gamma+2)\ldots.(\gamma+n-1)}\frac{x^n}{n!}$$

$$= \frac{(\alpha)_n(\beta)_n}{(\gamma)_n}\frac{x^n}{n!}.$$

To test the convergence of the series for other values of $a,b,c,$ let us apply D'Alembert's ratio test. We see that

$$\lim_{n\to\infty}\left|\frac{u_{n+1}}{u_n}\right| = \lim_{n\to\infty}\left|\frac{(\alpha)_{n+1}(\beta)_{n+1}}{(\gamma)_{n+1}}\frac{x^{n+1}}{(n+1)!}\frac{(\gamma)_n n!}{(\alpha)_n(\beta)_n x^n}\right|$$

$$= \lim_{n\to\infty}\left|\frac{(\alpha+n)(\beta+n)}{(\gamma+n)}\frac{x}{n+1}\right|.$$

Therefore, the series converges absolutely if $|x| < 1$ and diverges if $|x| > 1$. If $|x| = 1$, i.e., on the circle of convergence, the test fails. In this case, let us compare this series with the series

$$\Sigma_{n=1}^{\infty}\frac{1}{n^{1+\delta}}, \text{ where } 2\delta = Re(\gamma-\alpha-\beta) > 0$$

$$\lim_{n\to\infty}\left|\frac{(\alpha)_n(\beta)_n}{(\gamma)_n}\frac{n^{1+\delta}}{n!}\right| = \lim_{n\to\infty}\left|\frac{(\alpha)_n}{(\mathrm{n}-1)!\mathrm{n}^{\alpha}}\frac{(\beta)_n}{(\mathrm{n}-1)!\mathrm{n}^{\beta}}\frac{(\mathrm{n}-1)!\mathrm{n}^{\gamma}}{(\gamma)_n}\frac{(\mathrm{n}-1)!\mathrm{n}^{1+\delta}}{\mathrm{n}!\mathrm{n}^{\gamma-\alpha-\beta}}\right|$$

$$= \lim_{n\to\infty}\left|\frac{(\alpha)_n(\beta)_n}{(\gamma)_n}\frac{n^{1+\delta}}{n!}\right|$$

$$= \left|\frac{1}{\Gamma(\alpha)}\frac{1}{\Gamma(\beta)}\Gamma(\gamma)\right|.\lim_{n\to\infty}\left|\frac{1}{n^{\gamma-\alpha-\beta}}\right|$$

$$= \frac{\Gamma(\gamma)}{\Gamma(\alpha)\Gamma(\beta)}\lim_{n\to\infty}\left|\frac{1}{n^{Re(\gamma-\alpha-\beta)}}\right|.$$

We conclude that the series is absolutely convergent when $Re(\gamma-\alpha-\beta)>0$.

Hence, the hypergeometric series Equation (10.1), when γ is neither zero nor a negative integer, is absolutely convergent within the circle of convergence $|x|<1$ and divergent outside it. On the circle of convergence, the series is absolutely convergent if $Re(\gamma-\alpha-\beta)>0$.

When the series Equation (10.1) converges, it is supposed to converge to a function $F(\alpha,\beta,\gamma:x)$, which is called the *hypergeometric function* and is written as

$$F(\alpha,\beta,\gamma:x)=\sum_{n=0}^{\infty}\frac{(\alpha)_n(\beta)_n}{(\gamma)_n}\frac{x^n}{n!}.$$

Its notation may be written as ${}_2F_1$: the left suffix 2 and the right suffix 1 indicate, respectively, the number of parameters in the numerator and denominator of the function expression, i.e.,

$${}_2F_1(\alpha,\beta,\gamma:x)=\sum_{n=0}^{\infty}\frac{(\alpha)_n(\beta)_n}{(\gamma)_n}\frac{x^n}{n!}.$$

10.1.3 Symmetric Property of Hypergeometric Functions

A hypergeometric function does not change if the parameters α,β are interchanged keeping γ fixed. Thus,

$$F(\alpha,\beta,\gamma:x)=F(\beta,\alpha,\gamma:x).$$

Proof: We have, by definition, $F(\alpha,\beta,\gamma:x)=\sum_{n=0}^{\infty}\frac{(\alpha)_n(\beta)_n}{(\gamma)_n}\frac{x^n}{n!}$

$$F(\beta,\alpha,\gamma:x)=\sum_{n=0}^{\infty}\frac{(\beta)_n(\alpha)_n}{(\gamma)_n}\frac{x^n}{n!}.$$

$$\therefore F(\alpha,\beta,\gamma:x)=F(\beta,\alpha,\gamma:x).$$

10.2 Differentiation of Hypergeometric Functions

Show that $\frac{d}{dx}F(\alpha,\beta,\gamma:x)=\frac{\alpha\beta}{\gamma}F(\alpha+1,\beta+1,\gamma+1:x)$, and deduce that

(i) $\frac{d^n}{dx^n}F(\alpha,\beta,\gamma:x)=\frac{(\alpha)_n(\beta)_n}{(\gamma)_n}F(\alpha+n,\beta+n,\gamma+n:x)$

(ii) $\left[\frac{d^n}{dx^n}F(\alpha,\beta,\gamma:x)\right]_{x=0}=\frac{(\alpha)_n(\beta)_n}{(\gamma)_n}.$

Proof: By definition, we have

$$F(\alpha,\beta,\gamma:x)=\sum_{n=0}^{\infty}\frac{(\alpha)_n(\beta)_n}{(\gamma)_n}\frac{x^n}{n!}.$$

Differentiating with respect to x,

$$\frac{d}{dx}F(\alpha,\beta,\gamma:x)=\sum_{n=1}^{\infty}\frac{(\alpha)_n(\beta)_n}{(\gamma)_n}\frac{x^{n-1}}{(n-1)!}=\sum_{m=0}^{\infty}\frac{(\alpha)_{m+1}(\beta)_{m+1}}{(\gamma)_{m+1}}\frac{x^m}{(m)!}$$

[taking m such that $n=m+1$, i.e., $m=n-1$ so that when $n=1$, $m=0$ and $n=\infty$, $m=\infty$]

$$=\sum_{m=0}^{\infty}\frac{\alpha(\alpha+1)_m\beta(\beta+1)_m}{\gamma(\gamma+1)_m}\frac{x^m}{(m)!}$$

$$=\frac{\alpha\beta}{\gamma}\sum_{m=0}^{\infty}\frac{(\alpha+1)_m(\beta+1)_m}{(\gamma+1)_m}\frac{x^m}{(m)!}=\frac{\alpha\beta}{\gamma}F(\alpha+1,\beta+1,\gamma+1:x)$$

$$\frac{d}{dx}F(\alpha,\beta,\gamma:x)=\frac{\alpha\beta}{\gamma}F(\alpha+1,\beta+1,\gamma+1:x) \tag{10.2}$$

Deduction (i): For every positive integer, we must show that

$$\frac{d^n}{dx^n}F(\alpha,\beta,\gamma:x)=\frac{(\alpha)_n(\beta)_n}{(\gamma)_n}F(\alpha+n,\beta+n,\gamma+n:x). \quad (10.3)$$

Since $\alpha=(\alpha)_1, \beta=(\beta)_1, and\ \gamma=(\gamma)_1$ Equation (10.2) shows that Equation (10.3) is true for $n=1$. Now, we assume that Equation (10.3) is true for $n=m$.

$$\frac{d^m}{dx^m}F(\alpha,\beta,\gamma:x)=\frac{(\alpha)_m(\beta)_m}{(\gamma)_m}F(\alpha+m,\beta+m,\gamma+m:x) \quad (10.4)$$

Differentiating Equation (10.4) with respect of *x*, we get

$$\frac{d^{m+1}}{dx^{m+1}}F(\alpha,\beta,\gamma:x)=\frac{d}{dx}\left[\frac{d^m}{dx^m}F(\alpha,\beta,\gamma:x)\right]=\frac{d}{dx}\left[\frac{(\alpha)_m(\beta)_m}{(\gamma)_m}F(\alpha+m,\beta+m,\gamma+m:x)\right]$$

$$=\frac{(\alpha)_m(\beta)_m}{(\gamma)_m}\frac{(\alpha+m)(\beta+m)}{(\gamma+m)}F(\alpha+m+1,\beta+m+1,\gamma+m+1:x).$$

[Using Equation (10.2), by replacing $\alpha,\beta,\gamma\ by\ \alpha+m,\beta+m,\gamma+m$, respectively]

$$\frac{d^{m+1}}{dx^{m+1}}F(\alpha,\beta,\gamma:x)=\frac{(\alpha)_{m+1}(\beta)_{m+1}}{(\gamma)_{m+1}}F(\alpha+m+1,\beta+m+1,\gamma+m+1:x). \quad (10.5)$$

Equation (10.3) is true for $n=m$, *and hence, it is also true form* $m+1$.

Hence by mathematical induction Equation (10.3) is true for all integer value of n.

Deduction (ii):
Putting $x=0$ *in* Equation (ii), we get

$$\left[\frac{d^n}{dx^n}F(\alpha,\beta,\gamma:x)\right]_{x=0}$$
$$=\frac{(\alpha)_n(\beta)_n}{(\gamma)_n}\left[F(\alpha+n,\beta+n,\gamma+n:x)\right]_{x=0}$$
$$=\frac{(\alpha)_n(\beta)_n}{(\gamma)_n}F(\alpha+n,\beta+n,\gamma+n:0)=\frac{(\alpha)_n(\beta)_n}{(\gamma)_n}.1$$
$$=\frac{(\alpha)_n(\beta)_n}{(\gamma)_n}$$

10.3 An Integral Formula for a Hypergeometric Function

$$F(\alpha,\beta,\gamma:x)=\frac{\Gamma(\gamma)}{\Gamma(\beta)\Gamma(\gamma-\beta)}\int_0^1 t^{\beta-1}(1-t)^{\gamma-\beta-1}(1-xt)^{-\alpha}\,dt$$

or,

$$F(\alpha,\beta,\gamma:x)=\frac{1}{B(\beta,\gamma-\beta)}\int_0^1 t^{\beta-1}(1-t)^{\gamma-\beta-1}(-xt)^{-\alpha}\,dt,\text{ if } \gamma>\beta>0.$$

Proof: We know

$$F(\alpha,\beta,\gamma:x)=\sum_{n=0}^{\infty}\frac{(\alpha)_n(\beta)_n}{(\gamma)_n}\frac{x^n}{n!}=\sum_{n=0}^{\infty}(\alpha)_n\frac{\Gamma(\beta+n)}{\Gamma(\beta)}\frac{\Gamma(\Gamma(\gamma))}{\Gamma(\gamma+n)}\frac{x^n}{n!}$$
$$=\frac{\Gamma(\gamma)}{\Gamma(\beta)}\sum_{n=0}^{\infty}(\alpha)_n\frac{\Gamma(\beta+n)}{\Gamma(\gamma-\beta)}\frac{\Gamma(\gamma-\beta)}{\Gamma(\gamma+n)}\frac{x^n}{n!}$$
$$=\frac{\Gamma(\gamma)}{\Gamma(\beta)\Gamma(\gamma-\beta)}\sum_{n=0}^{\infty}(\alpha)_n\frac{\Gamma(\beta+n)\quad\Gamma(\gamma-\beta)}{\Gamma(\beta+n+\gamma-\beta)}\frac{x^n}{n!}$$
$$=\frac{\Gamma(\gamma)}{\Gamma(\beta)\Gamma(\gamma-\beta)}\sum_{n=0}^{\infty}(\alpha)_n\left\{\int_0^1 t^{\beta+n-1}(1-t)^{\gamma-\beta-1}dt\right\}\frac{x^n}{n!}$$
$$\left[\because\frac{\Gamma(\alpha)\quad\Gamma(\beta)}{\Gamma(\alpha+\beta)}=B(\alpha,\beta)=\int_0^1 t^{\alpha-1}(1-t)^{\beta-1}dt\right]$$

$$=\frac{\Gamma(\gamma)}{\Gamma(\beta)\Gamma(\gamma-\beta)}\left\{\int_0^1 t^{\beta-1}(1-t)^{\gamma-\beta-1}\right\}\left(\sum_{n=0}^{\infty}(\alpha)_n\frac{(xt)^n}{n!}\right)dt$$

$$=\frac{\Gamma(\gamma)}{\Gamma(\beta)\Gamma(\gamma-\beta)}\left\{\int_0^1 t^{\beta-1}(1-t)^{\gamma-\beta-1}\right\}(1-xt)^{-\alpha}\,dt \tag{10.6}$$

[since the general term of the expansion of $(1-xt)^{-\alpha}=\frac{(-\alpha)(-\alpha-1)(-\alpha-2)\ldots(-\alpha-n+1)}{n!}.(-xt)^n$

$$=(-1)^n\frac{(\alpha)(\alpha+1)(\alpha+2)\ldots(\alpha+n-1)}{n!}.(-1)^n(xt)^n=(\alpha)_n\frac{(xt)^n}{n!}].$$

Also,

$$B(\beta,\gamma-\beta)=\frac{\Gamma(\beta)\,\Gamma(\gamma-\beta)}{\Gamma(\gamma)};\quad\therefore\frac{\Gamma(\gamma)}{\Gamma(\beta)\Gamma(\gamma-\beta)}=\frac{1}{B(\beta,\gamma-\beta)}. \tag{10.7}$$

Using Equation (10.7) in Equation (10.6), we get

$$\therefore F(\alpha,\beta,\gamma:x)=\frac{1}{B(\beta,\gamma-\beta)}\left\{\int_0^1 t^{\beta-1}(1-t)^{\gamma-\beta-1}\right\}(1-xt)^{-\alpha}\,dt. \tag{10.8}$$

Hence, Equations (10.6) and (10.8) are the required results.

Gauss Theorem 10.3.1.

$$F(\alpha,\beta,\gamma:1)=\frac{\Gamma(\gamma)}{\Gamma(\gamma-\alpha)}\frac{\Gamma(\gamma-\beta-\alpha)}{\Gamma(\gamma-\beta)}$$

Proof: From Section 10.3 and putting $x=1$, *we get*

$$F(\alpha,\beta,\gamma:x)=\frac{\Gamma(\gamma)}{\Gamma(\beta)\Gamma(\gamma-\beta)}\int_0^1 t^{\beta-1}(1-t)^{\gamma-\beta-1}(1-xt)^{-\alpha}\,dt$$

For $x=1$,

$$F(\alpha,\beta,\gamma:1)=\frac{\Gamma(\gamma)}{\Gamma(\beta)\Gamma(\gamma-\beta)}\int_0^1 t^{\beta-1}(1-t)^{\gamma-\beta-1}(1-t)^{-\alpha}\,dt$$
$$=\frac{\Gamma(\gamma)}{\Gamma(\beta)\Gamma(\gamma-\beta)}\int_0^1 t^{\beta-1}(1-t)^{\gamma-\beta-\alpha-1}\,dt$$
$$=\frac{\Gamma(\gamma)}{\Gamma(\beta)\Gamma(\gamma-\beta)}\frac{\Gamma(\beta)\Gamma(\gamma-\beta-\alpha)}{\Gamma(\beta+\gamma-\beta-\alpha)}\quad\left[\int_0^1 t^{p-1}(1-t)^{q-1}\,dt=B(p,q)=\frac{\Gamma(p)\Gamma(q)}{\Gamma(p+q)}\right]$$
$$=\frac{\Gamma(\gamma)}{\Gamma(\gamma-\beta)}\frac{\Gamma(\gamma-\beta-\alpha)}{\Gamma(\gamma-\alpha)}.$$

Vandermonde's Theorem 10.3.2.

$$F(-n,\beta,\gamma:1)=\frac{(\gamma-\beta)_n}{(\gamma)_n}$$

Proof: By Gauss theorem, with $\alpha=-n$, *we get*

$$F(-n,\beta,\gamma:1)=\frac{\Gamma(\gamma)}{\Gamma(\gamma+n)}\frac{\Gamma(\gamma-\beta+n)}{\Gamma(\gamma-\beta)}$$
$$=\frac{\Gamma(\gamma)(\gamma-\beta+n-1)(\gamma-\beta+n-2)\ldots(\gamma-\beta)\Gamma(\gamma-\beta).}{(\gamma+n-1)(\gamma+n-2)\ldots.(\gamma)\Gamma(\gamma)\Gamma(\gamma-\beta)}$$
$$=\frac{(\gamma-\beta)_n}{(\gamma)_n}.$$

Kummer's Theorem 10.3.3.

$$F(\alpha,\beta,\beta-\alpha+1:-1)=\frac{\Gamma(\beta-\alpha+1)}{\Gamma(\beta+1)}\frac{\Gamma(\beta/2+1)}{\Gamma\left(\frac{\beta}{2}-\alpha+1\right)}$$

Proof: We know $F(\alpha,\beta,\gamma:x)=\frac{\Gamma(\gamma)}{\Gamma(\beta)\Gamma(\gamma-\beta)}\int_0^1 t^{\beta-1}(1-t)^{\gamma-\beta-1}(1-xt)^{-\alpha}\,dt$

Putting $x = -1\, and\, \gamma = \beta - \alpha + 1$, we get

$$F(\alpha,\beta,\beta-\alpha+1,:-1) = \frac{\Gamma(\beta-\alpha+1,)}{\Gamma(\beta)\Gamma(1-\alpha)}\int_0^1 t^{\beta-1}(1-t)^{-\alpha}(1+t)^{-\alpha}\,dt$$

$$= \frac{\Gamma(\beta-\alpha+1,)}{\Gamma(\beta)\Gamma(1-\alpha)}\int_0^1 t^{\beta-1}(1-t^2)^{-\alpha}\,dt.$$

Putting $t^2 = u, 2tdt = du \quad \therefore dt = \frac{1}{2\sqrt{u}}du$

$$= \frac{\Gamma(\beta-\alpha+1,)}{\Gamma(\beta)\Gamma(1-\alpha)}\int_0^1 u^{\frac{1}{2}(\beta-1)}(1-u)^{-\alpha}\frac{1}{2\sqrt{u}}du$$

$$= \frac{\Gamma(\beta-\alpha+1,)}{2\Gamma(\beta)\Gamma(1-\alpha)}\int_0^1 u^{(\beta/2-1)}(1-u)^{1-\alpha-1}du$$

$$= \frac{\Gamma(\beta-\alpha+1,)}{2\Gamma(\beta)\Gamma(1-\alpha)}\frac{\Gamma(\beta/2,)\Gamma(1-\alpha,)}{\Gamma(\beta/2-\alpha+1,)} \quad [\because \int_0^1 u^{p-1}(1-u)^{q-1}du = \frac{\Gamma(p,)\Gamma(q,)}{\Gamma(p+q,)}]$$

$$= \frac{\Gamma(\beta-\alpha+1,)}{\beta\Gamma(\beta)}\frac{\beta/2\Gamma(\beta/2,)}{\Gamma(\beta/2-\alpha+1,)}$$

$$= \frac{\Gamma(\beta-\alpha+1,)}{\Gamma(\beta+1)}\frac{\Gamma\left(\frac{\beta}{2}+1,\right)}{\Gamma(\beta/2-\alpha+1,)}.$$

Note: Sometimes, it is possible to reduce the solution to solving the second-order linear differential equation

$$x(1-x)y'' + \{\gamma - (\alpha+\beta+1)x\}y' - \alpha\beta\gamma = 0 \qquad (10.9)$$

in which $\alpha, \beta\, and\, \gamma$ are constants. The second-order linear differential equation, known as a *hypergeometric differential equation*, is of great importance because the solution of many problems of mathematical physics (study of the flow of compressible fluids) reduces to this type of equation. It can be shown that any ordinary linear differential equation of the second order, whose only singular points are regular singular points, one of which is the point at infinity, can be transformed into the form of Equation (10.9). For this, it is desirable to investigate the nature of the solutions of this equation.

Convergent of the Hypergeometric Series

The hypergeometric series has the form

$$1+\frac{\alpha\beta}{\gamma}x+\frac{\alpha(\alpha+1)\beta(\beta+1)}{1.2\gamma(\gamma+1)}x^2 +\frac{\alpha(\alpha+1)(\alpha+2)\beta(\beta+1)(\beta+2)}{1.2.3\gamma(\gamma+1)(\gamma+2)}x^3+\cdots\ldots \tag{10.10}$$

Let us denote the term having x^n

$$a_n=\frac{\alpha(\alpha+1)(\alpha+2)\beta(\beta+1)(\beta+2)\ldots(\beta+n-1).}{.\gamma(\gamma+1)(\gamma+2)\ldots(\gamma+n-1)..}\frac{x^n}{n!}$$

and

$$a_{n+1}=\frac{\alpha(\alpha+1)(\alpha+2)\beta(\beta+1)(\beta+2)\ldots(\beta+n-1)(\beta+n).}{.\gamma(\gamma+1)(\gamma+2)\ldots(\gamma+n-1).(\gamma+n).}\frac{x^{n+1}}{(n+1)!}$$

$$\therefore \frac{a_{n+1}}{a_n}=\frac{(\beta+n)}{(\gamma+n).}\frac{x^{n+1}}{(n+1)!}\frac{n!}{x^n}=\frac{(\alpha+n)(\beta+n)}{(\gamma+n).}\frac{x}{(n+1)}$$

$$=\frac{(1+\frac{\alpha}{n})(1+\frac{\beta}{n})}{(1+\frac{1}{n})(1+\frac{\gamma}{n})}x.$$

Here, $\frac{a_{n+1}}{a_n}\to x \text{ as } n\to\infty$.

Thus, for $-1<x<1$, the series converges. The solution of the hypergeometric equation converges within unit circle.

If $x=1$, we find that $\frac{a_n}{a_{n+1}}=1+\frac{1+\gamma-\alpha-\beta}{n}+O\left(\frac{1}{n^2}\right)$.

Hence, $\gamma>\alpha+\beta$; the series Equation (10.10) converges at $x=1$.

If $x=-1$, we find that $\frac{a_n}{a_{n+1}}=1-\frac{\alpha+\beta-1-\gamma}{n}+O\left(\frac{1}{n^2}\right)$.

Hence, $\gamma+1>\alpha+\beta$; the series Equation (10.10) converges at $x=-1$.

Relation: Integrals of a Hypergeometric Equation

Theorem 10.3.4. Prove that, in $0 \leq x \leq 1$,

$$F(\alpha,\beta,\gamma;x) = \frac{\Gamma(\gamma)}{\Gamma(\gamma-\alpha)} \frac{\Gamma(\gamma-\alpha-\beta)}{\Gamma(\gamma-\beta)} F(\alpha,\beta,\alpha+\beta-\gamma+1;1-x)$$
$$+\frac{\Gamma(\gamma)}{\Gamma(\alpha)} \frac{\Gamma(\alpha+\beta-\gamma)}{\Gamma(\beta)} (1-x)^{\gamma-\alpha-\beta} F(\gamma-\alpha,\gamma-\beta,\gamma-\alpha-\beta+1;1-x).$$

Proof: Since any solution can be linearly expressible in terms of two independent solutions, we have

$$F(\alpha,\beta,\gamma;x) = AF(\alpha,\beta,\alpha+\beta-\gamma+1;1-x)$$
$$+B(1-x)^{\gamma-\alpha-\beta} F(\gamma-\alpha,\gamma-\beta,\gamma-\alpha-\beta+1;1-x) \tag{10.11}$$

where A *and* B are constants. All the series are convergent in

$$0 \leq x \leq 1, \textit{if } (\alpha+b) < \gamma < 1.(\alpha+b) < \gamma < 1$$

Putting $x = 1$ *in* Equation (10.11), we get

$$F(\alpha,\beta,\gamma;1) = AF(\alpha,\beta,\alpha+\beta-\gamma+1;0)$$
$$= A\left(1+\frac{\alpha\beta}{\alpha+\beta-\gamma+1}\times 0+0+\cdots\right) = A.$$

Putting $x = 0$ *in* Equation (10.11), we get

$$F(\alpha,\beta,\gamma;0) = AF(\alpha,\beta,\alpha+\beta-\gamma+1;1)$$
$$+BF(\gamma-\alpha,\gamma-\beta,\gamma-\alpha-\beta+1;1).$$

$\because F(\alpha,\beta,\gamma;0) = 1$, and we have

$$1 = AF(\alpha,\beta,\alpha+\beta-\gamma+1;1) + BF(\gamma-\alpha,\gamma-\beta,\gamma-\alpha-\beta+1;1).$$

Substituting the value of A, we get

$$1 = F(\alpha,\beta,\gamma;1)F(\alpha,\beta,\alpha+\beta-\gamma+1;1) \\ + BF(\gamma-\alpha,\gamma-\beta,\gamma-\alpha-\beta+1;1).$$

Using $F(\alpha,\beta,\gamma;1) = \frac{\Gamma(\gamma)}{\Gamma(\gamma-\alpha)} \frac{\Gamma(\gamma-\alpha-\beta)}{\Gamma(\gamma-\beta)}$ in the above equation, we obtain

$$\frac{\Gamma(\gamma)}{\Gamma(\gamma-\alpha)} \frac{\Gamma(\gamma-\alpha-\beta)}{\Gamma(\gamma-\beta)} \frac{\Gamma(1-\gamma)}{\Gamma[1-(\gamma-\alpha)]} \frac{\Gamma[1-(\gamma-\alpha-\beta)]}{\Gamma[1-(\gamma-\beta)]} \\ + B\frac{\Gamma(1-\gamma)}{\Gamma(1-\alpha)} \frac{\Gamma[1-(\alpha+\beta-\gamma)]}{\Gamma(1-\beta)} = 1.$$

$$\therefore \frac{\Gamma(\gamma)}{\Gamma(\gamma-\alpha)\Gamma[1-(\gamma-\beta)]} \frac{\Gamma(\gamma-\alpha-\beta)}{\Gamma(\gamma-\beta)]} \frac{\Gamma(1-\gamma)}{} \frac{\Gamma[1-(\gamma-\alpha-\beta)]}{\Gamma[1-(\gamma-\beta)]} \\ + B\frac{\Gamma(1-\gamma)}{\Gamma(1-\alpha)} \frac{\Gamma[1-(\alpha+\beta-\gamma)]}{\Gamma(1-\beta)} = 1.$$

We have $F(1-\gamma)F(\gamma) = \frac{\pi}{\sin(\pi\gamma)}$.
Then,

$$1 = \frac{\dfrac{\pi}{\sin(\pi\gamma)} \quad \dfrac{\pi}{\sin(\pi(\gamma-\alpha-\beta))}}{\dfrac{\pi}{\sin(\pi(\gamma-\alpha))} \quad \dfrac{\pi}{\sin(\pi(\gamma-\beta))}} + B\frac{\Gamma(1-\gamma)}{\Gamma(1-\alpha)} \frac{\Gamma[1-(\alpha+\beta-\gamma)]}{\Gamma(1-\beta)}$$

$$= -\frac{\sin(\pi(\gamma-\alpha))\sin(\pi(\gamma-\beta))}{\sin(\pi\gamma)\sin(\pi(\gamma-\alpha-\beta))} + B\frac{\Gamma(1-\gamma)}{\Gamma(1-\alpha)} \frac{\Gamma[1-(\alpha+\beta-\gamma)]}{\Gamma(1-\beta)}$$

$$B\frac{\Gamma(1-\gamma)}{\Gamma(1-\alpha)} \frac{\Gamma[1-(\alpha+\beta-\gamma)]}{\Gamma(1-\beta)} = 1 + \frac{sin(\pi(\gamma-\alpha))sin(\pi(\gamma-\beta))}{sin(\pi\gamma)sin(\pi(\gamma-\alpha-\beta))}$$

$$= \frac{sin(\pi\gamma)sin(\pi(\gamma-\alpha-\beta)) + sin(\pi(\gamma-\alpha))sin(\pi(\gamma-\beta))}{sin(\pi\gamma)sin(\pi(\gamma-\alpha-\beta))}$$

$$= \frac{1}{2}\frac{cos\pi(\alpha+\beta-2\gamma) - cos\pi(\alpha+\beta) + cos\pi(\beta-\alpha) - cos\pi(2\gamma-\alpha-\beta)}{sin(\pi\gamma)sin(\pi(\alpha+\beta-\gamma))}$$

or, $$B\frac{\Gamma(1-\gamma)}{\Gamma(1-\alpha)}\frac{\Gamma\left[1-(\alpha+\beta-\gamma)\right]}{\Gamma(1-\beta)}=\frac{\cos\pi(\beta-\alpha)-\cos\pi(\alpha+\beta)}{2\sin(\pi\gamma)\sin(\pi(\alpha+\beta-\gamma))}$$

$$=\frac{\sin\pi\beta\,\sin\pi\alpha}{\sin(\pi\gamma)\sin(\pi(\alpha+\beta-\gamma))}$$

$$\frac{\dfrac{\pi}{\Gamma(\alpha)\Gamma(1-\alpha)}\;\dfrac{\pi}{\Gamma(\alpha)\Gamma(1-\alpha)}}{\dfrac{\pi}{\Gamma(\gamma)\Gamma(1-\gamma)}\;\dfrac{\pi}{\Gamma(\alpha+\beta-\gamma)\Gamma(1-\alpha-\beta+\gamma)}}\quad\left[\because \sin(\pi z)=\frac{\pi}{F(1-z)F(z)}\right]$$

or, $$B\frac{\Gamma(1-\gamma)}{\Gamma(1-\alpha)}\frac{\Gamma\left[1-(\alpha+\beta-\gamma)\right]}{\Gamma(1-\beta)}$$

$$=\frac{\Gamma(\gamma)\Gamma(1-\gamma)\Gamma(\alpha+\beta-\gamma)\Gamma(1-\alpha-\beta+\gamma)}{\Gamma(\alpha)\Gamma(1-\alpha)\Gamma(\beta)\Gamma(1-\beta)}.$$

$$\therefore B=\frac{\Gamma(\gamma)\Gamma(\alpha+\beta-\gamma)}{\Gamma(\alpha)\Gamma(\beta)}.$$

Putting the values of A *and* B *in* Equation (10.11), *we get*

$$F(\alpha,\beta,\gamma;x)=\frac{\Gamma(\gamma)\Gamma(\gamma-\alpha-\beta)}{\Gamma(\gamma-\alpha)\Gamma(\gamma-\beta)}F(\alpha,\beta,\alpha+\beta-\gamma+1;1-x)$$

$$+\frac{\Gamma(\gamma)\Gamma(\alpha+\beta-\gamma)}{\Gamma(\alpha)\Gamma(\beta)}(1-x)^{\gamma-\alpha-\beta}\,F(\gamma-\alpha,\gamma-\beta,\gamma-\alpha-\beta+1;1-x).$$

(10.12)

Replacing x *by* $\frac{1}{x}$ in Equation (10.12), we obtain

$$F\left(\alpha,\beta,\gamma;\frac{1}{x}\right)=\frac{\Gamma(\gamma)\Gamma(\gamma-\alpha-\beta)}{\Gamma(\gamma-\alpha)\Gamma(\gamma-\beta)}F\left(\alpha,\beta,\alpha+\beta-\gamma+1;1-\frac{1}{x}\right)$$

$$+\frac{\Gamma(\gamma)\Gamma(\alpha+\beta-\gamma)}{\Gamma(\alpha)\Gamma(\beta)}(1-\frac{1}{x})^{\gamma-\alpha-\beta}F\left(\gamma-\alpha,\gamma-\beta,\gamma-\alpha-\beta+1;1-\frac{1}{x}\right).$$

We know that $F\left(\alpha,\beta,\gamma;1-\frac{1}{x}\right)=(x)^{\alpha}F(\alpha,\gamma-\beta,\gamma;1-x)$.

$$\therefore F\left(\alpha,\beta,\gamma;\frac{1}{x}\right)=\frac{\Gamma(\gamma)\Gamma(\gamma-\alpha-\beta)}{\Gamma(\gamma-\alpha)\Gamma(\gamma-\beta)}(x)^{\alpha}F(\alpha,\alpha-\gamma+1,\alpha+\beta-\gamma+1;1-x)$$
$$+\frac{\Gamma(\gamma)\Gamma(\alpha+\beta-\gamma)}{\Gamma(\alpha)\Gamma(\beta)}(x)^{\beta}(x-1)^{\gamma-\alpha-\beta}F(\gamma-\alpha,1-\alpha,\gamma-\alpha-\beta;1-x).$$
(10.13)

where $1<x<2$ and $1>\gamma>(\alpha+\beta)$. These relations are typical for a large x, *which* exist between the solutions of the hypergeometric equation

$$x(1-x)y''+\{\gamma-(\alpha+\beta+1)x\}y'-\alpha\beta\gamma=0.$$

If we change the independent variable in this equation to any one of $1-x,\frac{1}{x},\frac{1}{1-x},\frac{x-1}{x},\frac{x}{x-1}$, the equation transforms to one of the same type. Equation (10.13) therefore has different types of solutions, $y={}_2F_1(\alpha,\beta,\gamma;x)$ *and* $y=x^{1-\gamma}{}_2F_1(\alpha-\gamma+1,\beta-\gamma+1,2-\gamma;x)$. Each is convergent within the unit circle.

10.4 Transformation of $F(\alpha,\beta,\gamma;x)$

We shall consider the following two types of transformations: simple transformation and quadratic transformation.

10.4.1 Simple Transformations

We have to show that $F\left(\alpha,\gamma-\beta,\gamma;\frac{x}{1-x}\right)$ is a simple transformation of $F(\alpha,\beta,\gamma;x)$. We have

$$(1-x)^{-\alpha}F\left(\alpha,\gamma-\beta,\gamma;\frac{x}{1-x}\right)=\sum_{n=0}^{\infty}\frac{(\alpha)_n(\gamma-\beta)_n(-1)^n x^n}{(\gamma)_n n!(1-x)^{n+\alpha}}$$

$$=\sum_{n=0}^{\infty}\frac{(\alpha)_n(\gamma-\beta)_n(-1)^n x^n}{(\gamma)_n n!}\sum_{m=0}^{\infty}\frac{(\alpha+n)_m}{m!}x^m$$

$$\left[\because(1-x)^{-n-\alpha}=\sum_{m=0}^{\infty}\frac{(\alpha+n)_m x^m}{m!}\right].$$

Since $(\alpha+n)_m = \dfrac{(\alpha)_{m+n}}{(\alpha)_n}$,

$$\therefore (1-x)^{-\alpha} F\left(\alpha,\gamma-\beta,\gamma;\frac{x}{1-x}\right) = \sum_{n=0}^{\infty}\sum_{m=0}^{\infty} \frac{(\alpha)_n(\gamma-\beta)_n(\alpha)_{m+n}(-1)^n x^{n+m}}{(\alpha)_n(\gamma)_n n!m!}$$

$$= \sum_{n=0}^{\infty}\sum_{m=0}^{n} \frac{(\gamma-\beta)_n(\alpha)_{m+n}(-1)^n x^{n+m}}{(\gamma)_n n!m!}$$

$$= \sum_{n=0}^{\infty}\sum_{m=0}^{n} \frac{(\gamma-\beta)_n(\alpha)_m(-1)^n x^m}{(\gamma)_n n!(m-n)!}$$

$$= \sum_{n=0}^{\infty}\sum_{m=0}^{n} \frac{(-n)!(\gamma-\beta)_n(\alpha)_m x^m}{(\gamma)_n n!m!} \left[\because (\text{m-n})! = \frac{(-1)^n m!}{(-n)_m}, 0 \le n \le m\right].$$

Since $\sum_{m=0}^{n} \dfrac{(-n)!(\gamma-\beta)_n}{(\gamma)_n n!}$ is a terminating hypergeometric series $F(-m,\gamma-\beta,\gamma;1)$, we can write

$$(1-x)^{-\alpha} F\left(\alpha,\gamma-\beta,\gamma;\frac{x}{1-x}\right) = \sum_{m=0}^{\infty} F(-m,\gamma-\beta,\gamma;1)\frac{(\alpha)_m x^m}{m!}.$$

We know $F(-n,\beta,\gamma;\alpha) = \dfrac{(\gamma-\beta)_n}{(\alpha)_n}$. We obtain

$$(1-x)^{-\alpha} F\left(\alpha,\gamma-\beta,\gamma;\frac{x}{1-x}\right) = \sum_{m=0}^{\infty} \frac{\Gamma(\gamma)\Gamma(\beta+m)}{\Gamma(\gamma+m)\Gamma(\beta)}\frac{(\alpha)_m x^m}{m!}$$

$$= \sum_{m=0}^{\infty} \frac{(\alpha)_m(\beta)_m}{(\gamma)_m m!} x^m = F(\alpha,\beta,\gamma;x).$$

$$\therefore (1-x)^{-\alpha} F\left(\alpha,\gamma-\beta,\gamma;\frac{x}{1-x}\right) = F(\alpha,\beta,\gamma;x), \tag{10.14}$$

which is valid under the condition $|x| < 1$ *and* $\left|\dfrac{x}{1-x}\right| < 1$.

Applying the transformation Equation (10.14) to the hypergeometric function, we get

$$F(\alpha,\beta,\gamma;x)=(1-x)^{-\alpha}(1-x)^{\gamma-\beta}F(\gamma-\alpha,\gamma-\beta,\gamma;x)$$
$$=(1-x)^{\gamma-\alpha-\beta}F(\gamma-\alpha,\gamma-\beta,\gamma;x),$$

which is valid when $|x|<1$.

There exist 24 such solutions, obtainable by suitable transformations, which are known as *Kummer's solutions* of the hypergeometric differential equation.

10.4.2 Quadratic Transformations

In a hypergeometric differential equation, we substitute $\gamma=2\beta$, and we get

$$x(1-x)\frac{d^2y}{dx^2}+\{2\beta-(\alpha+\beta+1)x\}\frac{dy}{dx}-\alpha\beta y=0. \quad (10.15)$$

One of its solutions is $y=F(\alpha,\beta,2\beta;x)$.

Let us transfer the variable x by $x=\frac{4t}{(1+t)^2}$

(where $\frac{dx}{dt}=\frac{4(1+t)^2-4t.2(1+t)}{1+t)^4}=\frac{4(1-t)}{(1+t)^3}$)

$$\frac{dx}{dy}=\frac{4(1+t)^2-4t.2(1+t)}{(1+t)^4}\frac{dt}{dy}=\frac{4(1+t)\{1+t-2t\}}{(1+t)^4}\frac{dt}{dy}=\frac{4(1-t)}{(1+t)^3}\frac{dt}{dy}$$

$$\therefore\frac{dy}{dx}=\frac{(1+t)^3}{4(1-t)}\frac{dy}{dt} \text{ and } \frac{d^2y}{dx^2}=\frac{d}{dx}\left(\frac{dz}{dx}\right)=\frac{d}{dx}\left[\frac{(1+t)^3}{4(1-t)}\frac{dy}{dt}\right]=\frac{d}{dt}\left[\frac{(1+t)^3}{4(1-t)}\frac{dy}{dt}\right]\frac{dt}{dx}$$
$$=\frac{(1+t)^6}{16(1-t)^2}\frac{d^2y}{dt^2}+\frac{(1+t)^5(4-2t}{16(1-t)^3}\frac{dy}{dt}.$$

Substituting these in Equation (10.15), we obtain

$$t(1-t)(1+t)^2\frac{d^2y}{dt^2}+2(1+t)\left(\beta-2\alpha t+\beta t^2-t^2\right)\frac{dy}{dt}-4\alpha\beta\left(1-t\right)y=0. \quad (10.16)$$

One of its solutions is

$$y = F\left(\alpha, \beta, 2\beta; \frac{4t}{(1+t)^2}\right). \tag{10.17}$$

If we replace $y = (1+t)^{2\alpha}$ in Equation (ii), then

$$\frac{dy}{dx} = (1+t)^{2\alpha}\frac{dy}{dt} + 2\alpha(1+t)^{2\alpha-1}y$$

$$\frac{d^2y}{dx^2} = (1+t)^{2\alpha}\frac{d^2y}{dt^2} + 4\alpha(1+t)^{2\alpha-1}\frac{dy}{dt} + 2\alpha(2\alpha-1)(1+t)^{2\alpha-2}y.$$

We obtain

$$t(1-t^2)\frac{d^2y}{dt^2} + 2[\beta - (2\alpha - \beta + 1)t^2]\frac{dy}{dt} - 2\alpha t(1 + 2\alpha - 2\beta)y = 0, \tag{10.18}$$

one of whose solutions can be written as

$$y = (1+t)^{-2\alpha}F\left(\alpha, \beta, 2\beta; \frac{4t}{(1+t)^2}\right). \tag{10.19}$$

We observe that Equation (10.15) remains unchanged by changing x by $-x$. If we substitute $v = t^2$ in that equation, we get

$$v(1-v)\frac{d^2y}{dv^2} + \left[\beta + \frac{1}{2} - \left(2\alpha - \beta + \frac{3}{2}\right)v\right]\frac{dy}{dv} - \alpha(\alpha - 2\beta + \frac{1}{2})y = 0.$$

Clearly, this equation is identical with Equation (10.18) and under the condition $|v| < 1$, we can write its general solution in the form

$$y = AF\left(\alpha, \alpha - \beta + \frac{1}{2}, \beta + \frac{1}{2}; v\right)$$
$$+ Bv^{\frac{1}{2}-\beta}F\left(\alpha - \beta + \frac{1}{2}, \alpha + 1 - 2\beta, \frac{3}{2} - \beta; v\right). \tag{10.20}$$

From the above discussion, it is obvious that for some constants A *and* B, when $|t|<1$ and $\left|\frac{4t}{(1+t)^2}\right|<1$ and 2β is neither zero nor a negative integer, the equation below holds:

$$(1+t)^{-2\alpha} F(\alpha,\beta,2\beta;\frac{4t}{(1+t)^2})$$
$$= AF\left(\alpha,\alpha-\beta+\frac{1}{2},\beta+\frac{1}{2};t^2\right)$$
$$+Bt^{1-2\beta} F\left(\alpha-\beta+\frac{1}{2},\alpha+1-2\beta,\frac{3}{2}-\beta;t^2\right). \tag{10.21}$$

Putting $t=0$ in this equation, we observe that the last term is not analytic due to the presence of the factor $t^{1-2\beta}$. Hence, B *must* be zero. We get

$$AF\left(\alpha,\alpha-\beta+\frac{1}{2},\beta+\frac{1}{2};t^2\right)=(1+t)^{-2\alpha} F\left(\alpha,\beta,2\beta;\frac{4t}{(1+t)^2}\right). \tag{10.22}$$

Again, putting $t=0$ *in* Equation (10.22), we get $A=1$.

Thus, if 2β is neither zero nor a negative integer, and if both $|t|<1$ and $\left|\frac{4t}{(1+t)^2}\right|<1$ hold, then

$$(1+t)^{-2\alpha} F\left(\alpha,\beta,2\beta;\frac{4t}{(1+t)^2}\right)=F\left(\alpha,\alpha-\beta+\frac{1}{2},\beta+\frac{1}{2};t^2\right). \tag{10.23}$$

10.5 Hypergeometric Equation

The differential equation of the form

$$x(1-x)y''+\{\gamma-(\alpha+\beta+1)x\}y'-\alpha\beta\gamma=0$$

is known as a *Gauss equation* or hypergeometric equation where α,β,γ are parameters.

Solution: The given equation is

$$x(1-x)y''+\{\gamma-(\alpha+\beta+1)x\}y'-\alpha\beta\gamma=0$$

or,
$$y''+\frac{\{\gamma-(\alpha+\beta+1)x\}}{x(1-x)}y'-\frac{\alpha\beta}{x(1-x)}y=0. \tag{10.24}$$

The points $x=0$ *and* $x=1$ are singular points of the differential equation (10.24). Let us put $x=\frac{1}{t}$ in Equation (10.24), we get

$$t^2(t-1)\frac{d^2y}{dt^2}+\{2t(t-1)-t[\gamma t-(\alpha+\beta+1)]\}\frac{dy}{dt}-\alpha\beta y=0. \tag{10.24a}$$

This equation has singularity at $t=0$ *and* $t=1$. So, Equation (10.24a) has singularity at $x=1$ *and* $x=\infty$. Therefore, Equation (10.24) has three singularities at $x=0$, $x=1$, *and* $x=\infty$.

Comparing the above equation with

$$y''+P(x)y'+Q(x)y=0,$$

Here, $P(x)=\frac{\{\gamma-(\alpha+\beta+1)x\}}{x(1-x)}$ and $Q(x)=-\frac{\alpha\beta}{x(1-x)}$.

$xP(x)=\frac{\{\gamma-(\alpha+\beta+1)x\}}{(1-x)}$ *and* $x^2Q(x)=-\frac{\alpha\beta\gamma x}{(1-x)}$ are analytic at $x=0$.

So, $x=0$ is a regular singular point of Equation (10.24).

Let the series solution of Equation (10.24a)

$$y=\sum_{n=0}^{\infty}c_nx^{k+n} \text{ where } c_0\neq 0. \tag{10.25}$$

Note: We can also write this equation,by simple manipulation to the following forms:

$$y''+\frac{\{\gamma-(\alpha+\beta+1)x\}}{x(1-x)}y'-\frac{\alpha\beta}{x(1-x)}y=0$$

or,

$$x^2y'' + x(1+x+x^2+\cdots)\{\gamma-(\alpha+\beta+1)x\}y' - \alpha\beta x(1+x+x^2+\cdots) = 0 \tag{i}$$

and,
$$(x-1)^2y'' - (x-1)\{\gamma-\alpha-\beta-1-\gamma(x-1)+..\}y' + \alpha\beta(x-1)\{1-(x-1)+..\}y = 0 \tag{ii}$$

and,
$$y'' + \left[\frac{1+\alpha+\beta}{x} + \frac{\gamma-\alpha-\beta-1}{x^2}(1+\frac{1}{x}+\frac{1}{2!x^2}+\cdots)\right]y' - \frac{\alpha\beta}{x^2}(1+\frac{1}{x}+\frac{1}{2!x^2}+\cdots)y = 0. \tag{iii}$$

It follows that the points $x=0, x=1,$ *and* $x=\infty$ are the regular singular points of Equation (10.24a).

Differentiating with respect to x, we get

$$y' = \sum_{n=0}^{\infty}(k+n)c_n x^{k+n-1}$$
$$y'' = \sum_{n=0}^{\infty}(k+n)(k+n-1)c_n x^{k+n-2}. \tag{10.26}$$

Substituting these values in Equation (10.24), we get

$$x(1-x)\sum_{n=0}^{\infty}(k+n)(k+n-1)c_n x^{k+n-2} + \{\gamma-(\alpha+\beta+1)x\}\sum_{n=0}^{\infty}(k+n)c_n x^{k+n-1} - \alpha\beta\sum_{n=0}^{\infty}c_n x^{k+n} = 0$$

or,
$$\sum_{n=0}^{\infty}(k+n)(k+n-1)c_n x^{k+n-1} - \sum_{n=0}^{\infty}(k+n)(k+n-1)c_n x^{k+n} + \gamma\sum_{n=0}^{\infty}(k+n)c_n x^{k+n-1} - (\alpha+\beta+1)\sum_{n=0}^{\infty}(k+n)c_n x^{k+n} - \alpha\beta\sum_{n=0}^{\infty}c_n x^{k+n} = 0$$

or, $$\sum_{n=0}^{\infty} c_n\left[(k+n)(k+n-1)+\gamma(k+n)\right]x^{k+n-1}$$
$$-\sum_{n=0}^{\infty} c_n\left[(k+n)(k+n-1)+(\alpha+\beta+1)(k+n)+\alpha\beta\right]x^{k+n}=0$$

or, $$\sum_{n=0}^{\infty} c_n(k+n)(k+n+\gamma-1)x^{k+n-1}$$
$$-\sum_{n=0}^{\infty} c_n(k+n+\alpha)(k+n+\beta)]x^{k+n}=0. \tag{10.27}$$

Equating the coefficient of the lowest power of x, i.e., x^{k-1}, we get

$$c_0 k(k+\gamma-1)=0$$

Since $c_0 \neq 0$, $k(k+\gamma-1)=0$, which is an indicial equation.
The roots of indicial equation are

$$k=0, k=1-\gamma,$$

which are unequal and do not differ by an integer.

To obtain the recurrence relation, we equate to zero the coefficient of x^{k+n-1}:

$$c_n(k+n)(k+n+\gamma-1)-c_{n-1}(k+n-1+\alpha)(k+n-1+\beta)=0$$
$$c_n=\frac{(k+n-1+\alpha)(k+n-1+\beta)}{(k+n)(k+n+\gamma-1)}c_{n-1}. \tag{10.28}$$

Case I(a): When $k=0$.
Putting $n=1,2,3,\ldots$ successively in Equation (10.28):

$$c_1=\frac{(1-1+\alpha)(1-1+\beta)}{(1)(1+\gamma-1)}c_0=\frac{\alpha\beta}{\gamma}c_0$$
$$c_2=\frac{(\alpha+1)(\beta+1)}{2(\gamma+1)}c_1=\frac{(\alpha+1)(\beta+1)}{2(\gamma+1)}\frac{\alpha\beta}{\gamma}c_0=\frac{\alpha(\alpha+1)\beta(\beta+1)}{1.2\gamma(\gamma+1)}c_0$$

$$c_3 = \frac{(\alpha+2)(\beta+2)}{3(\gamma+2)} c_2 = \frac{\alpha(\alpha+1)(\alpha+2)\beta(\beta+1)(\beta+2)}{1.2.3\gamma(\gamma+1)(\gamma+2)} c_0,$$

and so on.

Putting these values of c's in Equation (10.25) for = 0 and replacing c_0 *by* a, we get

$$y = a\left[1 + \frac{\alpha\beta}{\gamma}x + \frac{\alpha(\alpha+1)\beta(\beta+1)}{1.2\gamma(\gamma+1)}x^2 + \frac{\alpha(\alpha+1)(\alpha+2)\beta(\beta+1)(\beta+2)}{1.2.3\gamma(\gamma+1)(\gamma+2)}x^3 + \cdots\ldots\right] \tag{10.29}$$

If we take $a = 1$ in Equation (10.29), the right-hand side of Equation (10.29) is called a hypergeometric series and is represented by ${}_2F_1(\alpha,\beta,\gamma,x)$. This is ${}_2F_1(\alpha,\beta,\gamma,x)$, the solution of Equation (10.24).

$$\therefore y = {}_2F_1(\alpha,\beta,\gamma,x).$$

This series is known as a *hypergeometric series of the first kind* and is denoted by ${}_2F_1(\alpha,\beta,\gamma,x)$.

Case I(b): When $k = 1-\gamma, c_n$ in Equation (10.28) reduces to

$$\begin{aligned} c_n &= \frac{(1-\gamma+n-1+\alpha)(1-\gamma+n-1+\beta)}{(1-\gamma+n)(1-\gamma+n+\gamma-1)} c_{n-1} \\ &= \frac{(-\gamma+n+\alpha)(-\gamma+n+\beta)}{n(1-\gamma+n)} c_{n-1} = \frac{(\alpha'+n-1)(\beta'+n-1)}{n(\gamma'+n-1)} c_{n-1} \end{aligned} \tag{10.30}$$

where, $$\gamma' = 2-\gamma,\ \alpha' = \alpha-\gamma+1,\ \beta' = \beta-\gamma+1. \tag{10.31}$$

Putting $n = 1,2,3,\ldots$ successively in Equation (10.30) gives

$$c_1 = \frac{\alpha'\beta'}{1.\gamma'} c_0,\ c_2 = \frac{(\alpha'+1)(\beta'+1)}{1.2.(\gamma'+1)} \frac{\alpha'\beta'}{1.\gamma'} c_0 = \frac{\alpha'(\alpha'+1)\beta'(\beta'+1)}{1.2.\gamma'(\gamma'+1)} c_0$$

$$c_3 = \frac{\alpha'(\alpha'+1)(\alpha'+2)\beta'(\beta'+1)(\beta'+2)}{1.2.3\gamma'(\gamma'+1)(\gamma'+2)}c_0,$$

and so on.

Putting $k = 1-\gamma$ the values of c's and replacing c_0 by b in Equation (10.25), we get

$$y = bx^{1-\gamma}\left[1+\frac{\alpha'\beta'}{1.\gamma'}x+\frac{\alpha'(\alpha'+1)\beta'(\beta'+1)}{1.2.\gamma'(\gamma'+1)}x^2+\frac{\alpha'(\alpha'+1)(\alpha'+2)\beta'(\beta'+1)(\beta'+2)}{1.2.3\gamma'(\gamma'+1)(\gamma'+2)}x^3+\cdots.\right] \quad (10.32)$$

If we take $b=1$ *in* Equation (10.32), the series on the right-hand side of Equation (10.32) would be

$$y = x^{1-\gamma}\left[1+\frac{\alpha'\beta'}{1.\gamma'}x+\frac{\alpha'(\alpha'+1)\beta'(\beta'+1)}{1.2.\gamma'(\gamma'+1)}x^2+\frac{\alpha'(\alpha'+1)(\alpha'+2)\beta'(\beta'+1)(\beta'+2)}{1.2.3\gamma'(\gamma'+1)(\gamma'+2)}x^3+\ldots\right]$$

$$= x^{1-\gamma}\,{}_2F_1(\alpha',\beta',\gamma',x),$$

i.e., $y = x^{1-\gamma}\,{}_2F_1(\alpha',\beta',\gamma',x) = x^{1-\gamma}\,{}_2F_1(\alpha-\gamma+1,\beta-\gamma+1,2-\gamma,x)$ (10.33)

is another independent solution of Equation (10.24).

The series on the right-hand side of Equation (10.33) is known as the *hypergeometric series of the second kind* and is denoted by ${}_2F_1(\alpha-\gamma+1,\beta-\gamma+1,2-\gamma,x)$.

Hence, the general solution is

$$y = a\,{}_2F_1(\alpha,\beta,\gamma,x)+bx^{1-\gamma}\,{}_2F_1(\alpha-\gamma+1,\beta-\gamma+1,2-\gamma,x).$$

Case II: When $x=1$, putting $x=1-t$, Equation (10.24) reduces to

$$t(1-t)\frac{d^2y}{dt^2}+\left[(\alpha+\beta-\gamma+1)-(\alpha+\beta+1)t\right]\frac{dy}{dt}-\alpha\beta y=0,$$

which is identical with Equation (10.24), where γ has been replaced by $\alpha+\beta-\gamma+1$ *and* x *by* $t=(1-x)$. Hence, it follows from Equations (10.29) and (10.33) that the required solution near $x=1$ is

$$\begin{aligned} y &= {}_2F_1(\alpha,\beta,\alpha+\beta-\gamma+1,x) \\ &= (1-x)^{1-\gamma}\,{}_2F_1(\gamma-\alpha,\gamma-\beta,\gamma-\alpha-\beta+1:1-x). \end{aligned}$$

Case III: Solution near $x=\infty$.

Let us put $y=\sum_{n=0}^{\infty} c_n x^{-k-n}$ in Equation (10.24), then

$$c_{n+1}=\frac{(n+\alpha)(n+\alpha-\gamma+1)}{(n+1)(n+\alpha-\beta+1)}c_n$$

from which
$$\begin{aligned} c_1 &= \frac{\alpha(\alpha-\gamma+1)}{1.(\alpha-\beta+1)}c_0, \\ c_2 &= \frac{(\alpha+1)(\alpha-\gamma+2)}{2.(\alpha-\beta+1)}\frac{\alpha(\alpha-\gamma+1)}{1.(\alpha-\beta+1)}c_0 \\ &= \frac{\alpha(\alpha+1)(\alpha-\gamma+2)}{2.1.(\alpha-\beta+1)}\frac{(\alpha-\gamma+1)}{1.(\alpha-\beta+1)}c_0. \\ \therefore c_n &= \frac{(\alpha)_n(\alpha-\gamma+1)_n}{n!.(\alpha-\beta+1)_n}c_0. \end{aligned}$$

Taking $c_0=1$, we obtain the solution

$$y_1=x^{-\alpha}\,{}_2F_1\left(\alpha,\alpha-\gamma+1,\alpha-\beta+1:\frac{1}{x}\right).$$

Also, from symmetry, it follows that the other solution is

$$y_2=x^{-\beta}\,{}_2F_1\left(\beta,\beta-\gamma+1,\beta-\alpha+1:\frac{1}{x}\right).$$

The general solution is $y = Ay_1 + By_2$

$$= Ax^{-\alpha}\,{}_2F_1\left(\alpha,\alpha-\gamma+1,\alpha-\beta+1:\frac{1}{x}\right)$$
$$+ Bx^{-\beta}\,{}_2F_1\left(\beta,\beta-\gamma+1,\beta-\alpha+1:\frac{1}{x}\right).$$

10.6 Confluent Hypergeometric Series

The hypergeometric differential equation is

$$x(1-x)\frac{d^2y}{dx^2}+\left[\gamma-(1+\alpha+\beta)x\right]\frac{dy}{dx}-\alpha\beta y=0. \tag{10.34}$$

If we replace x by $\frac{x}{\beta}$, the above equation takes the form

$$x\left(1-\frac{x}{\beta}\right)\frac{d^2y}{dx^2}+\left\{\gamma-\left(1+\frac{\alpha+1}{\beta}\right)x\right\}\frac{dy}{dx}-\alpha y=0. \tag{10.35}$$

Its solution will be

$$\lim_{\beta\to\infty} F(\alpha,\beta,\gamma;\frac{x}{\beta}). \tag{10.36}$$

When $\beta \to \infty$, Equation (ii) reduces to

$$x\frac{d^2y}{dx^2}+(\gamma-x)\frac{dy}{dx}-\alpha y=0. \tag{10.37}$$

Equation (10.37) is known as a *confluent hypergeometric equation or Kummer's equation.*

Now, $$\lim_{\beta\to\infty}\frac{(\beta)_r}{\beta^r}=\frac{\beta(\beta+1)(\beta+2)\ldots.(\beta+r-1)}{\beta\beta\beta\ldots.r\,times}$$
$$=\lim_{\beta\to\infty}\left(1+\frac{1}{\beta}\right)\left(1+\frac{2}{\beta}\right)\ldots.\left(1+\frac{r-1}{\beta}\right)=1.$$

$$\therefore \lim_{\beta\to\infty}F\left(\alpha,\beta,\gamma;\frac{x}{\beta}\right)=\lim_{\beta\to\infty}\sum_{n=0}^{\infty}\frac{(\alpha)_n(\beta)_n}{n!(\gamma)_n}\left(\frac{x}{\beta}\right)^n$$
$$=\sum_{n=0}^{\infty}\frac{(\alpha)_n(\beta)_n}{n!(\gamma)_n}x^n\lim_{\beta\to\infty}\frac{(\beta)_n}{\beta^n}=\sum_{n=0}^{\infty}\frac{(\alpha)_n}{(\gamma)_n}\frac{x^n}{n!}.$$

The function Equation (10.37) reduces to the series

$$\sum_{n=0}^{\infty}\frac{(\alpha)_n}{(\gamma)_n}\frac{x^n}{n!}, \tag{10.38}$$

and the series is denoted by $F(\alpha,\gamma;x)$ and is called a *confluent hypergeometric function.*

The point $x=0$ is a regular singular point of the differential Equation (10.38). If γ is not an integer, the general solution of Equation (10.38) is

$$y=AF(\alpha,\gamma;x)+Bx^{1-\gamma}F(\alpha-\gamma+1,2-\gamma;x). \tag{10.39}$$

In an exceptional case, when $\gamma=1$, we have

$$y=F(\alpha,1;x). \tag{10.40}$$

We obtained simply putting $\gamma=1$ in $F(\alpha,\gamma;x)$. An integral form of this function is

$$F(\alpha,\gamma;x)=\frac{1}{B(\alpha,\gamma-\alpha)}\int_0^1(1-t)^{\gamma-\alpha-1}t^{\alpha-1}e^{xt}dt.$$

10.6.1 Solution of the Confluent Hypergeometric Differential Equation when $x = 0$ *and* γ is not an integer

The differential equation is

$$x\frac{d^2y}{dx^2} + (\gamma - x)\frac{dy}{dx} - \alpha y = 0. \tag{10.41}$$

Let us consider its solution

$$y = \sum_{n=0}^{\infty} c_n x^{k+n} \text{ where } c_0 \neq 0. \tag{10.42}$$

Differentiating with respect to x, we get

$$y' = \sum_{n=0}^{\infty} (k+n) c_n x^{k+n-1}$$

$$y'' = \sum_{n=0}^{\infty} (k+n)(k+n-1) c_n x^{k+n-2} \tag{10.43}$$

Substituting these values in Equation (10.41), we get

$$x\sum_{n=0}^{\infty}(k+n)(k+n-1)c_n x^{k+n-2}$$
$$+\{\gamma - x\}\sum_{n=0}^{\infty}(k+n)c_n x^{k+n-1} - \alpha\sum_{n=0}^{\infty} c_n x^{k+n} = 0$$

or,

$$\sum_{n=0}^{\infty}\left[(k+n)(k+n-1) + \gamma(k+n)\right]c_n x^{k+n-1}$$
$$-\sum_{n=0}^{\infty}\left[(k+n)c_n + \alpha\right]x^{k+n} = 0, \tag{10.44}$$

which is identical, and the coefficients of various powers of x are zero, i.e., equating the coefficients of x^{k-1}(the lowest power of x) $= 0$, we get $c_0\left[k(k-1) + \gamma k\right] = 0$ or $k(k-1) + \gamma k = 0$ since $c_0 \neq 0$.

$\therefore k = 0$ *and* $k = 1 - \gamma$ are the roots of an indicial equation.

Equating the coefficient of x^{k+r} in Equation (10.44), we get

$$c_{r+1}[(k+r)(k+r+1) + \gamma(k+r+1) - c_r(k+r+\alpha) = 0$$

$$c_{r+1} = \frac{(k+r+\alpha)}{(k+r+1)(k+r+\gamma)}c_r. \tag{10.45}$$

Case I: When $k=0$,

$$c_{r+1}=\frac{(r+\alpha)}{(r+1)(r+\gamma)}c_r. \tag{10.46}$$

Putting $r=0,1,2\ldots$ in Equation (10.46), we obtain

$$c_1=\frac{\alpha}{\gamma}c_0,\ c_2=\frac{(1+\alpha)}{2(1+\gamma)}c_1=\frac{(\alpha+1)}{1.2(\gamma+1)}\frac{\alpha}{\gamma}c_0\ c_m$$
$$=\frac{\alpha(\alpha+1)(\alpha+m-1)}{1.2\ldots m\,\gamma(\gamma+1)\ldots(\gamma+m-1)}c_0=\frac{(\alpha)_m}{m!(\gamma)_m}c_0.$$

Substituting these values in Equation (10.42),

$$y=c_0\left[1+\frac{\alpha}{\gamma}x+\frac{(\alpha+1)}{1.2(\gamma+1)}\frac{\alpha}{\gamma}x^2+\cdots\right]=c_0\sum_{m=0}^{\infty}\frac{(\alpha)_m}{m!(\gamma)_m}x^m.$$

Taking $c_0=1$, we have $y=\sum_{m=0}^{\infty}\frac{(\alpha)_m}{m!(\gamma)_m}x^m=F(\alpha,\gamma,x)$, which is called a *confluent hypergeometric function of the first kind.*

Case II. Putting $k=1-\gamma$, in Equation (10.45) we get

$$c_{r+1}=\frac{(k+r+\alpha)}{(k+r+1)(k+r+\gamma)}c_r=c_{r+1}=\frac{(1-\gamma+r+\alpha)}{(1-\gamma+r+1)(1-\gamma+r+\gamma)}c_r$$
$$=\frac{(1-\gamma+\alpha)+r}{(2-\gamma+r)(1+r)}c_r$$

$$c_{r+1}=\frac{\alpha'+r}{(\gamma'+r)(r+1)}c_r \text{ where } \alpha'=(1-\gamma+\alpha) \textit{ and } \gamma'=2-\gamma.$$

Putting $r=0,1,2,\ldots$ in above c_{r+1}

$$\therefore c_1=\frac{\alpha'}{1.\gamma'}c_0,\ c_2=\frac{\alpha'+1}{1.2.(\gamma'+1)}c_1=\frac{\alpha'+1}{1.2.(\gamma'+1)}\frac{\alpha'}{1.\gamma'}c_0,\ldots..c_m$$
$$=\frac{\alpha'(\alpha'+1)\ldots..(\alpha'+m-1)}{1.2..\gamma'(\gamma'+m-1)}c_0=\frac{(\alpha')_m}{m!(.\gamma')_m}c_0.$$

Substituting these values in Equation (10.42), we get

$$y = x^{1-\gamma}\left[c_0 + c_1 x + c_2 x^2 + \cdots\right]$$
$$= c_0 x^{1-\gamma}\left[1 + \frac{\alpha'}{1.\gamma'}x + \frac{\alpha'+1}{1.2.(\gamma'+1)}x^2 + \cdots\right]$$
$$= c_0 x^{1-\gamma} F(\alpha',\gamma';x) = c_0 x^{1-\gamma} F(1-\gamma+\alpha, 2-\gamma; x)$$

Putting $c_0 = 1$, we have

$$y = x^{1-\gamma}\left[c_0 + c_1 x + c_2 x^2 + \cdots\right]$$
$$= x^{1-\gamma} F(1-\gamma+\alpha, 2-\gamma; x),$$

which is called the *confluent hypergeometric function of the second kind.*

Thus, the general solution of the confluent hypergeometric differential equation is:

$$y = AF(\alpha,\gamma,x) + Bx^{1-\gamma} F(1-\gamma+\alpha, 2-\gamma; x).$$

10.6.2 Differentiation of Confluent Hypergeometric Functions

Show that $\frac{d}{dx}F(\alpha,\beta,x) = \frac{\alpha}{\beta}F(\alpha+1,\beta+1,x)$, and deduce that

(i) $\frac{d^n}{dx^n}F(\alpha,\beta,x) = \frac{(\alpha)_n}{(\beta)_n}F(\alpha+n,\beta+n,x)$

(ii) $\left[\frac{d^n}{dx^n}F(\alpha,\beta,x)\right]_{x=0} = \frac{(\alpha)_n}{(\beta)_n}.$

Proof: By definition, $F(\alpha,\beta,x) = \sum_{r=0}^{\infty}\frac{(\alpha)_r}{r!(\gamma)_r}x^r.$

Differentiating with respect to x, we have

$$\frac{d}{dx}F(\alpha,\beta,x) = \frac{d}{dx}\sum_{r=0}^{\infty}\frac{(\alpha)_r}{r!(\beta)_r}x^r = \sum_{r=1}^{\infty}\frac{(\alpha)_r}{r!(\beta)_r}rx^{r-1}\;[\text{term with r} = 0 \text{ vanishes}]$$
$$= \sum_{k=0}^{\infty}\frac{(\alpha)_{k+1}}{k!(\beta)_{k+1}}x^k\;[\text{let } r = k+1, \textit{ ie } k = r-1]$$
$$= \sum_{k=0}^{\infty}\frac{\alpha(\alpha+1)_k}{k!\beta(\gamma+1)_k}x^k = \frac{\alpha}{\beta}\sum_{k=0}^{\infty}\frac{(\alpha+1)_k}{k!(\gamma+1)_k}x^k = \frac{\alpha}{\beta}F(\alpha+1,\beta+1,x).$$

Solution of (i) and (ii) are similar as Section 10.2.

10.6.3 Integral Representation for Confluent Hypergeometric Function

Show that, $F(\alpha,\beta;x)=\dfrac{\Gamma(\beta)}{\Gamma(\alpha)\Gamma(\beta-\alpha)}\int_0^1 t^{\alpha-1}(1-t)^{\beta-\alpha-1}e^{xt}dt$

or, $F(\alpha,\beta,;x)=\dfrac{1}{B(\alpha,\beta-\alpha)}\int_0^1 t^{\alpha-1}(1-t)^{\beta-\alpha-1}e^{xt}dt$, if $\beta>\alpha>0$

Proof: We know

$$F(\alpha,\beta;x)=\sum_{n=0}^{\infty}\frac{(\alpha)_n}{(\beta)_n}\frac{x^n}{n!}=\sum_{n=0}^{\infty}\frac{\Gamma(\alpha+n)}{\Gamma(\alpha)}\frac{\Gamma(\beta)}{\Gamma(\beta+n)}\frac{x^n}{n!}$$

$$=\frac{\Gamma(\beta)}{\Gamma(\alpha)}\sum_{n=0}^{\infty}\frac{\Gamma(\alpha+n)}{\Gamma(\beta-\alpha)}\frac{\Gamma(\beta-\alpha)}{\Gamma(\beta-\alpha+\alpha+n)}\frac{x^n}{n!}$$

$$=\frac{\Gamma(\beta)}{\Gamma(\alpha)\Gamma(\beta-\alpha)}\sum_{n=0}^{\infty}\frac{\Gamma(\alpha+n)\quad\Gamma(\beta-\alpha)q}{\Gamma(\beta-\alpha+\alpha+n)}\frac{x^n}{n!}$$

$$=\frac{\Gamma(\beta)}{\Gamma(\alpha)\Gamma(\beta-\alpha)}\sum_{n=0}^{\infty}\left\{\int_0^1 t^{\alpha+n-1}(1-t)^{\beta-\alpha-1}dt\right\}\frac{x^n}{n!}$$

$$\left[\because\frac{\Gamma(\alpha)\ \Gamma(\beta)}{\Gamma(\alpha+\beta)}=B(\alpha,\beta)=\int_0^1 t^{\alpha-1}(1-t)^{\beta-1}dt\right]$$

$$=\frac{\Gamma(\beta)}{\Gamma(\alpha)\Gamma(\beta-\alpha)}\left\{\int_0^1 t^{\alpha-1}(1-t)^{\beta-\alpha-1}\right\}\left(\sum_{n=0}^{\infty}\frac{(xt)^n}{n!}\right)dt$$

$$=\frac{\Gamma(\beta)}{\Gamma(\alpha)\Gamma(\beta-\alpha)}\left\{\int_0^1 t^{\alpha-1}(1-t)^{\gamma-\beta-1}\right\}e^{xt}dt. \tag{10.47}$$

Also, $B(\alpha,\beta-\alpha)=\dfrac{\Gamma(\alpha)\ \Gamma(\beta-\alpha)}{\Gamma(\alpha+\beta-\alpha)}$;

$$\therefore \frac{\Gamma(\beta)}{\Gamma(\alpha)\Gamma(\beta-\alpha)} = \frac{1}{B(\alpha,\beta-\alpha)}. \tag{10.48}$$

Using Equation (10.48) in Equation (10.47), we get

$$\therefore F(\alpha,\beta;x) = \frac{1}{B(\alpha,\beta-\alpha)}\left\{\int_0^1 t^{\alpha-1}(1-t)^{\beta-\alpha-1}\right\}e^{xt}dt. \tag{10.49}$$

Hence, Equations (10.47) and (10.49) are the required results.

10.6.4 Theorem (Kummer's)

Proof: We know that

$$F(\alpha,\beta;x) = \frac{(\beta)}{\Gamma(\alpha)\Gamma(\beta-\alpha)}\left\{\int_0^1 t^{\alpha-1}(1-t)^{\beta-\alpha-1}\right\}e^{xt}dt. \tag{10.50}$$

Replacing α *by* $\beta-\alpha$ *and* x *by* $-x$ in Equation (i), we get

$$F(\beta-\alpha,\beta;-x) = \frac{\Gamma(\beta)}{\Gamma(\beta-\alpha)\Gamma(\beta-(\beta-\alpha))}\left\{\int_0^1 t^{\beta-\alpha-1}(1-t)^{\beta-(\beta-\alpha)-1}\right\}e^{-xt}dt$$

$$= \frac{\Gamma(\beta)}{\Gamma(\beta-\alpha)\Gamma(\alpha)}\left\{\int_0^1 t^{\beta-\alpha-1}(1-t)^{\alpha-1}\right\}e^{-xt}dt \quad [\text{let } u = 1-t, dt = -du] =$$

$$= -\frac{\Gamma(\beta)}{\Gamma(\beta-\alpha)\Gamma(\alpha)}\left\{\int_1^0 (1-u)^{\beta-\alpha-1}u^{\alpha-1}\right\}e^{-xt}du$$

$$= \frac{\Gamma(\beta)}{\Gamma(\beta-\alpha)\Gamma(\alpha)}\left\{\int_0^1 (1-u)^{\beta-\alpha-1}u^{\alpha-1}\right\}e^{-x(1-u)}du$$

$$= e^{-x}\frac{\Gamma(\beta)}{\Gamma(\beta-\alpha)\Gamma(\alpha)}\left\{\int_0^1 (1-u)^{\beta-\alpha-1}u^{\alpha-1}\right\}e^{xu}du$$

$$= e^{-x}F(\alpha,\beta;x)$$

$$\therefore F(\beta-\alpha,\beta;-x) = e^{-x}F(\alpha,\beta;x).$$

by Equation (10.50)

$$\therefore F(\alpha,\beta;x) = e^{x}F(\beta-\alpha,\beta;-x).$$

10.7 Contiguous Hypergeometric Functions

Definition 10.7.1. According to Gauss, the function $F(\alpha',\beta',\gamma';x)$ is said to be *contiguous* to $F(\alpha,\beta,\gamma;x)$ when it is increased or decreased by one and only one of the parameters α,β,γ by unity.

There exist six hypergeometric functions contiguous to $F(\alpha,\beta,\gamma;x)$. These are the following:

(i) $F_{\alpha+}=F(\alpha+1,\beta,\gamma;x)$
(ii) $F_{\beta+}=F(\alpha,\beta+1,\gamma;x)$
(iii) $F_{\gamma+}=F(\alpha,\beta,\gamma+1;x)$
(iv) $F_{\alpha-}=F(\alpha-1,\beta,\gamma;x)$
(v) $F_{\beta-}=F(\alpha,\beta-1,\gamma;x)$
(vi) $F_{\gamma-}=F(\alpha,\beta,\gamma-1;x)$

10.7.1 Contiguity Relationship

$$(\alpha-\beta)F(\alpha,\beta,\gamma;x)=\alpha F(\alpha+1,\beta,\gamma;x)-\beta F(\alpha,\beta+1,\gamma;x),$$

$$\text{i.e., } (\alpha-\beta)F(\alpha,\beta,\gamma;x)=\alpha F_{\alpha+}-\beta F_{\beta+}.$$

Proof: We have, by definition,

$$\begin{aligned}\alpha F(\alpha+1,\beta,\gamma;x)-\beta F(\alpha,\beta+1,\gamma;x)&=\sum_{r=0}^{\infty}\frac{\alpha(\alpha+1)_r(\beta)_r}{(\gamma)_r r!}x^r-\sum_{r=0}^{\infty}\frac{(\alpha)_r\beta(\beta+1)_r}{(\gamma+1)_r r!}x^r\\&=\sum_{r=0}^{\infty}\frac{(\alpha+r)(\alpha)_r(\beta)_r}{(\gamma)_r r!}x^r-\sum_{r=0}^{\infty}\frac{(\alpha)_r(\beta+1)(\beta)_r}{(\gamma+1)_r r!}x^r\\&=\sum_{r=0}^{\infty}\left[(\alpha+r)-(\beta+r)\right]\frac{(\alpha)_r(\beta)_r}{(\gamma)_r r!}x^r=(\alpha-\beta)\sum_{r=0}^{\infty}\frac{(\alpha)_r(\beta)_r}{(\gamma)_r r!}x^r\\&=(\alpha-\beta)F(\alpha,\beta,\gamma;x).\end{aligned}$$

It can be easily proved
$(\alpha-\beta)\alpha F(\alpha,\beta+1,\gamma;x)+\beta(\alpha+\beta-1)F(\alpha,\beta;x)-\beta F(\alpha,\beta-1;x)=0$
by similar proceeding of above proof.

Example 10.7.1. Prove that

(i) $e^x={}_1F_1(\alpha;\alpha;x)$
(ii) $(1-x)^{-n}={}_2F_1(\alpha,\beta;\beta;x)$
(iii) $\log(1+x)=x\,{}_2F_1(1,1;2;-x)$

Solution: (i) We have, by definition,

$$ {}_1F_1(\alpha;\beta;x)=\sum_{r=0}^{\infty}\frac{(\alpha)_r}{r!(\beta)_r}x^r=1+\frac{\alpha}{\beta}.\frac{x}{1!}+\frac{\alpha(\alpha+1}{\beta(\beta+1)}.\frac{x^2}{2!}+\ldots\ldots\ldots \quad \text{(i)}$$

Replacing β *by* α in Equation (i), we get

$$ {}_1F_1(\alpha;\alpha;x)=1+\frac{\alpha}{\alpha}.\frac{x}{1!}+\frac{\alpha(\alpha+1}{\alpha(\alpha+1)}.\frac{x^2}{2!}+\ldots\ldots\ldots\ldots $$

$$ =1+\frac{x}{1!}+\frac{x^2}{2!}+\cdots=e^x. $$

(ii) We have, by definition,

$$ {}_2F_1(\alpha,\beta;\gamma;x)=1+\frac{\alpha\beta}{\gamma}.\frac{x}{1!}+\frac{\alpha(\alpha+1)\beta(\beta+1)}{\gamma(\gamma+1)}.\frac{x^2}{2!}+\cdots\ldots\ldots $$

Replacing γ *by* β, we get

$$ {}_2F_1(\alpha,\beta;\beta;x)=1+\frac{\alpha\beta}{\beta}.\frac{x}{1!}+\frac{\alpha(\alpha+1)\beta(\beta+1)}{\beta(\beta+1)}.\frac{x^2}{2!}+\ldots\ldots\ldots $$

$$ =1+\frac{\alpha}{1}.\frac{x}{1!}+\frac{\alpha(\alpha+1)}{1}.\frac{x^2}{2!}+\ldots. $$

$$ =1+.\alpha.x+\alpha(\alpha+1)\frac{x^2}{2!}+\cdots.+\ldots $$

$$ =1+(-\alpha)(-x)+\frac{(-\alpha)(-\alpha-1)}{1}.\frac{(-x)^2}{2!} $$

$$ +\frac{(-\alpha)(-\alpha-1)(-\alpha-2)}{3!}(-x)^2+\cdots\ldots=(1-x)^{-\alpha} $$

(iii) We have, by definition,

$$ {}_2F_1(\alpha,\beta;\gamma;x)=1+\frac{\alpha\beta}{\gamma}.\frac{x}{1!}+\frac{\alpha(\alpha+1)\beta(\beta+1)}{\gamma(\gamma+1)}.\frac{x^2}{2!}+\ldots\ldots\ldots $$

Replacing $\alpha, \beta, \gamma, and\, x$ by 1, 1, 2, and $-x$, respectively, in Equation (i), we get

$$ {}_2F_1(1,1;2;-x) = 1 + \frac{1.1}{2}.\frac{-x}{1!} + \frac{1(1+1).1.(1+1)}{2(2+1)}.\frac{(-x)^2}{2!} + \dots\dots\dots $$

Multiplying both sides by x, we get

$$ x\,{}_2F_1(1,1;2;-x) = x + \frac{1.1}{2}.\frac{-x}{1!}.x + \frac{1(1+1).1.(1+1)}{2(2+1)}.\frac{(-x)^2}{2!}.x $$

$$ = x - \frac{x^2}{2} + .\frac{x^3}{3} - \dots\dots\dots\dots\dots. = \log(1-x). $$

10.8 Generalized Hypergeometric Series

We studied in the preceding Article 10.7 the hypergeometric function $F(\alpha, \beta; \gamma; x)$, which has two parameters α, β in the numerator and one parameter γ in the denominator. We can now think of a function that has any number of parameters in the numerator and denominator. Consider the series

$$ \sum_{n=0}^{\infty} \frac{(\alpha_1)_n (\alpha_2)_n \dots (\alpha_p)_n \cdot x^n}{(\beta_1)_n (\beta_2)_n \dots (\beta_q)_n \cdot n!}. \tag{10.51} $$

If this series converged to a function F, we call F *the generalized hypergeometric function*, and it is in the form of ${}_pF_q(\alpha_1, \alpha_2, \dots\alpha_p; \beta_1, \beta_2, \dots\beta_q; x)$, which can also be written as

$$ {}_pF_q\left[\begin{matrix} \alpha_1, \alpha_2, \dots\alpha_p; x \\ \beta_1, \beta_2, \dots\beta_q; \end{matrix}\right] \text{ or briefly as } {}_pF_q. $$

The series i is called the *generalized hypergeometric series.*

We will investigate under the conditions that series i be converged if no denominator parameter should be zero or negative. Next, if any numerator parameter is zero or a negative integer, the series terminates.

10.8.1 Convergence of a Generalized Hypergeometric Function

Let us apply the D'Alembert's ratio test of convergence. We have

$$\left|\frac{u_{n+1}}{u_n}\right| = \frac{\left|\prod_1^p(\alpha_i)_{n+1}x^{n+1}\right|\left|\prod_1^q(\beta_j)_n n!\right|}{\left|\prod_1^p(\alpha_i)_n(n+1)!\right|\left|\prod_1^q(\beta_j)_{n+1}x^n\right|}$$

$$= \left|\frac{(\alpha_1+n)(\alpha_2+n)\ldots(\alpha_p+n).}{(n+1)(\beta_1+n)(\beta_2+n)\ldots.(\beta_q+n).}x\right|.$$

It is clear that (i) if $p \le q$, the limit of $\left|\frac{u_{n+1}}{u_n}\right|$ as $n \to \infty$ is zero, whatever the value of x. Thus, when $p \le q$, the series is convergent absolutely. (ii) If $p = q+1$, the limit of $\left|\frac{u_{n+1}}{u_n}\right|$ as $n \to \infty$ is $|x|$; hence, the series converges absolutely within the circle of convergence $|x|<1$ and diverges when $|x|>1$. (iii) When $p>q+1$, the limit $\left|\frac{u_{n+1}}{u_n}\right|$ as $n \to \infty$ is infinity and the series diverges.

(iv) If $p = q+1\, and\, |x| = 1$, then comparing series (i) with the series $\sum_{n=1}^{\infty}\frac{1}{n^{1+\delta}}$ where $\delta = \frac{1}{2}Re\left(\sum_{j=1}^{q}\beta_j - \sum_{i=1}^{p}\alpha_i\right)$, we find that the series under consideration is absolutely convergent if $Re\left(\sum_{j=1}^{q}\beta_j - \sum_{i=1}^{p}\alpha_i\right) > 0$.

We summarize as follows:

(i) If $p \le q$, the series is convergent absolutely for every finite x.
(ii) If $p = q+1$, the series converges absolutely within the circle of convergence $|x|<1$ and diverges when $|x|>1$.
(iii) If $p = q+1\, and\, |x| = 1$, the series under consideration is absolutely convergent if $Re\left(\sum_{j=1}^{q}\beta_j - \sum_{i=1}^{p}\alpha_i\right) > 0$.
(iv) When $p > q+1$, the series diverges for $x \ne 0$, and for $x = 0$, its value is one.

10.8.2 Differential Equations Satisfied by ${}_pF_q$

We know that the differential equation satisfied by the hypergeometric function $y = F(\alpha, \beta; \gamma; x)$, *is*

$$\left[\theta(\theta+\gamma-1)-x(\theta+\alpha)(\theta+\beta)\right]y=0 \tag{10.52}$$

where $\theta=(\frac{d}{dx})$.

We can obtain the differential equation satisfied by

$$y={}_pF_q\begin{bmatrix}\alpha_1,\alpha_2,\ldots\alpha_p;x\\ \beta_1,\beta_2,\ldots\beta_q;\end{bmatrix}=\sum_{n=0}^{\infty}\frac{(\alpha_1)_n(\alpha_2)_n\cdots(\alpha_p)_n\cdot x^n}{(\beta_1)_n(\beta_2)_n\cdots(\beta_q)_n\cdot n!}.$$

$\because \theta(x^n)=nx^{n-1}$. We have

$$\theta\prod_{j=1}^{q}(\theta+\beta_j-1)y=\sum_{n=1}^{\infty}\frac{n\prod_{j=1}^{q}(n+\beta_j-1)\prod_{i=1}^{p}(\alpha_i)_n}{\prod_{j=1}^{q}(\beta_j)_n}\frac{x^n}{n!}$$

$$=\sum_{n=1}^{\infty}\frac{\prod_{i=1}^{p}(\alpha_i)_n\quad x^n}{\prod_{j=1}^{q}(\beta_j)_n(n-1)!}$$

$$=\sum_{n=1}^{\infty}\frac{\prod_{i=1}^{p}(\alpha_i)_{n+1}\quad x^{n+1}}{\prod_{j=1}^{q}(\beta_j)_n n!}$$

$$=\sum_{n=1}^{\infty}\frac{\prod_{i=1}^{p}(\alpha_i+n)\prod_{i=1}^{p}(\alpha_i)_n}{\prod_{j=1}^{q}(\beta_j)_n}\frac{x^{n+1}}{n!}$$

$$=x\prod_{i=1}^{p}(\theta+\alpha_i)y.$$

From this, when no β_j is a non – positive integer, the function $y={}_pF_q$ is a solution of the differential equation

$$\left[\theta\prod_{j=1}^{q}(\theta+\beta_j-1)-x\prod_{i=1}^{p}(\theta+\alpha_i)\right]y=0. \tag{10.53}$$

Clearly, the solution $y_0={}_pF_q\begin{bmatrix}\alpha_1,\alpha_2,\ldots\alpha_p;x\\ \beta_1,\beta_2,\ldots\beta_q;\end{bmatrix}$ of Equation (10.53) is valid in the neighborhood of origin, when $p\le q+1$. We shall find the other q solutions in the same neighborhood.

Now, we replace α *by* α_1, β *by* α_2 and γ *by* β_1 in Equation (10.52) to bring this with the notation of Equation (10.53). Then, we may write the general solution of Equation (10.52) in the form:

$$y=Ay_0+by_1 \tag{10.54}$$

where $y_0 = {}_2F_1(\alpha_1,\alpha_2;\beta_1;x)$
and $y_1 = x^{1-\beta}\ {}_pF_q(\alpha_1-\beta_1+1,\alpha_2-\beta_1+1;2-\beta_1;x)$ with the condition that β_1 is not an integer.

Generalizing this solution, we write Equation (10.53) in the form

$$y=\sum_{n=0}^{\infty} A_n y_n \tag{10.55}$$

where $A_n s$ are arbitrary constants and

$$y_0 = {}_pF_q(\alpha_1,\alpha_2,\ldots\alpha_p;\beta_1,\beta_2,\ldots\beta_q;x),$$

and for $n=1,2,\ldots q$,

$$\begin{aligned} y_n &= x^{1-\beta_n}\left[\begin{matrix}\alpha_1-\beta_n+1,\alpha_2-\beta_n+1,\ldots,\alpha_p-\beta_n+1; \\ \beta_1-\beta_n+1,\beta_{n-1}-\beta_n+1,2-\beta_n,\beta_{n+1}-\beta_n+1,\ldots\beta_q-\beta_n+1;x\end{matrix}\right. \\ &= \sum_{k=0}^{\infty}\frac{\prod_{i=1}^{p}(\alpha_i-\beta_n+1)_k}{\prod_{j=1}^{q}(\beta_j-\beta_n+1)_k(2-\beta_n)_k}x^{k+1-\beta_n}\end{aligned}$$

with the condition that β_n is not an integer.

We will prove that y_n satisfies Equation (10.53).

Using the definition and properties of θ, we have

$$\begin{aligned} &\theta\prod_{j=1}^{q}(\theta+\beta_j-1)y_n \\ &= \sum_{k=1}^{\infty}\frac{(k+1-\beta_n)\prod_{j=1}^{q}(k+1-\beta_n+\beta_j-1)\prod_{i=1}^{p}(\alpha_i-\beta_n+1)_k\ x^{k+1-\beta_n}}{\prod_{j=1}^{q}(\beta_j-\beta_n+1)_k(2-\beta_n)_k} \\ &= \sum_{k=0}^{\infty}\frac{\prod_{i=1}^{p}(\alpha_i-\beta_n+1)_k}{\prod_{j=1}^{q}(\beta_j-\beta_n+1)_{k-1}(2-\beta_n)_{k-1}}x^{k+1-\beta_n}\end{aligned}$$

because $\prod_{j=1}^{q}(k-\beta_m+\beta_j)$ cancels with the last term of the product $\prod_{j=1}^{q}(\beta_j-\beta_m+1)_k$, which is equal to it, and $(k+1-\beta_n)$ in the numerator cancels with the last term of $(2-\beta_n)_k$. Also, the term with $k=0$ is omitted above because, in the case $j = n$, we shall have k in the numerator, which makes the term zero. Now, shifting the index k *to* $(k+1)$, we get

$$\theta\prod_{j=1}^{q}(\theta+\beta_j-1)y_n=\sum_{k=0}^{\infty}\frac{\prod_{i=1}^{p}(\alpha_i-\beta_n+1)_{k+1}}{\prod_{j=1}^{q}(\beta_j-\beta_n+1)_k(2-\beta_n)_k}x^{k+2-\beta_n}$$

$$=x\sum_{k=0}^{\infty}\frac{\prod_{i=1}^{p}(\alpha_i-\beta_n+1+k)\prod_{i=1}^{p}(\alpha_i-\beta_n+1)_k}{\prod_{j=1}^{q}(\beta_j-\beta_n+1)_k}\frac{x^{k+1-\beta_n}}{(2-\beta_n)_k}$$

$$=x\prod_{i=1}^{p}(\theta+\alpha_i)y_n\quad[\because(\theta+\alpha)x^n=(n+\alpha)x^n].$$

Hence, $y_n, n = 1,2\ldots q$, satisfies the generalized hypergeometric Equation (ii). If any β_n is an integer, we get a logarithmic term in the general solution.

10.8.3 Properties of $_pF_q$

Here, we will establish three properties of the function $_pF_q$, which are theorems and are known after the name of the mathematicians who are associated with them.

Saalschutz' Theorem 10.8.1.
When $p = q+1$ *and* $Re\left(\sum_{j=1}^{q}\beta_j-\sum_{i=1}^{p}\alpha_i\right)>0, |x| = 1$, the generalized hypergeometric series converges absolutely to a function $_{q+1}F_q$. In a particular case, we will find a function $_3F_2$ in terms of the factorial functions.

We know $F(\gamma-\alpha,\gamma-\beta;\gamma;x)=(1-x)^{\alpha+\beta-\gamma}F(\alpha,\beta;\gamma;x)$.

It can be written as

$$\sum_{n=0}^{\infty}\frac{(\gamma-\alpha)_n(\gamma-\beta)_n}{(\gamma)_n}\frac{x^n}{n!}$$

$$=\left[\sum_{n=0}^{\infty}\frac{(\gamma-\alpha-\beta)_n}{n!}x^n\right]\left[\sum_{m=0}^{\infty}\frac{(\alpha)_m(\beta)_m}{(\gamma)_m}\frac{x^m}{m!}\right]$$

$$=\sum_{n=0}^{\infty}\sum_{m=0}^{\infty}\frac{(\gamma-\alpha-\beta)_n(\alpha)_m(\beta)_m}{(\gamma)_m m!n!}x^{m+n}$$

$$=\sum_{n=0}^{\infty}\sum_{m=0}^{\infty}\frac{(\gamma-\alpha-\beta)_{n-m}(\alpha)_m(\beta)_m}{(\gamma)_m m!(n-m)!}x^n$$

[replacing n by $n-m$ we get $(\gamma-\alpha-\beta)_{n-m}=\dfrac{(-1)^m(\gamma-\alpha-\beta)_n}{(1-\gamma+\alpha+\beta-n)_n}$ *and* $(n-m)!=\dfrac{(-1)^m n!}{(-n)_m}$].

We have

$$\sum_{n=0}^{\infty}\frac{(\gamma-\alpha)_n(\gamma-\beta)_n}{(\gamma)_n}\frac{x^n}{n!}=\sum_{n=0}^{\infty}\sum_{m=0}^{\infty}\frac{\dfrac{(-1)^m(\gamma-\alpha-\beta)_n}{(1-\gamma+\alpha+\beta-n)_n}(\alpha)_m(\beta)_m}{(\gamma)_m m!\dfrac{(-1)^m n!}{(-n)_m}}x^n$$

$$=\sum_{n=0}^{\infty}\sum_{m=0}^{\infty}\frac{(-1)^m(\gamma-\alpha-\beta)_n(\alpha)_m(\beta)_m(\beta)_m}{(\gamma)_m m!(-1)^m n!(1-\gamma+\alpha+\beta-n)_n}x^n$$

$$=\sum_{n=0}^{\infty}{}_3F_2\left[\begin{matrix}-n,\alpha,\beta;\\ \gamma,1-\gamma+\alpha+\beta-n;\end{matrix}\;1\right]\frac{(\gamma-\alpha-\beta)_n}{n!}x^n.$$

Comparing the coefficients of x^n, we get

$$ {}_3F_2\left[\begin{matrix}-n,\alpha,\beta;\\ \gamma,1-\gamma+\alpha+\beta-n;\end{matrix}\;1\right]=\frac{(\gamma-\alpha)_n(\gamma-\beta)_n}{(\gamma)_n(\gamma-\alpha-\beta)_n}. \tag{10.56}$$

In this equation, n is a non-negative integer, and α,β,γ are independent of n.

We observe that in Equation (10.56), ${}_3F_2$ has the parameters such that the sum of the denominator parameters exceeds the sum of the numerator parameters by unity. If any ${}_{q+1}F_q$ satisfies this property, then it is called *Saalschutzian*.

We will find another *Saalschutzian* in terms of factorial functions.

We have ${}_2F_1\left[\begin{matrix}\alpha,\beta;\\ 1+\alpha-\beta;\end{matrix}\;1\right]=(1-x)^{-\alpha}\;{}_2F_1\left[\begin{matrix}\frac{1}{2}\alpha,\frac{1}{2}+\frac{1}{2}\alpha-\beta;\\ 1+\alpha-\beta;\end{matrix}\;-\frac{4x}{(1-x)^2}\right]$

[from quadratic transformation].

Now,

$$\sum_{n=0}^{\infty}\frac{(\alpha)_n(\beta)_n\cdot}{(1+\alpha-\beta)_n}\frac{x^n}{n!}$$

$$=\sum_{m=0}^{\infty}\frac{(\frac{1}{2}\alpha)_m(\frac{1}{2}+\frac{1}{2}\alpha-\beta)_m\cdot}{(1+\alpha-\beta)_m(1-x)^{\alpha+2m}}\frac{2^{2m}x^m}{m!}$$

$$=\sum_{m=0}^{\infty}\frac{(\frac{1}{2}\alpha)_m(\frac{1}{2}+\frac{1}{2}\alpha-\beta)_m(-1)^m\cdot}{(1+\alpha-\beta)_m}\frac{2^{2m}x^m}{m!}\sum_{n=0}^{\infty}\frac{(\alpha+2m)_n}{n!}x^n$$

$$=\sum_{m=0}^{\infty}\sum_{n=0}^{\infty}\frac{(\frac{1}{2}\alpha)_m(\frac{1}{2}+\frac{1}{2}\alpha-\beta)_m(-1)^m\cdot}{(1+\alpha-\beta)_m}\frac{2^{2m}}{m!}\frac{(\alpha+2m)_n}{n!}x^{n+m}$$

$$=\sum_{m=0}^{\infty}\sum_{n=0}^{\infty}\frac{(\frac{1}{2}\alpha)_m(\frac{1}{2}+\frac{1}{2}\alpha-\beta)_m(-1)^m\cdot}{(1+\alpha-\beta)_m}\frac{2^{2m}}{m!}\frac{(\alpha)_{n+2m}}{n!(\alpha)_{2m}}x^{n+m}$$

$[\because(\alpha+2m)_n=\frac{(\alpha)_{n+2m}}{(\alpha)_{2m}}]$, and applying $(\alpha)_{2m}=2^{2m}(\frac{\alpha}{2})_m(\frac{\alpha}{2}+\frac{1}{2})_m$, we get

$$\sum_{n=0}^{\infty}\frac{(\alpha)_n(\beta)_n\cdot}{(1+\alpha-\beta)_n}\frac{x^n}{n!}=\sum_{m=0}^{\infty}\sum_{n=0}^{\infty}\frac{(\frac{1}{2}\alpha)_m(\frac{1}{2}+\frac{1}{2}\alpha-\beta)_m(-1)^m\cdot}{(1+\alpha-\beta)_m m!n!}\frac{2^{2m}}{2^{2m}(\frac{\alpha}{2})_m(\frac{\alpha}{2}+\frac{1}{2})_m}\frac{(\alpha)_{n+2m}}{(\alpha)_{2m}}x^{n+m}$$

$$=\sum_{m=0}^{\infty}\sum_{n=0}^{\infty}\frac{(\frac{1}{2}+\frac{1}{2}\alpha-\beta)_m(-1)^m\cdot}{(1+\alpha-\beta)_m m!n!}\frac{1}{(\frac{\alpha}{2}+\frac{1}{2})_m}(\alpha)_{n+2m}x^{n+m}$$

$$=\sum_{m=0}^{\infty}\sum_{n=0}^{\infty}\frac{(\frac{1}{2}+\frac{1}{2}\alpha-\beta)_m(-1)_m}{(1+\alpha-\beta)_m m!}\frac{1}{(\frac{\alpha}{2}+\frac{1}{2})_m}\frac{(\alpha)_{n+m}}{(n-m)!}x^n$$

$$=\sum_{m=0}^{\infty}\sum_{n=0}^{\infty}\frac{(\frac{1}{2}+\frac{1}{2}\alpha-\beta)_m(-n)_m}{(1+\alpha-\beta)_m}\frac{1}{(\frac{\alpha}{2}+\frac{1}{2})_m}\frac{(\alpha)_n(\alpha+n)_m}{m!n!}x^n$$

$$=\sum_{n=0}^{\infty}{}_3F_2\left[\begin{matrix}-n,\alpha+n,\frac{1}{2}+\frac{1}{2}\alpha-\beta;\\ 1+\alpha-\beta,\frac{\alpha}{2}+\frac{1}{2};\end{matrix}\;1\right]\frac{(\alpha)_n}{n!}x^n.$$

Hence, euating the coefficients of x^n, we get

$$ {}_3F_2\left[\begin{matrix} -n, \alpha+n, \frac{1}{2}+\frac{1}{2}\alpha-\beta; \\ 1+\alpha-\beta, \frac{\alpha}{2}+\frac{1}{2}; \end{matrix} 1\right] = \frac{(\beta)_n \cdot}{(1+\alpha-\beta)_n}. \tag{10.57} $$

where α, β are independent of n.

Whipples's Theorem 10.8.2.

Here, we will describe two transformations, one for terminating and the other for non-terminating generalized hypergeometric series, obtained by F.J.W. Whipple in 1927.

First, we find a terminating generalized hypergeometric series.

Consider the hypergeometric function

$$ {}_2F_1\left[\begin{matrix} \beta, \gamma; \\ \beta+\gamma; \end{matrix} t(1-x+xt)\right] = \sum_{n=0}^{\infty} \frac{(\beta)_n (\gamma)_n \cdot (1-x+xt)^n t^n}{(\beta+\gamma)_n \quad n!}. $$

By the binomial theorem, we get

$$ (1-x+xt)^n = \sum_{m=0}^{n} \frac{n!(1-x)^{n-m}(xt)^m}{m!(n-m)!}. $$

$$ \therefore {}_2F_1\left[\begin{matrix} \beta, \gamma; \\ \beta+\gamma; \end{matrix} t(1-x+xt)\right] = \sum_{n=0}^{\infty} \frac{(\beta)_n (\gamma)_n \cdot t^n}{(\beta+\gamma)_n \; n!} \sum_{m=0}^{n} \frac{n!(1-x)^{n-m}(xt)^m}{m!(n-m)!} $$

$$ = \sum_{n=0}^{\infty} \sum_{m=0}^{n} \frac{(\beta)_n (\gamma)_n \cdot (1-x)^{n-m} x^m t^{n+m}}{(\beta+\gamma)_n \quad m!(n-m)!}. $$

Replacing $n \, by \, n-m$ and the integral part of $\frac{1}{2} n \, by \left[\frac{1}{2} n\right]$, we get

$$
{}_2F_1\left[\begin{matrix}\beta,\gamma;\\ \beta+\gamma;\end{matrix} t(1-x+xt)\right]=\sum_{n=0}^{\infty}\sum_{m=0}^{\left[\frac{1}{2}n\right]}\frac{(\beta)_{n-m}(\gamma)_{n-m}\cdot}{(\beta+\gamma)_{n-m}}\frac{(1-x)^{n-2m}x^m t^n}{m!(n-2m)!}
$$

$$
=\sum_{n=0}^{\infty}\sum_{m=0}^{\left[\frac{1}{2}n\right]}\frac{(-n)_{2m}(\beta)_n(\gamma)_n\cdot}{(\beta+\gamma)_n(1-\beta-n)_m(1-\gamma-n)_m}\frac{(1-\beta-\gamma-n)_m(-1)^m(1-x)^{n-2m}x^m t^n}{m!n!}
$$

$$
=\sum_{n=0}^{\infty}\sum_{m=0}^{\left[\frac{1}{2}n\right]}\frac{2^{2m}(-\frac{1}{2}n)_m(-\frac{1}{2}n+\frac{1}{2})_m(\beta)_n(\gamma)_n\cdot}{(\beta+\gamma)_n(1-\beta-n)_m(1-\gamma-n)_m}\frac{(1-\beta-\gamma-n)_m(-1)^m(1-x)^{n-2m}x^m t^n}{m!n!}
$$

$$
=\sum_{n=0}^{\infty}(1-x)^n\,{}_3F_2\left[\begin{matrix}-\frac{1}{2}n,\frac{1}{2}n+\frac{1}{2},1-\beta-\gamma-n;;\\ 1-\beta-n,1-\gamma-n;\end{matrix}\frac{4x}{(1-x)^2}\right]\frac{(\beta)_n(\gamma)_n\cdot t^n}{(\beta+\gamma)_n\ n!}. \tag{10.58}
$$

Another expansion of

$$
{}_2F_1\left[\begin{matrix}\beta,\gamma;\\ \beta+\gamma;\end{matrix} t(1-x+xt)\right]=\sum_{n=0}^{\infty}\frac{(\beta)_n(\gamma)_n\cdot}{(\beta+\gamma)_n}\frac{[1-x(1-t)]^n t^n}{n!}.
$$

Using the binomial theorem $[1-x(1-t)]^n=\sum_{m=0}^{\infty}\frac{n!(-1)^m(1-t)^m(x)^m}{m!(n-m)!}$, and putting in the above equation,

$$
{}_2F_1\left[\begin{matrix}\beta,\gamma;\\ \beta+\gamma;\end{matrix} t(1-x+xt)\right]=\sum_{n=0}^{\infty}\sum_{m=0}^{n}\frac{(\beta)_n(\gamma)_n\cdot}{(\beta+\gamma)_n}\frac{(-1)^m(1-t)^m(x)^m t^n}{m!(n-m)!}
$$

$$
\therefore\ {}_2F_1\left[\begin{matrix}\beta,\gamma;\\ \beta+\gamma;\end{matrix} t(1-x+xt)\right]
$$

$$
=\sum_{n=0}^{\infty}\sum_{m=0}^{n}\frac{(\beta)_{n+m}(\gamma)_{n+m}\cdot}{(\beta+\gamma)_{n+m}}\frac{(-1)^m(1-t)^m(x)^m t^n}{m!n!}\quad\text{(replacing } n \text{ by } n+m)
$$

$$
=\sum_{n=0}^{\infty}\sum_{m=0}^{n}\frac{(\beta+m)_n(\gamma+m)_n\cdot}{(\beta+\gamma+m)_n}\frac{t^n}{n!}
$$

$$
=\sum_{m=0}^{\infty}\frac{(\beta)_m(\gamma)_m(-1)^m x^m(1-t)^m}{(\beta+\gamma)_m}\frac{t^m}{m!}
$$

$$
=\sum_{m=0}^{\infty}{}_2F_1\left[\begin{matrix}\beta+m,\gamma+m;\\ \beta+\gamma+m;\end{matrix}\right]\frac{(\beta)_m(\gamma)_m(-1)^m x^m(1-t)^m}{(\beta+\gamma)_m}\frac{t^m}{m!}.
$$

Now,

$$
{}_2F_1\left[\begin{matrix}\beta,\gamma; \\ \beta+\gamma;\end{matrix} t(1-x+xt)\right]=\sum_{m=0}^{\infty}(1-t)^{-m}\,{}_2F_1\left[\begin{matrix}\beta,\gamma; \\ \beta+\gamma+m;\end{matrix} t\right]\frac{(\beta)_m(\gamma)_m(-x)^m(1-t)^m}{(\beta+\gamma)_m}\frac{t^m}{m!}
$$

$$
=\sum_{m=0}^{\infty}\sum_{n=0}^{\infty}\frac{(\beta)_n(\gamma)_n}{(\beta+\gamma+m)_n}\frac{t^n}{n!}\frac{(\beta)_m(\gamma)_m(-x)^m}{(\beta+\gamma)_m}\frac{t^m}{m!}
$$

$$
=\sum_{m=0}^{\infty}\sum_{n=0}^{\infty}\frac{(\beta)_n(\gamma)_n}{(\beta+\gamma)_{n+m}}\frac{(\beta)_m(\gamma)_m(-x)^m}{m!n!}t^{m+n}
$$

$$
=\sum_{n=0}^{\infty}\sum_{m=0}^{\infty}\frac{(\beta)_{n-m}(\gamma)_{n-m}}{(\beta+\gamma)_n}(\beta)_m(\gamma)_m(-x)^m\frac{t^n}{(n-m)!m!}\ \text{(replacing } m+n \text{ by } n)
$$

$$
=\sum_{n=0}^{\infty}\sum_{m=0}^{\infty}\frac{(\beta)_m(\gamma)_m(-n)_m(x)^m}{(1-\beta-n)_m(1-\gamma-n)_m(\beta+\gamma)_n}\frac{t^n}{(n!m!}\ \text{(replacing } n-m \text{ by } n)
$$

$$
=\sum_{n=0}^{\infty}{}_3F_2\left[\begin{matrix}-n,\beta,\gamma; \\ 1-\beta-n,1-\gamma-n;\end{matrix} x\right]\frac{(\beta)_n(\gamma)_n}{(\beta+\gamma)_n}\frac{t^n}{n!}.
$$

$$
(10.59)
$$

Comparing the coefficient of t^n in the right-hand members of Equations (10.58) and (10.59), we get

$$
{}_3F_2\left[\begin{matrix}-n,\beta,\gamma; \\ 1-\beta-n,1-\gamma-n;\end{matrix} x\right]
$$

$$
=(1-x)^n\,{}_3F_2\left[\begin{matrix}-\frac{1}{2}n,-\frac{1}{2}n+\frac{1}{2},1-\beta-\gamma-n;; \\ 1-\beta-n,1-\gamma-n;\end{matrix} \frac{4x}{(1-x)^2}\right]. \quad (10.60)
$$

Here, n is a non-negative integer, and β,γ are independent of n. Equation (10.60) is known as *Whipple's theorem* for terminating the generalized hypergeometric series ${}_3F_2$.

For non-terminating ${}_3F_2$, we get the result

$$
{}_3F_2\left[\begin{matrix}\alpha,\beta,\gamma; \\ 1-\alpha-\beta,1+\alpha-\gamma;\end{matrix} x\right]
$$

$$
=(1-x)^{-\alpha}\,{}_3F_2\left[\begin{matrix}\frac{1}{2}\alpha,\frac{1}{2}\alpha+\frac{1}{2},1+\alpha-\beta-\gamma;; \\ 1+\alpha-\beta,1+\alpha-\gamma;\end{matrix} \frac{-4x}{(1-x)^2}\right]. \quad (10.61)
$$

Dixon's Theorem 10.8.3.

The series ${}_{3-q+1}F_q\left[\begin{matrix}\alpha_0,,\alpha_1,,\ldots\ldots\alpha_q,;\\ \beta_1,\beta_2,\ldots\ldots\beta_q;\end{matrix}x\right]$ is said to be *well poised.*

If $1+\alpha_0,=\alpha_1,+\beta_1,=\alpha_1,+\beta_1,=\cdots=\alpha_q,+\beta_q,.$

We observe that in Whipple's theorems (10.8.2), with the above condition on the left sides, are well poised. We will write here the summation of a well-poised ${}_3F_2$ with a unit argument, which is first obtained by Dixon. Now,

$$ {}_3F_2\left[\begin{matrix}\alpha,\beta,\gamma;\\ 1+\alpha-\beta,1+\alpha-\gamma;\end{matrix}1\right]=\sum_{m=0}^{\infty}\frac{(\alpha)_m(\beta)_m(\gamma)_m}{(1+\alpha-\beta)_m(1+\alpha-\gamma)_m}. $$

By deduction, we get

$$ {}_3F_2\left[\begin{matrix}\alpha,\beta,\gamma;\\ 1+\alpha-\beta,1+\alpha-\gamma;1\end{matrix}\right] $$

$$ =\frac{\Gamma(1+\alpha-\beta)\Gamma(1+\alpha-\gamma)}{\Gamma(1+\alpha-\beta-\gamma)\Gamma(1+\alpha)}\times\sum_{n=0}^{\infty}{}_3F_2\left[\begin{matrix}-n,\alpha;\\ 1+\alpha+n;\end{matrix}1\right]\frac{(\beta)_n(\gamma)_n}{(1+\alpha)_n n!}. $$

Applying Kummer's theorem with $\beta=-n,$ we get

$$ {}_3F_2\left[\begin{matrix}\alpha,\beta,\gamma;\\ 1+\alpha-\beta,1+\alpha-\gamma;1\end{matrix}\right] $$

$$ =\frac{\Gamma(1+\alpha-\beta)\Gamma(1+\alpha-\gamma)\Gamma\left(1+\frac{1}{2}\alpha\right)\Gamma\left(1+\frac{1}{2}\alpha-\beta-\gamma\right)}{\Gamma(1+\alpha-\beta-\gamma)\Gamma(1+\alpha)\Gamma\left(1+\frac{1}{2}\alpha-\beta\right)\Gamma\left(1+\frac{1}{2}\alpha-\gamma\right)}. $$

If $\alpha, \beta, and\ \gamma$ are such that each of its gamma functions exists herein, then

$$
{}_3F_2\left[\begin{matrix}\alpha,\beta,\gamma;\\ 1+\alpha-\beta,1+\alpha-\gamma;1\end{matrix}\right]
=\frac{\Gamma\left(1+\frac{1}{2}\alpha\right)\Gamma(1+\alpha-\beta)\Gamma(1+\alpha-\gamma)\Gamma\left(1+\frac{1}{2}\alpha-\beta-\gamma\right)}{\Gamma(1+\alpha)\Gamma\left(1+\frac{1}{2}\alpha-\beta\right)\Gamma\left(1+\frac{1}{2}\alpha-\gamma\right)\Gamma(1+\alpha-\beta-\gamma)}.
\tag{10.62}
$$

We can present an alternative form of Equation (10.62), if we apply the following formulae:

$$
\Gamma(1+x)=x\Gamma(x) \text{ and } \Gamma(x)\Gamma(1-x)=\frac{\pi}{sin\pi x}.
$$

By calculation, we get Equation (vii) in the form:

$$
{}_3F_2\left[\begin{matrix}\alpha,\beta,\gamma;\\ 1+\alpha-\beta,1+\alpha-\gamma;1\end{matrix}\right]
=\frac{cos\frac{1}{2}\pi\alpha\sin\left(\beta-\frac{1}{2}\alpha\right)\Gamma(1-\alpha)\Gamma\left(\beta-\frac{1}{2}\alpha\right)\Gamma(1+\alpha-\gamma)\Gamma\left(1+\frac{1}{2}\alpha-\beta-\gamma\right)}{\sin\pi(\beta-\alpha)\Gamma\left(1-\frac{1}{2}\alpha\right)\Gamma(\beta-\alpha)\Gamma\left(1+\frac{1}{2}\alpha-\gamma\right)\Gamma(1+\alpha-\beta-\gamma)}.
\tag{10.63}
$$

10.9 Integrals Involving Generalized Hypergeometric Functions

We will obtain a simple integral for ${}_pF_q$ by extending the method in Article 10.3 and Equation (10.6). We know that if n is a non negative integer and $Re(a_j) > Re(a_i) > 0\ for\ i=1,2\ldots p\ and\ j=1,2,\ldots q;\ then$

$$
\frac{(a)_n}{(b)_n}=\frac{\Gamma(b)}{\Gamma(a)\Gamma(b-a)}\int_0^1 t^{a+n-1}(1-t)^{b-a-1}\,dt.
$$

Also, we have

$$ {}_pF_q\left[\begin{matrix} a_1,a_2,\dots a_p;x \\ b_1,b_2,\dots b_q; \end{matrix}\right] = \sum_{n=0}^{\infty} \frac{(a_1)_n(a_2)_n\dots(a_i)_n\dots(a_p)_n}{(b_1)_n(b_2)_n\dots(b_j)_n\dots(b_q)_n}\frac{x^n}{n!} $$

$$ =\sum_{n=0}^{\infty}\frac{(a_1)_n(a_2)_n\dots(a_{i-1})_n(a_{i+1})_n\dots(a_p)_n}{(b_1)_n(b_2)_n\dots(b_{j-1})_n(b_{j+1})_n\dots(b_q)_n}\frac{x^n}{n!}\frac{\Gamma(b_j)}{\Gamma(a_i)\Gamma(b_j-a_i)}\left\{\int_0^1 t^{(a_i+n-1)}(1-t)^{b_j-a_i-1}dt\right\}] $$

$$ =\frac{\Gamma(b_j)}{\Gamma(a_i)\Gamma(b_j-a_i)}\int_0^1 t^{(a_i-1)}(1-t)^{b_j-a_i-1}\times\sum_{n=0}^{\infty}\frac{(a_1)_n(a_2)_n\dots(a_{i-1})_n(a_{i+1})_n\dots(a_p)_n}{(b_1)_n(b_2)_n\dots(b_{j-1})_n(b_{j+1})_n\dots(b_q)_n}\frac{(xt)^n}{n!}dt $$

$$ =\frac{\Gamma(b_j)}{\Gamma(a_i)\Gamma(b_j-a_i)}\int_0^1 t^{(a_i-1)}(1-t)^{b_j-a_i-1}{}_{p-1}F_{q-1}dt $$

when ${}_{p-1}F_{q-1} = {}_{p-1}F_{q-1}\left[\begin{matrix} a_1a_2\dots a_{i-1}a_{i+1}\dots a_p; \\ b_1b_2\dots b_{j-1}b_{j+1}\dots b_q; \end{matrix} xt\right]$.

Thus, if $p\le q+1, Re(b_j)>Re(a_i)>0, \textit{and none of } b_1b_2\dots b1_q$ is zero or a negative integer, we have for $|x|<1$

$$ {}_pF_q\left[\begin{matrix} a_1a_2\dots a_p; \\ b_1b_2\dots b_q; \end{matrix} x\right] = \frac{\Gamma(b_j)}{\Gamma(a_i)\Gamma(b_j-a_i)}\int_0^1 t^{(a_i-1)}(1-t)^{b_j-a_i-1}{}_{p-1}F_{q-1}\left[\begin{matrix} a_1a_2\dots a_{i-1}a_{i+1}\dots a_p; \\ b_1b_2\dots b_{j-1}b_{j+1}\dots b_q; \end{matrix} xt\right]dt. $$

If $p\le q$, the condition $|x|<1$ *may be omitted*
For easy calculation:

$$ \frac{(a)_n}{(b)_n}=\frac{(\Gamma(a+n)\Gamma(b)}{\Gamma(b+n)\Gamma(a)}=\frac{\Gamma(b)}{\Gamma(a)\Gamma(b-a)}\frac{\Gamma(a+n)\Gamma(b-a)}{\Gamma(a+n)} $$

$$ =\frac{\Gamma(b)}{\Gamma(a)\Gamma(b-a)}B(a+n,b-a) $$

$$ =\frac{\Gamma(b)}{\Gamma(a)\Gamma(b-a)}\int_0^1 t^{a+n-1}(1-t)^{b-a-1}dt. $$

We also know that $(1-t)^{-\alpha}=1+\alpha t+\frac{\alpha(\alpha+1)}{2!}t^2+\frac{\alpha(\alpha+1)(\alpha+2)}{3!}t^3$ $+\cdots\dots=\sum_{n=0}^{\infty}\frac{(\alpha)_n}{n!}t^n$.
Hence, in Article10.3 and Equation (10.6), we get

$$ F(\alpha,\beta,\gamma:x)=\frac{\Gamma(\gamma)}{\Gamma(\beta)\Gamma(\gamma-\beta)}\left\{\int_0^1 t^{\beta-1}(1-t)^{\gamma-\beta-1}\right\}(1-xt)^{-\alpha}dt]. $$

10.10 Some Special Generalized Hypergeometric Functions

We consider some particular cases of function ${}_pF_q$.

Putting $p=q=0$, we have

${}_0F_0(-;-;x)=\sum_{n=0}^{\infty}\frac{x^n}{n!}$, which is an exponential series.

If $p=1, q=0$, we have

${}_1F_0(\alpha;-;x=\sum_{n=0}^{\infty}\frac{(\alpha)_n x^n}{n!}$, which is a binomial series, convergent for $|x|<1$.

If $p=0, q=1$, we have

$$ {}_0F_1(-;\beta;x)=\sum_{n=0}^{\infty}\frac{x^n}{(\beta)_n n!}. $$

If we take, in this case, $\beta=1+n, x=\frac{-z^2}{4}$ and multiply both sides by $\frac{(\frac{1}{2}z)^n}{\Gamma(1+n)}$, we get ($n$ is not a negative integer)

$$ \frac{(\frac{1}{2}z)^n}{\Gamma(1+n)}\,{}_0F_1\left(-;1+n;\frac{-z^2}{4}\right)=\sum_{m=0}^{\infty}\frac{(\frac{1}{2}z)^n}{\Gamma(1+n)}\frac{(\frac{-z^2}{4})^m}{(1+n)_m m!} $$

$$ =\sum_{m=0}^{\infty}\frac{(-1)^m}{\Gamma(1+n)}\frac{(\frac{z}{2})^{2m+n}}{(1+n)_m m!} $$

$$ =\sum_{m=0}^{\infty}\frac{(-1)^n}{\Gamma(1+n+m)}\frac{(\frac{z}{2})^{2m+n}}{m!}=J_n(z), $$

which is a Bessel's function $J_n(x)$.

Now, we consider the properties of the function ${}_1F_1$, which can be directly obtained from those of ${}_pF_q$ as we will get two of special properties.

Now, the function ${}_1F_1(\alpha;\beta;x) = \sum_{n=0}^{\infty} \frac{(\alpha)_n x^n}{(\beta)_n n!}$, which is analytic for all finite x, when β is not zero or a negative integer, is known as the *Pochhammer–Barnes confluent hypergeometric functions.*

(a) Now, from the properties of ${}_pF_q$, the confluent hypergeometric differential equation is

$$x\frac{d^2y}{dx^2} + (\beta - x)\frac{dy}{dx} - \alpha y = 0, \tag{10.64}$$

and its general solution in the neighborhood of $x = 0$ if β is a non-negative integral, is

$$y = A\,{}_1F_1(\alpha;\beta;x) + Bx^{1-\beta}\,{}_1F_1(\alpha+1-\beta;2-\beta;x). \tag{10.65}$$

(b) Three contiguous function relations between ${}_1F_1$ and its two contiguous functions are

$$(\alpha-\beta+1)\,{}_1F_1(\alpha;\beta;x) = \alpha\,{}_1F_1(\alpha+1;\beta;x) - (\beta-1)\,{}_1F_1(\alpha;\beta-1;x) \tag{10.66}$$

$$\beta(\alpha+x)\,{}_1F_1(\alpha;\beta;x) = \alpha\beta\,{}_1F_1(\alpha+1;\beta;x) - (\alpha-\beta)\,{}_1F_1(\alpha;\beta+1;x) \tag{10.67}$$

$$\beta\,{}_1F_1(\alpha;\beta;x) = \alpha\beta\,{}_1F_1(\alpha-1;\beta;x) + x\,{}_1F_1(\alpha;\beta+1;x) \tag{10.68}$$

(c) If $Re(\beta) > Re(\alpha) > 0$, the hypergeometric differential equations

$${}_1F_1(\alpha;\beta;x) = \frac{\Gamma(\beta)}{\Gamma(\alpha)\Gamma(\beta-\alpha)}\int_0^1 e^{xt}t^{\alpha-1}(1-t)^{\beta-\alpha-1}. \tag{10.69}$$

We may write Equation (10.64) as

$$x(1-x)\frac{d^2y}{dx^2} + \{\gamma - (1+\alpha+\beta)x\}\frac{dy}{dx} - \alpha\beta y = 0. \tag{10.70}$$

One of its solutions is

$$y = {}_2F_1(\alpha;\beta;\gamma;x).$$

Let us replace x *by* x/β in Equation (vii). We get

$$x(1-x/\beta)\frac{d^2y}{dx^2}+\left\{\gamma-\left(1+\frac{1+\alpha}{\beta}\right)x\right\}\frac{dy}{dx}-\alpha y=0. \quad (10.71)$$

Let $\beta \to \infty$. We find that one of its solutions,

$$x\frac{d^2y}{dx^2}+\{\gamma-x\}\frac{dy}{dx}-\alpha y=0, \quad (10.72)$$

is $y=\log_{\beta\to\infty} {}_2F_1(\alpha;\beta;\gamma;x/\beta)=\log_{\beta\to\infty}\sum_{n=0}^{\infty}\frac{(\alpha)_n(\beta)_n}{(\gamma)_n}\frac{x^n}{n!\,\beta^n}.$

Here, $\log_{\beta\to\infty}, \frac{(\beta)_n}{\beta^n}=\log_{\beta\to\infty}\beta\left(1+\frac{1}{\beta}\right)\left(1+\frac{2}{\beta}\right)\ldots.\left(1+\frac{n-1}{\beta}\right)=1.1\ldots1=1.$

Therefore, the solution of Equation (ix) reduces to

$$y=\log_{\beta\to\infty}\sum_{n=0}^{\infty}\frac{(\alpha)_n}{(\gamma)_n}\frac{x^n}{n!}={}_1F_1(\alpha;\gamma;x).$$

Kummer's Theorem 10.10.1.

We derive some special properties possessed by the confluent hypergeometric function ${}_1F_1(\alpha;\beta;x)$ in the following:

(i) We can write

$$e^{-x}\,{}_1F_1(\alpha;\beta;x)=\sum_{n=0}^{\infty}\frac{(-1)^n x^n}{n!}\sum_{m=0}^{\infty}\frac{(\alpha)_m}{(\beta)_m}\frac{x^m}{m!}$$

$$=\sum_{n=0}^{\infty}\sum_{m=0}^{\infty}\frac{(-1)^n(\alpha)_m}{(\beta)_m}\frac{x^{n+m}}{n!m!}=\sum_{n=0}^{\infty}\sum_{m=0}^{n}\frac{(-1)^{n-m}(\alpha)_m}{(\beta)_m}\frac{x^n}{(n-m)!m!}$$

$$=\sum_{n=0}^{\infty}\sum_{m=0}^{n}\frac{(-1)^n\,(\alpha)_m}{(\beta)_m}\frac{(-n)_m\,x^n}{n!m!}$$

$$=\sum_{n=0}^{\infty}\sum_{m=0}^{n}\frac{(-n)_m(\alpha)_m}{(\beta)_m}\frac{(-x)^n}{n!m!}$$

$$=\sum_{n=0}^{\infty}{}_2F_1(-n,\alpha;\beta;x)\frac{(-x)^n}{n!}$$

$$=\sum_{n=0}^{\infty}\frac{(\beta-\alpha)_n}{(\beta)_n}\frac{(-x)^n}{n!}={}_1F_1(\beta-\alpha;\beta;-x).$$

If β is neither zero nor negative, we get

$$ {}_1F_1(\alpha;\beta;x)=e^x\,{}_1F_1(\beta-\alpha;\beta;-x), \tag{10.73}$$

which is known as *Kummer's first theorem.*

(ii) Putting $\beta=2\alpha$ in Equation (x), we get

$$ {}_1F_1(\alpha;2\alpha;x)=e^x\,{}_1F_1(\alpha;2\alpha;-x)$$
$$\text{or}\quad e^{-x/2}\,{}_1F_1(\alpha;2\alpha;x)=e^{x/2}\,{}_1F_1(\alpha;2\alpha;-x).$$

This shows that $e^{-x/2}\,{}_1F_1(\alpha;2\alpha;x)$, and therefore also $e^{-x}\,{}_1F_1(\alpha;2\alpha;2x)$ is an even function of x.

We have now

$$e^{-x}\,{}_1F_1(\alpha;2\alpha;2x)=\sum_{n=0}^{\infty}\frac{(-1)^n x^n}{n!}\sum_{m=0}^{\infty}\frac{(\alpha)_m}{(2\alpha)_m}\frac{2^m x^m}{m!}$$

$$=\sum_{n=0}^{\infty}\sum_{m=0}^{\infty}\frac{(-1)^n}{n!}\frac{(\alpha)_m}{(2\alpha)_m}\frac{2^m x^{m+n}}{m!}$$

$$=\sum_{n=0}^{\infty}\sum_{m=0}^{\infty}\frac{(-1)^{n-m}}{(n-m)!}\frac{(\alpha)_m}{(2\alpha)_m}\frac{2^m x^n}{m!}$$

$$=\sum_{n=0}^{\infty}\sum_{m=0}^{\infty}\frac{(-n)_m\,(\alpha)_m}{(2\alpha)_m}\frac{2^m(-x)^n}{m!n!}$$

$$=\sum_{n=0}^{\infty}{}_2F_1(-n,\alpha;2\alpha;2)\frac{(-x)^n}{n!}.$$

Since the left member of this equation is an even function of x we get the following results for k, a non negative integer:

$$_2F_1(-2k-1,\alpha;2\alpha;2)=0$$

$$e^{-x}\,{}_1F_1(\alpha;2\alpha;2x)=\sum_{k=0}^{\infty}{}_2F_1(-2k,\alpha;2\alpha;2)\frac{(x)^{2k}}{(2k)!}. \qquad (10.74)$$

We know that $y={}_1F_1(\alpha;\beta;x)$ is one solution of the differential equation

$$x\frac{d^2y}{dx^2}+\{\beta-x\}\frac{dy}{dx}-\alpha y=0.$$

Substituting $\beta=2\alpha\,, x=2z\ and\ y=e^x w,$ we see that $\mathrm{w}=e^{-x}\,{}_1F_1(\alpha;2\alpha;2z)$ is one of the solutions of the differential equation

$$z\frac{d^2w}{dx^2}+2\alpha\frac{dw}{dx}-zw=0. \qquad (10.75)$$

Again, replacing $z\ by\ \delta$ and substituting $\delta=\frac{1}{4}x^2$, Equation (xii) becomes

$$\delta^2\frac{d^2w}{d\delta^2}+\left(\alpha+\frac{1}{2}\right)\delta\frac{dw}{d\delta}-\delta w=0,$$

which, in terms of $\theta\equiv\delta\frac{d}{d\delta}$, may be written as

$$\left[\theta\left(\theta+\alpha+\frac{1}{2}\right)-\delta\right]w=0. \qquad (10.76)$$

Comparing it with Equation (10.65), we find that $w={}_0F_1\left(-;\beta+\frac{1}{2}\ ;\frac{1}{4}x^2\right)$ is the solution of Equation (10.76).

The general solution of Equation (10.75), under the condition that $\alpha + \frac{1}{2}$ is not an integer, i.e., 2α is not an odd integer, is

$$w = A\,{}_0F_1\left(-;\alpha + \frac{1}{2};\frac{1}{4}x^2\right) + B(x^2)^{\frac{1}{2}-\alpha}\,{}_0F_1\left(-;\frac{3}{2}-\alpha;\frac{1}{4}x^2\right). \tag{10.77}$$

We have seen also that $w = e^{-x}\,{}_1F_1(\alpha;2\alpha;2z)$ also satisfies the differential Equation (10.76).

Therefore, for some A *and* B, there exists the identity

$$e^{-x}\,{}_1F_1(\alpha;2\alpha;2z) = A\,{}_0F_1\left(-;\alpha + \frac{1}{2};\frac{1}{4}x^2\right) + B(x)^{1-2\alpha}\,{}_0F_1\left(-;\frac{3}{2}-\alpha;\frac{1}{4}x^2\right).$$

Analytically, the first term of the above equation requires $B = 0$, then $A = 0$. Therefore, we get

$$e^{-x}\,{}_1F_1(\alpha;2\alpha;2z) = {}_0F_1\left(-;\alpha + \frac{1}{2};\frac{1}{4}x^2\right), \tag{10.78}$$

The result Equation (10.78) is known as *Kummer's second theorem.* Using Equations (10.78) and (10.74), we find

$$\sum_{k=0}^{\infty} {}_2F_1(-2k,\alpha;2\alpha;2)\frac{(x)^{2k}}{(2k)!} = \sum_{k=0}^{\infty} \frac{(x)^{2k}}{2^{2k}k!(\alpha+\frac{1}{2})_k}$$

$$= \sum_{k=0}^{\infty} \frac{(\frac{1}{2})_k (x)^{2k}}{2^{2k}(\frac{1}{2})_k k!(\alpha+\frac{1}{2})_k}$$

$$= \sum_{k=0}^{\infty} \frac{(\frac{1}{2})_k (x)^{2k}}{2^{2k}(2k)!(\alpha+\frac{1}{2})_k}.$$

This gives

$${}_2F_1(-2k,\alpha;2\alpha;2) = \frac{(\frac{1}{2})_k}{(\alpha+\frac{1}{2})_k}. \quad (10.79)$$

Ramanujan's Theorem 10.10.2.

If we use the result of Dixon's theorem and substitute $\alpha = -2n$, $\beta = \alpha$, $\gamma = 1-\beta-2n$ in the result of Dixon's Theorem 10.8.3, Equation (10.63)

$${}_2F_1\left(\begin{matrix} -2n,\alpha,1-\beta-2n; \\ 1-\alpha-2n,\beta; \end{matrix} 1\right) = \frac{(2n)!(\alpha)_n(\beta-\alpha)_n}{n!(\alpha)_{2n}(\beta)_n}. \quad (10.80)$$

We will use this result to get a theorem obtained by the Indian mathematician S. Ramanujan. Let us consider

$$
{}_2F_1\left[\begin{matrix}\alpha;\\ \beta;\end{matrix} x\right] {}_2F_1\left[\begin{matrix}\alpha;\\ \beta;\end{matrix} -x\right] = \sum_{n=0}^{\infty}\frac{(\alpha)_n}{(\beta)_n}\frac{x^n}{n!}\sum_{m=0}^{\infty}\frac{(-1)^m(\alpha)_m}{(\beta)_m}\frac{x^m}{m!}
$$

$$
=\sum_{n=0}^{\infty}\sum_{m=0}^{\infty}\frac{(\alpha)_m}{(\beta)_m}\frac{(\alpha)_n}{(\beta)_n}\frac{(-1)^m x^{m+n}}{m!n!}
$$

$$
=\sum_{n=0}^{\infty}\sum_{m=0}^{\infty}\frac{(\alpha)_m}{(\beta)_m}\frac{(\alpha)_{n-m}}{(\beta)_{n-m}}\frac{(-1)^m x^n}{(n-m)!n!}
$$

$$
=\sum_{n=0}^{\infty}\sum_{m=0}^{\infty}\frac{(\alpha)_m}{(\beta)_m}\frac{(\alpha)_n(1-\beta-m)_m}{(\beta)_{n-m}(1-\alpha-n)_m}\frac{(-n)_m x^n}{m!n!}
$$

$$
=\sum_{n=0}^{\infty}{}_3F_2\left(\begin{matrix}-n,\alpha,1-\beta-n;\\ 1-\alpha-n,\beta;\end{matrix}1\right)\frac{(\alpha)_m}{(\beta)_m}\frac{x^n}{n!}
$$

$$
=\sum_{n=0}^{\infty}\frac{n!(\alpha)_{n/2}(\beta-\alpha)_{n/2}}{(n/2)!(\beta)_{n/2}(\alpha)_n}\frac{(\alpha)_n}{(\beta)_n}\frac{x^n}{n!}
$$

$$
=\sum_{n=0}^{\infty}\frac{(\alpha)_n(\beta-\alpha)_n}{(\beta)_{2n}}\frac{x^{2n}}{n!}
$$

$$
=\sum_{n=0}^{\infty}\frac{(\alpha)_n(\beta-\alpha)_n}{(\beta)_n(\frac{1}{2}\beta)_n(\frac{1}{2}\beta+\frac{1}{2})_n}\frac{1}{n!}(\frac{x^2}{4})^n.
$$

Hence,

$$
{}_2F_1\left[\begin{matrix}\alpha;\\ \beta;\end{matrix} x\right] {}_2F_1\left[\begin{matrix}\alpha;\\ \beta;\end{matrix} -x\right] = {}_2F_3\left(\begin{matrix}\alpha,\beta-\alpha,;\\ \beta,\frac{1}{2}\beta,\frac{1}{2}\beta+\frac{1}{2};\end{matrix}\frac{x^2}{4}\right). \tag{10.81}
$$

10.11 Barnes Type Contour Integrals

The contour integral

$$I_{p,q} = \frac{1}{2\pi i}\int_B \frac{[\Pi_{n=1}^{p}\Gamma(\alpha_n + s)]\Gamma(-s)(-x)^q}{\Pi_{i=1}^{q}\Gamma(\beta_i + s)} ds \qquad (10.82)$$

where the path of integration B is the imaginary axis, modified, if necessary, by loops to make the poles Γ(-s) lie to the right of the path, and the poles of $\Gamma(\alpha_n + s), n = 1,2\ldots p$ *to* the left of the path, is known as *Barnes type contour integral.*

We will see below that this integral can be connected with the generalized hypergeometric series.

If $p = q+1$ and no α_n is zero or a negative, the Barnes type integral $I_{p,q}$ is an analytic function of x in the cut plane $[\arg(-x)] < \pi$. If $p = q$ and no α_n is zero or a negative integer, the same integral $I_{p,q}$ is an analytic function of x in the open half plane $[\arg(-x)] < \pi/2$.

We will therefore find the relation connecting (i) ${}_{q+1}F_p$ with the Barnes integral $I_{q+1,p}$ and (ii) ${}_pF_q$ with the Barnes integral $I_{q,,p}$.

(i) Let us write

$$\Phi(s) = \frac{\Gamma(-s)(-x)^s \Pi_{n=1}^{p}\Gamma(\alpha_n + s)}{\Pi_{i=1}^{q}\Gamma(\beta_i + s)} \qquad (10.83)$$

and consider the integral $\frac{1}{2\pi i}\int_{B_n + C_n} \Phi(s)ds$ (10.84)

over the closed path $(B_n + C_n)$ where n is a non-negative integer, and B_n is Barnes path terminated at $s = \left(n + \frac{1}{2}\right)i$ *and* $s = -\left(n + \frac{1}{2}\right)$, and C_n is the semicircle $s = \left(n + \frac{1}{2}\right)e^{i\theta}$, $-\frac{1}{2\pi} \le \theta \le \frac{1}{2\pi}$. As n tends to infinity, the path B_n tends to Barnes path B. Since we are moving round the contour in the clockwise direction, the value of the integral Equation (iii) will be the negative of the sum of the residues of $f(s)$ at the simple poles $s = k, k = 0, 1\ldots.n$, *lying within* the contour. To find the sum of residues,

since $\Gamma(-s)\Gamma(1+s)=\dfrac{-\pi}{\sin\pi s}$,
we may write, for the sake of convenience,

$$\Phi(s)=\frac{-\pi(-x)^s}{\sin\pi s\,\Gamma(1+s)}\frac{\Pi_{n=1}^{p}\Gamma(\alpha_n+s)}{\Pi_{i=1}^{q}\Gamma(\beta_i+s)}.$$

Then, the negative residue of $\Phi(s)$ at $s=k$ is

$$\lim_{s\to k}\frac{\pi(s-k)(-x)^s}{\sin\pi s\,\Gamma(1+s)}\frac{\Pi_{n=1}^{p}\Gamma(\alpha_n+s)}{\Pi_{i=1}^{q}\Gamma(\beta_i+s)}=\frac{(-1)^k(-x)^s}{\Gamma(1+k)}\frac{\Pi_{n=1}^{p}\Gamma(\alpha_n+s)}{\Pi_{i=1}^{q}\Gamma(\beta_i+s)}$$

$$=\frac{\Pi_{n=1}^{p}\Gamma(\alpha_n)(\alpha_n)_k}{\Pi_{i=1}^{q}\Gamma(\beta_i)(\beta_i)_k}\frac{x^k}{k!}$$

$$\frac{1}{2\pi i}\int_{B_n+C_n}\Phi(s)ds=\frac{1}{2\pi i}\int_{B_n}\Phi(s)ds+\frac{1}{2\pi i}\int_{C_n}\Phi(s)ds$$

$$=\frac{\Pi_{n=1}^{p}\Gamma(\alpha_n)}{\Pi_{i=1}^{q}\Gamma(\beta_i)}\sum\frac{\Pi_{n=1}^{p}(\alpha_n)_k}{\Pi_{i=1}^{q}(\beta_i)_k}\frac{x^k}{k!}\qquad(10.85)$$

When $n\to\infty$, $\dfrac{1}{2\pi i}\int_{B_n}\Phi(s)ds\to I_{p,q}$, and when $n\to\infty$, $\dfrac{1}{2\pi i}\int_{C_n}f(s)ds$ $\to 0$ under the conditions $p=q+1$, $[x]<1$, $|\arg(-x)|\le\pi-\delta$, $\delta>0$, and no α_n or β_i is zero or a negative integer.

Under the above conditions, the following equation holds:

$$\frac{1}{2\pi i}\int(-x)^s\frac{\Gamma(-s)\Pi_{n=1}^{q+1}\Gamma(\alpha_n+s)}{\Pi_{i=1}^{q}\Gamma(\beta_i+s)}$$

$$=\frac{\Pi_{n=1}^{q+1}\Gamma(\alpha_n)}{\Pi_{i=1}^{q}\Gamma(\beta_i)}\,{}_{q+1}F_q\left[\begin{matrix}\alpha_1,\alpha_2,\ldots\alpha_{q+1};\\ \beta_1,\beta_2,\ldots\beta_q;\end{matrix}\;x\right].\qquad(10.86)$$

(i) If $p=q$ and no α_n is zero or a negative integer, the Barnes integral $I_{p,q}$ is analytic in the open half plane $|\arg(-x)|<\frac{1}{2}\pi$, i.e., $Re(x)<0$. We conclude that

when $n\to\infty, \frac{1}{2\pi i}\int_{B_n} f(s)ds\to 0.$

If $Re(x)<0$ and α_n or β_i is zero or negative integer,

$$\frac{1}{2\pi i}\int (-x)^s\frac{\Gamma(-s)\Pi_{n=1}^{q+1}\Gamma(\alpha_n+s)}{\Pi_{i=1}^{q}\Gamma(\beta_i+s)}=\frac{\Pi_{n=1}^{q}\Gamma(\alpha_n)}{\Pi_{i=1}^{q}\Gamma(\beta_i)}\,{}_q^0F_q\left[\begin{matrix}\alpha_1,\alpha_2,\ldots\alpha_{q;}\\ \beta_1,\beta_2,..\beta_{q;}\end{matrix}x\right].$$

(10.87)

10.12 Example

Example 10.12.1. Show that if $L\{F(t)\}$ denotes Laplace transform of the function $f(t)$, then

$$L\left\{t^\gamma\,{}_pF_q\left[\begin{matrix}\alpha_1,\alpha_2,\ldots\alpha_{q;}\\ \beta_1,\beta_2,..\beta_{q;}\end{matrix}xt\right]\right\}=\frac{\Gamma(1+\gamma)}{s^{1+\gamma}}\,{}_{p+1}F_q\left[\begin{matrix}1+\gamma,\alpha_1,\alpha_2,\ldots\alpha_{q;}\\ \beta_1,\beta_2,..\beta_{q;}\end{matrix}x/s\right].$$

Solution: We have

$$f(s)=L\{F(t)\}=\int_0^\infty e^{-st}F(t)dt.$$

Taking $F(t)=t^\gamma\,{}_pF_q\left[\begin{matrix}\alpha_1,\alpha_2,\ldots\alpha_{q;}\\ \beta_1,\beta_2,..\beta_{q;}\end{matrix}xt\right]$, we get

$$f(s)=\int_0^\infty e^{-st}t^\gamma\,{}_pF_q\left[\begin{matrix}\alpha_1,\alpha_2,\ldots\alpha_{p;}\\ \beta_1,\beta_2,..\beta_{q;}\end{matrix}xt\right]dt$$

$$=\int_0^\infty e^{-st}t^\gamma\sum_{n=0}^{\infty}\frac{\Pi_{i=1}^{p}\Gamma(\alpha_i)_n}{\Pi_{j=1}^{q}\Gamma(\beta_j)_n}\frac{x^n t^n}{n!}dt.$$

Assuming the validity of term-by-term integration, we have

$$f(s)=\sum_{n=0}^{\infty}\frac{\Pi_{i=1}^{p}\Gamma(\alpha_i)_n}{\Pi_{j=1}^{q}\Gamma(\beta_j)_n}\frac{x^n}{n!}\int_0^{\infty}e^{-st}t^{\gamma+n}dt=\sum_{n=0}^{\infty}\frac{\Pi_{i=1}^{p}\Gamma(\alpha_i)_n}{\Pi_{j=1}^{q}\Gamma(\beta_j)_n}\frac{x^n}{n!}\frac{\Gamma(\gamma+n+1)}{s^{\gamma+n+1}}$$

$$=\frac{\Gamma(\gamma+1)}{s^{\gamma+1}}\,{}_{p+1}F_q\left[\begin{matrix}1+\gamma,\alpha_1,\alpha_2,\ldots\alpha_{p;}\\ \beta_1,\beta_2,..\beta_{q;}\end{matrix}\; x/s\right].$$

Example 10.12.2. Show that ${}_1F_1(\alpha;\beta;x)=\frac{1}{\Gamma(\alpha)}\int_0^{\infty}e^{-t}t^{\alpha-1}\,{}_0F_1(-;\beta;xt)$.

Solution: *Right side* $=\frac{1}{\Gamma(\alpha)}\int_0^{\infty}e^{-t}t^{\alpha-1}\sum_{n=0}^{\infty}\frac{x^nt^n}{\Gamma(\beta_j)_n\;n!}\mathrm{dt}$

$$=\frac{1}{\Gamma(\alpha)}\sum_{n=0}^{\infty}\int_0^{\infty}e^{-t}t^{\alpha+n-1}dt$$

$$=\sum_{n=0}^{\infty}\frac{x^n\Gamma(\alpha+n)}{n!\Gamma(\alpha)(\beta)_n}=\sum_{n=0}^{\infty}\frac{x^n(\alpha)_n}{n!(\beta)_n}={}_1F_1(\alpha;\beta;x).$$

Example 10.12.3. The error function $\mathrm{erf}(x)$ *is* defined by

$$\mathrm{erf}(x)=\frac{2}{\sqrt{\pi}}\int_0^x e^{-t^2}dt.$$

Show that $\mathrm{erf}(x)=\frac{2x}{\sqrt{\pi}}\,{}_1F_1(\frac{1}{2},\frac{3}{2};-x^2)$.

Solution: We can write $\mathrm{erf}(x)=\frac{2}{\sqrt{\pi}}\int_0^x {}_0F_0(-;-;-t^2)dt$.

Using $p=q=0,\alpha=\beta=1,\gamma=-1,m=2,$ the result in Article 2.3 Equation (2.4), we get

$$\mathrm{erf}(x)=\frac{2x}{\sqrt{\pi}}\,{}_2F_2\left[\begin{matrix}\frac{1}{2},1;\\ 1,\frac{3}{2};\end{matrix}\;-t^2\right]=\frac{2x}{\sqrt{\pi}}\,{}_1F_1(\frac{1}{2},\frac{3}{2};-x^2).$$

Example 10.12.4. Prove that

$$(\beta)_k \frac{d^k}{d^k x}\left[e^{-x}\,{}_1F_1(\alpha,\beta;x)\right]=(-1)^k(\beta-\alpha)_k e^{-x}\,{}_1F_1(\alpha,\beta+k;x).$$

Solution: Using Kummer's first theorem, we can write

$$(\beta)_k \frac{d^k}{d^k x}\left[e^{-x}\,{}_1F_1(\alpha,\beta;x)\right]=(\beta)_k \frac{d^k}{d^k x}\sum_{n=0}^{\infty}\frac{(\beta-\alpha)_n}{(\beta)_n}\frac{(-x)^n}{n!}.$$

Assuming the validity of term by term integration, we have

$$=(\beta)_k \sum_{n=0}^{k}\frac{(\beta-\alpha)_n}{(\beta)_n}\frac{(-1)^n n(n-1)\ldots(n-k+1)(x)^{n-k}}{(\beta)_n n!}$$

$$=(\beta)_k \sum_{n=0}^{k}\frac{(\beta-\alpha)_n}{(\beta)_n}\frac{(-1)^n (x)^{n-k}}{(\beta)_n (n-k)!}$$

$$=(\beta)_k \sum_{n=0}^{k}\frac{(\beta-\alpha)_{n+k}}{(\beta)_n}\frac{(-1)^{n+k} (x)^{n}}{(\beta)_{n+k}\, n!}$$

$$=(\beta)_k \sum_{n=0}^{k}\frac{(\beta-\alpha)_{n+k}}{(\beta)_n}\frac{(\beta-\alpha+k)_n (-1)^{n+k} (-x)^{n}}{\beta_k(\beta+k)_n\, n!}$$

$$=(-1)^k(\beta-\alpha)_k \sum_{n=0}^{k}\frac{}{(\beta)_n}\frac{(\beta-\alpha+k)_n (-x)^{n}}{(\beta+k)_n\, n!}$$

$$=(-1)^k(\beta-\alpha)_k\,{}_1F_1(\beta-\alpha+k,\beta+k;x)$$

$$=(-1)^k(\beta-\alpha)_k e^{-x}\,{}_1F_1(\alpha,\beta+k;x).$$

Example 10.12.5. Prove that $P_n(x)={}_2F_1\left(-n,n+1;1;\frac{1-x}{2}\right)$.

Solution:

$$P_n(x)=\frac{1}{n!2^n}\frac{d^n}{dx^n}(x^2-1)^n, \text{ by Rodrigues formula}$$

$$=\frac{(-1)^n}{n!2^n}\frac{d^n}{dx^n}(1-x^2)^n$$

$$=\frac{(-1)^n}{n!}\frac{d^n}{dx^n}\left[(1-x)^n\frac{(1+x)^n}{2^n}\right]=\frac{(-1)^n}{n!}\frac{d^n}{dx^n}\left[(1-x)^n\left\{1-\frac{1-x}{2}\right\}^n\right]$$

$$=\frac{(-1)^n}{n!}\frac{d^n}{dx^n}\left[(1-x)^n\left\{1-n\frac{1-x}{2}+\frac{n(n-1)}{2!}\left(\frac{1-x}{2}\right)^2\ldots..\right\}\right]$$

$$=\frac{(-1)^n}{n!}\frac{d^n}{dx^n}\left[(1-x)^n-\frac{n}{2}(1-x)^{n+1}+\frac{n(n-1)}{2!2^2}(1-x)^{n+2}-\ldots.\right]$$

$$=\frac{(-1)^n}{n!}\left[(-1)^n n!-(-1)^n\frac{n}{2}(n+1)!(1-x)+(-1)^n\frac{n(n-1)}{2!2^2}\frac{(n+2)!}{2!}(1-x)^2-\ldots.\right]$$

$$[\textit{using this}, \frac{d^n}{dx^n}(a-bx)^m=(-1)^n b^n\frac{m!}{(m-n)!}(a-bx)^{(m-n)}]$$

$$=\frac{(-1)^{2n}}{n!}\left[n!-\frac{n}{2}(n+1).n!(1-x)+\frac{n(n-1)n(n+1)n!}{2!2^2.2!}+\cdots..\right]$$

$$=1+\frac{(-n)(n+1)}{1.1!}\left(\frac{1-x}{2}\right)+\frac{-n(-n+1)(n+1)(n+2)}{2.1.2!}\left(\frac{1-x}{2}\right)^2+\cdots\ldots$$

$$={}_2F_1\left(-n,n+1;1;\frac{1-x}{2}\right), \text{ by definition.}$$

10.13 Exercise

1. Show that

$${}_0F_1\left[\begin{matrix}-;\\ \alpha;\end{matrix}\,x\right]{}_0F_1\left[\begin{matrix}-;\\ \beta;\end{matrix}\,-x\right]={}_0F_3\left(\begin{matrix}\frac{1}{2}\alpha+\frac{1}{2}\beta,\frac{1}{2}\alpha+\frac{1}{2}\beta-\frac{1}{2};\\ \alpha,\beta,\alpha+\beta-1;\end{matrix}\,4x\right).$$

2. If $J_n(x)$ is defined by

$$J_n(x)=\frac{(\frac{x}{2})^n}{\Gamma(1+n)}\,{}_0F_1(-,1+n;-\frac{x^3}{4}),$$

n being a non-negative integer, prove that

$$\int_0^\infty e^{-t^2}t^{2\alpha-n-1}J_n(xt)\,dt=\frac{\Gamma(\alpha)x^n}{2^{n+1}\Gamma(1+n)}\,{}_1F_1(\alpha,1+n;-\frac{x^3}{4}).$$

3. Show that if L^{-1} denotes inverse Laplace transform, then

$$L^{-1}\left\{\frac{1}{S}\,{}_pF_{q+1}\left[\begin{matrix}\alpha_1,\alpha_2,\ldots\alpha_{p;}\\ S+1,\beta_1,\beta_2,..\beta_{q;}\end{matrix}\;x\right]\right\}$$

$$={}_pF_{q+1}\left[\begin{matrix}\alpha_1,\alpha_2,\ldots\alpha_{p;}\\ 1,\beta_1,\beta_2,..\beta_{q;}\end{matrix}\;x\left(1-e^{-t}\right)\right]$$

and $L^{-1}\left(\dfrac{S^k}{(S-x)^{k+1}}\right)={}_1F_1\left[\begin{matrix}k+1;\\ 1;\end{matrix}\;xt\right].$

4. Show that if $Re(\beta>0)$ and if n is a non-negative integer, then

$${}_1F_1\left[\begin{matrix}-\frac{1}{2}n(n+1);\\ \beta+\frac{1}{2};\end{matrix}\;1\right]=\frac{2^n(\beta)_n}{(2\beta)_n}.$$

5. Prove that

$$L\left\{ {}_pF_q\left[\begin{matrix}\alpha_1,\alpha_2,\ldots\alpha_{q};\\ \beta_1,\beta_2,..\beta_{q};\end{matrix} xt\right],s\right\} = \frac{1}{s}\,{}_{p+1}F_q\left[\begin{matrix}1,\alpha_1,\alpha_2,\ldots\alpha_{p};\\ \beta_1,\beta_2,..\beta_{q};\end{matrix} x/s\right]$$

where $L\{f(t),s\} = \int_0^\infty e^{-st} f(t)dt$.

6. If n is a non-negative integer and α,β,γ are independent of n, prove that

$${}_3F_q\left[\begin{matrix}-n,\alpha,\beta;\\ \gamma,1-\gamma+\alpha+\beta-n;\end{matrix} 1\right] = \frac{(\gamma-\alpha)_n(\gamma-\beta)_n}{(\gamma)_n(\gamma-\alpha-\beta)_n}.$$

7. Solve the Legendre equation $(1-x^2)\frac{d^2y}{dx^2} - 2x\frac{dy}{dx} + n(n+1)y = 0$ by changing it to a hypergeometric equation.

11

Bessel Functions

Abstract

Bessel functions were introduced by German mathematician and astronomer Friedrich Wilhelm Bessel (1784–1846) in his 1826 paper. They are solutions to a second-order differential equation and are known as Bessel functions of the first kind of order n. Bessel functions are widely used in physics and engineering to model phenomena such as wave propagation, heat transfer, and fluid dynamics. In this chapter, we study their generating functions, recurrence relations, and orthogonality properties.

Keywords: Bessel function, orthogonality, modified Bessel functions, Bessel function zeros, Bessel series, Neumann functions

11.1 Introduction

Bessel function was introduced by German mathematician and astronomer *Friederich Wilhelm Bessel (1784–1846)* whose paper on Bessel functions appeared in 1826, while discussing a problem in astonomy. The Bessel functions have received the most extensive treatment among all the special functions. We shall be discussing only the introductory results of these functions in this chapter. In the preceding chapter we dissussed how Bessel function is the solution of a second order differential equation and it is a multiple of the function ${}_0F_1$. Actually we define $J_n(x)$ as *the Bessel function of the first kind of order* n. The equation

$$x^2y'' + xy' + (x^2 - n^2)y = 0, \tag{11.1}$$

where n is a non-negative constant, is known as the Bessel's equation of order n. The function occurring in its solution are called Bessel function. In this chapter properties of Bessel's functions such as orthogonality, recurrence relations, generating function are studied.

Dipankar De. Ordinary Differential Equations and Special Functions, (373–430) © 2026 Scrivener Publishing LLC

11.2 Bessel's Equation

One of the most important differential equations in applied mathematics is

$$x^2y'' + xy' + (x^2 - n^2)y = 0, \tag{11.1}$$

which is known as *Bessel's equation of order n.* In particular, situations are called *Bessel functions of order n.* Many physical problems involving vibrations or heat conduction in cylindrical regions give rise to these equations.

Let its solution be

$$y = \sum_{m=0}^{\infty} c_m x^{k+m}. \tag{11.2}$$

$\therefore y' = \sum_{m=0}^{\infty} c_m (k+m)x^{k+m-1}$ and $y'' = \sum_{m=0}^{\infty} c_m (k+m)(k+m-1)x^{k+m-2}$.
Substituting these values in Equation (11.1), we get

$$x^2 \sum_{m=0}^{\infty} c_m (k+m)(k+m-1)x^{k+m-2} + x\sum_{m=0}^{\infty} c_m (k+m)x^{k+m-1}$$
$$+ (x^2 - n^2)\sum_{m=0}^{\infty} c_m x^{k+m} = 0$$

or,

$$\sum_{m=0}^{\infty} c_m \left\{(k+m)(k+m-1) + (k+m) - n^2\right\} x^{k+m} + \sum_{m=0}^{\infty} c_m x^{k+m+2} = 0.$$

But the bracket expression above is the coefficient of the smallest power of x, and it can be written as

$$\begin{aligned}(k+m)(k+m-1) + (k+m) - n^2 &= (k+m)^2 - (k+m) + (k+m) - n^2 \\ &= (k+m)^2 - n^2 \\ &= (k+m+n)(k+m-n).\end{aligned}$$

The above identity becomes

$$\sum_{m=0}^{\infty} c_m \{(k+m+n)(k+m-n)\} x^{k+m} + \sum_{m=0}^{\infty} c_m x^{k+m+2} = 0.$$

Equating to zero the coefficient of x^{k+m} gives an indicial equation (Putting $m = 0$):

$$c_0(k+n)(k-n) = 0, c_0 \neq 0.$$

Its roots are k= $n, -n$.
This gives

$$y = x^k c_0 (1 - \frac{1}{(k+2+n)(k+2-n)} x^2$$

$$+ \frac{1}{(k+4+n)(k+4-n)} \frac{1}{(k+2+m)(k+2-n)} x^4 - \cdots \ldots\ldots).$$

We obtain the indicial equation $k^2 - n^2 = 0$, which implies that $k = n. -n$.
For $k = n$,

$$y_1 = x^n c_0 \left(1 - \frac{1}{4(n+1)} x^2 + \frac{1}{4^2 . 2!(n+1)(n+2)} x^4 - \right.$$
$$\left. \frac{1}{4^3 . 3!(n+1)(n+2)(n+3)} x^6 + \cdots \ldots\ldots \right). \tag{11.3}$$

For $k = -n$,

$$y_2 = x^{-n} c_0 \left(1 - \frac{1}{4(-n+1)} x^2 + \frac{1}{4^2 . 2!(-n+1)(-n+2)} x^4 - \right.$$
$$\left. \frac{1}{4^3 . 3!(-n+1)(-n+2)(-n+3)} x^6 + \cdots \ldots\ldots \right). \tag{11.4}$$

Case I. When n is not an integral or zero, the complete solution of Equation (11.1) is $y = c_1 y_1 + c_2 y_2$.

If we take $c_0 = \frac{1}{2^n} \frac{1}{\Gamma(n+1)}$, then the solution given by Equation (11.3) is called *the Bessel function of the first kind of order n*. Thus,

$$J_n(x) = \left(\frac{x}{2}\right)^n \left\{ \begin{array}{l} \frac{1}{\Gamma(n+1)} - \frac{1}{1!\Gamma(n+2)} \left(\frac{x}{2}\right)^2 + \frac{1}{2!\Gamma(n+3)} \left(\frac{x}{2}\right)^4 \\ \qquad - \frac{1}{3!\Gamma(n+4)} \left(\frac{x}{2}\right)^6 + \cdots . \end{array} \right\} \quad n > 0.$$

$$\therefore J_n(x) = \Sigma_{r=0}^{\infty} (-1)^r \left(\frac{x}{2}\right)^{n+2r} \frac{1}{r!\Gamma(n+r+1)} \tag{11.5}$$

and corresponding to Equation (11.4), we have

$$J_{-n}(x) = \Sigma_{r=0}^{\infty}(-1)^r \left(\frac{x}{2}\right)^{-n+2r} \frac{1}{r!\Gamma(-n+r+1)}, \tag{11.6}$$

which is called *the Bessel function of the first kind of order* $-n$.

Hence, the complete solution of Bessel's Equation (11.1) may be expressed in the form

$$y = AJ_n(x) + BJ_{-n}(x). \tag{11.7}$$

Case II. When $n=0$, $y_1 = y_2$, and

The complete solution of Equation (11.1), which reduces to Bessel's equation of order zero.

Case III. When n is an integral,

y_2 fails to give a solution for the positive value of n, and y_1 fails to give a solution for negative values of n. Thus, aother independent integral of the Bessel's Equation (11.1) is needed to form its general solution. We now proceed to find an independent solution of Equation (11.1), when n is an integer.

Let $\mathrm{y} = \mathrm{u}(x)J_n(x)$ be the solution of Equation (11.1).

Substituting the values of y,

$$y' = u'J_n + uJ_n'$$

$$y'' = u''J_n + 2u'J_n' + uJ_n''$$

in Equation (11.1)

$$x^2(u''J_n + 2u'J_n' + uJ_n'') + x(u'J_n + uJ_n') + (x^2 - n^2)\mathrm{u}J_n = 0$$

or,

$$u(x^2J_n'' + xJ_n' + (x^2 - n^2)J_n) + x^2u''J_n + 2x^2u'J_n' + xu'J_n = 0. \tag{11.8}$$

Now, since J_n is a solution of Equation (11.1),

$$x^2J_n'' + xJ_n' + (x^2 - n^2)J_n = 0.$$

Equation (11.8) reduces to $x^2u''J_n + 2x^2u'J_n' + xu'J_n = 0$

or,
$$\frac{u''}{u'} + \frac{2J_n'}{J_n} + \frac{1}{x} = 0,$$

dividing by $x^2u'J_n$

or,
$$\frac{d}{dx}(logu') + 2\frac{d}{dx}(logJ_n) + \frac{d}{dx}(logx) = 0$$

or,
$$\frac{d}{dx}[(logu') + 2(logJ_n) + (logx)] = 0.$$

Integrating, we get $logu'J_n^2x = logB.$

$$\therefore u'J_n^2x = B \Rightarrow u' = \frac{B}{J_n^2x}.$$

$$\therefore u = B\int\frac{dx}{J_n^2x} + A.$$

Thus, $y = u(x)J_n(x) = AJ_n(x) + BJ_n(x)\int\frac{dx}{J_n^2x}.$

Hence, the complete solution of Bessel's equation is

$$y = AJ_n(x) + BY_n(x) \tag{11.9}$$

where,
$$Y_n(x) = J_n(x)\int\frac{dx}{J_n^2x}. \tag{11.10}$$

$Y_n(x)$ is called the *Bessel function of the second kind of order* n or *Neumann*[1] *function.*

[1]Named after the German mathematician and physicist *Carl Neumann* (1832–1925) whose work was on potential theory.

Putting $k=-n+r$, i.e., $r=k+n$, *also* $\Gamma(k+1)=k!$, where k is an integer, Equation (11.5) becomes

$$J_{-n}(x)=\sum_{k=0}^{\infty}(-1)^{k+n}\left(\frac{x}{2}\right)^{2k+n}\frac{1}{(k+n)!k!}$$

$$=(-1)^n\sum_{k=0}^{\infty}(-1)^k\left(\frac{x}{2}\right)^{2k+n}\frac{1}{\Gamma(k+n+1)k!}.$$

Hence, $$J_{-n}(x)=(-1)^n J_n(x). \tag{11.11}$$

11.3 Recurrence Formulae for $J_n(x)$

The following recurrence formulae can be derived from the series expression for $J_n(x)$:

1. $\frac{d}{dx}[x^n J_n(x)]=x^n J_{n-1}(x)$
2. $\frac{d}{dx}[x^{-n} J_n(x)]=-x^{-n} J_{n+1}(x)$
3. $J_n(x)=\frac{x}{2n}[J_{n-1}(x)+J_{n+1}(x)]$
4. $J_n'(x)=\frac{1}{2}[J_{n-1}(x)-J_{n+1}(x)]$
5. $J_n'(x)=\frac{n}{x}J_n(x)-J_{n+1}(x)$
6. $J_{n+1}(x)=\frac{2n}{x}J_n(x)-J_{n-1}(x)$

Proofs: 1. We know that

$$J_n(x)=\sum_{r=0}^{\infty}(-1)^r\left(\frac{x}{2}\right)^{n+2r}\frac{1}{r!\Gamma(n+r+1)}. \tag{11.12}$$

Multiplying Equation (11.4) by x^n, we get

$$x^n J_n(x) = x^n \sum_{r=0}^{\infty} (-1)^r \left(\frac{x}{2}\right)^{n+2r} \frac{1}{r!\Gamma(n+r+1)}$$

$$= \sum_{r=0}^{\infty} (-1)^r x^{2(n+r)} \frac{1}{2^{n+2r} r!\Gamma(n+r+1)}.$$

$$\therefore \frac{d}{dx}[x^n J_n(x)] = \sum_{r=0}^{\infty} (-1)^r 2(n+r) x^{2(n+r)-1} \frac{1}{2^{n+2r} r!\Gamma(n+r+1)}$$

$$= x^n \sum_{r=0}^{\infty} (-1)^r \left(\frac{x}{2}\right)^{n-1+2r} \frac{1}{r!\Gamma(n-1+r+1)} = x^n J_{n-1}(x).$$

2. We know

$$J_n(x) = \sum_{r=0}^{\infty} (-1)^r \left(\frac{x}{2}\right)^{n+2r} \frac{1}{r!\Gamma(n+r+1)}$$

$$= \left(\frac{x}{2}\right)^n \left\{ \begin{array}{l} \dfrac{1}{\Gamma(n+1)} - \dfrac{1}{1!\Gamma(n+2)}\left(\dfrac{x}{2}\right)^2 + \dfrac{1}{2!\Gamma(n+3)}\left(\dfrac{x}{2}\right)^4 \\ \qquad - \dfrac{1}{3!\Gamma(n+4)}\left(\dfrac{x}{2}\right)^6 + \cdots \end{array} \right\} =. \tag{11.13}$$

Multiplying both sides of Equation (11.5) by x^{-n}, we get

$$x^{-n} J_n(x) = \sum_{r=0}^{\infty} (-1)^r \left(\frac{x}{2}\right)^{2r} \frac{1}{r!\Gamma(n+r+1)}$$

$$\frac{d}{dx}[x^{-n} J_n(x)] = \sum_{r=0}^{\infty} (-1)^r \frac{2r x^{2r-1}}{2^{n+2r} r!\Gamma(n+r+1)}$$

$$= -x^{-n} \sum_{r=0}^{\infty} (-1)^{r-1} \frac{x^{n+1+2(r-1)}}{2^{n+1+2(r-1)} (r-1)!\Gamma(n+r+1)}$$

$$= -x^{-n} \sum_{k=0}^{\infty} \frac{(-1)^k (\frac{x}{2})^{n+1+2k}}{k!\Gamma(n+r+1)}$$

$$= -x^{-n} J_{n+1}(x).$$

3. From (1), $\frac{d}{dx}[x^n J_n(x)] = x^n J_{n-1}(x)$

or,
$$x^n\ J_n'(x) + nx^{n-1} J_n(x) = x^n J_{n-1}(x)$$

or,
$$J_n'(x) + \left(\frac{n}{x}\right) J_n(x) = J_{n-1}(x). \tag{i}$$

From (2), $\frac{d}{dx}[x^{-n} J_n(x)] = -x^{-n} J_{n+1}(x)$

or,
$$x^{-n} J_n{}'(x) - nx^{-n-1} J_n(x) = -x^{-n} J_{n+1}(x)$$

or,
$$-J_n{}'(x) + \frac{n}{x} J_n(x) = J_{n+1}(x). \tag{ii}$$

Adding Equations (i) and (ii), we obtain

$$2\frac{n}{x} J_n(x) = J_{n-1}(x) + J_{n+1}(x).$$

$$\therefore J_n(x) = \frac{x}{2n}[J_{n-1}(x) + J_{n+1}(x)].$$

(4) Subtracting Equation (ii) from Equation (i), we obtain

$$2J_n'(x) = J_{n-1}(x) - J_{n+1}(x)$$

or, $$J_n'(x) = \frac{1}{2}(J_{n-1}(x) - J_{n+1}(x)).$$

(5) From (4), $J_n'(x) = \frac{1}{2}(J_{n-1}(x) - J_{n+1}(x))$.

Substituting from (3), $J_{n-1}(x) = \frac{2n}{x} J_n(x) - J_{n+1}(x)$.
From above, we get,

$$\begin{aligned} J_n'(x) &= \frac{1}{2}(J_{n-1}(x) - J_{n+1}(x)) \\ &= \frac{1}{2}\left[\frac{2n}{x} J_n(x) - J_{n+1}(x)\right] - \frac{1}{2} J_{n+1}(x) \\ &= \frac{n}{x} J_n(x) - J_{n+1}(x). \end{aligned}$$

(6) From (3), we have

$$J_n(x) = \frac{x}{2n}[J_{n-1}(x) + J_{n+1}(x)]$$

or,

$$\frac{2n}{x} J_n(x) = [J_{n-1}(x) + J_{n+1}(x)] \text{ [multiplying both sides by } \frac{2n}{x}\text{]}$$

or, $$J_{n+1}(x) = \frac{2n}{x} J_n(x) - J_{n-1}(x).$$

11.4 Expansion of J_o, J_1, and $J_{\frac{1}{2}}$

(a) We have from Equation (11.4)

$$J_n(x) = \Sigma_{r=0}^{\infty}(-1)^r\left(\frac{x}{2}\right)^{n+2r}\frac{1}{r!\Gamma(n+r+1)}$$

$$=\left(\frac{x}{2}\right)^n\left\{\begin{array}{l}\dfrac{1}{\Gamma(n+1)}-\dfrac{1}{1!\Gamma(n+2)}\left(\dfrac{x}{2}\right)^2+\dfrac{1}{2!\Gamma(n+3)}\left(\dfrac{x}{2}\right)^4\\ \qquad -\dfrac{1}{3!\Gamma(n+4)}\left(\dfrac{x}{2}\right)^6+\cdots.\end{array}\right\}$$

$$J_0(x)=\left(\frac{x}{2}\right)^n\left\{\begin{array}{l}\dfrac{1}{\Gamma(1)}-\dfrac{1}{1!\Gamma(2)}\left(\dfrac{x}{2}\right)^2+\dfrac{1}{2!\Gamma(3)}\left(\dfrac{x}{2}\right)^4\\ \qquad -\dfrac{1}{3!\Gamma(4)}\left(\dfrac{x}{2}\right)^6+\cdots.\end{array}\right\}\text{[putting n=0]}$$

$$=1-\frac{1}{1!}\left(\frac{x}{2}\right)^2+\frac{1}{(2!)^2}\left(\frac{x}{2}\right)^4-\frac{1}{(3!)^2}\left(\frac{x}{2}\right)^6+\cdots$$

(b)

$$J_1(x)=\left(\frac{x}{2}\right)^n\left\{\frac{1}{\Gamma(2)}-\frac{1}{1!\Gamma(3)}\left(\frac{x}{2}\right)^2+\frac{1}{2!\Gamma(4)}\left(\frac{x}{2}\right)^4-\frac{1}{3!\Gamma(5)}\left(\frac{x}{2}\right)^6+\cdots.\right\}\text{[putting n = 1]}$$

$$J_1(x)=\left(\frac{x}{2}\right)^1\left\{\frac{1}{\Gamma(2)}-\frac{1}{1!\Gamma(3)}\left(\frac{x}{2}\right)^2+\frac{1}{2!\Gamma(4)}\left(\frac{x}{2}\right)^4-\frac{1}{3!\Gamma(5)}\left(\frac{x}{2}\right)^6+\cdots.\right\}$$

$$=\frac{x}{2}\left\{1-\frac{1}{1!2!}\left(\frac{x}{2}\right)^2+\frac{1}{2!3!}\left(\frac{x}{2}\right)^4-\frac{1}{3!4!}\left(\frac{x}{2}\right)^6+\cdots.\right\}.$$

The values of $J_0(x)$ *and* $J_1(x)$ are (four decimal places of interval 0.1)

0.0	0.1	0.2	0.3	0.4	0.5
$J_0(x)=1.0000$	0.9975	0.9900	0.9776	0.9604	0.9385
$J_1(x)=0.0000$	0.0499	0.0995	0.1483	0.1960	0.2423

With the help of these values, we can draw the graphs of $J_0(x)$ *and* $J_1(x)$ [shown in Figure 11.1] for $x > 0$. Their close resemblance to $\cos x$ *and* $\sin x$ is interesting.

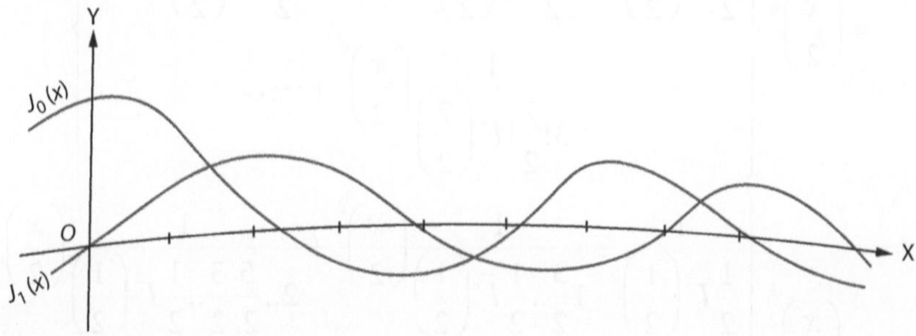

Figure 11.1 The Bessel polynomial.

(c) Value of $J_{\frac{1}{2}},(x)$

We know that

$$J_n(x) = \left(\frac{x}{2}\right)^n \left\{ \begin{array}{l} \dfrac{1}{\Gamma(n+1)} - \dfrac{1}{1!\Gamma(n+2)}\left(\dfrac{x}{2}\right)^2 + \dfrac{1}{2!\Gamma(n+3)}\left(\dfrac{x}{2}\right)^4 \\ \qquad - \dfrac{1}{3!\Gamma(n+4)}\left(\dfrac{x}{2}\right)^6 + \cdots . \end{array} \right\}.$$

Putting $n = \frac{1}{2}$, we have

$$J_{\frac{1}{2}},(x) = \left(\frac{x}{2}\right)^{\frac{1}{2}} \left\{ \begin{array}{l} \dfrac{1}{\Gamma\left(\frac{1}{2}+1\right)} - \dfrac{1}{1!\Gamma\left(\frac{1}{2}+2\right)}\left(\dfrac{x}{2}\right)^2 + \dfrac{1}{2!\Gamma\left(\frac{1}{2}+3\right)}\left(\dfrac{x}{2}\right)^4 \\ \qquad - \dfrac{1}{3!\Gamma\left(\frac{1}{2}+4\right)}\left(\dfrac{x}{2}\right)^6 + \cdots . \end{array} \right\}$$

$$= \left(\frac{x}{2}\right)^n \left\{ \begin{array}{l} \dfrac{1}{\Gamma\left(\frac{3}{2}\right)} - \dfrac{1}{1!\Gamma\left(\frac{5}{2}\right)}\left(\dfrac{x}{2}\right)^2 + \dfrac{1}{2!\Gamma\left(\frac{7}{2}\right)}\left(\dfrac{x}{2}\right)^4 \\ \qquad - \dfrac{1}{3!\Gamma\left(\frac{9}{2}\right)}\left(\dfrac{x}{2}\right)^6 + \cdots . \end{array} \right\}$$

$$=\left(\frac{x}{2}\right)^{\frac{1}{2}}\left\{\begin{array}{l}\dfrac{1}{\frac{1}{2}\Gamma\left(\frac{1}{2}\right)}-\dfrac{1}{1!\frac{3}{2}!\Gamma\left(\frac{3}{2}\right)}\left(\frac{x}{2}\right)^2+\dfrac{1}{2!\frac{5}{2}!\Gamma\left(\frac{5}{2}\right)}\left(\frac{x}{2}\right)^4\\ \qquad-\dfrac{1}{3!\frac{7}{2}!\Gamma\left(\frac{7}{2}\right)}\left(\frac{x}{2}\right)^6+\cdots..\end{array}\right\}$$

$$=\left(\frac{x}{2}\right)^{\frac{1}{2}}\left\{\begin{array}{l}\dfrac{1}{\frac{1}{2}\Gamma\left(\frac{1}{2}\right)}-\dfrac{1}{1\frac{3}{2}..\frac{1}{2}\Gamma\left(\frac{1}{2}\right)}\left(\frac{x}{2}\right)^2+\dfrac{1}{2..\frac{5}{2.}\frac{3}{2}..\frac{1}{2}\Gamma\left(\frac{1}{2}\right)}\left(\frac{x}{2}\right)^4\\ \qquad-\dfrac{1}{3!.2.\frac{5}{2.}\frac{3}{2}..\frac{1}{2}\Gamma\left(\frac{1}{2}\right)}\left(\frac{x}{2}\right)^6+\cdots..\end{array}\right\}$$

$$=\frac{\sqrt{x}}{\sqrt{2}}\frac{1}{\Gamma\left(\frac{1}{2}\right)}\left\{\frac{2}{1}-\frac{2x^2}{3!}+\frac{2x^4}{5!}-\frac{2x^6}{7!}+\cdots..\right\}$$

Multiplying the right side by $\frac{x}{2}$ and the outside by $\frac{2}{x}$, we get

$$=\frac{\sqrt{2}}{\sqrt{x}}\frac{1}{\sqrt{\pi}}\left\{\frac{x}{\mathbf{1}!}-\frac{x^3}{3!}+\frac{x^5}{5!}-\frac{x^7}{7!}+\cdots..\right\}$$

$$\boldsymbol{J}_{\frac{1}{2},}(\boldsymbol{x})=\frac{\sqrt{2}}{\sqrt{x}}\frac{1}{\sqrt{\pi}}\sin x. \tag{11.14}$$

Similarly, taking $n=-\frac{1}{2}$ in the $J_{n,}(x)$ series, we get

$$J_{-\frac{1}{2}}(x)=\left(\frac{x}{2}\right)^{-\frac{1}{2}}\left\{\begin{array}{l}\dfrac{1}{\Gamma\left(-\frac{1}{2}+1\right)}-\dfrac{1}{1!\Gamma\left(-\frac{1}{2}+2\right)}\left(\frac{x}{2}\right)^2+\dfrac{1}{2!\Gamma\left(-\frac{1}{2}+3\right)}\left(\frac{x}{2}\right)^4\\ \qquad-\dfrac{1}{3!\Gamma\left(-\frac{1}{2}+4\right)}\left(\frac{x}{2}\right)^6+\cdots..\end{array}\right\}$$

$$= \left(\frac{\sqrt{2}}{\sqrt{x}}\right)\left\{\begin{array}{l}\frac{1}{\Gamma\left(\frac{1}{2}\right)} - \frac{1}{1!\Gamma\left(\frac{3}{2}\right)}\left(\frac{x}{2}\right)^2 + \frac{1}{2!\Gamma\left(\frac{5}{2}\right)}\left(\frac{x}{2}\right)^4 \\ -\frac{1}{3!\Gamma\left(\frac{7}{2}\right)}\left(\frac{x}{2}\right)^6 + \cdots..\end{array}\right\}$$

$$= \left(\frac{\sqrt{2}}{\sqrt{x}}\right)\left\{\begin{array}{l}\frac{1}{\Gamma\left(\frac{1}{2}\right)} - \frac{1}{1!\frac{1}{2}\Gamma\left(\frac{1}{2}\right)}\left(\frac{x}{2}\right)^2 + \frac{1}{2!\frac{3}{2..}\frac{1}{2}\Gamma\left(\frac{1}{2}\right)}\left(\frac{x}{2}\right)^4 \\ -\frac{1}{3!\frac{5}{2}\frac{3..}{.2}\frac{1}{2}\Gamma\left(\frac{1}{2}\right)}\left(\frac{x}{2}\right)^6 + \cdots..\end{array}\right\}$$

$$= \left(\frac{\sqrt{2}}{\sqrt{x}}\right)\frac{1}{\Gamma\left(\frac{1}{2}\right)}\left\{\begin{array}{l}1 - \frac{1}{1!\frac{1}{2}}\left(\frac{x}{2}\right)^2 + \frac{1}{2\frac{3}{2..}\frac{1}{2}}\left(\frac{x}{2}\right)^4 \\ -\frac{1}{3.2\frac{5}{2}\frac{3..}{.2}\frac{1}{2}}\left(\frac{x}{2}\right)^6 + \cdots..\end{array}\right\}$$

$$= \left(\frac{\sqrt{2}}{\sqrt{x}}\right)\frac{1}{\sqrt{\pi}}\left\{1 - \frac{x^2}{2!} + \frac{x^4}{4!} - \frac{x^6}{6!} + \cdots..\right\}$$

$$= \frac{\sqrt{2}}{\sqrt{\pi x}}\cos x$$

(d) $\left[J_{\frac{1}{2},}(x)\right]^2 + \left[J_{-\frac{1}{2}}(x)\right]^2 = \left(\frac{\sqrt{2}}{\sqrt{x}}\frac{1}{\sqrt{\pi}}\sin x\right)^2 + \left(\frac{\sqrt{2}}{\sqrt{\pi x}}\cos x\right)^2 = \left(\frac{\sqrt{2}}{\sqrt{x}}\frac{1}{\sqrt{\pi}}\right)^2$
$[sin^2 x + \cos^2 x] = \frac{2}{\pi x}$.

Example 11.4.1. Express $J_5(x)$ in terms of $n = 1,2,3,4$ of $J_1(x)$ and $J_0(x)$.

Solution: We know that

$$J_{n+1}(x) = \frac{2n}{x}J_n(x) - J_{n-1}(x).$$

Putting $n = 1, 2, 3, 4$ successively, we get

$$J_2(x) = \frac{2}{x} J_1(x) - J_0(x) \tag{i}$$

$$J_3(x) = \frac{4}{x} J_2(x) - J_1(x) \tag{ii}$$

$$J_4(x) = \frac{6}{x} J_3(x) - J_2(x) \tag{iii}$$

$$J_5(x) = \frac{8}{x} J_4(x) - J_3(x) \tag{iv}$$

Putting the value $J_2(x)$ *from Equation (i) in* Equation (ii), *we have*

$$\begin{aligned} J_3(x) &= \frac{4}{x} J_2(x) - J_1(x) \\ &= \frac{4}{x}[\frac{2}{x} J_1(x) - J_0(x)] - J_1(x) \\ &= \frac{8}{x^2} J_1(x) - \frac{4}{x} J_0(x) - J_1(x) = (\frac{8}{x^2} - 1) J_1(x) - \frac{4}{x} J_0(x). \end{aligned}$$

Substituting the value of $J_3(x)$ from Equation (ii) in Equation (iii), we get

$$\begin{aligned} J_4(x) &= \frac{6}{x} J_3(x) - J_2(x) \\ &= \frac{6}{x}\left[\frac{8}{x^2} - 1) J_1(x) - \frac{4}{x} J_0(x)\right] - [\frac{2}{x} J_1(x) - J_0(x)] \\ &= (\frac{48}{x^3} - \frac{8}{x}) J_1(x) - \left(\frac{24}{x^2} - 1\right) J_0(x). \end{aligned}$$

Finally, we have

$$\begin{aligned} J_5(x) &= \frac{8}{x} J_4(x) - J_3(x) \\ &= \frac{8}{x}\left[(\frac{48}{x^3} - \frac{8}{x}) J_1(x) - \left(\frac{24}{x^2} - 1\right) J_0(x)\right] - \left[(\frac{8}{x^2} - 1) J_1(x) - \frac{4}{x} J_0(x)\right] \\ &= (\frac{384}{x^4} - \frac{72}{x^2} + 1) J_1(x) - \left(\frac{192}{x^3} - \frac{12}{x}\right) J_0(x). \end{aligned}$$

Example 11.4.2. Prove that (a) $J_{\frac{5}{2}}(x)=\sqrt{\left(\frac{2}{\pi x}\right)}\left\{\frac{3-x^2}{x^2}\sin x-\frac{3}{x}\cos x\right\}$

(b) $J_{\frac{3}{2}}(x)=\sqrt{\left(\frac{2}{\pi x}\right)}\left\{\frac{1}{x}\sin x-\cos x\right\}$

(c) $J_{-\frac{3}{2}}(x)=\sqrt{\left(\frac{2}{\pi x}\right)}\left\{-\sin x-\frac{1}{x}\cos x\right\}.$

Solution: (a) We know that

$$J_{n+1}(x)=\frac{2n}{x}J_n(x)-J_{n-1}(x). \qquad \text{(i)}$$

Putting $n=\frac{1}{2}$ in the above equation, we get

$$J_{\frac{3}{2}}(x)=\frac{2.\frac{1}{2}}{x}J_{\frac{1}{2}}(x)-J_{\frac{1}{2}-1}(x)=\frac{1}{x}J_{\frac{1}{2}}(x)-J_{-\frac{1}{2}}(x).$$

Putting the values of $J_{\frac{1}{2}}(x)$ *and* $J_{-\frac{1}{2}}(x)$, we get

$$J_{\frac{3}{2}}(x)=\frac{1}{x}J_{\frac{1}{2}}(x)-J_{-\frac{1}{2}}(x)=$$
$$=\frac{1}{x}\frac{\sqrt{2}}{\sqrt{x}}\frac{1}{\sqrt{\pi}}\sin x-\frac{\sqrt{2}}{\sqrt{\pi x}}\cos x=\frac{\sqrt{2}}{\sqrt{\pi x}}\left(\frac{\sin x}{x}-\cos x\right).$$

Putting $n=\frac{3}{2}$ in the above Equation (i),

$$J_{\frac{5}{2}}(x)=\frac{2.\frac{3}{2}}{x}J_{\frac{3}{2}}(x)-J_{\frac{3}{2}-1}(x)=\frac{3}{x}J_{\frac{3}{2}}(x)-J_{\frac{1}{2}}(x)$$
$$=\frac{3}{x}\frac{\sqrt{2}}{\sqrt{\pi x}}\left(\frac{\sin x}{x}-\cos x\right)-\frac{\sqrt{2}}{\sqrt{\pi x}}\sin x$$
$$=\frac{\sqrt{2}}{\sqrt{\pi x}}\left(\left(\frac{3}{x^2}-1\right)\sin x-\frac{3}{x}\cos x\right).$$

(b) We know that

$$J_{\frac{1}{2}}(x) = \frac{\sqrt{2}}{\sqrt{x}}\frac{1}{\sqrt{\pi}}\sin x,$$

$$J_{-\frac{1}{2}}(x) = \frac{\sqrt{2}}{\sqrt{x}}\frac{1}{\sqrt{\pi}}\cos x,$$

and we know

$$J_{n+1}(x) + J_{n-1}(x) = \frac{2n}{x}J_n(x). \tag{ii}$$

Replacing n *by* $\frac{1}{2}$ in Equation (ii), we get $J_{-\frac{1}{2}}(x) + J_{\frac{3}{2}}(x) = \frac{1}{x}J_{\frac{1}{2}}(x)$

or, $$J_{\frac{3}{2}}(x) = \frac{1}{x}J_{\frac{1}{2}}(x) - J_{-\frac{1}{2}}(x) = \frac{1}{x}\frac{\sqrt{2}}{\sqrt{x}}\frac{1}{\sqrt{\pi}}\sin x - \frac{\sqrt{2}}{\sqrt{x}}\frac{1}{\sqrt{\pi}}\cos x$$
$$= \frac{\sqrt{2}}{\sqrt{x}}\frac{1}{\sqrt{\pi}}\left(\frac{\sin x}{x} - \cos x\right).$$

(c) Replacing $n = -\frac{1}{2}$ in Equation (ii), we have

$$J_{-\frac{3}{2}}(x) = -\frac{1}{x}J_{\frac{1}{2}}(x) - J_{-\frac{1}{2}}(x) = -\frac{1}{x}\frac{\sqrt{2}}{\sqrt{x}}\frac{1}{\sqrt{\pi}}\cos x - \frac{\sqrt{2}}{\sqrt{x}}\frac{1}{\sqrt{\pi}}\sin x$$
$$= \frac{\sqrt{2}}{\sqrt{x}}\frac{1}{\sqrt{\pi}}\left(-\frac{\cos x}{x} - \sin x\right).$$

Example 11.4.3. Prove that

(a) $J_n''(x) = \frac{1}{4}[J_{n-2}(x) - 2\,J_n(x) + J_{n+2}(x)]$

(b) $\dfrac{d}{dx}[xJ_n(x)J_{n+1}(x)] = x[J_n^2(x) - J_{n+1}^2(x)].$

Solution:

(a) We know that

$$J_n'(x) = \frac{1}{2}[J_{n-1}(x) - J_{n+1}(x)]. \tag{i}$$

Differentiating with respect to x, we get

$$J_n''(x) = \frac{1}{2}[J_{n-1}'(x) - J_{n+1}'(x)]. \tag{ii}$$

Changing n to (n-1) in Equation (i), we have

$$J_{n-1}'(x) = \frac{1}{2}[J_{n-2}(x) - J_n(x)]. \tag{iii}$$

Changing n to (n+1) in Equation (i), we have

$$J_{n+1}'(x) = \frac{1}{2}[J_n(x) - J_{n+2}(x)]. \tag{iv}$$

Substituting $J_{n-1}'(x)$ and $J_{n+1}'(x)$ in Equation (ii), we have

$$\begin{aligned} J_n''(x) &= \frac{1}{2}[J_{n-1}'(x) - J_{n-1}'(x)] \\ &= \frac{1}{2}\left[\left\{\frac{1}{2}[J_{n-2}(x) - J_n(x)]\right\} - \left\{\frac{1}{2}[J_n(x) - J_{n+2}(x)]\right\}\right] \\ &= \frac{1}{4}[J_{n-2}(x) - 2J_n(x) + J_{n+2}(x)]. \end{aligned}$$

(b)

$$\frac{d}{dx}\left[xJ_n(x)J_{n+1}(x)\right] = \left[J_n(x)J_{n+1}(x)\right] + x[J_n(x)J_{n+1}'(x) + J_{n+1}(x)J_n'(x)] \tag{i}$$

We know that

$$J_n'(x) = \frac{n}{x} J_n(x) - J_{n+1}(x). \tag{ii}$$

Changing n to n+1 in $J_n'(x) = J_{n-1}(x) - \frac{n}{x} J_n(x)$

$$J_{n+1}'(x) = J_n(x) - \frac{n+1}{x} J_{n+1}(x). \tag{iii}$$

Now, substituting from Equations (iii) and (ii) in Equation (i), we get

$$\begin{aligned}\frac{d}{dx}\left[x J_n(x) J_{n+1}(x)\right] &= \left[J_n(x) J_{n+1}(x)\right)\right] \\ &\quad + x\left[J_n(x)\left\{J_n(x) - \frac{n+1}{x} J_{n+1}(x)\right\} + J_{n+1}(x)\left\{J_{n-1}(x) - \frac{n}{x} J_n(x)\right\}\right] \\ &= x[j_n^2(x) - j_{n+1}^2(x)].\end{aligned}$$

Example 11.4.4. Prove that $\int x j_0^2(x) dx = \frac{1}{2} x^2 [j_0^2(x) + j_1^2(x)]$.

Solution:

$$\begin{aligned}\int x j_0^2(x)\, dx &= \int j_0^2(x) x\, dx \\ &= \frac{x^2}{2} j_0^2(x) - \int 2 J_0(x) j_0'(x). \frac{x^2}{2} dx \\ &= \frac{x^2}{2} j_0^2(x) + \int J_0(x) J_1(x). x^2 dx \, [\text{by using } \frac{d}{dx}[x^{-n} J_n(x)] \\ &= -x^{-n} J_{n+1}(x), \text{putting n} = 0] \\ &= \frac{x^2}{2} j_0^2(x) + \int x J_1(x) \frac{d}{dx} x J_1(x) dx \left[applying\ \frac{d}{dx} x J_1(x) = x J_0(x) \right] \\ &= \frac{x^2}{2} j_0^2(x) + \frac{1}{2}[x J_1(x)]^2 = \frac{1}{2} x^2 [j_0^2(x) + j_1^2(x)].\end{aligned}$$

Example 11.4.5. Prove that

(a) $J_n(x) J_{-n}'(x) - J_n'(x) J_{-n}(x) = -\dfrac{2 \sin n\pi}{x\pi}$

(b) $\frac{d}{dx}\left(\frac{J_{-n}(x)}{J_n(x)}\right) = -\frac{2\sin n\pi}{\pi x J_n^2}$.

Solution: (a) We know that $J_n(x)$ *and* $J_{-n}(x)$ are the solutions of Bessel's equation.

$$y'' + (\frac{1}{x})y' + \left(\frac{1-n^2}{x^2}\right)y = 0.$$

$$\therefore J_n(x)'' + (\frac{1}{x})J_n(x)' + \left(\frac{1-n^2}{x^2}\right)J_n(x) = 0 \quad \text{(i)}$$

$$and\ J_{-n}(x)'' + (\frac{1}{x})J_{-n}(x)' + \left(\frac{1-n^2}{x^2}\right)J_{-n}(x) = 0. \quad \text{(ii)}$$

Multiplying Equation *(i) by* $J_{-n}(x)$ and Equation (ii) by $J_n(x)$ and subtracting, we have

$$J_{-n}(x)J_n''(x) - J_{-n}''(x)J_n(x) + \frac{1}{x}\left[J_{-n}(x)J_n'(x) - J_{-n}'(x)J_n(x)\right] = 0. \quad \text{(iii)}$$

Let

$$J_{-n}(x)\, J_n'(x) - J_{-n}'(x)J_n(x) = \vartheta. \quad \text{(iv)}$$

Differentiating Equation (iv) with respect to x, we get

$$J_{-n}(x)\, J_n''(x) + J_n'(x)J_{-n}'(x) - \left(J_{-n}'(x)J_n(x) + J_{-n}'(x)J_n'(x)\right) = \vartheta'$$
$$or\ J_{-n}(x)\, J_n''(x) - J_{-n}'(x)J_n(x) = \vartheta'. \quad \text{(v)}$$

Using Equations (iv) and (v) in Equation (iii), we get

$$\vartheta' + \frac{1}{x}\vartheta = 0$$

or,
$$\frac{d\vartheta}{dx}+\frac{1}{x}\vartheta=0.$$

Integrating, we have $\log\vartheta+\log x=\log c$

or,
$$\vartheta x=c.$$

$$\therefore \vartheta=\frac{c}{x}$$

$$J_{-n}(x)\,J_n'(x)-J_{-n}'(x)J_n(x)=\frac{c}{x} \qquad \text{(vi) [By Equation (iv)]}$$

$$Now,\ J_n(x)=\frac{1}{2^n\Gamma(n+1)}\left\{\begin{array}{l} x^n-\dfrac{x^{n+2}}{4(n+1)}+\dfrac{x^{n+4}}{4.8.(n+1)(n+2)} \\ -\dfrac{x^{n+6}}{4.8..16.(n+1)(n+2)(n+3)}+\cdots.. \end{array}\right\}$$

$$and\ J_{-n}(x)=\frac{1}{2^{-n}\Gamma(-n+1)}\left\{\begin{array}{l} x^{-n}-\dfrac{x^{-n+2}}{4(1-n)}+\dfrac{x^{-n+4}}{4.8.(1-n)(2-n)} \\ -\dfrac{x^{-n+6}}{4.8..16.(1-n)(2-n)(3-n)}+\cdots.. \end{array}\right\}.$$

Using these values in Equation (vi) gives

$$\frac{1}{2^n\Gamma(n+1)}\left\{nx^{n-1}-\frac{(n+2)x^{n+1}}{4(n+1)}+\frac{(n+4)x^{n+3}}{4.8.(n+1)(n+2)}-\cdots\right\}$$
$$\times\frac{1}{2^{-n}\Gamma(-n+1)}\left\{x^{-n}-\frac{x^{-n+2}}{4(1-n)}+\frac{x^{-n+4}}{4.8.(1-n)(2-n)}-\cdots\right\}-$$
$$\frac{1}{2^{-n}\Gamma(-n+1)}\left\{-nx^{-n-1}-\frac{(2-n)x^{-n+1}}{4(1-n)}+\frac{(4-n)x^{-n+3}}{4.8.(1-n)(2-n)}-\cdots\right\}$$
$$\times\frac{1}{2^n\Gamma(n+1)}\left\{x^n-\frac{x^{n+2}}{4(n+1)}+\frac{x^{n+4}}{4.8.(n+1)(n+2)}-\right\}=\frac{c}{x}.$$

Comparing the coefficient of $\frac{1}{x}$ from both sides, we get

$$\frac{1}{2^n\Gamma(n+1)}\frac{n}{2^{-n}\Gamma(-n+1)}+\frac{1}{2^n\Gamma(n+1)}\frac{n}{2^{-n}\Gamma(-n+1)}=c$$

or, $$c=\frac{2}{2^n\Gamma(n+1)}\frac{n}{2^{-n}\Gamma(-n+1)}=\frac{2}{\Gamma(n)}\frac{1}{\Gamma(-n+1)}=\frac{2\sin n\pi}{\pi}.$$

$$\therefore J_n(x)\,J'_{-n}(x)-J'_n(x)J_{-n}(x)=-\frac{c}{x}.$$

$$J_n(x)\,J'_{-n}(x)-J'_n(x)J_{-n}(x)=-\frac{2\sin n\pi}{x\pi}. \quad \text{(vii)}$$

(b) Dividing Equation (vii) by $j_n^2(x)$, we get

$$\frac{J_n(x)\,J'_{-n}(x)-J'_n(x)J_{-n}(x)}{j_n^2}=-\frac{2\sin n\pi}{x\pi\, j_n^2}$$

or, $$\frac{d}{dx}\left(\frac{J_{-n}(x)}{J_n(x)}\right)=-\frac{2\sin n\pi}{x\pi\, j_n^2}$$

Example 11.4.6. Prove that $J_n(x)=(-2)^n x^n\frac{d^nJ_0(x)}{d(x^2)^n}$.

Solution: We know that $J_0(x)$ is a solution of Bessel's equation of order zero (n=0).

$$\therefore \frac{d^2y}{dx^2}+\left(\frac{1}{x}\right)\frac{dy}{dx}+y=0. \quad \text{(i)}$$

Let $x^2 = X. \therefore x = \sqrt{X}$

and, $$\frac{dX}{dx} = 2x = 2\sqrt{X}. \tag{ii}$$

Now, $$\frac{dy}{dx} = \frac{dy}{dX}\frac{dX}{dx} = 2\sqrt{X}\frac{dy}{dX} \text{ by Equation (ii),} \tag{iii}$$

and, $$\frac{d}{dx} = 2\sqrt{X}\frac{d}{dX} \tag{iiia}$$

$$\frac{d^2y}{dx^2} = \frac{d}{dx}\frac{dy}{dx} = 2\sqrt{X}\frac{d}{dX}\left(2\sqrt{X}\frac{dy}{dX}\right) \text{ by Equations (iii) and (iiia)}$$

$$= 4\sqrt{X}\left[\frac{1}{2\sqrt{X}}\frac{dy}{dX} + \sqrt{X}\frac{d}{dX}(\frac{dy}{dX})\right] = 4X\frac{d^2y}{dX^2} + 2\frac{dy}{dX}. \tag{iv}$$

Putting the values of $x, \frac{d^2y}{dx^2}, and \frac{dy}{dx}$ from Equations (iii) and (iv) in Equation (i), we get

$$4X\frac{d^2y}{dX^2} + 2\frac{dy}{dX} + \left(\frac{1}{\sqrt{X}}\right)2\sqrt{X}\frac{dy}{dX} + y = 0$$

or, $$4X\frac{d^2y}{dX^2} + 4\frac{dy}{dX} + y = 0. \tag{v}$$

Differentiating Equation (v) n times with respect to X by the Leibnitz theorem, we get

$$4\left[X\frac{d^{n+2}y}{dX^2} + n.1\frac{d^{n+1}y}{dX^2}\right] + 4\frac{d^{n+1}y}{dX^2} + \frac{d^ny}{dX^2} = 0$$

or, $$4X\frac{d^{n+2}y}{dX^{n+2}}+4(n+1)\frac{d^{n+1}y}{dX^{n+1}}+\frac{d^{n}y}{dX^{n}}=0 \tag{vi}$$

Since $J_0(x)$ is the solution of Equation (i), we take

$$Y=\frac{d^n y}{dX^n}=\frac{d^n J_0(x)}{d(x^n)^2} \quad \text{therefore,} \quad \frac{dY}{dX}=\frac{d^{n+1}y}{dX^{n+1}} \quad and \quad \frac{d^2Y}{dX^2}=\frac{d^{n+2}y}{dX^{n+2}}. \tag{vii}$$

Equation (vi) gives us

$$4X\frac{d^2Y}{dX^2}+4(n+1)\frac{dY}{dX}+Y=0. \tag{viii}$$

But $J_n(x)$ is the solution of Bessel's equation

$$\frac{d^2y}{dx^2}+\frac{1}{x}\frac{dy}{dx}+(1-\frac{n^2}{x^2})y=0 \tag{ix}$$

Let $y=x^n z$ so that

$$J_n(x)=x^n z. \tag{x}$$

Differentiating Equation (x), $\frac{dy}{dx}=x^n\frac{dz}{dx}+nx^{n-1}z$

and $\frac{d^2y}{dx^2}=x^n\frac{d^2z}{dx^2}+2nx^{n-1}\frac{dz}{dx}+n(n-1)x^{n-2}z.$

Using the above values of y, $\frac{dy}{dx}$, *and* $\frac{d^2y}{dx^2}$ *in* Equation (ix), *we get*

$$x^n\frac{d^2z}{dx^2}+2nx^{n-1}\frac{dz}{dx}+n(n-1)x^{n-2}z+\frac{1}{x}\left(x^n\frac{dz}{dx}+nx^{n-1}z\right)$$
$$+\left(1-\frac{n^2}{x^2}\right)x^n z=0$$

or,
$$x^n \frac{d^2z}{dx^2} + (2n+1)\frac{x^n}{x}\frac{dz}{dx} + x^n z = 0,$$

or,
$$\frac{d^2z}{dx^2} + (2n+1)\frac{1}{x}\frac{dz}{dx} + z = 0. \tag{xi}$$

Using Equations (ii), (iii), and (iv) (after replacing *y by z here*), Equation (xi) gives

$$\left(4X\frac{d^2z}{dX^2} + 2\frac{dz}{dX}\right) + (2n+1)\frac{1}{x}2\sqrt{X}\frac{dz}{dX} + z = 0$$

or,
$$4X\frac{d^2z}{dX^2} + 4(n+1)\frac{dz}{dX} + z = 0\ (since\ \sqrt{X} = x) \tag{xii}$$

Comparing Equations (viii) and (xii), we have $z = kY$ where k is a constant to be determined. Using Equations (vii) and (x), we have $x^n z = kx^n Y$ or

$$J_n(x) = kx^n \frac{d^n J_0(x)}{d(x^2)^n}. \tag{xiii}$$

But $J_0(x) = \Sigma_{r=0}^{\infty} \frac{(-1)^r x^{2r}}{(2^r r!)^2} = \Sigma_{r=0}^{\infty} \frac{(-1)^r X^r}{(2^r r!)^2}$.

Differentiating both sides with respect to x^2 n times (since $x^2 = X$), we have

$$\frac{d^n J_0(x)}{d(x^n)^2} = \Sigma_{r=0}^{\infty} \frac{(-1)^r}{(2^r r!)^2} \frac{d^n}{d(X)^n} X^r$$
$$= \frac{(-1)^n}{(2^n n!)^2} n! + \text{terms involving } X \text{ (i.e., } x^2). \tag{xiv}$$

Using the value of $J_n(x)$ in Equation (xiii), we get $J_n(x)=kx^n\dfrac{d^nJ_0(x)}{d(x^2)^n}$

$$\left(\frac{x}{2}\right)^n\frac{1}{\Gamma(n+1)}\left\{1-\frac{1}{1!(n+1)}\left(\frac{x}{2}\right)^2++\cdots..\right\}$$
$$=kx^n\left[\frac{(-1)^n}{(2^nn!)^2}n!\ +\ \text{terms involving } X\right].$$

Equating the coefficient of x^n from both sides, we get

$$\frac{1}{n!2^n}=\text{k}\frac{(-1)^n}{(2^nn!)^2}n!$$
$$\therefore\frac{1}{2^n}=k\frac{(-1)^n}{(2^n)^2}.\therefore k=(-1)^n2^n=(-2)^n. \qquad \text{(xv)}$$

Putting this value of k in Equation (xiii), we get

$$J_n(x)=kx^n\frac{d^nJ_0(x)}{d(x^2)^n}=(-2)^nx^n\frac{d^nJ_0(x)}{d(x^2)^n}.$$

Example 11.4.7. (a) $J_n(x)=\dfrac{x^n}{2^{n-1}\Gamma(n)}\int_0^{\pi/2}sin\theta cos^{2n-1}\theta J_0(xsin\theta)d\theta$

(b) (i) $J_0'=-J_1$
(ii) $J_2-J_0=2J_0''$

Solution: We have

$$J_n(x)=\Sigma_{r=0}^{\infty}\frac{(-1)^r}{r!\Gamma(n+r+1)}\left(\frac{x}{2}\right)^{2r+n}. \qquad \text{(i)}$$

Putting $x=xsin\,\theta, n=0$,

$$J_0(xsin\,\theta)=\Sigma_{r=0}^{\infty}\frac{(-1)^r}{r!\Gamma(r+1)}\left(\frac{xsin\,\theta}{2}\right)^{2r}. \qquad \text{(ii)}$$

$$Let\ I = \int_0^{\pi/2} sin\theta cos^{2n-1}\theta J_0(xsin\theta)d\theta. \qquad \text{(iii)}$$

$$\therefore I = \int_0^{\pi/2} sin\theta cos^{2n-1}\theta \left(\sum_{r=0}^{\infty} \frac{(-1)^r}{r!\Gamma(r+1)} \left(\frac{xsin\,\theta}{2} \right)^{2r} \right) d\theta, \text{ using Equation (ii)}$$

$$= \left(\sum_{r=0}^{\infty} \frac{(-1)^r x^{2r}}{r!2^{2r}\Gamma(r+1)} \right) \int_0^{\pi/2} cos^{2n-1}\theta sin^{2r+1}\theta d\theta$$

$$= \left(\sum_{r=0}^{\infty} \frac{(-1)^r x^{2r}}{r!2^{2r}\Gamma(r+1)} \right) \frac{\Gamma(n)\Gamma(r+1)}{\Gamma(r+n+1)} \text{ (using Gamma function)}$$

$$= \frac{\Gamma(n)}{2} \sum_{r=0}^{\infty} \frac{(-1)^r}{r!\Gamma(r+1)} \left(\frac{x}{2} \right)^{2r}$$

$$\frac{x^n}{2^{n-1}\Gamma(n)} I = \left(\frac{x}{2} \right)^n \sum_{r=0}^{\infty} \frac{(-1)^r}{r!\Gamma(r+1)} \left(\frac{x}{2} \right)^{2r} = \sum_{r=0}^{\infty} \frac{(-1)^r}{r!\Gamma(r+1)} \left(\frac{x}{2} \right)^{2r+n}.$$

$$\therefore \frac{x^n}{2^{n-1}\Gamma(n)} \int_0^{\pi/2} sin\theta cos^{2n-1}\theta J_0(xsin\theta)d\theta = J_n(x) \text{[using Equations (i) and (iii)]}.$$

(b) (i) We know that the recurrence relation is $xJ_n' = nJ_n - xJ_{n+1}$.

Replacing $n\ by\ 0\ in$ Equation (i), we have $xJ_0' = -xJ_1$, or, $J_0' = -J_1$.

(b) (ii) We know that the recurrence relation is

$$2J_n' = J_{n-1} - J_{n+1}. \qquad \text{(ii)}$$

Differentiating Equation (ii) with respect to x, we get

$$2J_n'' = J_{n-1}' - J_{n+1}'. \qquad \text{(iii)}$$

Replacing $n\ by\ n-1\ and\ n+1$ successively in Equation (ii), we obtain

$$2J_{n-1}' = J_{n-2} - J_n \qquad \text{(iv)}$$

$$2J_{n+1}' = J_n - J_{n+2}. \qquad \text{(v)}$$

Putting *the values of* J'_{n-1} and J'_{n+1} in Equation (iii), we get

$$4J''_n = J_{n-2} - J_n - (J_n - J_{n+2})$$

$$= J_{n-2} - 2J_n + J_{n+2} \qquad \text{(vi)}$$

Replacing n *by* 0 in Equation (vi), we obtain $4J''_0 = J_{-2} - 2J_0 + J_2 = (-1)^2 J_2 - 2J_0 + J_2$ $[\because J_{-n} = -1)^n J_n$

$$4J''_0 = 2J_2 - 2J_0$$

$$2J''_0 = J_2 - J_0.$$

11.5 Generating Function for $J_n(x)$

Theorem 11.5.1.

Prove that $e^{\frac{1}{2}x(t-\frac{1}{t})} = \sum_{n=-\infty}^{\infty} t^n J_n(x)$.

Proof: We have

$$e^{\frac{1}{2}x(t-\frac{1}{t})} = e^{\frac{1}{2}xt} . e^{-\frac{x}{2t}}$$

$$= \left[1 + \left(\frac{xt}{2}\right) + \frac{1}{2!}\left(\frac{xt}{2}\right)^2 + \frac{1}{3!}\left(\frac{xt}{2}\right)^3 + ..\right]$$

$$\times \left[1 - \left(\frac{x}{2t}\right) + \frac{1}{2!}\left(\frac{x}{2t}\right)^2 - \frac{1}{3!}\left(\frac{x}{2t}\right)^3 + \cdots\right]$$

With the coefficient of t^n on the above product, we get

$$= \frac{1}{n!}\left(\frac{x}{2}\right)^n - \frac{1}{(n+1)!}\left(\frac{x}{2}\right)^{n+2} + \frac{1}{2!(n+2)!}\left(\frac{x}{2}\right)^{n+4} - \cdots \ldots \ldots . = J_n(x).$$

As all the integral powers of t, both positive and negative, occur, we have

$$e^{\frac{1}{2}x(t-\frac{1}{t})} = J_0(x) + tJ_1(x) + t^2 J_2(x) + t^3 J_3(x) + \cdots . + t^{-1} J_{-1}(x)$$

$$+ t^{-2} J_{-2}(x) + t^{-3} J_{-3}(x) + \cdots$$

$$= \sum_{n=-\infty}^{\infty} t^n J_n(x).$$

This shows that the Bessel function of various orders can be derived as the coefficients of different powers of t in the expression of $e^{\frac{1}{2}x(t-\frac{1}{t})}$. For this reason, it is known as the *generating function of Bessel function.*

Example 11.5.1. Trigonometry expansion involving Bessel functions:

(a) $cos\,(xsin\theta) = J_0 + 2cos2\theta J_2 + 2cos4\theta J_4 + \cdots.$
(b) $sin\,(xsin\theta) = 2\,sin\theta J_1 + 2sin3\theta J_2 + \cdots.$
(c) $cosx = J_0 - 2J_2 + 2J_4 + \cdots. = J_0(x) + 2\sum_{n=1}^{\infty}(-1)^n J_{2n}(x)$
(d) $sinx = 2J_1\text{-}2J_3 + 2J_5 \ldots. = 2\sum_{n=0}^{\infty}(-1)^n J_{2n+1}(x)$

Solution: We know that

$$e^{\frac{1}{2}x(t-\frac{1}{t})} = J_0 + (t-t^{-1})J_1 + (t^2 - t^{-2})J_2 + (t^3 - t^{-3})J_3 \ldots + \ldots\ldots \quad \text{(i)}$$

Let $t = e^{i\theta}$, so that $t^n = e^{in\theta}$ and $t^{-n} = e^{-in\theta}$.
From Equation (i), we get

$$e^{\frac{1}{2}x(e^{i\theta}-e^{-i\theta})} = J_0 + (e^{i\theta} - e^{-i\theta})J_1 + (e^{2i\theta} + e^{-2i\theta})J_2 + (e^{3i\theta} - e^{-3i\theta})J_3 \ldots + \ldots \quad \text{(ii)}$$

$$\because cosn\theta = \frac{\mathbf{1}}{\mathbf{2}}(e^{ni\theta} + e^{-ni\theta})\ and\ \ sinn\theta = \frac{\mathbf{1}}{\mathbf{2i}}(e^{ni\theta} - e^{-ni\theta}).$$

Equation (ii) gives

$$e^{(\text{xsin}\theta)} = J_0 + 2isin\theta\ J_1 + 2\cos 2\theta J_2 + 2isin3\theta J_3 \ldots. +$$

or,
$$cos\,(xsin\theta) + i\,sin\,(xsin\theta) = (J_0 + 2\cos 2\theta J_2 + \cdots.)$$
$$+2i(sin\theta\ J_1 + + sin3\theta J_3 \ldots + \ldots). \quad \text{(iii)}$$

Equating the real and imaginary part, we get

$$cos\,(xsin\theta) = J_0 + 2cos2\theta J_2 + 2cos4\theta J_4 + \cdots. \quad \text{(iv)}$$

$$and\ sin\,(xsin\theta) = 2\,sin\theta J_1 + 2sin3\theta J_2 + \cdots\ \ proved. \quad \text{(v)}$$

Replacing θ *by* $\theta = \frac{\pi}{2} - \theta$ in Equation (iv), we get

$$\cos\left(x\sin(\frac{\pi}{2} - \theta) \right) = J_0 + 2\cos 2(\frac{\pi}{2} - \theta) J_2 + 2\cos 4(\frac{\pi}{2} - \theta) J_4 + \cdots.$$

or,
$$\cos(x\cos\theta) = J_0 - 2\cos 2\theta J_2 + 2\cos 4\theta J_4 + \cdots. \quad \text{(vi)}$$

Replacing θ *by* $\theta = (\frac{\pi}{2} - \theta)$ in Equation (v), we get

$$\sin\left(x\sin(\frac{\pi}{2} - \theta) \right) = 2\sin(\frac{\pi}{2} - \theta) J_1 + 2\sin 3(\frac{\pi}{2} - \theta) J_2 + \cdots$$

$$\sin(x\cos\theta) = 2\cos\theta J_1 - 2\cos 3\theta J_3 + 2\cos 5\theta J_5 + \cdots \ldots \quad \text{(vii)}$$

Replacing θ *by* $\theta = 0$ in Equations (vi) and (vii), we get

$$\cos(x) = J_0 - 2J_2 + 2J_4 + \cdots. = J_0(x) + 2\sum_{n=1}^{\infty} (-1)^n J_{2n}(x)$$

and $\sin(x) = 2J_1 - 2J_3 + 2J_5 + \cdots. = 2\sum_{n=0}^{\infty} (-1)^n J_{2n+1}(x)$.

Example 11.5.2. Bessel Integrals

Show that

(a) $J_n(x) = \frac{1}{\pi}\int_0^{\pi} \cos(n\theta - x\sin\theta)\, d\theta$, n being an integer.

(b) $J_0(x) = \frac{1}{\sqrt{\pi}}\int_0^{\pi} \cos(x\cos\theta)\, d\theta = \frac{1}{\pi}\int_0^{\pi} \cos(x\sin\theta)\, d\theta$.

(c) $J_0^2 + 2J_1^2 + 2J_2^2 + 2J_3^2 + \cdots.. = 1$.

(d) $J_0(x) = 1 - \frac{x^2}{2^2} + \frac{x^4}{2^2 4^2} - .. \sum_{R=0}^{\infty} \frac{(-1)^r x^{2r}}{(2^r r!)^2}$.

(e) $J_n(-x) = (-1)^n J_n(x)$.

(f) $J_n(x) = \frac{1}{\sqrt{\pi}\Gamma(n+1/2)} (\frac{x}{2})^n \int_0^{\pi} \cos(x\sin\theta) \cos^{2n}\theta\, d\theta$.

Solution: We know that

$$e^{\frac{1}{2}x(t-\frac{1}{t})} = J_0(x) + tJ_1(x) + t^2J_2(x) + t^3J_3(x) + \cdots + t^{-1}J_{-1}(x) + t^{-2}J_{-2}(x)$$
$$+ t^{-3}J_{-3}(x) + \ldots\ldots.$$

Since $J_{-n}(x) = (-)^n J_n(x) \qquad [J_{-1}(x) = -J_1(x), J_{-2}(x) = (-1)^2 J_2(x), \ldots\ldots],$

$$\therefore e^{\frac{1}{2}x(t-\frac{1}{t})} = J_0(x) + \left(t - \frac{1}{t}\right)J_1(x) + \left(t^2 + \frac{1}{t^2}\right)J_2(x)$$
$$+ \left(t^3 - \frac{1}{t^3}\right)J_3(x) + \cdots\ldots \qquad \text{(i)}$$

Putting $t = \cos\theta + i\,sin\theta,$
$t^p = \cos p\theta + i\,sinp\theta$ and $t^{-p} = \cos p\theta - i\,sinp\theta$
$t^p + \frac{1}{t^p} = 2\cos p\theta$ and $t^p - \frac{1}{t^p} = 2\text{i}\sin p\theta.$
Putting in Equation (i) for $i = 1,2,3,\ldots,$ we get

$$e^{\frac{1}{2}x\left(t-\frac{1}{t}\right)} = e^{\frac{1}{2}x(2\text{isin},)} = \cos(xsin\theta) + isin(x\cos\theta)$$
$$= J_0(x) + \left(t - \frac{1}{t}\right)J_1(x) + \left(t^2 + \frac{1}{t^2}\right)J_2(x) + \left(t^3 - \frac{1}{t^3}\right)J_3(x) + \cdots\ldots$$
$$= J_0(x) + 2\text{i}\sin\theta J_1(x) + 2\cos 2\theta\, J_2(x) + 2\text{i}\sin 3\theta\, J_3(x)$$
$$+ 2\cos 4\theta J_4(x) + \ldots.$$
$$= J_0(x) + 2[\cos 2\theta\, J_2(x) + \cos 4\theta J_4(x) + \cdots] + 2i[\sin\theta\, J_1(x)$$
$$+ \sin 3\theta\, J_3(x)].$$

$$\therefore \cos(xsin\theta) = J_0(x) + 2[\cos 2\theta\, J_2(x) + \cos 4\theta J_4(x) + \cdots] \qquad \text{(ii)}$$

$$sin(x\cos\theta) = 2[\sin\theta\, J_1(x) + \sin 3\theta\, J_3(x)], \qquad \text{(iii)}$$

which is known as *Jacobi series.*

Multiplying Equation (ii) by $\cos n\theta$ and Equation (iii) by $\sin n\theta$

and integrating each of the resulting expressions between 0 and π, we obtain

$$\frac{1}{\pi}\int_0^{\pi}\cos(xsin\theta)\cos n\theta\, d\theta = \begin{cases} J_n(x), n\ is\ even\ or\ 0 \\ 0, n\ odd \end{cases},$$

and,
$$\frac{1}{\pi}\int_0^{\pi}\sin(xsin\theta)\sin n\theta\, d\theta = \begin{cases} 0, n\ is\ even \\ ,J_n(x), n\ odd \end{cases}.$$

$$\therefore J_n(x) = \frac{1}{\pi}\int_0^{\pi}\cos(xsin\theta)\cos n\theta\, d\theta + \frac{1}{\pi}\int_0^{\pi}\sin(xsin\theta)\,\mathrm{sincos}\, n\theta\, d\theta$$
$$= \frac{1}{\pi}\int_0^{\pi}\cos(n\theta - xsin\theta)\, d\theta.$$

This is Bessel's original definition of $J_n(x)$ on planetary motion.

(b) Putting $\theta = (\frac{1}{2}\pi - \varphi)$ in Equation (ii),

$$\cos\left(xsin(\frac{1}{2}\pi - \varphi)\right) = J_0(x) + 2\left[\cos 2(\frac{1}{2}\pi - \varphi)J_2(x) + \cos 4(\frac{1}{2}\pi - \varphi)J_4(x) + \cdots\right]$$

or, $$\cos(xcos\varphi) = J_0(x) + 2[-\sin 2\,\varphi)J_2(x) + \cos 4\varphi J_4(x) + \cdots].$$

Integrating with respect to φ, we obtain

$$\int_0^{\pi}\cos(xcos\varphi)d\varphi = \int_0^{\pi}J_0(x) + 2[-\sin 2\,\varphi)J_2(x) + \cos 4\varphi J_4(x) + \cdots]\, d\varphi$$
$$= \lfloor J_0(x)\pi + 2[0.J_2(x) + 0.J_4(x) + \cdots]\rfloor = \pi J_0(x)$$

$$J_0(x) = \frac{1}{\pi}\int_0^{\pi}\cos(xcos\varphi)d\varphi. \tag{iia}$$

Similarly, putting $\theta = (\frac{1}{2}\pi - \varphi)$ in Equation (iii), we get

$$\frac{1}{\pi}\int_0^{\pi} \cos(x \sin \theta)\, d\theta = J_0(x).$$

(c) Squaring Equations (ii) and (iii) and integrating with respect to θ from 0 to π (m,n being integers), using the following in it, we get

$$\int_0^{\pi} \cos(m\theta) \cos n\theta \, d\theta = \int_0^{\pi} \sin m, \ \sin n\theta \, d\theta = 0, (m \neq n)$$

$$\int_0^{\pi} \cos^2 n\theta \, d\theta = \int_0^{\pi} \sin^2 n\theta \, d\theta = \frac{\pi}{2}$$

$$\int_0^{\pi} [J_0(x)]^2 d\theta + 4\int_0^{\pi} \cos^2 2\theta J_2(x)^2 \, d\theta + 4\int_0^{\pi} \cos^2 4\theta J_4(x)^2 + \cdots$$

$$= \int_0^{\pi} \cos^2(x sin\theta)\, d\theta$$

or, $$[J_0(x)]^2 \pi + 4J_2(x)^2 \frac{\pi}{2} + 4J_4(x)^2 \frac{\pi}{2} + \cdots + \cdots = \int_0^{\pi} \cos^2(x sin\theta)\, d\theta \quad \text{(iii)}$$

and,

$$4\left[\int_0^{\pi} \sin^2\theta J^2{}_1(x)\, d\theta\right] + 4\int_0^{\pi}\left[\sin^2 3\theta \, J_3{}^2(x) d\theta\right] + \cdots = \int_0^{\pi} sin^2(x \cos \theta) d\theta$$

or, $$4\left[J^2{}_1(x)\frac{\pi}{2}\right] + 4\left[\frac{\pi}{2} J_3{}^2(x)\right] + \cdots = \int_0^{\pi} sin^2(x \sin \theta) d\theta. \quad \text{(iv)}$$

Adding Equations (iii) and (iv), we obtain

$$[J_0(x)]^2 \pi + 2\pi\left[J^2{}_1\left(J^2{}_1(x)x\right) + J^2{}_1(x) + J^2{}_1(x) + \cdots\right] = \pi$$

or, $$J^2_0+2\left[J^2_1(x)+J^2_2(x)+J^2_3(x)+\cdots\right]=1.$$

(d) From Equation (iia)

$$J_0(x)=\frac{1}{\pi}\int_0^{\pi}\cos(x\cos\varphi)d\varphi$$
$$=\frac{1}{\pi}\int_0^{\pi}\left[1-\frac{x^2\cos^2\varphi}{2!}+\frac{x^4\cos^4\varphi}{4!}-\right]d\varphi. \qquad \text{(v)}$$

But

$$\int_0^{\pi}\cos^{2n}\theta d\theta=\frac{1.3.5\ldots.(2n-1)}{2.4.6\ldots.(2n)}\pi. \qquad \text{(vi)}$$

Using Equation (vi) in Equation (v), we get

$$J_0(x)=\frac{1}{\pi}\int_0^{\pi}\left[1-\frac{x^2\cos^2\varphi}{2!}+\frac{x^4\cos^4\varphi}{4!}-\right]d\varphi$$
$$=\frac{1}{\pi}\left[\pi+\frac{x^2}{2!}\frac{1}{2}\pi+\frac{x^4}{4!}\frac{1.3}{2.4}\pi+\right]=\left[1+\frac{x^2}{2!}\frac{1}{2}+\frac{x^4}{4!}\frac{1.3}{2.4}+\frac{x^6}{6!}\frac{1.3.5}{2.4.6}+\cdots\right]$$
$$J_0(x)=\sum_{r=0}^{\infty}\frac{(-1)^r x^{2r}}{(2^r r!)^2}.$$

(e) We have

$$\sum_{n=-\infty}^{\infty}J_n(x)\,z^n=\exp\left\{\frac{x}{2}\left(z-\frac{1}{z}\right)\right\}. \qquad \text{(i)}$$

Replacing x *by* $-x$ *in* Equation (i), *we get*

$$\sum_{n=-\infty}^{\infty}J_n(-x)\,z^n=\exp\left\{\frac{-x}{2}\left(z-\frac{1}{z}\right)\right\}=\exp\left\{\frac{x}{2}\left(-z-\frac{1}{-z}\right)\right\}$$
$$=\sum_{n=-\infty}^{\infty}J_n(x)(-z)^n$$

by Equation (i)

or, $$\sum_{n=-\infty}^{\infty} J_n(-x)\, z^n = \sum_{n=-\infty}^{\infty} J_n(x)(-1)^n (z)^n.$$

Equating the coefficient of z^n, we get $J_n(-x) = (-1)^n\ J_n(x)$.

(f) Let $I = \int_0^{\pi} \cos(x\sin\theta) \cos^{2n} \theta d\theta$.

$\therefore I = 2\int_0^{\pi/2} \cos(x\sin\theta) \cos^{2n} \theta d\theta$, since $\cos(x\sin\theta) \cos^{2n}\theta$ is an even function

$$= 2\int_0^{\pi/2} \cos^{2n}\theta \left[1 - \frac{x^2\sin^2\theta}{2!} + \frac{x^4\sin^4\theta}{4!} - \cdots\right] d\theta \qquad \text{(ii)}$$

But

$$\int_0^{\pi/2} \cos^p \theta \sin^q \theta d\theta = \frac{\Gamma(\frac{p+1}{2})\Gamma(\frac{q+1}{2})}{2\Gamma(\frac{p+q+2}{2})}. \qquad \text{(iii)}$$

Using Equation (iii), Equation (ii) is in the form

$$I = 2\int_0^{\pi/2} \cos^{2n}\theta \left[1 - \frac{x^2\sin^2\theta}{2!} + \frac{x^4\sin^4\theta}{4!} - \cdots\right] d\theta$$

$$= 2\left[\frac{\Gamma(n+\frac{1}{2})\Gamma(\frac{1}{2})}{2\Gamma(n+1)} - \frac{x^2}{2!}\frac{\Gamma(n+\frac{1}{2})\Gamma(\frac{3}{2})}{2\Gamma(n+2)} + \frac{x^4}{4!}\frac{\Gamma(n+\frac{1}{2})\Gamma(\frac{5}{2})}{2\Gamma(n+3)} - \cdots\right]$$

$$= \Gamma(n+\frac{1}{2})\left[\frac{\sqrt{\pi}}{\Gamma(n+1)} - \frac{x^2}{2!}\frac{\frac{1}{2}\sqrt{\pi}}{(n+1)\Gamma(n+1)} + \frac{x^4}{4!(n+2)}\frac{\frac{3}{2}\cdot\frac{1}{2}\sqrt{\pi}}{(n+1)\Gamma(n+1)} - \cdots\right]$$

$$= \Gamma(n+\frac{1}{2})\frac{\sqrt{\pi}}{\Gamma(n+1)}\left[1 - \frac{x^2}{4\Gamma(n+1)} + \frac{x^4}{4.8(n+2)(n+1)} - \cdots\right]. \qquad \text{(iv)}$$

Also,

$$J_n(x) = \frac{x^n}{2^n}\frac{1}{\Gamma(n+1)}\left[1 - \frac{x^2}{4\Gamma(n+1)} + \frac{x^4}{4.8(n+2)(n+1)} - \cdots\right]. \qquad \text{(v)}$$

Multiplying both sides of Equation (iv) by $\frac{1}{\sqrt{\pi}\Gamma\left(n+\frac{1}{2}\right)}(\frac{x}{2})^n$, we get

$$\frac{1}{\sqrt{\pi}\Gamma(n+\frac{1}{2})}(\frac{x}{2})^n I=\frac{1}{\sqrt{\pi}\Gamma(n+\frac{1}{2})}(\frac{x}{2})^n\Gamma(n+\frac{1}{2})$$

$$\frac{\sqrt{\pi}}{\Gamma(n+1)}\left[1-\frac{x^2}{4\Gamma(n+1)}+\frac{x^4}{4.8(n+2)(n+1)}-\ldots.\right]$$

$$=\frac{x^n}{2^n}\frac{1}{\Gamma(n+1)}\left[1-\frac{x^2}{4\Gamma(n+1)}+\frac{x^4}{4.8(n+2)(n+1)}-\ldots.\right]=J_n(x).$$

Using Equations (i), (v), and (vi), we get

$$\frac{1}{\sqrt{\pi}\Gamma\left(n+\frac{1}{2}\right)}(\frac{x}{2})^n\int_0^{\pi}\cos(x\sin\theta)\cos^{2n}\theta d\theta=J_n(x).$$

Example 11.5.3. If $a>0$, *prove that* $\int_0^{\infty}e^{-ax}J_0(bx)=\frac{1}{\sqrt{a^2+b^2}}$.

Solution: We know that $J_0(x)=\frac{1}{\pi}\int_0^{\pi}\cos(x\sin\theta)\,d\theta$.

$$\therefore J_0(bx)=\frac{1}{\pi}\int_0^{\pi}\cos(bx\sin\theta)\,d\theta. \qquad \text{(i)}$$

$$\therefore \int_0^{\infty}e^{-ax}J_0(bx)dx=\int_0^{\infty}e^{-ax}\left\{\frac{1}{\pi}\int_0^{\pi}\cos(bx\sin\theta)\,d\theta\right\}dx,$$

using Equation (i)

$$=\frac{1}{\pi}\int_0^{\infty}\left\{\int_0^{\pi}e^{-ax}\cos(bx\sin\theta)\,d\theta\right\}dx$$

$$=\frac{1}{\pi}\int_0^{\pi}\left\{\int_0^{\infty}e^{-ax}\cos(bx\sin\theta)dx\right\}d\theta,$$

interchanging the order of integration

$$=\frac{1}{\pi}\int_0^{\pi}\left\{\int_0^{\infty}e^{-ax}\frac{e^{ibxsin\theta}+e^{-ibxsin\theta}}{2}dx\right\}d\theta$$

$$=\frac{1}{2\pi}\int_0^{\pi}\left\{\int_0^{\infty}e^{-(a-ibsin\theta)x}+e^{-(a+ibsin\theta)x}dx\right\}d\theta$$

$$=\frac{1}{2\pi}\int_0^{\pi}\left[\frac{e^{-(a-ibsin\theta)x}}{-(a-ibsin\theta)}+\frac{e^{-(a+ibsin\theta)x}}{-(a+ibsin\theta)}\right]_0^{\infty}d\theta$$

$$=\frac{1}{2\pi}\int_0^{\pi}\left[\frac{1}{(a-ibsin\theta)}+\frac{1}{(a+ibsin\theta)}\right]d\theta$$

$$=\frac{1}{2\pi}\int_0^{\pi}\left[\frac{2a}{a^2+b^2sin^2\theta}\right]d\theta=\frac{a}{\pi}\int_0^{\pi}\left[\frac{1}{a^2+b^2sin^2\theta}\right]d\theta$$

$$=\frac{a}{\pi}\int_0^{\pi}\left[\frac{cosec^2\theta}{a^2cosec^2\theta+b^2}\right]d\theta$$

$$=\frac{a}{\pi}\int_0^{\pi}\left[\frac{cosec^2\theta}{a^2(1+cot^2\theta)+b^2}\right]d\theta$$

$$=\frac{2a}{\pi}\int_0^{\pi/2}\left[\frac{cosec^2\theta}{(a^2+b^2)+a^2cot^2\theta}\right]d\theta.$$

Putting $cot\theta=z, dz=-cosec^2\theta d\theta$

$$=\frac{2a}{\pi}\int_0^{\infty}\left[\frac{1}{(a^2+b^2)+a^2z^2}\right]dz=\frac{2}{\pi a}\int_0^{\infty}\left[\frac{1}{\frac{(a^2+b^2)}{a^2}+z^2}\right]dz$$

$$=\frac{2}{\pi a}\int_0^{\infty}\left[\frac{1}{\left[\sqrt{\frac{(a^2+b^2)}{a^2}}\right]^2+z^2}\right]dz=\frac{2}{\pi a}\frac{a}{\sqrt{(a^2+b^2)}}\left[tan^{-1}\frac{z}{\sqrt{\frac{(a^2+b^2)}{a^2}}}\right]_0^{\infty}$$

$$=\frac{2}{\pi a}\frac{a}{\sqrt{(a^2+b^2)}}\cdot\left(\frac{\pi}{2}-0\right)=\frac{1}{\sqrt{(a^2+b^2)}}.$$

Example 11.5.4. Prove that

(a) $J_n(x) = \dfrac{(x/2)^n}{\sqrt{\pi}\Gamma(n+1)} \int_{-1}^{1} (1-t^2)^{n-1/2} e^{ixt} dt, \, if \; n \geq -1/2$

(b) $J_n(x) = \dfrac{(x)^n}{2^{n-1}\Gamma\left(\frac{1}{2}\right)\Gamma(n+1/2)} \int_0^1 (1-t^2)^{n-1/2} \cos xt \, dt.$

Solution: Let

$$I = \int_{-1}^{1} (1-t^2)^{n-1/2} e^{ixt} dt \qquad \text{(i)}$$

$$I = \int_{-1}^{1} (1-t^2)^{n-1/2} \left\{ \sum_{r=0}^{\infty} \frac{(ixt)^r}{r!} \right\} dt.$$

$$\therefore I = \sum_{r=0}^{\infty} \frac{(ix)^r}{r!} \int_{-1}^{1} (1-t^2)^{n-1/2} t^r dt. \qquad \text{(ii)}$$

If r is odd, the integrand in the above integral is an odd function. Then, it vanishes; whereas if r is even, the integrand is an even function. Putting $r = 2m$, by the property of definite integral, we get $\int_{-1}^{1} (1-t^2)^{n-1/2} t^r dt = 2 \int_0^1 (1-t^2)^{n-1/2} t^{2m} dt$.

Then, from Equation (ii),

$$I = 2\sum_{m=0}^{\infty} \frac{(ix)^{2m}}{(2m)!} \int_0^1 (1-t^2)^{n-1/2} t^{2m} dt$$

$$= \sum_{m=0}^{\infty} \frac{(i^2)^m (x)^{2m}}{(2m)!} \int_0^1 (1-t^2)^{n-1/2} t^{2m-1} 2t dt.$$

Putting $z = t^2$, so that $2tdt = dz$,

$$I = \sum_{m=0}^{\infty} \frac{(-1)^m (x)^{2m}}{(2m)!} \int_0^1 (1-z)^{n-1/2} (z^{1/2})^{2m-1} dz$$

$$= \sum_{m=0}^{\infty} \frac{(-1)^m (x)^{2m}}{(2m)!} \int_0^1 (1-z)^{\left(n+\frac{1}{2}\right)-1} z^{\left(m+\frac{1}{2}\right)-1} dz$$

$$= \sum_{m=0}^{\infty} \frac{(-1)^m (x)^{2m}}{(2m)!} \times \frac{\Gamma\left(n+\frac{1}{2}\right)\Gamma(m+\frac{1}{2})}{\Gamma(n+m+1)} \text{ using Gamma function}$$

and $n+\frac{1}{2} > 0. m+\frac{1}{2} > 0, \text{i.e.,} n > -\frac{1}{2}, m > -1/2$

$$= \Gamma\left(n+\frac{1}{2}\right) \sum_{m=0}^{\infty} \frac{(-1)^m (x)^{2m}}{(2m)!\Gamma(n+m+1)} \times \frac{(2m)!\sqrt{\pi}}{2^m m!} [\because \Gamma\left(m+\frac{1}{2}\right)$$

$$= \frac{(2m)!\sqrt{\pi}}{2^{2m} m!}]$$

$$= \Gamma\left(n+\frac{1}{2}\right)\sqrt{\pi}\left(\frac{x}{2}\right)^{-n} \sum_{m=0}^{\infty} \frac{(-1)^m}{m!\Gamma(n+m+1)} \times \left(\frac{x}{2}\right)^{2m+n}$$

$$= \Gamma\left(n+\frac{1}{2}\right)\sqrt{\pi}\left(\frac{x}{2}\right)^{-n} j_n(x) \text{ [by definition of } j_n(x),\ j_n].$$

$$\therefore j_n(x) = \frac{(x/2)^n}{\sqrt{\pi \Gamma(n+1/2)}} I$$

$$= \frac{(x/2)^n}{\sqrt{\pi \Gamma(n+1/2)}} \int_{-1}^{1} (1-t^2)^{n-1/2} e^{ixt} dt. \qquad \text{(iii)}$$

(b) $e^{ixt} = \cos xt + i \sin xt.$

Let, $I = \int_{-1}^{1} (1-t^2)^{n-1/2} e^{ixt} dt = \int_{-1}^{1} (1-t^2)^{n-1/2} \cos xt \, dt$

$+i \int_{-1}^{1} (1-t^2)^{n-1/2} \sin xt \, dt = 2 \int_{-1}^{1} (1-t^2)^{n-1/2} \cos xt \, dt + 0. \qquad \text{(iv)}$

Using Equations (iv) and (iii), we get

$$J_n(x) = \frac{(x/2)^n}{\sqrt{\pi}\Gamma(n+1/2)} I = \frac{(x/2)^n}{\sqrt{\pi}\Gamma(n+1/2)} 2 \int_{-1}^{1} (1-t^2)^{n-1/2} \cos xt \, dt$$

$$= \frac{(x)^n}{2^{n-1}\Gamma\left(\frac{1}{2}\right)\Gamma(n+\frac{1}{2})} \int_{0}^{1} (1-t^2)^{n-1/2} \cos xt \, dt.$$

Example 11.5.5. Use the generating function to show that $\sum_{n=-\infty}^{\infty} J_n(-x) = (-1)^n J_n(x)$.

Solution: We have

$$\sum_{n=-\infty}^{\infty} J_n(x)z^n = exp\left\{\frac{x}{2}\left(z - \frac{1}{z}\right)\right\} \quad \text{(i)}$$

Replacing $x\ by - x\ in$ Equation (i), we obtain $\sum_{n=-\infty}^{\infty} J_n(-x)z^n = exp\left\{\frac{-x}{2}\left(z - \frac{1}{z}\right)\right\} = exp\left\{\frac{x}{2}\left(-z - \frac{1}{-z}\right)\right\}.$

$$\sum_{n=-\infty}^{\infty} J_n(-x)z^n = \sum_{n=-\infty}^{\infty} J_n(x)(-z)^n,$$

by Equation (i)

$$or\ \sum_{n=-\infty}^{\infty} J_n(-x)z^n = \sum_{n=-\infty}^{\infty} J_n(x)(-1)^n(z)^n. \quad \text{(ii)}$$

Equating the coefficients of z^n from both sides of Equation (ii), we have

$$J_n(-x) = (-1)^n J_n(x).$$

11.6 Modified Bessel Functions

The Bessel equation of order *n is*

$$x^2\frac{d^2y}{dx^2} + x\frac{dy}{dx} + (x^2 - n^2)y = 0. \quad (11.15)$$

Putting $x = it, we\ get\ an\ equation$

$$t^2\frac{d^2y}{dt^2} + t\frac{dy}{dx} - (t^2 + n^2)y = 0. \quad (11.16)$$

This is known as *modified Bessel's equation of order n.*

We can obtain two linearly independent solutions of Equation (ii), which is denoted by $J_n(it)\, and\, G_{-n}(it)$. We take the solution of Equation (ii) in the following form:

$$I_n(t) = e^{-in\pi/2} J_n(it) \tag{11.17}$$

and,

$$K_n(t) = e^{in\pi/2} G_{-n}(it) \tag{11.18}$$

where $I_n(t)\, and\, K_n(t)$ are real functions of t and are known as *modified Bessel functions of the first and second kinds of order n*, where n is not an integer. We can write its series solutions by employing the usual procedure in the form

$$y = AI_n(t) + BK_n(t).$$

Now,

$$I_n(t) = e^{-in\pi/2} \sum_{r=0}^{\infty} \frac{(-1)^r}{r!\Gamma(n+r+1)} \left(\frac{it}{2}\right)^{n+2r}$$

$$= \sum_{r=0}^{\infty} \frac{1}{\Gamma(r+1)\Gamma(n+r+1)} \left(\frac{t}{2}\right)^{n+2r} \tag{11.19}$$

$[\because n \text{ is even, so } \sin n\frac{\pi}{2} = 0]$.

$$K_n(t) = \frac{\pi}{2\sin(n\pi)} [I_{-n}(t) - I_n(t)] \tag{11.20}$$

$$[\because J_{-n}(x) = (-1)^n J_n(x)]$$

where

$$K_{-n}(t) = K_n(t) \tag{11.20a}$$

and, $$I_{-n}(t) = I_n(t) \tag{11.20b}$$

where n is an integer.

From Equation (11.19),

$$I_n(t) = \Sigma_{r=0}^{\infty} \frac{1}{\Gamma(r+1)\Gamma(n+r+1)} \left(\frac{t}{2}\right)^{n+2r}$$

$$= \frac{t^n}{2^n} \frac{1}{\Gamma(n+1)} \Sigma_{r=0}^{\infty} \frac{1}{\Gamma(r+1)\Gamma(n+r+1)} \left(\frac{t^2}{4}\right)^r$$

$$= \frac{t^n}{2^n} \frac{1}{\Gamma(n+1)} \; 0F_1\left(-, n+1, \frac{t^2}{4}\right).$$

Comparing with Bessel's solution gives the relation

$$I_n(t) = (i)^{-n} J_n(it).$$

11.6.1 Recurrence Relations for Modified Bessel Functions

Theorem 11.6.1. Prove that

I. $tI_n' = nI_n' + tI_{n+1}$

II. $\frac{2n}{t} I_n = I_{n-1} - I_{n+1}$

III. $2^2 I_n'' = 2I_{n-1}' + 2I_{n+1}'$.

Proof I: We have $xJ_n' = nJ_n - xJ_{n+1}$

or, $$x\frac{d}{dx} J_n(x) = nJ_n(x) - xJ_{n+1}(x). \tag{11.21}$$

Putting $x = it$, $\therefore \frac{dx}{dt} = i, \frac{d}{dt} = \frac{dx}{dt}\frac{d}{dx} = i\frac{d}{dx}$, we obtain

$$it\frac{d}{dt} J_n(it)\frac{dt}{dx} = nJ_n(it) - itJ_{n+1}(it).$$

Multiplying both sides by $e^{-in\pi/2}$,

$$e^{-in\pi/2} it \frac{d}{dt} J_n(it) \frac{1}{i} = n e^{-in\pi/2} J_n(it) - ite^{-\frac{in\pi}{2} - (\frac{i\pi}{2})} J_{n+1}(it) e^{\frac{i\pi}{2}}$$

or, $$t \frac{d}{dt} I_n(t) = nI_n(t) + tI_{n+1}(t)$$

or, $$tI_n' = nI_n + tI_{n+1}.$$

Proof II: We have $\frac{2n}{x} J_n = J_{n-1} + J_{n+1}.$

Putting $x = it$, $\frac{2n}{it} J_n(it) = J_{n-1}(it) + J_{n+1}(it).$

Multipling both sides by $e^{-in\pi/2}$, we get

$$\frac{2n}{it} e^{-in\pi/2} J_n(it) = e^{-\frac{in\pi}{2} + i\pi/2} J_{n-1}(it) e^{-i\pi/2} + e^{-\frac{in\pi}{2} - i\pi/2} J_{n+1}(it) e^{i\pi/2}$$

or, $$\frac{2n}{it} I_n(t) = -iI_{n-1}(t) + iI_{n+1}(t)$$

or, $$\frac{2n}{t} I_n(t) = I_{n-1}(t) - I_{n+1}(t).$$

Proof III: We have $2^2 J_n'' = 2J_{n-1}' - 2J_{n+1}'$

or, $$2^2 \frac{d^2}{dx^2} J_n(x) = 2\frac{d}{dx} J_{n-1}(x) - 2\frac{d}{dx} J_{n+1}(x).$$

Putting $x = it$, $\frac{d}{dx} = \frac{dt}{dx}\frac{d}{dt} = \frac{1}{i}\frac{d}{dt}$ and $\frac{d^2}{dx^2} = \frac{d}{dx}\left(\frac{1}{i}\frac{d}{dt}\right) = \frac{1}{i}\frac{d}{dt}\left(\frac{1}{i}\frac{d}{dt}\right)$
$= -\frac{d^2}{dt^2}$

$$2^2\frac{d^2}{dx^2}J_n(it) = 2\frac{1}{i}\frac{d}{dt}J_{n-1}(it) - 2\frac{d}{dx}J_{n+1}(it)$$

or, $$-2^2\frac{d^2}{dt^2}J_n(it) = 2\frac{1}{i}\frac{d}{dt}J_{n-1}(it) - 2\frac{1}{i}\frac{d}{dt}J_{n+1}(it).$$

Multiplying both sides by $e^{-ni\pi/2}$, we get

$$-2^2\frac{d^2}{dt^2}e^{-ni\pi/2}J_n(it) = 2\frac{1}{i}\frac{d}{dt}e^{-\frac{ni\pi}{2}+i\pi/2}J_{n-1}(it)e^{-i\pi/2}$$
$$-2\frac{1}{i}\frac{d}{dt}e^{-\frac{ni\pi}{2}-i\pi/2}J_{n+1}(it)e^{i\pi/2}$$

or, $$-2^2\frac{d^2}{dt^2}I_n(t) = 2\frac{1}{i}\frac{d}{dt}I_{n-1}(t)(-i) - 2\frac{1}{i}\frac{d}{dt}I_{n+1}(t).i$$

or, $$-2^2 I_n''(t) = -2I_n'(t) - 2I_{n+1}'(t)$$

or, $$2^2 I_n''(t) = 2I_n'(t) + 2I_{n+1}'(t).$$

Note 1. *Hypergeometric Function*: The power series

$$1+\frac{a.b}{c}t+\frac{a(a+1)b(.b+1)}{c(c+1)}\frac{t^2}{2!}+\frac{a(a+1)(a+2)b(.b+1)(b+2)}{c(c+1)(c+2)}\frac{t^3}{3!}$$

$$+\ldots\ldots=\frac{(a)_n(b)_n}{(c)_n}\frac{t^n}{n!}={}_2F_1(a,b,c,t)$$

2. *Factorial Function*: We define $(\alpha)_n$ by the equation

$$(\alpha)_n=\alpha(\alpha+1)(\alpha+2)\ldots(\alpha+n-1), n\geq 1.$$

11.7 Equations Reducible to Bessel Equation

In many problems, we come across such differential equations, which can easily be reduced to Bessel equation and, therefore, can be solved by means of Bessel equations.

Example 11.7.1. Reduce the following differential equation to Bessel's form.

(a)

$$x^2\frac{d^2y}{dx^2}+x\frac{dy}{dx}+(k^2x^2-n^2)y=0=0 \tag{i}$$

(b)

$$x\frac{d^2y}{dx^2}+a\frac{dy}{dx}+k^2xy=0=0 \tag{ii}$$

Solution: Put $z=kx$, so that $\frac{dy}{dx}=\frac{dy}{dz}\frac{dz}{dx}=\frac{dy}{dz}k=k\frac{dy}{dz}$

$$\frac{d^2y}{dx^2}=\frac{d}{dx}\frac{dy}{dx}=\frac{d}{dx}k\frac{dy}{dz}=\frac{dz}{dx}\frac{d}{dz}k\frac{dy}{dz}=k^2\frac{d^2y}{dz^2}.$$

Now, Equation (i) is formed:

$$x^2k^2\frac{d^2y}{dz^2}+xk\frac{dy}{dz}+(k^2x^2-n^2)y=0$$

$$or, \qquad z^2\frac{d^2y}{dz^2}+z\frac{dy}{dz}+(z^2-n^2)y=0,$$

which is Bessel's form.

Its solution is $y=c_1J_n(z)+c_2J_{-n}(z)$

$=c_1J_n(kx)+c_2J_{-n}(kx)$, where n is a non-integer.

$=c_1J_n(kx)+c_2Y_n(kx)$, where n is an integer.

(*b*) Putting $y=x^nz$

$$\frac{dy}{dx}=x^n\frac{dz}{dx}+nx^{n-1}z$$

$$\frac{d^2y}{dx^2}=\frac{d}{dx}\frac{dy}{dx}=\frac{d}{dx}\left(x^n\frac{dz}{dx}+nx^{n-1}z\right)=x^n\frac{d^2z}{dx^2}+2nx^{n-1}\frac{dz}{dx}+n(n-1)x^{n-2}z.$$

Putting in Equation (ii), we obtain

$$x(x^n\frac{d^2z}{dx^2}+2nx^{n-1}\frac{dz}{dx}+n(n-1)x^{n-2}z)+a(x^n\frac{dz}{dx}+nx^{n-1}z)+k^2x.x^nz=0$$

$$\text{or,} \qquad x^{n+1}\frac{d^2z}{dx^2}+x^n(2n+a)\frac{dz}{dx}+x^{n-1}[n(n-1)+na+k^2x^2]z=0.$$

Dividing the above equation by x^{n-1} and putting $2n+a=1$, *we obtain*

$$x^2\frac{d^2z}{dx^2}+x\frac{dz}{dx}+[k^2x^2-n^2]z=0$$

$$[since, n^2-n+na=n^2-n+n(1-2n)=-n^2].$$

It is the same differential equation as Equation (a), so its solution is
$z=c_1J_n(kx)+c_2Y_n(kx)$, where n is an integral.
$=c_1J_n(kx)+c_2J_{-n}(kx)$, where n is a non-integral
Hence, the solution of Equation (ii) is
$y=x^nz=x^n[c_1J_n(kx)+c_2J_{-n}(kx)]$, where n is a non–integral.
$=x^n[c_1J_n(kx)+c_2Y_n(kx)]$, where n is an integral, *where* $n=\frac{1}{2}(1-a)$.

11.8 Orthogonality of Bessel Functions

We shall state the orthogonal property in the following theorem

Lemma 11.8.1. If $J_n(x)$ is the solution of the Bessel equation of order n,

$$x^2y''+xy'+(x^2-n^2)y=0, \tag{11.22}$$

then $J_n(\lambda x)$ is the solution of

$$x^2y''+xy'+(\lambda^2x^2-n^2)y=0 \tag{11.23}$$

Proof: Let z *be a variable, and* $J_n(z)$ satisfies the Bessel equation

$$z^2J_n''(z)+zJ_n'(z)+(z^2-n^2)J_n(z)=0. \tag{11.24}$$

Let us take $z = \lambda x$. *Then*, we have

$$\frac{dJ_n}{dz} = \frac{dJ_n}{dx}\frac{dx}{dz} = \frac{1}{\lambda}\frac{dJ_n(z)}{dx} \tag{11.25}$$

$$\frac{d^2J_n}{dz^2} = \frac{1}{\lambda}\frac{d}{dx}\left[\frac{dJ_n(z)}{dx}\right]\frac{dx}{dz} = \frac{1}{\lambda^2}\frac{d^2J_n(z)}{dx^2}. \tag{11.26}$$

Using Equations (11.25) and (11.26) in Equation (11.24), we get

$$(\lambda x)^2\frac{1}{\lambda^2}\frac{d^2J_n(\lambda x)}{dx^2} + \lambda x\frac{1}{\lambda}\frac{dJ_n(\lambda x)}{dx} + (\lambda^2x^2 - n^2)J_n(\lambda x) = 0$$

or, $$(x)^2\frac{d^2J_n(\lambda x)}{dx^2} + x\frac{dJ_n(\lambda x)}{dx} + (\lambda^2x^2 - n^2)J_n(\lambda x) = 0.$$

$$\therefore x^2y'' + xy' + (\lambda^2x^2 - n^2)y = 0. \tag{11.27}$$

Theorem 11.8.1. Prove that

$$\int_0^a xJ_n(\lambda_i x)\,J_n(\lambda_j x)dx = \begin{cases} 0 & \text{if } i \neq j \text{ (different roots)} \\ \frac{a^2}{2}J_{n+1}^2(\lambda_i a). & i = j \text{ (equal roots)} \end{cases}$$

where λ_i, λ_j are the roots of $J_n(x) = 0$

Proof: **Case I.** Let $i \neq j$, i.e., *let* λ_i, λ_j be unequal roots of Bessel equation $J_n(\lambda a) = 0$.

$$\therefore J_n(\lambda_i a) = 0 \text{ and } J_n(\lambda_j a) = 0. \tag{11.28}$$

Let, $$u(x) = \mathrm{J}_n(\lambda_i x) \text{ and } v(x) = \mathrm{J}_n(\lambda_j x). \tag{11.29}$$

We know that the solutions of the equation are

$$x^2 y'' + xy' + [\lambda^2 x^2 - n^2]y = 0, \tag{11.30}$$

$$x^2 u'' + xu' + [\lambda_i^2 x^2 - n^2]u = 0, \tag{11.31}$$

and, $$x^2 v'' + xv' + [\lambda_j^2 x^2 - n^2]v = 0. \tag{11.32}$$

Multiplying Equation (11.31) by $\frac{v}{x}$ and Equation (11.32) by $\frac{u}{x}$ and then subtracting, we get

$$x(u''v - v''u) + (u'v - v'u) + (\lambda_i^2 - \lambda_j^2)xuv = 0$$

or, $$\frac{d}{dx}[x(u'v - v'u)] = (\lambda_j^2 - \lambda_i^2)xuv \tag{11.33}$$

Now, integrating both sides from 0 to a, we obtain

$$(\lambda_j^2 - \lambda_i^2)\int_0^a xuv\, dx = |x(u'v - v'u)|_0^a = a(u'v - v'u)_{x=a}. \tag{11.34}$$

Using Equation (11.29), Equation (11.34) reduces to

$$(\lambda_j^2 - \lambda_i^2)\int_0^a x\mathrm{J}_n(\lambda_i x)\mathrm{J}_n(\lambda_j x)\, dx = \left| x\left\{(\mathrm{J}_n'(\lambda_i x)\, \mathrm{J}_n(\lambda_j x) - \mathrm{J}_n'(\lambda_j x)\ \mathrm{J}_n(\lambda_i x)\right\}\right|_0^a$$
$$= a\left[\mathrm{J}_n'(\lambda_i a)\, \mathrm{J}_n(\lambda_j a) - \mathrm{J}_n'(\lambda_j a)\ \mathrm{J}_n(\lambda_i a)\right] = 0 \text{ [using Equation (11.28)]}.$$

Since $\lambda_i \neq \lambda_j$, we get

$$\int_0^a xJ_n(\lambda_i x)J_n(\lambda_j x)\,dx = 0 \text{ when } i \neq j. \tag{11.35}$$

Case II: Let $i = j$. Multiplying Equation (11.31) by $2u'$, we have

$$2x^2u'u'' + 2xu'u' + 2\left[\lambda_i^2x^2 - n^2\right]uu' = 0$$

or,
$$\frac{d}{dx}\left[x^2(u')^2 - n^2u^2 + \lambda_i^2x^2u^2\right] - 2\lambda_i^2xu^2 = 0$$

$$2\lambda_i^2xu^2 = \frac{d}{dx}\left[x^2(u')^2 - n^2u^2 + \lambda_i^2x^2u^2\right]. \tag{11.36}$$

Integrating Equation (11.36) with respect to x from 0 *to a*,

$$2\lambda_i^2\int_0^a xu^2dx = \left[x^2(u')^2 - n^2u^2 + \lambda_i^2x^2u^2\right]_0^a. \tag{11.37}$$

Using Equations (11.22) and (11.23) and noting that $J_n(0) = 0$, we have

$$2\lambda_i^2\int_0^a xu^2dx = a^2[\{J_n'(\lambda_i x)\}^2]_{x=a}. \tag{11.38}$$

Using recurrence relation, we have

$$\frac{d}{dx}J_n(x) = \frac{n}{x}J_n(x) - J_{n+1}(x). \tag{11.39}$$

Replacing x by $\lambda_i x$ in Equation (11.39), we have

$$\frac{d}{d\lambda_i x} J_n(\lambda_i x) = \frac{n}{x} J_n(\lambda_i x) - J_{n+1}(\lambda_i x)$$

or,
$$\frac{1}{\lambda_i}\frac{d}{dx} J_n(\lambda_i x) = \frac{n}{\lambda_i x} J_n(\lambda_i x) - J_{n+1}(\lambda_i x)$$

$$J_n'(\lambda_i x) = \frac{n}{x} J_n(\lambda_i x) - \lambda_i J_{n+1}(\lambda_i x).$$

$$\therefore\ [\{J_n'(\lambda_i x)\}^2]_{x=a} = \left[\left\{\frac{n}{x} J_n(\lambda_i x\ ) - \lambda_i J_{n+1}(\lambda_i x)\right\}^2\right]_{x=a} = \{0 - \lambda_i J_{n+1}(\lambda_i a)\}^2$$

$= \lambda_i{}^2 J_{n+1}^2(\lambda_i a)$ [by Equation (11.28)

Using this value in Equation (11.38) and dividing both sides of the resulting equation by $2\lambda_i{}^2$, we get

$$\int_0^a x J_n^2(\lambda_i a) dx = \frac{a^2}{2} J_{n+1}^2(\lambda_i a). \tag{11.40}$$

Combining Equations (11.35) and (11.40), we have

$$\int_0^a x J_n(\lambda_i x) J_n(\lambda_j x) dx = \frac{a^2}{2} J_{n+1}^2(\lambda_i a)\delta_j^i. \tag{11.41}$$

11.8.1 Fourier–Bessel Expansion for $f(x)$

If we assume that a function $f(x)$, which is a continuous function having finite number of oscillations in the interval $(0,\alpha)$, then we can write

$$f(x) = c_1 J_n(\alpha_1 x) + c_2 J_n(\alpha_2 x) + \cdots . + c_n J_n(\alpha_n x) + \cdots = \sum_{k=1}^{\infty} c_k J_n(\alpha_k x) \tag{11.42}$$

where $n > -1$ and $\alpha_1, \alpha_2, \ldots..$ are the positive roots of $J_n(\alpha_n x) = 0$.

Then, the series Equation (11.42) is known as the *Fourier–Bessel series.*

If we assume that the series Equation (11.42) can be integrated term-by-term over the interval (o, α), the coefficient $c_n (n = 1, 2, ..)$ can be obtained below:

$$\int_0^a xf(x) J_n(\alpha_n x) dx = c_j \int_0^a x J_n^2(\alpha_j x) dx = c_j \frac{a^2}{2} J_{n+1}^2(a\alpha_j)$$

$$\therefore c_j = \frac{2}{a^2 J_{n+1}^2(a\alpha_j)} \int_0^a xf(x) J_n(\alpha_j x) dx.$$

Proof: *Multiplying both sides of* Equation (11.42) *by* $xJ_n(\alpha_j x)$,

$$xf(x) J_n(\alpha_j x) = \sum_{j=1}^{\infty} c_j x J_n(\alpha_j x) J_n(\alpha_j x). \tag{11.43}$$

Integrating both sides of Equation (11.43) with respect to x from 0 to a, we get

$$\int_0^a xf(x) J_n(\alpha_j x) dx = \sum_{j=1}^{\infty} c_j \int_0^a x J_n(\alpha_j x) J_n(\alpha_j x) dx. \tag{11.44}$$

From the orthogonality property of Bessel functions, we have

$$\int_0^a x J_n(\alpha_i x) J_n(\alpha_j x)\, dx = \begin{cases} 0 i \neq j \\ \frac{a^2}{2} [J_{n+1}(\alpha_j a)]^2 . i = j \end{cases}. \tag{11.45}$$

Using Equation (11.45), Equation (11.44) reduces to

$$\int_0^a xf(x) J_n(\alpha_j x)\, dx = c_j \frac{a^2}{2} \left[J_{n+1}(\alpha_j x) \right]^2 . \tag{11.46}$$

Replacing *by* Equation (11.43), we have

$$c_i \frac{a^2}{2}[J_{n+1}(\alpha_i x)]^2 = \int_0^a x f(x) J_n(\alpha_i x)\, dx$$

or,

$$c_i = \frac{2}{a^2 J_{n+1}^2(a\alpha_i)} \int_0^a x f(x) J_n(\alpha_i x) dx.$$

11.9 Zeros of Bessel Functions

Here, we require a knowledge of the zeros of Bessel function. We shall show that the Bessel function $J_n(x)$ has an infinite number of zeros if n is real.

Poisson's integral of $J_n(x)$ has the form

$$J_n(x) = \frac{2\left(\frac{x}{2}\right)^n}{\sqrt{\pi}\,\Gamma(n+\frac{1}{2})} \int_0^{\frac{\pi}{2}} \cos(x\cos\theta)\, \sin^{2n}\theta d\theta.$$

Taking $-\frac{1}{2} < n < \frac{1}{2}$ and let

$$(2m+1)\cos\theta = t,\ x = (m+\frac{1}{2})\pi, \text{ where } m = 0,1,2\ldots.$$

We get

$$J_n\left(m\pi + \frac{\pi}{2}\right) = \frac{2\left[\left(\frac{m}{2}+\frac{1}{4}\right)\pi\right]^n}{\sqrt{\pi\Gamma(n+\frac{1}{2})}} \int_0^{2m+1} \cos\left(\frac{\pi t}{2}\right)\left[1-\frac{t^2}{(2m+1)}\right]^{n-\left(\frac{1}{2}\right)} \frac{dt}{2m+1}.$$

The definite integral can be written in the form

$$\frac{1}{2}u_0 - u_1 + u_2 - \cdots . + (-1)^m u_m$$

where, $$u_r = (-1)^r \int_{2r-1}^{2r+1} cos\left(\frac{\pi t}{2}\right)\left[1-\frac{t^2}{(2m+1)^2}\right]^{n-\frac{1}{2}} \frac{dt}{2m+1}$$

$$= \int_{-1}^{1} \cos(\frac{\pi t}{2})\left[1-\frac{2r+t}{(2m+1)}\right]^{n-\frac{1}{2}} \frac{dt}{2m+1}$$

where t is replaced by $(2r+t)$. This shows that u_r is positive as $n < u_{r+1}$. Hence, $J_n\left(m\pi + \frac{\pi}{2}\right) > 0$ as m is even. <0 *as m* is odd.

Consequently, $J_n(x)$ changes sign as x increases from $\left(m-\frac{1}{2}\right)\pi$ *to* $\left(m+\frac{1}{2}\right)\pi$, and the function vanishes for some values of x in that interval. We have shown that when $-\frac{1}{2} < n < \frac{1}{2}$, $J_n(x)$ has an infinite number of zeros, and it can be shown by using the formulae

$$J_{\frac{1}{2}}(x)\text{=}(2/\pi x)\text{^}(1/2)\sin x, \; J_{\frac{1}{-2}}(x) = \left(\frac{2}{\pi x}\right)^{\frac{1}{2}} \cos x.$$

The result remains valid when $n = \pm\frac{1}{2}$

since $\frac{d}{dx}[x^{-n}J_n(x)] = -x^{-n}J_{n+1}(x)$.

Since $x^{-n}J_n(x)$ *and* $x^{-n}J_{n+1}(x)$ are both continuous functions of x, it follows from Rolle's theorem that there is at least one zero of $x^{-n}J_{n+1}(x)$ between each consecutive pair of zeros of $x^{-n}J_n(x)$.

$$\frac{1}{x}\frac{d}{dx}[x^n J_n(x)] = x^{n-1}J_{n-1}(x).$$

A similar way shows that there is at least one zero of $x^n J_{n-1}(x)$ between each consecutive pair of zeros of $x^n J_n(x)$.

Hence, we have established that when n is real, $J_n(x)$ has infinite number of zeros.

Again, the asymptotic expansion of $J_n(x)$ is as follows:

$$J_n(x) = \sqrt{\frac{2}{\pi x}}\cos\left(x - \frac{n\pi}{2} - \frac{\pi}{4}\right)\sum_{s=0}^{\infty}\frac{(-1)^s(n,2s)}{(2x)^{2s}}$$
$$-\sin\left(x - \frac{n\pi}{2} - \frac{\pi}{4}\right)\sum_{s=0}^{\infty}\frac{(-1)^s(n,2s+1)}{(2x)^{2s+1}}$$

where $(n,s) = \dfrac{\Gamma(n+\frac{1}{2}+s)}{s\Gamma(n+\frac{1}{2}-s)}$.

This expansion shows that the large positive zeros is approximated to those of

$$\cos\left(x - \frac{n\pi}{2} - \frac{\pi}{4}\right)$$

$x = \left(s - \frac{n}{2} - \frac{1}{4}\right)\pi$, where x is a large positive integer. If we take $s = 11, n = 0$, we find that $J_0(x) = 0$ when $x \cong 10.75\pi = 33.77$.

11.10 Ber and Bei Functions

In many problems, we come across such differential equation, which can be reduced to a Bessel equation and, therefore, can be solved by means of Bessel functions.

Consider the differential equation

$$x^2\frac{d^2y}{dx^2}+x\frac{dy}{dx}+(k^2x^2-n^2)y=0. \quad \text{(i)}$$

Put $n=0, and\ k^2=-i,$ The Equation (i) reduced to

$$x\frac{d^2y}{dx^2}+\frac{dy}{dx}-ixy=0 \quad (11.47)$$

which occurs in certain problems of electrical engineering.

So that its particular solution is

$$y=J_0(kx)=J_0\left[(-i)^{\frac{1}{2}}x\right]=J_0\left[(i)^{\frac{3}{2}}x\right].$$

Replacing x by $i^{\frac{3}{2}}$x in the series for $J_0(x)$ in Equation (i), we get

$$\begin{aligned}y&=1-\frac{i^3x^2}{2^2}+\frac{i^6x^4}{(2!)^2 2^4}-\cdots..\\&=\left[1-\frac{x^4}{2^2.4^2}+\frac{x^8}{2^2.4^26^2.8^2.}-\cdots..\right]\\&+\text{i}\left[\frac{x^2}{2^2}-\frac{x^6}{2^2.4^26^2.8^2.}+\frac{x^{10}}{2^2.4^26^2.8^2.10^2.}-\cdots\right].\end{aligned} \quad (11.48)$$

The series in the above brackets are taken to define Bessel real (or Ber) and Bessel imaginary (or Bei) functions.

Thus,

$$berx = 1 - \frac{x^4}{2^2.4^2} + \frac{x^8}{2^2.4^26^2.8^2.} - \cdots = \sum_{m=1}^{\infty}(-1)^m \frac{x^{4m}}{2^2.4^26^2.8^2..(4m)^2.}$$

$$beix = \frac{x^2}{2^2} - \frac{x^6}{2^2.4^26^2.8^2.} + \frac{x^{10}}{2^2.4^26^2.8^2.10^2.} - \cdots$$

$$= \sum_{m=1}^{\infty}(-1)^m \frac{x^{4m-2}}{2^2.4^26^2.8^2..(4m-2)^2.}$$

$\therefore \quad y = berx + ibeix$ is a solution of Equation (11.47).

11.11 Exercise

1. Prove that (*i*) $\int J_3(x)dx = c - J_2(x) - \frac{2}{x}J_1(x)$

   (ii) $\int xJ_0^2(x)dx = \frac{1}{2}x^2[J_0^2(x) + J_1^2(x)]$.

2. Prove that

   (i) $\frac{d}{dx}[J_n^2(x)] = \frac{x}{2n}[J_{n-1}^2(x) - J_{n+1}^2(x)]$

   (ii) $\frac{d}{dx}[J_n^2(x) + J_{n+1}^2(x)] = 2\left[\frac{n}{2}J_n^2(x) - \frac{n+1}{n}J_{n+1}^2(x)\right]$.

3. Establish the Jacobi series: (i) $\cos(x\cos\theta) = J_0 - 2J_2\cos 2\theta + 2J_4\cos 4\theta - \cdots$.

   (ii) $\sin(x\cos\theta) = 2\left[J_{1\cos\theta} - J_3\cos 3\theta + J_5\cos 5\theta - \cdots\right]$.

4. Prove that $J_n(x) = \frac{x^n}{2^{n-1}\Gamma(n)}\int_0^{\frac{\pi}{2}} \sin\theta\, \cos^{2n-1}\theta\, J_0(x\sin\theta)d\theta$,

   where $n > -\frac{1}{2}$.

5. Prove that

   (i) $J_{n+3} + J_{n+5} = \frac{2}{x}(n+4)J_{n+4}$

   (ii) $4J_n'' = J_{n-2} - 2J_n + J_{n+2}$

   (iii) $J_2 - J_0 = J_0''$.

6. Prove that (i) $\frac{d}{dx}J_0(x) = -J_1(x)$.
7. Show that $\int_0^x x^3 J_0(x)dx = x^3 J_1(x) - 2x^2 J_2(x)$.
8. Show that $\int_0^1 J_0\{x(t-x)\}^{1/2} = 2\sin\left(\frac{t}{2}\right)$.
9. (i) Show that $\int_0^\infty \frac{J_n(x)}{x}dx = \frac{1}{n}$.

   (ii) Prove that $\int_0^{\pi/2} \sqrt{\pi x}\, j_{\frac{1}{2}}(2x)dx = 1$.
10. Prove the following recurrence relations for the Bessel functions and modified Bessel functions:

    (i) $4J_n''(x) = J_{n-2}(x) - 2J_n(x) + J_{n+2}(x)$

    (ii) $xI_n'(x) = xI_{n-1}(x) - nI_n(x)$.
11. Prove that $\int_0^1 J_0\left(\sqrt{\{x(t-x)\}}\right)dx = 2sin\frac{1}{2}t$.
12. Prove that for Re $n > -\frac{1}{2}$,

$$J_n(z) = \frac{2(\frac{1}{2}z)^n}{\Gamma\left(\frac{1}{2}\right)\Gamma(n+\frac{1}{2})}\int_0^{\pi/2} sin^{2n}\theta\ \cos(zcos\theta)\ d\theta.$$

13. Show that $\frac{d}{dx}\left[\frac{J_{-n}(x)}{J_n(x)}\right] = \frac{2\sin n\pi}{\pi x J_n^2(x)}$.
14. Prove that $\frac{d}{dx}[J_n^2 + J_{n+1}^2] = \frac{2}{x}[nJ_n^2 - (n+1)J_{n+1}^2]$.
15. Using the generating function, show that $J_n(-x) = (-1)^n J_n(x)$.
16. Prove that (i) $\frac{d}{dx}[x^{-n}J_n] = -x^{-n}J_{n+1}$

    (ii) $\frac{d}{dx}[x^n J_n] = x^n J_{n-1}$.
17. Prove that (i) $\frac{d}{dx}J_n^2(x) = \frac{x}{2n}[J_{n-1}^2(x) - J_{n+1}^2(x)]$

    (ii) $\frac{d}{dx}[J_n^2(x) + J_{n+1}^2(x)] = 2\left[\frac{n}{2}J_n^2(x) - \frac{(n+1)}{x}J_{n+1}^2(x)\right]$.

18. If $\alpha_1, \alpha_2 \ldots \alpha_n$ *are the positive roots of* $J_0(x) = 0$, *show that*

$$\frac{1}{2} = \sum_{n=1}^{\infty} \left\{ \frac{J_0(\alpha_n x)}{\alpha_n J_1(\alpha_n x)} \right\}.$$

19. Show that (i) $x^n J_n(x)$ is a solution of the equation $xy'' + (1-2n)y' + xy = 0$
    (ii) $x^{-n} J_n(x)$ is a solution of the equation $xy'' + (1+2n)y' + xy = 0$.
20. For Bessel function $J_n(x)$, prove that

$$J_n(x) = \frac{1}{\pi} \int_0^{\pi} \cos(n\theta - x\sin\theta)\, d\theta,$$

and hence, show that

$$J_0(x) = \frac{2}{\pi} \int_0^1 \frac{\cos xt\, dt}{\sqrt{(1-t^2)}} d\theta.$$

12
Legendre Polynomials

Abstract
The Legendre differential equation is crucial in boundary value problems involving spherical symmetry, due to the orthogonality of its solutions and their applicability in solving Laplace's equation. Legendre polynomials are widely used in potential theory, quantum mechanics, and geophysics.

In this chapter, we study the properties, orthogonality, generating functions, and recurrence relations of Legendre polynomials, along with their role in solving physical problems with spherical symmetry.

Keywords: Legendre's differential equation, generating function, recurrence relations, orthogonality, Rodrigues formula

12.1 Introduction

The French mathematician Adrien-Marie Legendre (1752–1833) made important contributions to number theory, special functions, calculus of variations, and elliptic integrals.

The gravitational potential ρ at any point P, of a unit mass at another point A, is $\rho = (a^2 - 2arcos\theta + r^2)^{-1/2}$ where r is the distance of P from the origin O, and θ is the angle between $OA\ and\ OP$. Sometimes, we need to expand ρ in terms of $r\ and\ r^{-1}$.

This expansion originates an important class of functions, which we will study in this chapter. The differential equation's importance in applied mathematics, particularly in boundary value problem for spheres, will be studied here.

In this chapter, we study the properties of these functions in greater detail, that is, the properties – orthogonality, generating functiond, recurrence relations, Fourier - legendre expansion and also the properties of Associated legendre polynomials aqnd legendre polynomials of second kind.

Dipankar De. Ordinary Differential Equations and Special Functions, (431–496) © 2026 Scrivener Publishing LLC

12.2 Legendre's Equation

The differential equation of the form

$$(1-x^2)\frac{d^2y}{dx^2}-2x\frac{dy}{dx}+n(n+1)y=0, \tag{12.1}$$

where n is a positive integer, is known as *Legendre's equation.*

Let the series solution of Equation (i) be

$$y=\sum_{m=0}^{\infty} c_m x^{k-m}. \tag{12.2}$$

Differentiating Equation (12.2) and putting these values of $\frac{d^2y}{dx^2}$, $\frac{dy}{dx}$ *and* y in Equation (12.1), we get

$$(1-x^2)\sum_{m=0}^{\infty} c_m(k-m)(k-m-1)x^{k-m-2}-2x\sum_{m=0}^{\infty} c_m(k-m)x^{k-m-1}$$
$$+n(n+1)\sum_{m=0}^{\infty} c_m x^{k-m}=0$$

or,

$$\sum_{m=0}^{\infty} c_m(k-m)(k-m-1)x^{k-m-2}$$
$$-\sum_{m=0}^{\infty} c_m x^{k-m}\{(k-m)(k-m-1)-n(n+1)+2(k-m)\}=0. \tag{12.3}$$

Now,

$$(k-m-1)-n(n+1)+2(k-m)=(k-m)^2-(k-m)+2(k-m)-n(n+1)$$
$$=(k-m)^2+(k-m)-n^2-n=(k-m+n)(k-m-n)+(k-m-n)$$
$$=(k-m-n)(k-m+n+1).$$

Equation (12.3) becomes

$$\sum_{m=0}^{\infty} c_m(k-m)(k-m-1)x^{k-m-2}$$
$$+\sum_{m=0}^{\infty} c_m x^{k-m}(k-m-n)(k-m+n+1)=0. \tag{12.4}$$

For an indicial equation, equating to zero of the coefficient of the highest power of x^k (*here* $m=0$), we get

$$c_0(k-n)(k+n+1)=0 \text{ or } (k-n)(k+n+1)=0\ (\because c_0 \neq 0). \quad (12.5)$$

The roots of Equation (12.5) are $k=n,-(n+1)$.

They are unequal and differ by integer.

Now, the next lower power of is x^{k-1}, and since their coefficients in Equation (12.4) are equal to zero, we get

$$c_1\,(k-1-n)(k+n)=0. \quad (12.6)$$

For $k=n$ and $-(n+1)$, no factors with c_1 in Equation (12.5) is zero. So, $c_1=0$.

Now, equating the coefficient of x^{k-m} in Equation (12.4) to zero (putting m=$m-2$):

$$c_{m-2}(k-m+2)(k-m+1)-c_m(k-m-n)(k-m+n+1)=0.$$

$$\therefore c_m=\frac{(k-m+2)(k-m+1)}{(k-m-n)(k-m+n+1)}c_{m-2}.$$

(12.7)

Here, all

$$c_3=c_5=c_7\ldots=0\ (\text{since } c_1=0) \quad (12.8)$$

[which hold for both $k=n$ *and* $k=-(n+1)$].

Now, to obtain $c_2,c_4,c_6\ldots$

Case I: When $k=n$, *then* Equation (12.7) *becomes*

$$c_m=-\frac{(n-m+2)(n-m+1)}{m(2n-m+1)}c_{m-2}=\ldots \quad (12.9)$$

For, $m=2,4,6,\ldots$ in Equation (12.9), we get

$$c_2=-\frac{n(n-1)}{2(2n-1)}c_0,\ c_4=-\frac{(n-2)(n-3)}{4(2n-3)}c_2=\frac{(n-2)(n-3)}{4(2n-3)}\frac{n(n-1)}{2(2n-1)}c_0$$

$$=\frac{n(n-1)(n-2)(n-3)}{2.4(2n-1)(2n-3)}c_0,$$

and so on.

For $k=n$, Equation (12.2) is

$$y=\sum_{m=0}^{\infty} c_m x^{n-m} = c_0x^n + c_1x^{n-1} + c_2x^{n-2} + c_3x^{n-3} + c_4x^{n-4} + c_5x^{n-5} + c_6x^{n-6} + \cdots. \tag{12.10}$$

Using Equation (12.8) and putting the values of $c_1, c_2 c_3, c_4 c_5, c_6 \ldots$.

$$\begin{aligned} &= c_0x^n + c_2x^{n-2} + c_4x^{n-4} + c_6x^{n-6} + \cdots. \\ &= c_0\left[x^n - \frac{n(n-1)}{2(2n-1)}x^{n-2} + \frac{n(n-1)(n-2)(n-3)}{2.4(2n-1)(2n-3)}x^{n-4} + \cdots..\right] \\ &= a\left[x^n - \frac{n(n-1)}{2(2n-1)}x^{n-2} + \frac{n(n-1)(n-2)(n-3)}{2.4(2n-1)(2n-3)}x^{n-4} + \cdots..\right]. \end{aligned} \tag{12.11}$$

Case II: When $k=-(n+1)$, *then* Equation (12.7) becomes

$$c_m = \frac{(n+m-1)(n+m)}{m(2n+m+1)} c_{m-2}. \tag{12.12}$$

Putting $m=2,4,6,\ldots$ in Equation (12.12), we have

$$\begin{aligned} c_2 &= \frac{(n+1)(n+2)}{2(2n+3)}c_0,\ c_4 = \frac{(n+3)(n+4)}{4(2n+5)}c_2 \\ &= \frac{(n+3)(n+4)}{4(2n+5)}\frac{(n+1)(n+2)}{2(2n+3)}c_0 = \frac{(n+1)(n+2)(n+3)(n+4)}{2.4\big(2n+3)(2n+5)\big)}c_0, \end{aligned}$$

and so on.

For $k=-(n+1)$, Equation (12.2) *gives*

$$y=\textstyle\sum_{m=0}^{\infty} c_m x^{-n-m-1} = c_0x^{-n-1} + c_1x^{-n-2} + c_2x^{-n-3} + c_3x^{-n-4} + c_4x^{-n-5} + c_5x^{-n-6} + c_6x^{-n-7} + \cdots. \tag{12.13}$$

Using Equation (12.8) and the above values of $c_2, c_4, c_6, c_8 \ldots$ in Equation (12.13), we get

$$y=c_0\left[x^{-n-1}+\frac{(n+1)(n+2)}{2(2n+3)}x^{-n-3}+\frac{(n+1)(n+2)(n+3)(n+4)}{2.4(2n+3)(2n+5))}x^{-n-5}\right.$$
$$\left.+\frac{(n+1)(n+2)(n+3)(n+4)(n+5)(n+6)}{2.4.6(2n+3)(2n+5))(2n+7)(2n+9))}x^{-n-7}+\cdots\right]$$

$$y=b\left[x^{-n-1}+\frac{(n+1)(n+2)}{2(2n+3)}x^{-n-3}+\frac{(n+1)(n+2)(n+3)(n+4)}{2.4(2n+3)(2n+5))}x^{-n-5}\right.$$
$$\left.+\frac{(n+1)(n+2)(n+3)(n+4)(n+5)(n+6)}{2.4.6(2n+3)(2n+5))(2n+7)(2n+9))}x^{-n-7}+\cdots\right]. \quad (12.14)$$

These are two independent solutions of Equation (12.1).

Now, if we put $=\frac{[1.3.5.,,,,(2n-1)]}{n!}$, the solution (Equation 12.11) is denoted by $P_n(x)$ and is called a *Legendre's function of the first kind or Legendre's polynomial of degree n*. Thus, $P_n(x)$ is a solution of Equation (12.1). Observe that Equation (xi) is a terminating series, and so, it gives rise to a polynomial of degree *n*.

Again, if we take $b=\frac{n!}{[1.3.5.,,,,(2n+1)]}$, the solution Equation (12.14) is denoted by $Q_n(x)$ and is called *Legendre's function of the second kind or Legendre's polynomial of degree n*. Thus, s is a solution of Equation (i). Since n is a positive integer, Equation (xiv) is an infinite or non-terminating series, and $Q_n(x)$ is not a polynomial. Thus, $P_n(x)\, and\, Q_n(x)$ are two linearly independent solutions of Equation (12.1). Hence, the general solution of Equation (12.1) is

$$y = AP_n(x) + B\,Q_n(x),$$

where $A\, and\, B$ are arbitrary constants.

12.3 Another Form of Legendre's Polynomial $P_n(x)$

Legendre's polynomial of degree *n* is defined by

$$P_n(x)=\frac{[1.3.5.,,,,(2n-1)]}{n!}\left[x^n-\frac{n(n-1)}{2(2n-1)}x^{n-2}+\frac{n(n-1)(n-2)(n-3)}{2.4(2n-1)(2n-3)}x^{n-4}+\cdots..\right]. \quad (12.15)$$

The general term t_r of Polynomial (12.15) is given by

$$t_r = \frac{1.3.5.,,,,(2n-1)}{n!}(-1)^r \frac{n(n-1)(n-2)(n-3)\dots(n-2r+1)}{2.4\dots.2r(2n-1)(2n-3)..(2n-2r+1)} x^{n-2r}. \tag{12.16}$$

Now, $$t_r \frac{1.3.5.,,,,(2n-1)}{n!} = \frac{(2n)!}{2.4\dots.2n} = \frac{(2n)!}{2^n.}. \tag{12.17}$$

$$n(n-1)(n-2)(n-3)\dots(n-2r+1)$$
$$= \frac{n(n-1)(n-2)(n-3)\dots(n-2r+1)(n-2r)..3.2.1}{(n-2r)..3.2.1} = \frac{n!}{(n-2r)!}. \tag{12.18}$$

and, $$2.4.6\dots.2r = 2^r\, r! \tag{12.18a}$$

$$(2n-1)(2n-3)..(2n-2r+1) = \frac{(2n)\big(2n-1(2n-2)\big)(2n-3)..(2n-2r)(2n-2r+1)}{(2n)(2n-2)..(2n-2r)}$$
$$= \frac{(2n)\big(2n-1(2n-2)\big)(2n-3)..(2n-2r+2)(2n-2r+1)(2n-2r)\dots 3.2.1}{(2n)(2n-2)..(2n-2r+2)[(2n-2r)\dots 3.2.1]}$$
$$= \frac{(2n)!}{(2n-2r)!\{(2n)(2n-2)..(2n-2r+2)\}} = \frac{(2n)!}{(2n-2r)![2^r n(n-1)(n-2)\dots.(n-r+1)]}$$
$$= \frac{(2n)!(n-r)\dots 3.2.1}{(2n-2r)![2^r n(n-1)(n-2)\dots.(n-r+1)\dots 3.2.1]} = \frac{(2n)!(n-r)!}{(2n-2r)!2^r n!}. \tag{12.19}$$

Using (12.17), (12.18), (12.18a), (12.19) in (12.16), we get

$$t_r = \frac{[1.3.5.,,,,(2n-1)]}{n!}(-1)^r \frac{n(n-1)(n-2)(n-3)\dots(n-2r+1)}{2.4\dots.2r(2n-1)(2n-3)..(2n-2r+1)} x^{n-2r}$$

$$= \frac{\frac{(2n)!}{2^n.n!}(-1)^r}{n!} \frac{n!}{(n-2r)!\,2^r r!} \frac{x^{n-2r}}{\frac{(2n)!(n-r)!}{(2n-2r)!2^r n!}}$$

$$= (-1)^r \frac{(2n-2r)!}{2^n r!(n-2r)!(n-r)!} x^{n-2r}. \tag{12.20}$$

Since Equation (12.15) is a polynomial of degree n, r must be chosen so that $n-2r \geq 0$, i.e., $r \leq \frac{1}{2}n$. Thus, if n is even, r *goes from* 0 *to* $[\frac{1}{2}n]$, and while n is odd, r *goes from* 0 *to* $\frac{1}{2}(n-1)$. Therefore, r *goes from* 0 *to* $[\frac{1}{2}n]$,

that is, $\left[\frac{1}{2}n\right] = \begin{cases} \frac{1}{2}n, \textit{if } n \textit{ is even} \\ \frac{1}{2}(n-1), \textit{if } n \textit{ is odd} \end{cases}$.

The Legendre's polynomial of degree n is given by

$$P_n(x) = \sum_{r=0}^{\left[\frac{1}{2}n\right]} (-1)^r \frac{(2n-2r)!}{2^n r!(n-2r)!(n-r)!} x^{n-2r}. \quad (12.21)$$

12.3.1 Legendre's Function of the First Kind or Legendre's Polynomial of Degree n

The Legendre's polynomial of degree n is given by

$$P_n(x) = \frac{[1.3.5.,,,,(2n-1)]}{n!}\left[x^n - \frac{n(n-1)}{2(2n-1)}x^{n-2} + \frac{n(n-1)(n-2)(n-3)}{2.4(2n-1)(2n-3)}x^{n-4} + \cdots..\right].$$

(12.22)

$$P_n(x) = \sum_{r=0}^{\left[\frac{1}{2}n\right]} (-1)^r \frac{(2n-2r)!}{2^n r!(n-2r)!(n-r)!} x^{n-2r}. \quad (12.23)$$

12.3.2 Legendre's Function of the Second Kind

$$Q_n(x) = \frac{n!}{[1.3.5.,,,,(0-2n+1)]}\left[x^{-n-1} + \frac{(n+1)(n+2)}{2(2n+3)}x^{-n-3} + \frac{(n+1)(n+2)(n+3)(n+4)}{2.4(2n+3)(2n+5))}x^{-n-5} + \frac{(n+1)(n+2)(n+3)(n+4)(n+5)(n+6)}{2.4.6(2n+3)(2n+5))(2n+7)(2n+9))}x^{-n-7} + \cdots\right] \quad (12.24)$$

12.3.3 Determination of the First Few Legender's Polynomials $P_n(x)$

Putting $n = 0,1,2,3,4,\ldots$ in Equation (12.22), we have

$$P_0(x) = \frac{1}{0!}x^0 = 1,\ P_1(x) = \frac{1}{1!}x = x,\ P_2(x) = \frac{1.3}{2!}\left[x^2 - \frac{2.1}{2.3}.1\right]$$

$$= \frac{3}{2}\left[x^2 - \frac{.1}{.3}\right] = \frac{1}{2}[3x^2 - 1]$$

$$P_3(x) = \frac{1.3.5}{3!}\left[x^3 - \frac{3.2.}{2.5}.x\right] = \frac{1}{2}[5x^3 - 3x],$$

$$P_4(x) = \frac{1.3.5.7}{4!}\left[x^4 - \frac{4.3.}{2.7}.x^2 + \frac{4.3.2.1}{2.4.7.5}.1\right] = \frac{35}{8}\left[x^4 - \frac{6}{7}.x^2 + \frac{3}{35}\right]$$

$$= \frac{1}{8}\left[35x^4 - 30x^2 + 3\right].$$

12.4 Generating Function for Legendre's Polynomials

Show that $(1 - 2xt + t^2)^{-\frac{1}{2}} = \Sigma_{n=0}^{\infty} t^n P_n(x)$, $|x| \le 1, |t| < 1$

[Note: $(1 - 2xt + t^2)^{-\frac{1}{2}}$ is called the generating function for the Legendre's polynomial $P_n(x)$].

Proof: Since $|x| \le 1, |t| < 1$, we have

$$(1 - 2xt + t^2)^{-\frac{1}{2}} = [1 - t(2x - t)]^{-\frac{1}{2}} =$$

using $(1-z)^{-\frac{1}{2}} = 1 + \frac{1}{2}z + \frac{\frac{1}{2}.\frac{3}{2}}{2!}z^2 + \frac{\frac{1}{2}.\frac{3}{2}.\frac{5}{2}}{3!}z^3 + \cdots = 1 + \frac{2!}{(1!)^2.2^2}z + \frac{4!}{(2!)^2.2^4}z^2$
$+ \frac{6!}{(3!)^2.2^6}z^3 + \cdots$

$$(1-2xt+t^2)^{-\frac{1}{2}} = 1+\frac{2!}{(1!)^2.2^2}t(2x-t)+\frac{4!}{(2!)^2.2^4}t^2(2x-t)^2$$

$$+\frac{6!}{(3!)^2.2^6}t^3(2x-t)^3+\cdots.+\frac{(2n-2r)!}{[(n-r)!]^2 2^{(2n-2r)}}.t^{n-r}(2x-t)^{n-r}+\cdots..$$

$$+\frac{(2n)!}{(n!)^2 2^{2n}}.t^n(2x-t)^n.$$

The term of t^n in the term containing $t^{n-r}(2x-t)^{n-r}$

$$=\frac{(2n-2r)!}{[(n-r)!]^2 2^{(2n-2r)}}\times\frac{(n-r)!}{r!(n-2r)!}(-1)^r t^n(2x)^{n-2r}.$$

The coefficient of $t^n = \dfrac{(-1)^r(2n-2r)!}{2^n r!(n-r)!(n-2r)!}t^n.$

Collecting all terms in t^n, which will occur in the term containing $t^n(2x-t)^n$ and proceeding terms, we see that terms in t^n are

$$\Sigma_{r=0}^{N}\frac{(-1)^r(2n-2r)!}{2^n r!(n-r)!(n-2r)!}t^n(x)^{n-2r} = P_n(x)t^n \qquad (12.25)$$

where $N=\frac{1}{2}$n or $N=\frac{1}{2}(n-1)$ if n is even or odd, respectively.

$$\therefore (1-2xt+t^2)^{-\frac{1}{2}} = P_n(x)t^n. \qquad (12.26)$$

This shows that $P_n(x)$ is the coefficient of t^n in the expression of $(1-2xt+t^2)^{-\frac{1}{2}}$.

It is known as the *generating function of Legendre's polynomial.*

Alternative Method:

$$
\begin{aligned}
(1-2xt+t^2)^{-\frac{1}{2}} &= [1-t(2x-t)]^{-\frac{1}{2}} \\
&= 1+\frac{1}{2}t(2x-t)+\frac{1}{2}.\frac{3}{4}t^2(2x-t)^2+\cdots. \\
&\quad +\frac{1}{2}.\frac{3\ldots.}{4\ldots..}.\frac{(2n-3)}{2n-2}t^{n-1}(2x-t)^{n-1} \\
&\quad +\frac{1}{2}.\frac{3\ldots.}{4\ldots.}.\frac{(2n-1)}{2n}t^{n}(2x-t)^{n}+\ldots. \qquad (12.27)
\end{aligned}
$$

The coefficient of t^n in $\frac{1}{2}.\frac{3\ldots.}{4\ldots..}.\frac{(2n-1)}{2n}t^n(2x-t)^n$

$$
=\frac{1}{2}.\frac{3\ldots.}{4\ldots..}.\frac{(2n-1)}{2n}(2x)^n=\frac{1.3.5\ldots.(2n-1)}{2^n(1.2.3\ldots n)}2^n.x^n=\frac{1.3.5\ldots.(2n-1)}{n!}x^n
$$

(12.28)

Again, the coefficient of t^n in $\frac{1}{2}.\frac{3\ldots.}{4\ldots..}.\frac{(2n-3)}{2n-2}t^{n-1}(2x-t)^{n-1}$

$$
\begin{aligned}
&=\frac{-(1.3.5\ldots.(2n-3)}{2^{n-1}(1.2.3\ldots n-1)}\frac{2n-1}{n}\frac{n}{2n-1}[(n-1)2^{n-2}x^{n-2} \\
&=-\frac{1.3.5\ldots.(2n-3)(2n-1)}{n!}\frac{n(n-1)}{2(2n-1)}X^{n-2}
\end{aligned} \qquad (12.29)
$$

$$
\left[(n-1)th\ term\ of (2x-t)^{n-1}=\binom{n-1}{1}(2x)^{n-2}(-t)=-(n-1)2^{n-2}x^{n-2}\right]
$$

and so on. Using Equations (12.28) and (12.29), we see that the coefficient of t^n in expansion (12.27) is given by

$$
=\frac{1.3.5\ldots.(2n-1)}{n!}\left[x^n-\frac{n(n-1)}{2(2n-1)}x^{n-2}+\frac{n(n-1)(n-2)(n-3)}{2.4(2n-1)(2n-3)}x^{n-4}-\ldots\right],
$$

which is $P_n(x)$ by definition is a Legendre's polynomial.

We find that the coefficients of $t, t^2, \ldots.$ are $P_1(x), P_2(x)$ in the expansion of $(1-2xt+t^2)^{-\frac{1}{2}}$. Thus,

$$(1-2xt+t^2)^{-\frac{1}{2}} = P_0(x)+P_1(x)t+P_2(x)t^2+\cdots.+P_n(x)t^n$$
$$+\cdots. = \textstyle\sum_{n=0}^{\infty} P_n(x)t^n. \tag{12.29a}$$

Replacing *x by–x* in (12.27)

$$(1+2xt+t^2)^{\frac{1}{2}} = \sum_{n=0}^{\infty} t^n\ P_n(-x) \tag{12.29b}$$

Replacing *t by–t* in (12.27)

$$(1+2xt+t^2)^{\frac{1}{2}} = \sum_{n=0}^{\infty}(-1)^n t^n\ P_n(x) \tag{12.29c}$$

From (12.29a) and (12.29b) we have

$$\textstyle\sum_{n=0}^{\infty} t^n\ P_n(-x) = \sum_{n=0}^{\infty}(-1)^n t^n P_n(x)$$

$$\therefore P_n(-x) = P_n(x)$$

Note. (a) $P_n(1)=1$.

Putting $x=1$ *in* Equation (12.29a),

$$(1-2t+t^2)^{-\frac{1}{2}} = (1-t)^{-1} = (1+t+t^2+t^3+\cdots+t^n+..) = \sum_{n=0}^{\infty} P_n(1)t^n.$$

$$\therefore P_n(1)=1.$$

(b) $P_n(-1)=(-1)^n$.

Putting $x=-1$ *in* Equation (12.29a),

$$(1+2t+t^2)^{-\frac{1}{2}} = (1+t)^{-1} = (1-t+t^2-t^3+\cdots+(-1)^n t^n+..)$$
$$= \sum_{n=0}^{\infty} P_n(-1)t^n.$$

$$\therefore P_n(-1)=(-1)^n.$$

$$(c)\ P_n(0)=\begin{cases}(-1)^{\frac{n}{2}}\dfrac{1.3.5....(n-1)}{2.4.6.....n}, \text{when n is even} \\ 0, \textit{when n is odd}\end{cases}.$$

Putting $x=0$ *in* Equation (12.28),

$$P_0(x)+P_1(x)t+P_2(x)t^2+\cdots.+P_n(x)t^n+\cdots.=\sum_{n=0}^{\infty}P_n(0)t^n=(1+t^2)^{-\frac{1}{2}}$$

$$=1-\frac{1}{2}t^2+\frac{1.3}{2.2^2}t^4-..+(-1)^r\frac{1.3.5....(2r-1)}{2.4.6...2r}t^{2r}+\cdots.$$

Equating the coefficients of t^{2m}, we get $P_{2m}(0)=(-1)^m\dfrac{1.3.5....(2m-1)}{2.4.6...2m}$,

i.e., $P_n(0)=(-1)^{n/2}\dfrac{1.3.5....(n-1)}{2.4.6...n}$, if n is an even integer,
and $P_{2m+1}(0)=0$ (since no coefficients of t^{2m+1} are in the expansions),
i.e., $P_n(0)=0$, *when n* is odd.

12.5 Various Forms of $P_n(x)$

12.5.1 Rodrigues' Formula

We shall prove that

$$P_n(x)=\frac{1}{n!2^n}\frac{d^n}{dx^n}(x^2-1)^n. \tag{12.30}$$

Let $v=(x^2-1)^n$

$$v_1=\frac{dv}{dx}=n2x(x^2-1)^{n-1}$$

or,
$$(x^2-1)v_1=2nxv$$

or, $$(1-x^2)v_1 + 2nxv = 0. \tag{12.31}$$

Differentiating $(n+1)$ times and applying Leibnitz theorem, we get

$$(1-x^2)v_{n+2} + (n+1)(-2x)v_{n+1} + \frac{(n+1)}{2}n(-2)v_n + 2n[xv_{n+1} + (n+1)v_n] = 0$$

or, $$(1-x^2)\frac{d^2v_n}{dx^2} - 2x\frac{dv_n}{dx} + n(n+1)v_n = 0.$$

This is Legendre's equation, and the solution is cv_n, and its finite series solution is $P_n(x)$.

$$\therefore P_n(x) = cv_n = c\frac{d^n}{dx^n}(x^2-1)^n. \tag{12.32}$$

To determine c, put $x = 1$ *in* Equation (12.32):

$$1 = c\left[\frac{d^n}{dx^n}(x-1)^n(x+1)^n\right]_{x=1}$$
$$= c[n!(x+1)^n] + \textit{terms containing } (x-1) \textit{ and its power}]_{x=1}$$
$$= cn!2^n.$$
$$\therefore c = \frac{1}{n!2^n}.$$

Substituting this value of c in Equation (iii), we get $P_n(x) = \frac{1}{n!2^n}\frac{d^n}{dx^n}(x^2-1)^n$, which is known as the *Rodrigue's formula.*

Note (i) that using the Rodrigue's formula and expansion of $(x^2-1)^n$, *we* can easily establish $P_n(x) = \sum_{r=0}^{\left[\frac{1}{2}n\right]} (-1)^r \frac{(2n-2r)!}{2^n r!(n-2r)!(n-r)!} x^{n-2r}$.

(ii) All the roots of $P_n(x)=0$ are real and lie between $-1\, and +1$.

12.5.2 Murphy's Formula

Let us consider the function

$$(1-2xt+t^2)^{-\frac{1}{2}} = [(1-t)^2-2t(x-1)]^{-\frac{1}{2}}$$

$$= (1-t)^{-1}\left[1-\frac{2t(x-1)}{(1-t)^2}\right]^{-1/2}$$

$$= (1-t)^{-1}\sum_{m=0}^{\infty}\frac{(\frac{1}{2})_m.2^m t^m (x-1)^m}{m!(1-t)^{2m}}$$

$$= \sum_{n=0}^{\infty}\sum_{m=0}^{\infty}\frac{(\frac{1}{2})_m.2^m t^m (x-1)^m}{m!}.\frac{(2m+1)_n t^n}{n!}$$

$$= \sum_{n=0}^{\infty}\sum_{m=0}^{\infty}\frac{(\frac{1}{2})_m.(2m+1)_n 2^m (x-1)^m}{m!}.\frac{t^{n+m}}{n!}$$

$$= \sum_{n=0}^{\infty}\sum_{m=0}^{\infty}\frac{(\frac{1}{2})_m.\Gamma(n+2m+1)2^m (x-1)^m}{m!\Gamma(2m+1)}.\frac{t^{n+m}}{n!}$$

$$= \sum_{n=0}^{\infty}\sum_{m=0}^{\infty}\frac{\Gamma(\frac{1}{2})(\frac{1}{2})_m.(n+2m)!2^m (x-1)^m}{m!2^m\Gamma(m+1)\Gamma(m+1/2)}.\frac{t^{n+m}}{n!}$$

$$= \sum_{n=0}^{\infty}\sum_{m=0}^{\infty}\frac{(n+2m)!(x-1)^m}{2^m(m!)^2}.\frac{t^{n+m}}{n!}$$

$$= \sum_{n=0}^{\infty}\sum_{m=0}^{\infty}\frac{(n+m)!(x-1)^m}{2^m(m!)^2}.\frac{t^n}{(n-m)!}$$

$$= \sum_{n=0}^{\infty}\sum_{m=0}^{n}\frac{(-1)^m(-n)_m(n+1)_m(x-1)^m}{2^m(m!)^2}.\frac{t^n}{}. \tag{12.33}$$

But we know that

$$(1-2xt+t^2)^{-\frac{1}{2}} = \Sigma_{n=0}^{\infty} P_n(x)t^n. \tag{12.34}$$

Comparing the coefficient of t^n from Equations (i) and (ii), we get

$$P_n(x) = \sum_{m=0}^{n} \frac{(-1)^m(-n)_m(n+1)_m(x-1)^m}{2^m(m!)^2}$$

$$= {}_2F_1\left[-n, n+1; 1; \frac{1-x}{2}\right],$$

which is known as *Murphy's formula for* $P_n(x)$.

12.5.3 Bateman's Generating Relations

We know that

$$P_n(x) = \Sigma_{m=0}^{\infty} \frac{(n!)^2}{(m!)^2\{(n-m)!\}^2}\left\{\frac{1}{2}(x-1)\right\}^{n-m}\left\{\frac{1}{2}(x+1)\right\}^m.$$

Multiplying both sides by $\frac{t^n}{(n!)^2}$ and summing from $n=0$ *to* $n=\infty$, we get

$$\sum_{n=0}^{\infty}\frac{t^n}{(n!)^2}P_n(x) = \sum_{n=0}^{\infty}\sum_{m=0}^{\infty}\frac{t^n}{(m!)^2\{(n-m)!\}^2}\left\{\frac{1}{2}(x-1)\right\}^{n-m}\left\{\frac{1}{2}(x+1)\right\}^m$$

$$\Sigma_{n=0}^{\infty}\Sigma_{m=0}^{\infty}\frac{1}{(m!)^2\{(n)!\}^2}\left\{\frac{1}{2}(x-1)\right\}^n\left\{\frac{1}{2}(x+1)\right\}^m t^{n+m}$$

$$= \Sigma_{n=0}^{\infty}\frac{1}{(n!)^2}\left\{\frac{1}{2}t(x-1)\right\}^n \Sigma_{m=0}^{\infty}\frac{1}{(m!)^2}\left\{\frac{1}{2}(x+1)t\right\}^m$$

$$= {}_0F_1\left[-,1;1;\frac{(x-1)}{2}t\right]\ {}_0F_1\left[-,1;1;\frac{(x+1)}{2}t\right],$$

which is known as *Bateman's generating relation.*

Note: 1. Hypergeometric form: $P_n(x) = \dfrac{2^n(\frac{1}{2})_n x^n}{n!}\, {}_2F_1\left[\begin{matrix} -\frac{n}{2}, -\frac{n}{2}+\frac{1}{2}; \\ \frac{1}{2}-n; \end{matrix}\ \frac{1}{x^2}\right]$.

2. In terms of Bessel functions, another form of $P_n(x)$

$\sum \dfrac{P_n(x)t^n}{n!} = e^{xt} J_0\left\{t\sqrt{(1-x^2)}\right\}$, is also a generating relation of $P_n(x)$.

12.6 Recurrence Formulae for $P_n(x)$

(1) $(n+1)P_{n+1}(x) = (2n+1)xP_n(x) - nP_{n-1}(x)$
(2) $nP_n(x) = xP_n'(x) - P_{n-1}'(x)$
(3) $(2n+1)P_n(x) = P_{n+1}'(x) - P_{n-1}'(x)$
(4) $P_n'(x) = xP_{n-1}'(x) + nP_{n-1}(x)$

Proofs: (1) We know that

$$\therefore (1-2xt+t^2)^{-\frac{1}{2}} = \sum_{n=0}^{\infty} P_n(x)t^n. \tag{12.35}$$

Differentiating partially with respect to t, we get

$$-\frac{1}{2}(1-2xt+t^2)^{-\frac{3}{2}}(2t-2x) = \sum_{n=0}^{\infty} nP_n(x)t^{n-1}$$

or, $$(x-t)(1-2xt+t^2)^{-\frac{1}{2}} = (1-2xt+t^2)\sum_{n=0}^{\infty} nP_n(x)t^{n-1}$$

or, $$(x-t)\sum P_n(x)t^n = (1-2xt+t^2)\sum_{n=0}^{\infty} nP_n(x)t^{n-1}.$$

Equating the coefficients of t^n on both sides, we get

$$xP_n(x) - P_{n-1}(x) = (n+1)P_{n+1}(x) - 2xnP_n(x) + (n-1)P_{n-1}(x)$$

or,
$$(n+1)P_{n+1}(x) = (2n+1)xP_n(x) - nP_{n-1}(x).$$

(2) Differentiating (12.35) partially with respect to x,

$$-\frac{1}{2}(1-2xt+t^2)^{-\frac{3}{2}}(-2t) = \sum_{n=0}^{\infty} P_n'(x)t^n$$

or,
$$t(1-2xt+t^2)^{-\frac{3}{2}} = \Sigma P_n'(x)t^n. \tag{12.36}$$

Again, differentiating it with respect to t, we get

$$(1-2xt+t^2)^{-\frac{3}{2}}(x-t) = \Sigma\, nP_n(x)t^{n-1}. \tag{12.37}$$

Dividing Equation (12.37) by Equation (12.36), we get

$$\frac{x-t}{t} = \frac{\Sigma\, nP_n(x)t^{n-1}}{\Sigma\, P_n'(x)t^n}$$

or,
$$\Sigma\, nP_n(x)t^n = (x-t)\Sigma\, P_n'(x)t^n.$$

Equating the coefficients of t^n from both sides, we get $nP_n(x) = xP_n'(x) - P_{n-1}'(x)$.

(3) From (1), $(n+1)P_{n+1}(x)=(2n+1)xP_n(x)-nP_{n-1}(x)$.

Differentiating above with respect to x, we get

$$(n+1)P'_{n+1}(x)=(2n+1)P_n(x)+(2n+1)xP'_n(x)-nP'_{n-1}(x). \quad (12.38)$$

Substituting for $xP'_n(x)$ from (2) in Equation (12.38), we get

$$(n+1)P'_{n+1}(x)=(2n+1)P_n(x)+(2n+1)[nP_n(x)+P'_{n-1}(x)]-nP'_{n-1}(x)$$

or,
$$(2n+1)P_n(x)=P'_{n+1}(x)-P'_{n-1}(x).$$

(4) From Equation (12.38),

$$\begin{aligned}(n+1)P'_{n+1}(x)&=(2n+1)P_n(x)+(n+1)xP'_n(x)+n[xP'_n(x)-P'_{n-1}(x)]\\&=(2n+1)P_n(x)+(n+1)xP'_n(x)+n^2P_n(x)\\&=(n+1)xP'_n(x)+(n^2+2n+1)P_n(x)\end{aligned}$$

or,
$$P'_{n+1}(x)=xP'_n(x)+(n+1)P_n(x).$$

Replacing *n by (n−1)*, we get $P'_n(x)=xP'_{n-1}(x)+nP_{n-1}(x)$.
Other recurrence relations:

(5) $(1-x^2)P'_n(x)=n[(P_{n-1}(x)-xP_n(x)]$.
(6) $(1-x^2)P'_n(x)=(n+1)[(xP_n(x)-P_{n+1}(x)]$.

12.7 Christoffel's Summation Formula

Prove that $(2k+1)P_k(x)P_k(y)=\dfrac{l+1}{x-y}[P_{l+1}(x)P_l(y)-P_l(x)P_{l+1}(y)]$.

Proof: From recurrence relation (1), we have

$$(2k+1)xP_k(x) = (k+1)P_{k+1}(x) + kP_{k-1}(x) \tag{12.39}$$

and, $$(2k+1)yP_k(y) = (k+1)P_{k+1}(y) + kP_{k-1}(y). \tag{12.40}$$

Multiplying Equation (12.39) by $P_k(y)$ and Equation (12.40) by $P_k(x)$ and subtracting, we get

$$\begin{aligned}(2k+1)(x-y)P_k(x)P_k(y) &= (k+1)[P_{k+1}(x)P_k(y) - P_{k+1}(y)P_k(x)]\\ &+k[P_{k-1}(x)P_k(y) - P_{k-1}(y)P_k(x)].\end{aligned} \tag{12.41}$$

Replacing $k \, by \, 0,1,2,\ldots l-1,\ l$ successively in Equation (12.41) and adding all these equations, we get

$$(x-y)\sum_{k=0}^{l} P_k(x)P_k(y) = (l+1)[P_{l+1}(x)P_l(y) - P_{l+1}(y)P_l(x)]$$

or, $$\sum_{k=0}^{l} P_k(x)P_k(y) = \frac{(l+1)}{x-y}[P_{l+1}(x)P_l(y) - PP_n'(x) = P_{l+1}(y)P_l(x)]. \tag{12.41a}$$

Replacing $y \, by \, 1 \, in$ (12.41a) and use $P_l(1) = P_{l+1}(1) = P_k(1) = 1$. We have got required result.

12.7.1 Christoffel's Expansion

Show that $P_n'(x) = (2n-1)P_{n-1}(x) - (2n-5)P_{n-3}(x) + (2n-9)P_{n-5}(x) + \cdots$; the last term of the series is $3P_1(x)$ or P_0 according to whether n is even or odd

or, $$P_n'(x) = \sum_{r=0}^{|1/2(n-1)|}(2n-4r-1)P_{n-2r-1}(x), \text{ where } |\frac{1}{2}(\text{n-1})|$$

$$= \begin{cases} \frac{1}{2}(\text{n}-1), \text{if n is even} \\ \frac{1}{2}(\text{n}-2), \text{if n is odd} \end{cases}.$$

Solution: We have the recurrence relation $(2n+1)P_n = P'_{n+1} - P'_{n-1}$.
Replacing n *by* $n-1$ in the recurrence relation, we get

$$P'_n = (2n-1)P_{n-1} + P'_{n-2}. \tag{12.42}$$

Case I. When n is even. Replacing n *by* $n, n-2, n-4, \ldots 4, 2$ successively in Equation (12.42) and putting $P_0 = 1,\ P'_0 = 0,$ we get

$$\begin{aligned}
P'_n &= (2n-1)P_{n-1} + P'_{n-2} \\
P'_{n-2} &= (2n-5)P_{n-3} + P'_{n-4} \\
P'_{n-4} &= (2n-9)P_{n-5} + P'_{n-6} \\
&\ldots\ldots\ldots \quad \ldots\ldots \quad \ldots\ldots\ldots\ldots \\
P'_4 &= 7P_3 + P'_2 \quad P'_2 = 3P_1 + P'_0.
\end{aligned}$$

Adding these,

$$P'_n = (2n-1)P_{n-1} + (2n-5)P_{n-3} + P'_{n-4} + \cdots \ldots + 3P_1. \tag{12.43}$$

Case II. When n is odd.
Putting $n = n, n-2, n-4, \ldots \ldots 5, 3$ successively in Equation (i) and also using $P_0 = 1,\ P'_0 = 0,$ we get

$$\begin{aligned}
P'_n &= (2n-1)P_{n-1} + P'_{n-2} \\
P'_{n-2} &= (2n-5)P_{n-3} + P'_{n-4} \\
P'_{n-4} &= (2n-9)P_{n-5} + P'_{n-6} \\
&\ldots\ldots\ldots \quad \ldots\ldots \quad \ldots\ldots\ldots\ldots \\
P'_5 &= 9P_4 + P'_3 \\
P'_3 &= 5P_2 + P'_1 = 5P_2 + P_0.
\end{aligned}$$

Adding these,

$$P'_n = (2n-1)P_{n-1} + (2n-5)P_{n-3} + P'_{n-4} + \cdots \ldots + 5P_2 + P_0. \tag{12.44}$$

Combining Equations (12.43) and (12.44), we get

$$P_n'(x) = \sum_{r=0}^{|1/2(n-1)|} (2n-4r-1)\, P_{n-2r-1}(x).$$

12.8 Orthogonality of Legendre Polynomials

Prove that $\int_{-1}^{1} P_m(x)P_n(x)dx = \begin{cases} 0 & ,m \neq n \\ \dfrac{2}{2n+1}. & m = n \end{cases}$.

Solution: We know that the solutions of

$$(1-x^2)u'' - 2xu' + m(m+1)u = 0 \tag{12.45}$$

$$(1-x^2)v'' - 2xv' + n(n+1)v = 0 \tag{12.46}$$

are $P_m(x)$ and $P_n(x)$, respectively.

Multiplying Equation (12.45) by v and Equation (12.46) by u and subtracting, we get

$$(1-x^2)[u''v - v''u] - 2x(u'v - v'u) + [m(m+1) - n(n+1)]uv = 0$$

or,
$$\frac{d}{dx}[(1-x^2)(u'v - v'u)] + (m-n)(m+n+1)uv = 0.$$

Integrating from $-1\ to +1$, we get

$$(m-n)(m+n+1)\int_{-1}^{1} uv dx = [(1-x^2)(u'v - v'u)]_{-1}^{1} = 0.$$

$$\therefore \int_{-1}^{1} uvdx = 0.$$

$$\int_{-1}^{1} P_m(x).P_n(x) \quad dx = 0 \text{ where } m \neq n. \tag{12.47}$$

This is known as the *orthogonality property of Legendre's polynomials*.
When $m = n$, we have Rodrigue's formula.
When $m = n$, we have from Rodrigue's formula

$$\int_{-1}^{1} \{P_n(x)\}^2 dx = \int_{-1}^{1} \left[\frac{1}{n!2^n} \frac{d^n}{dx^n} (x^2 - 1)^n \right]^2 dx$$

or,

$$(n!2^n)^2 \int_{-1}^{1} \{P_n(x)\}^2 dx = \int_{-1}^{1} D^n (x^2 - 1)^n D^n (x^2 - 1)^n dx \qquad [\text{let } D \equiv \frac{d}{dx}]$$

$$= \left| D^n (x^2 - 1)^n D^{n-1} (x^2 - 1)^n \right|_{-1}^{1}$$

$$- \int_{-1}^{1} D^{n-1} (x^2 - 1)^n D^{n+1} (x^2 - 1)^n dx$$

$$(n!2^n)^2 \int_{-1}^{1} \{P_n(x)\}^2 dx = - \int_{-1}^{1} D^{n-1} (x^2 - 1)^n D^{n+1} (x^2 - 1)^n dx$$

(putting $x = \pm$, the first term of the right – hand side vanishes).

Integrating by parts $(n-1)$ times on the right – hand side, in a similar way, we finally get

$$= (-1)^n \int_{-1}^{1} (x^2 - 1)^n D^{2n} (x^2 - 1)^n dx$$

$$= (-1)^n \int_{-1}^{1} (2n)!(x^2 - 1)^n dx = 2.(2n)! \int_{0}^{1} (x^2 - 1)^n dx.$$

Substituting $x = sin\theta$, we get

$$(n!2^n)^2\int_{-1}^{1}[P_n(x)]^2dx = 2.(2n)!\int_0^{\frac{\pi}{2}} cos^{2n+1}\theta d\theta$$

$$= 2.(2n)!\frac{2n(2n-2)..4.2}{(2n+1)(2n-1)..2.1}$$

$$= 2.(2n)![2n(2n-2)..\frac{4.2]^2}{(2n+1)!} = \frac{2}{(2n+1)}(2^n n!)^2.$$

$$\therefore \int_{-1}^{1}[P_n(x)]^2dx = \frac{2}{(2n+1)}. \quad (12.48)$$

12.9 Fourier–Legendre's Expansion of $f(x)$

If $f(x)$ is a function defined from $x-1$ *to* $x = 1$, we can write

$$f(x) = \sum_{n=0}^{\infty} c_n P_n(x). \quad (12.49)$$

To determine the coefficient c_n, multiplying both sides by $P_n(x)$, and integrating from -1 *to* 1, then using Equations (iii) and (iv), we get

$$\int_{-1}^{1} f(x)P_n(x)dx = c_n\int_{-1}^{1} P_n^2(x)dx = \frac{2c_n}{2n+1}.$$

$$\therefore c_n = \left(n+\frac{1}{2}\right)\int_{-1}^{1} f(x)P_n(x)dx.$$

Equation (12.49) is known as the *Fourier–Legendre expansion of* $f(x)$.

Remark 12.9.1. Suppose that function $f(x)$ is continuous and has continuous derivatives in $[-1,1]$, then we prove that series (12.49) converges uniformly in $[-1,1]$, and series expansion (12.49) is unique.

12.9.1 Expansion of x^n in Legendre Polynomials

$$x^n = \frac{n!}{2^n}\sum_{k=0}^{\left[\frac{n}{2}\right]} \frac{(2n-4k+1)P_{n-2k}(x)}{k!(\frac{3}{2})_{n-k}} \quad \text{where } \left[\frac{n}{2}\right] = \begin{cases} \dfrac{n}{2}, \text{if } n \text{ is even} \\ \dfrac{n-1}{2}, \text{if } n \text{ is odd} \end{cases}$$

and $(a)_n = a(a+1)\ldots(a+n-1)$.

Proof: Let

$$x^n = c_n P_n(x) + c_{n-2}P_{n-2}(x) + c_{n-4}P_{n-4}(x) + \cdots c_r P_r(x) + \cdots \quad (12.50)$$

where $P_n(x), P_{n-2}(x),\ldots$ are all even or odd fumctions of x according to whether x^n is an even or odd function. Again, $P_{n-1}(x), P_{n-3}(x),\ldots$ cannot occur in the series because they are odd or even functions *of* x according to whether x^n is an even or odd function of x, respectively. Since $P_n(x)$ contains terms of degree n and lower, hence expansion (12.50) cannot contain any P with a suffix higher than n.

Multiplying both sides of Equation (12.50) by $P_m(x)$ and integrating from -1 *to* 1, we get

$$\int_{-1}^{1} x^n P_m(x)dx = \int_{-1}^{1} P_m(x)[c_n P_n(x) + c_{n-2}P_{n-2}(x) + c_{n-4}P_{n-4}(x) + \cdots c_r P_r(x) + \cdots].dx$$

$$= c_m \int_{-1}^{1} [P_m(x)]^2 dx \quad [\text{since } \int_{-1}^{1} x^n P_m(x) P_n(x) dx = \begin{cases} 0, \text{if } m \neq n \\ \dfrac{2}{2m+1}, \text{if } m = n \end{cases}]$$

$$\int_{-1}^{1} x^n P_m(x)dx = c_m \frac{2}{2m+1}. \quad (12.51)$$

$$\therefore c_m = \frac{2m+1}{2}\int_{-1}^{1} x^n P_m(x)dx$$

$$= \frac{2m+1}{2}\int_{-1}^{1} x^n \left[\frac{1}{2^m m!}\frac{d^m}{dx^m}(x^2-1)^m\right]dx \text{ (by Rodrigue's formula)}$$

$$= \frac{2m+1}{2^{m+1} m!}\int_{-1}^{1} x^n \left[\frac{d^m}{dx^m}(x^2-1)^m\right]dx$$

$$= \frac{2m+1}{2^{m+1} m!}\left[\left\{x^n \frac{d^{m-1}}{dx^{m-1}}(x^2-1)^m\right\}_{-1}^{1} - \int_{-1}^{1} nx^{n-1}\frac{d^{m-1}}{dx^{m-1}}(x^2-1)^m dx\right]$$

$$= -\frac{2m+1}{2^{m+1} m!}\int_{-1}^{1} nx^{n-1}\frac{d^{m-1}}{dx^{m-1}}(x^2-1)^m dx, \text{ [taking integration } (m-1) \textit{ times} \text{ by parts]}$$

$$= (-1)^m \frac{2m+1}{2^{m+1} m!} n.(n-1)\ldots(n-m+1)\int_{-1}^{1} x^{n-m}(x^2-1)^m dx$$

$$= (-1)^{2m} \frac{2m+1}{2^{m+1} m!} n.(n-1)\ldots(n-m+1)\int_{-1}^{1} x^{n-m}(1-x^2)^m dx$$

$$c_m = \frac{2m+1}{2^{m+1} m!} n.(n-1)\ldots(n-m+1)\int_{-1}^{1} x^{n-m}(1-x^2)^m dx. \quad (12.52)$$

In our deduction m can be one of the integers $n, n-2, n-4, \ldots$ only and, hence, $(n-m)$ can be one of the integers 0, 2, 4,... only.

Hence, $x^{n-m}(1-x^2)^m$ is an even function of x.

$$\therefore \int_{-1}^{1} x^{n-m}(1-x^2)^m dx = 2\int_{0}^{1} x^{n-m}(1-x^2)^m dx. \quad (12.53)$$

Now, Equation (12.52),

$$c_m = 2\frac{2m+1}{2^{m+1} m!} n.(n-1)\ldots(n-m+1)\int_{0}^{1} x^{n-m}(1-x^2)^m dx,$$

$$= \frac{2m+1}{2^{m+1} m!}\frac{n!}{(n-m)!}\int_{0}^{1} x^{n-m-1}(1-x^2)^m 2x dx$$

$$= \frac{2m+1}{2^{m+1} m!}\frac{n!}{(n-m)!}\int_{0}^{1} t^{(n-m-1)/2}(1-t)^m dt \qquad [\text{let } x^2 = t, dt = 2xdx]$$

$$= \frac{2m+1}{2^{m+1} m!}\frac{n!}{(n-m)!}\int_{0}^{1} t^{\frac{n-m+1}{2}-1}(1-t)^{(m+1)-1} dt$$

$$= \frac{2m+1}{2^{m+1} m!}\frac{n!}{(n-m)!}\frac{\Gamma(n-m+1)/2\,\Gamma(m)}{\Gamma[\frac{(n-m+1)}{2}+m]}$$

$$= \frac{2m+1}{2^{m+1}}\frac{n!}{(n-m)!}\frac{\Gamma(n-m+1)/2}{\frac{(n+m+1)}{2}\frac{(n+m-1)}{2}\frac{(n+m-3)}{2}\cdots\frac{(n-m+1)}{2}\Gamma[\frac{(n-m+1)}{2}}$$

$$= \frac{(2m+1)\; n(n-1)\ldots(n-m+2)(n-m+1)}{(n-m)!}\frac{(n-m)!}{(n+m+1)(n+m-1)\ldots(n-m+3)(n-m+1)}$$

$$c_m = \frac{n(n-1)\ldots(n-m+2)}{(n+m+1)(n+m-1)\ldots(n-m+3)}(2m+1). \qquad (12.54)$$

Putting $m = n, (n-2), (n-4)\ldots$ in Equation (v),

$$c_n = \frac{n(n-1)(n-2)..3.2}{(2n+1)(2n-1)....5.3}(2n+1) = \frac{n!}{3.5\ldots(2n-1)(2n+1)}(2n+1)$$

$$c_{n-2} = \frac{n(n-1)(n-2)..5.4}{(2n-1)(2n-3)\ldots7.5.}(2n-3) = \frac{n!}{3.5.7\ldots(2n-1)(2n+1)}\frac{(2n+1)}{2}(2n-3)$$

$$c_{n-4} = \frac{n!}{3.5.7\ldots(2n-1)(2n+1)}\frac{(2n+1)(2n-1))}{2.4}(2n-7).$$

Putting these values in Equation (i),

$$x^n = \frac{n!}{3.5\ldots(2n-1)(2n+1)}[(2n+1)P_n(x) + (2n-3)\frac{(2n+1)}{2}P_{n-2}(x) + (2n-7)\frac{(2n+1)(2n-1)}{2.4}P_{n-4}(x) + .. + \frac{1}{(n+1)}P_0(x)], \textit{if } n \textit{ is even} \qquad (12.55)$$

$$x^n = \frac{n!}{3.5\ldots(2n-1)(2n+1)}[(2n+1)P_n(x) + (2n-3)\frac{(2n+1)}{2}P_{n-2}(x) + (2n-7)\frac{(2n+1)(2n-1)}{2.4}P_{n-4}(x) + ..\frac{3P_1(x)}{(n+2)}], \textit{if } n \textit{ is odd.} \qquad (12.56)$$

That is,

$$x^n = \frac{n!}{2^n}\sum_{k=0}^{[\frac{n}{2}]}\frac{(2n-4k+1)P_{n-2k}(x)}{k!(\frac{3}{2})_{n-k}} \text{ where } \left[\frac{n}{2}\right] = \begin{cases} \frac{n}{2}, \textit{if } n \textit{ is even} \\ \frac{n-1}{2}, \textit{if } n \textit{ is odd} \end{cases} \qquad (12.57)$$

Corollary 12.9.1. Let

$$f(x) = a_0 + a_1 x + ax^2 + \cdots . + a_n x^n + \cdots, \tag{12.58}$$

where $a_0, a_1 \ldots . a_n$, are constants. Then, $f(x)$ can be expanded in Legendre polynomials in the form

$$f(x) = \sum_{n=0}^{\infty} c_n P_n(x)$$

where,

$$c_n = \frac{n!}{3.5. \ldots .(2n+1)} \left[a_n + \frac{(n+1)(n+2)}{2(2n+3)} a_{n+2} + \frac{(n+1)(n+2)(n+3)(n+4)}{2.4.(2n+3)(2n+5)} a_{n+2} + \cdots . \right].$$

Proof: Let

$$f(x) = a_0 + a_1 x + ax^2 + \cdots . + a_n x^n + \cdots, \tag{12.59}$$

and assume

$$f(x) = \sum_{n=0}^{\infty} c_n P_n(x). \tag{12.60}$$

Applying Equation (12.56) (above article), we have

$$a_n x^n = a_n \frac{n!}{3.5. \ldots .(2n+1)} [(2n+1) P_n(x) + \cdots]$$

$$a_{n+2} x^{n+2} = a_{n+2} \frac{(n+2)!}{3.5. \ldots .(2n+5)} [(2n+5) P_{n+2}(x) + (2n+1) \frac{(2n+5)}{2} P_n(x) + \cdots].$$

Putting these values in Equation (12.59) and equating the coefficients of $P_n(x)$ from Equations (12.59) and (12.60), we get

$$\begin{aligned} c_n &= \frac{n!}{3.5....(2n+1)}(2n+1)a_n + \frac{(n+2)!}{3.5....(2n+5)}(2n+1)\frac{(2n+5)}{2}a_{n+2} + \cdots ... \\ &= \frac{n!}{3.5....(2n-1)}\left[a_n + \frac{(n+1)(n+2)}{2(2n+3)}a_{n+2} + \frac{(n+1)(n+2)(n+3)(n+4)}{2.4(2n+3)(2n+5)}a_{n+2} + \cdots.\right]. \end{aligned}$$

Example 12.9.1. Show that $\int_{-1}^{1}(1-x^2)P_m'(x)dx = \begin{cases} 0, & when \quad m \neq n \\ \dfrac{2n(n+1)}{2n+1}, & when\, m = n \end{cases}$.

Solution: Integrating by parts,

$$\begin{aligned} \int_{-1}^{1}(1-x^2)P_m'(x)P_n'(x)dx &= [(1-x^2)P_m'(x)P_n(x)]_{-1}^{1} \\ &\quad -\int_{-1}^{1}\frac{d}{dx}[(1-x^2)P_m'(x)]P_n(x)dx \\ &= -\int_{-1}^{1}]P_n(x)[(1-x^2)P_m''(x) + (-2x)P_m'(x)]dx. \end{aligned} \tag{i}$$

Here, since $P_m(x)$ is a solution of Legendre's equation

$$(1-x^2)\frac{d^2y}{dx^2} - 2x\frac{dy}{dx} + m(m+1)y = 0,$$

so $(1-x^2)P_m''(x) - 2xP_m'(x) + m(m+1)P_m(x) = 0$.

Substituting these in Equation (i), we get

$$\begin{aligned} \int_{-1}^{1}(1-x^2)P_m'(x)P_n'(x)dx &= -\int_{-1}^{1}]P_n(x)[-m(m+1)]P_m(x)dx \\ &= m(m+1)\int_{-1}^{1}P_m(x)P_n(x)dx. \end{aligned}$$

When $m \neq n$, $\int_{-1}^{1}P_m(x)P_n(x)dx = 0$.

$$\therefore \int_{-1}^{1}(1-x^2)P_m'(x)P_n'(x)dx = 0.$$

When $m=n, \int_{-1}^{1} P_m(x)P_n(x)dx=\frac{2}{2n+1}$.

$$\therefore \int_{-1}^{1}(1-x^2)P_m'(x)P_n'(x)dx=n(n+1)\frac{2}{2n+1}=\frac{2n(n+1)}{(2n+1)}.$$

Example 12.9.2. Show that $\int_{-1}^{1} x^2P_{n-1}(x)P_{n+1}(x)dx=\frac{2n(n+1)}{(2n-1)(2n+1)(2n+3)}$.

Solution: We know that $(2n+1)xP_n=(n+1)P_{n+1}+nP_{n-1}$.

$$\therefore xP_{n-1}=\frac{1}{2n-1}[nP_n+(n-1)P_{n-2}] \text{ [replacing } by\ n-1]$$

$$\text{and } xP_{n+1}=\frac{1}{2n+3}[(n+2)P_{n+2}+(n+1)P_n] \text{ [replacing } n\ by\ n+1].$$

$$\therefore x^2P_{n-1}P_{n+1}=\frac{1}{(2n-1)(2n+3)}[n(n+2)P_nP_{n+2}+n(n+1)P_n^2$$
$$+(n-1)(n+2)P_{n-2}P_{n+2}+(n+1)(n-1)P_nP_{n-2}].$$

Integrating both sides from -1 *to* 1, we get

$$\int_{-1}^{1} x^2P_{n-1}(x)P_{n+1}(x)dx=\frac{n(n+1)}{(2n-1)(2n+3)}\int_{-1}^{1} P_n^2dx$$

(for orthogonality property all terms under integration are zero)

$$\int_{-1}^{1} x^2P_{n-1}(x)P_{n+1}(x)dx=\frac{n(n+1)}{(2n-1)(2n+3)}\frac{2}{(2n+1)}$$
$$=\frac{2n(n+1)}{(2n-1)(2n+1(2n+3)}.$$

Example 12.9.3. If $f(x)=0,-1<x\le 0$

$$=x, 0<x\le 1,$$

show that, $f(x)=\frac{1}{4}P_0(x)+\frac{1}{2}P_1(x)+\frac{5}{16}P_2(x)-\frac{3}{32}P_4(x)+\cdots.$

Solution: Let $f(x)=\sum_{n=0}^{\infty}c_nP_n(x)$.
We have

$$c_n=\left(n+\frac{1}{2}\right)\int_{-1}^{1}f(x)P_n(x)dx=\left(n+\frac{1}{2}\right)[\int_{-1}^{0}0.P_n(x)dx+\int_0^1 xP_n(x)dx]$$

$$=\left(n+\frac{1}{2}\right)\int_0^1 xP_n(x)dx$$

$$c_0=\frac{1}{2}\int_0^1 xP_0(x)dx=\frac{1}{2}\int_0^1 xdx=\frac{1}{4}$$

$$c_1=\frac{3}{2}\int_0^1 xP_1(x)dx=\frac{3}{2}\int_0^1 x^2dx=\frac{1}{2}$$

$$c_2=\frac{5}{2}\int_0^1 xP_2(x)dx=\frac{3}{2}\int_0^1 x\frac{3x^2-1}{2}dx=\frac{5}{4}\left|\frac{3x^4}{4}-\frac{x^2}{2}\right|_0^1=\frac{5}{16}$$

$$c_3=\frac{5}{2}\int_0^1 xP_3(x)dx=\frac{7}{2}\int_0^1 x\frac{5x^3-3x}{2}dx=\frac{7}{4}\left|\frac{5x^5}{5}-3\frac{x^3}{3}\right|_0^1=0,\text{ and so on.}$$

$$\therefore f(x)=\frac{1}{4}P_0(x)+\frac{1}{2}P_1(x)+\frac{5}{16}P_2(x)-\frac{3}{32}P_4(x)+\cdots.$$

Example 12.9.4. Show that for any function $f(x)$, for which the *nth* derivative is continuous,

$$\int_{-1}^{1}f(x)P_n(x)dx=\frac{1}{2^n n!}\int_{-1}^{1}(1-x^2)^n f^n(x)(x)dx.$$

Solution: Using Rodrigue's formula, $P_n(x) = \frac{1}{2^n n!}\frac{d^n}{dx^n}(x^2-1)^n$

$$\int_{-1}^{1} f(x)P_n(x)dx = \frac{1}{2^n n!}\int_{-1}^{1} f(x)\frac{d^n}{dx^n}(x^2-1)^n dx.$$

Integrating by parts:

$$\int_{-1}^{1} f(x)P_n(x)dx = \frac{1}{2^n n!}\left[\left| f(x)\frac{d^{n-1}}{dx^{n-1}}(x^2-1)^n\right|_{-1}^{1} - \int_{-1}^{1} f'(x)\frac{d^{n-1}}{dx^{n-1}}(x^2-1)^n dx\right]$$

$$= -\frac{1}{2^n n!}\int_{-1}^{1} f'(x)\frac{d^{n-1}}{dx^{n-1}}(x^2-1)^n dx$$

$$= (-1)^2\frac{1}{2^n n!}\int_{-1}^{1} f''(x)\frac{d^{n-2}}{dx^{n-2}}(x^2-1)^n dx$$

$$= (-1)^n\frac{1}{2^n n!}\int_{-1}^{1} f^n(x)(x^2-1)^n dx \text{ [integrating } (n-2) \text{ times by parts]}$$

$$= (-1)^{2n}\frac{1}{2^n n!}\int_{-1}^{1} f^n(x)(1-x^2)^n$$

$$= \frac{1}{2^n n!}\int_{-1}^{1} f^n(x)(1-x^2)^n.$$

Example 12.9.5. Prove that $\int_{-1}^{1} P_n(x)dx = \begin{cases} 2, \text{if } n=0 \\ 0, \text{if } n \geq 1 \end{cases}.$

Solution: Using Rodrigue's formula, we have

$$\int_{-1}^{1} P_n(x)dx = \frac{1}{2^n n!}\int_{-1}^{1}\frac{d^n}{dx^n}(x^2-1)^n dx$$

$$= \frac{1}{2^n n!}\left[\frac{d^{n-1}}{dx^{n-1}}(x^2-1)^n\right]_{-1}^{1} = \frac{1}{2^n n!}\left[\frac{d^{n-1}}{dx^{n-1}}(x-1)^n(x+1)^n\right]_{-1}^{1}.$$

Using Leibnitz theorem:

$$D^n(uv) = D^n u.v + \binom{n}{1} D^{n-1}u.Dv + \cdots + uD^n v$$

$$= \frac{1}{2^n n!}[D^{n-1}(x-1)^n.(x+1)^n + (n-1)D^{n-2}(x-1)^n.D(x+1)^n$$

$$+\ldots(x-1)^n.D^{n-1}(x+1)^n]$$

$$= \frac{1}{2^n n!}[n!(x-1).(x+1)^n + \cdots. + (x-1)^n(x+1)]_{-1}^{1} = 0.$$

When $n = 0, P_n(x) = 1. \therefore \int_{-1}^{1} P_n(x)dx = \int_{-1}^{1} dx = 2.$

Example 12.9.6. (i) If $m < n$, show that $\int_{-1}^{1} x^m P_n(x)dx = 0.$
Deduce that $\int_{-1}^{1} x^4 P_6(x)dx = 0.$

(ii) Prove that $\int_{-1}^{1} x^n P_n(x)dx = \dfrac{\Gamma\left(\frac{1}{2}\right)\Gamma(n+1)}{2^n \Gamma(n+\frac{3}{2})} = \dfrac{2^{n+1}(n!)^2}{(2n+1)!}$.

Solution: Using Rodrigue's formula, we have

$$\int_{-1}^{1} x^m P_n(x)dx = \frac{1}{2^n n!}\int_{-1}^{1} x^m \frac{d^n}{dx^n}(x^2-1)^n dx$$

$$= \frac{(-1)^2 m(m-1)}{2^n n!}\int_{-1}^{1} x^{m-2}\frac{d^{n-2}}{dx^{n-2}}(x^2-1)^n dx$$

1

$$= \frac{(-1)^m m(m-1)(m-2)..2.1}{2^n n!}\int_{-1}^{1} x^{m-m}\frac{d^{n-m}}{dx^{n-m}}(x^2-1)^n dx \ldots\ldots\ldots\ldots\ldots(\alpha)$$

[on continuing its integration up to (m−2) steps]

1

$$=\frac{(-1)^m m!}{2^n n!}\int_{-1}^{1}\frac{d}{dx}\left(\frac{d^{n-m-1}}{dx^{n-m-1}}(x^2-1)^n\right)dx$$

1

$$=\frac{(-1)^m m!}{2^n n!}\left(\frac{d^{n-m-1}}{dx^{n-m-1}}(x^2-1)^n\right)_{-1}^{1}=0.$$

For $m=4, n=6$ in the above result, we get the required result because m<n.

(ii) Here, $m=n$. Proceeding as above up to (α), we get

$$\int_{-1}^{1}x^n P_n(x)dx=\frac{(-1)^n n!}{2^n n!}\int_{-1}^{1}x^{n-n}\frac{d^{n-n}}{dx^{n-n}}(x^2-1)^n dx=\frac{(-1)^n}{2^n}\int_{-1}^{1}(x^2-1)^n dx$$

$$=\frac{1}{2^n}2\int_0^1(1-x^2)^n dx=\frac{1}{2^{n-1}}\int_0^{\pi/2}(cos^2\theta)^n cos\theta d\theta \text{ [putting } x=sin\theta]$$

$$=\frac{1}{2^{n-1}}\int_0^{\pi/2}(cos^{2n+1}\theta d\theta=\frac{1}{2^{n-1}}\frac{\Gamma(n+1)\Gamma(\frac{1}{2})}{2\Gamma(\frac{2n+3}{2})}\ [\because \int_0^{\pi/2}cos^p\theta sin^q\theta d\theta=\frac{\Gamma\left(\frac{p+1}{2}\right)\Gamma(\frac{q+1}{2})}{2\Gamma(\frac{p+q+2}{2})}]$$

$$\int_{-1}^{1}x^n P_n(x)dx=\frac{1}{2^{n-1}}\frac{n!\sqrt{\pi}}{\frac{2n+1}{2}.\frac{2n-1}{2}\ldots\frac{3}{2}\frac{1}{2}.\sqrt{\pi}}=\frac{n!}{2^{n-1}}\frac{2^{n+1}}{(2n+1)(2n-1)..3.1}$$

$$=\frac{2(n!)\ 2n(2n-2)\ldots4.2}{(2n+1)!}=\frac{2(n!)\ 2n.2(n-1)\ldots(2.2).(2.1)}{(2n+1)!}=\frac{2(n!)2^n.n!}{(2n+1)!}=\frac{2^{n+1}(n!)^2}{(2n+1)!}.$$

Example 12.9.7. (a) Show that all the roots of $P_n(x)=0$ are real and lie between -1 *and* 1.

(b) Prove that all the roots of $P_n(x)=0$ are distinct.

Solution: (a) Let

$$f(x)=(x^2-1)^n=(x-1)^n(x+1)^n. \qquad \text{(i)}$$

Here, $f(x)$ vanishes for $x=1$ *and* $x=-1$.

From Rolle's theorem, we know that $f'(x)$ must vanish at least at one value α of x lying between -1 *and* 1.

Now, $$f'(x)=n(x-1)^{n-1}(x+1)^n+n(x-1)^n(x+1)^{n-1}. \qquad \text{(ii)}$$

Here, Equation (ii) shows that $f'(x)$ vanishes at $x=1$ *and* $x=-1$ and also proves that $f'(x)$ vanishes at $x=\alpha$ where $-1<\alpha<1$.

Similarly, applying Rolle's theorem on $f'(x)$, we conclude that $f''(x)$ must be vanished at $x=\beta$, where $-1<\beta<\alpha$, and also at $x=\gamma$ such that $\alpha<\gamma<1$. Proceeding likewise, we conclude that $f^{(n)}(x)=0$ must have n real roots lying between -1 *and* 1.

Using Rodrigue's formula,

$$P_n(x)=\frac{1}{2^n n!}\frac{d^n}{dx^n}(x^2-1)^n=\frac{1}{2^n n!}f^{(n)}(x). \qquad \text{(iii)}$$

$$\text{Since } f^{(n)}(x)=0, \text{ it implies that } P_n(x)=0. \qquad \text{(iv)}$$

Hence, $f^{(n)}(x)=0$ has n real roots lying between -1 *and* 1. From Equation (iv), it shows that $P_n(x)=0$ has n real roots lying between -1 *and* 1.

(b) If possible, let the roots of $P_n(x)=0$ be not all distinct. At least two roots must be equal. Let α be a repeated root, then from the theory of equations, we have

$$P_n(x)=0 \quad \text{and} \quad P_n'(\alpha)=0. \qquad \text{(i)}$$

$$\therefore (1-x^2)P_n''(x)-2xP_n'(x)+n(n+1)P_n(x)=0. \qquad \text{(ii)}$$

Differentiating r times and using the Leibnitz theorem, Equation (ii) gives

$$(1-x^2)P_n^{r+2}(x)+r(-2x)P_n^{r+1}(x)+\frac{r(r-1)}{2}(-2)P_n^r(x)-2[xP_n^{r+1}(x)$$
$$+rP_n^r(x)]+n(n+1)P_n^r(x)=0$$

or,
$$(1-x^2)P_n^{r+2}(x)-2x(r+1)P_n^{r+1}(x)$$
$$-[r^2+r-n(n+1)]P_n^r(x). \quad \text{(iii)}$$

Putting $r=0$ *and* $x=\alpha$ in Equation (iii), and using Equation (i), we get

$$(1-\alpha^2)P_n''(\alpha)=0 \text{ or } P_n''(\alpha)=0. \quad \text{(iv)}$$

Again, putting $r=1$ *and* $x=\alpha$ in Equation (iii), and using Equations (i) and (iv), we get

$$(1-\alpha^2)P_n'''(\alpha)=0, \text{ or } P_n'''(\alpha)=0. \quad \text{(v)}$$

Similarly, we proceed for $r=2,3..n-2, x=\alpha,$ we get

$$\therefore P_n^{(n)}(x)=0 \quad \text{(vi)}$$

$$P_n(x)=\frac{1.3.5\ldots(2n-1)}{n!}\left[x^n-\frac{n(n-1)}{2(2n-1)}x^{n-2}+\cdots\right]$$

$P_n^{(n)}(x)=\dfrac{1.3.5\ldots(2n-1)}{n!}n!.$ It implies that

$$P_n^{(n)}(x)\neq 0. \quad \text{(vii)}$$

Equations (vi) and (vii) contrdict. It indicates that our assumption that there are at least two equal roots is wrong.

So, all roots of $P_n(x)=0$ are distinct.

Example 12.9.8. Prove that $\int_{-1}^{1} xP_n(x)P_{n-1}(x)dx=\dfrac{2n}{4n^2-1}.$

Solution: We know that $(2n+1)xP_n = (n+1)P_{n+1} + nP_{n-1}$

or,
$$xP_n = \frac{(n+1)}{2n+1}P_{n+1} + \frac{n}{2n+1}P_{n-1}. \qquad \text{(i)}$$

Multiplying both sides by $P_{n-1}(x)$ and then integrating from -1 *to* 1,

$$\int_{-1}^{1} xP_n(x)P_{n-1}(x)dx = \int_{-1}^{1}\frac{(n+1)}{2n+1}P_{n+1}P_{n-1}dx + \int_{-1}^{1}\frac{n}{2n+1}\{P_{n-1}\}^2 dx. \quad \text{(ii)}$$

But,
$$\int_{-1}^{1} P_n(x)P_{n-1}(x)dx = \begin{cases} 0, if\ m \neq n \\ 2/(2n+1), if\ m = n \end{cases}. \qquad \text{(iii)}$$

Using Equation (iii) in Equation (ii), we get

$$\int_{-1}^{1} xP_n(x)P_{n-1}(x)dx = \frac{(n+1)}{2n+1}\times 0 + \frac{n}{2n+1}\times 2/[2(n-1)+1]$$
$$= \frac{2n}{(2n+1)(2n-1)}.$$
$$\therefore \int_{-1}^{1} xP_n(x)P_{n-1}(x)dx = \frac{2n}{4n^2-1}$$

Example 12.9.9. (i) Express $x^4 + 2x^3 + 2x^2 - x - 3$ in terms of Legendre's polynomial.

(ii) Prove that $\dfrac{1-z^2}{(1-2xz+z^2)} = \sum_{n=0}^{\infty}(2n+1)z^n P_n.$

Solution: We have $P_0(x) = 1, P_1(x) = x, P_2(x) = (3x^2-1)/2, P_3(x) = (5x^3-3x)/2, P_4(x) = (35x^4 - 30x^2 + 3)/8.$

Thus,

$$x^4 = \frac{8}{35}P_4(x) + \frac{6}{7}x^2 - \frac{3}{35}$$
$$x^3 = \frac{2}{5}P_3(x) + \frac{3}{5}x$$
$$x^2 = \frac{1}{3} + \frac{2}{3}P_2(x) \qquad x = P_1(x)$$
$$\therefore x^4 - 2x^3 + 2x^2 - x - 3 = \frac{8}{35}P_4(x) + \frac{6}{7}x^2 - \frac{3}{35}$$
$$+2\left(\frac{2}{5}P_3(x) + \frac{3}{5}x\right) + 2\left(\frac{1}{3} + \frac{2}{3}P_2(x)\right) - P_1(x) - 3$$
$$= \frac{8}{35}P_4(x) + \frac{4}{5}P_3(x) + \frac{4}{3}P_2(x) - P_1(x) + \frac{6}{7}(\frac{1}{3} + \frac{2}{3}P_2(x))$$
$$+\frac{6}{5}P_1(x) + (\frac{2}{3} - \frac{3}{35} - 3)$$
$$= \frac{8}{35}P_4(x) + \frac{4}{5}P_3(x) + (\frac{4}{3} + \frac{12}{21})P_2(x) + \frac{1}{5}P_1(x) + \frac{224}{105}$$
$$= \frac{8}{35}P_4(x) + \frac{4}{5}P_3(x) + \frac{40}{21}P_2(x) + \frac{1}{5}P_1(x) + \frac{224}{105}.$$

(*ii*) We know that

$$(1 - 2xt + t^2)^{-\frac{1}{2}} = \sum_{n=0}^{\infty} P_n(x)t^n. \qquad \text{(i)}$$

Differentiating both sides with respect to t,

$$-\frac{1}{2}(-2x + 2t)(1 - 2xt + t^2)^{-\frac{3}{2}} = \sum_{n=0}^{\infty} nt^{n-1}P_n(x)$$

or,

$$(x - t)(1 - 2xt + t^2)^{-\frac{3}{2}} = \sum_{n=0}^{\infty} nt^{n-1}P_n(x). \qquad \text{(ii)}$$

Multiplying both sides by $2t$,

$$2t(x-t)(1-2xt+t^2)^{-\frac{3}{2}} = 2\sum_{n=0}^{\infty} nt^n P_n(x). \qquad \text{(iii)}$$

Adding Equations (i) and (iii),

$$\frac{1}{(1-2xt+t^2)^{\frac{1}{2}}} + \frac{2t(x-t)}{(1-2xt+t^2)^{\frac{3}{2}}} = \sum_{n=0}^{\infty} P_n(x)t^n + 2\sum_{n=0}^{\infty} nt^n P_n(x)$$

or,

$$\frac{1-2xt+t^2+2t(x-t)}{(1-2xt+t^2)^{\frac{3}{2}}} = \sum_{n=0}^{\infty}(2n+1)P_n(x)$$

or,

$$\frac{1-t^2}{(1-2xt+t^2)^{\frac{3}{2}}} = \sum_{n=0}^{\infty}(2n+1)P_n(x).$$

12.10 Associated Legendre's Functions

Theorem 12.10.1. If w is a solution of the Legendre's equation

$$(1-x^2)\frac{d^2y}{dx^2} - 2x\frac{dy}{dx} + n(n+1)y = 0, \qquad (12.61)$$

then show that $(1-x^2)^{m/2}\dfrac{d^m w}{dx^m}$ is a solution of the equation

$$(1-x^2)\frac{d^2y}{dx^2} - 2x\frac{dy}{dx} + \left\{n(n+1) - \frac{m^2}{1-x^2}\right\}y = 0. \qquad (12.62)$$

Proof: Since w is the solution of Equation (i), we get

$$(1-x^2)\frac{d^2w}{dx^2}-2x\frac{dw}{dx}+n(n+1)w=0 \tag{12.63}$$

Differentiating Equation (12.63) m times using Leibnitz theorem, we get

$$(1-x^2)\frac{d^{m+2}w}{dx^{m+2}}-2xm\frac{d^{m+1}w}{dx^{m+1}}+\frac{m(m-1)}{2}(-2)\frac{d^m w}{dx^m}$$
$$-2[x\frac{d^{m+1}w}{dx^{m+1}}+m\frac{d^m w}{dx^m}]+n(n+1)\frac{d^m w}{dx^m}=0$$

or, $$(1-x^2)\frac{d^{m+2}w}{dx^{m+2}}-2x(m+1)\frac{d^{m+1}w}{dx^{m+1}}+[n(n+1)-m(m+1)]\frac{d^m w}{dx^m}=0. \tag{12.64}$$

Let, $$v=\frac{d^m w}{dx^m}. \tag{12.65}$$

Then, Equation (12.64) becomes

$$(1-x^2)\frac{d^2v}{dx^2}-2x(m+1)\frac{dv}{dx}+[n(n+1)-m(m+1)]v=0. \tag{12.66}$$

Let, $$u=(1-x^2)^{m/2}v \text{ and} \tag{12.67}$$

$$v=(1-x^2)^{-m/2}u. \tag{12.68}$$

$$\therefore \frac{dv}{dx} = (1-x^2)^{-\frac{m}{2}}\frac{du}{dx} + (-\frac{m}{2})(1-x^2)^{-\frac{m}{2}-1}(-2x)u.$$

$$= (1-x^2)^{-\frac{m}{2}}\frac{du}{dx} + mux(1-x^2)^{-\frac{m}{2}-1}. \tag{12.69}$$

$$\frac{d^2v}{dx^2} = (1-x^2)^{-\frac{m}{2}}\frac{d^2u}{dx^2} - \frac{m}{2}(1-x^2)^{-\frac{m}{2}-1}(-2x)\frac{du}{dx}$$

$$+ mux[\left(-\frac{m}{2}-1\right)](1-x^2)^{-\frac{m}{2}-2}(-2x) + m(1-x^2)^{-\frac{m}{2}-1}\left[u + x\frac{du}{dx}\right]. \tag{12.70}$$

Substituting $v, \frac{dv}{dx}$ *and* $\frac{d^2v}{dx^2}$ in Equation (12.66), we have

$$(1-x^2)^{-\frac{m}{2}+1}\frac{d^2u}{dx^2} + mx(1-x^2)^{-\frac{m}{2}}\frac{du}{dx} + mx^2u(m+2)(1-x^2)^{-\frac{m}{2}-1}$$

$$+ mx(1-x^2)^{-\frac{m}{2}}\frac{du}{dx} + mu(1-x^2)^{-\frac{m}{2}} - 2(m+1)x(1-x^2)^{-\frac{m}{2}}\frac{du}{dx}$$

$$-2m(m+1)x^2u(1-x^2)^{-\frac{m}{2}-1}$$

$$+\{n(n+1) - m^2 + m\}(1-x^2)^{-\frac{m}{2}}u = 0$$

or,

$$(1-x^2)\frac{d^2u}{dx^2} - 2x\frac{du}{dx} + \left[n(n+1) - m^2 - m - \frac{2m(m+1)}{1-x^2}x^2 + m + \frac{m(m+2)}{1-x^2}x^2\right]u = 0$$

or, $$(1-x^2)\frac{d^2u}{dx^2} - 2x\frac{du}{dx}\left\{n(n+1) - \frac{m^2}{1-x^2}\right\}u = 0. \tag{12.71}$$

Using Equations (12.71), (12.67), and (12.65), we get $u=(1-x^2)^{\frac{m}{2}}v=$ $=(1-x^2)^{\frac{m}{2}}\dfrac{d^m w}{dx^m}$, which is a solution of Equation (12.71) and, hence, a solution of Equation (12.62).

Here, Equation (12.62) is called the *associated Legendre's equation.*

Remark 12.10.1. Since $P_n(x)$ *and* $Q_n(x)$ are solutions of Equation (12.63), we have

$P_n^m(x)=(1-x^2)^{\frac{m}{2}}\dfrac{d^m}{dx^m}P_n(x)$ *and* $Q_n^m(x)=(1-x^2)^{\frac{m}{2}}\dfrac{d^m}{dx^m}Q_n(x)$, which are also solutions of Equation (12.62). $P_n^m(x)$ *and* $Q_n^m(x)$ are called the associated Legendre's function of degree n and *order m of the first and second kinds,* respectively.

Remark 12.10.2. For $m>0,$ *we have* $P_n^m(x)=(1-x^2)^{\frac{m}{2}}\dfrac{d^m}{dx^m}P_n(x).$

Using Rodrigue's formula, $P_n^m(x)=(1-x^2)^{\frac{m}{2}}\dfrac{1}{2^n n!}\dfrac{d^{m+n}}{dx^{m+n}}(x^2-1)^n$ where the right-hand side is well defined for negative values of m *such that* $m+n>0$, i.e., $m\geq -n$.

$$\therefore P_n^m(x)=\begin{cases}(1-x^2)^{\frac{m}{2}}\dfrac{d^m}{dx^m}P_n(x) & \text{for } m\geq 0\\[2ex] (1-x^2)^{\frac{m}{2}}\dfrac{1}{2^n n!}\dfrac{d^{m+n}}{dx^{m+n}}(x^2-1)^n & \text{for } m\geq -n\end{cases}.$$

Example 12.10.1. Find the general solutions of the following equations:

$$\text{(a) } \varphi''(a)\varphi''+\cot\theta.\varphi'+\left\{n(n+1)-m^2/\sin^2\theta\right\}\varphi=0 \tag{i}$$

$$\text{(b) } \varphi''+\cot\theta.\varphi'+n(n+1)\varphi=0 \tag{ii}$$

Solution:

$$Put\ \cos\theta=\mu\ then\ \varphi'=\frac{d\varphi}{d\theta}=\frac{d\varphi}{d\mu}.\frac{d\mu}{d\theta}=-\sin\theta\frac{d\varphi}{d\mu}. \tag{iii}$$

$$\therefore \frac{d}{d\theta} \equiv -sin\theta \frac{d}{d\mu} \tag{iv}$$

Now, from Equation (iii),

$$\varphi'' = \frac{d}{d\theta}\frac{d\varphi}{d\theta} = -\frac{d}{d\theta}\left(sin\theta \frac{d\varphi}{d\mu} \right) = -cos\theta \frac{d\varphi}{d\mu} - sin\theta \frac{d}{d\theta}(\frac{d\varphi}{d\mu})$$

$$= -cos\theta \frac{d\varphi}{d\mu} - sin\theta \times -sin\theta \frac{d}{d\mu}(\frac{d\varphi}{d\mu}) \text{by (v)}$$

$$= -cos\theta \frac{d\varphi}{d\mu} + sin^2, \frac{d^2\varphi}{d\mu^2}$$

$$= -\mu \frac{d\varphi}{d\mu} + (1-\mu^2)\frac{d^2\varphi}{d\mu^2}.$$

Using values of φ', φ'' in Equation (i), we get

$$-\mu \frac{d\varphi}{d\mu} + (1-\mu^2)\frac{d^2\varphi}{d\mu^2} + \cot\theta.(-sin\theta \frac{d\varphi}{d\mu}) + \left\{ n(n+1) - \frac{m^2}{1-\mu^2} \right\}\varphi = 0$$

or,

$$(1-\mu^2)\frac{d^2\varphi}{d\mu^2} - 2\mu \frac{d\varphi}{d\mu} + \left\{ n(n+1) - \frac{m^2}{1-\mu^2} \right\}\varphi = 0, \tag{vi}$$

which is the associated Legendre's equation.

Hence, the general solution of Equation (vi) is

$$\varphi = AP_n^m(\mu) + BQ_n^m(\mu) = AP_n^m(cos\theta) + BQ_n^m(cos\theta) \tag{vii}$$

Note. Putting $cos\theta = \mu$ and doing as before, Equation (iia) or (iib) gives us

$$\varphi = AP_n(\mu) + BQ_n(\mu) = AP_n(cos\theta) + BQ_n(cos\theta).$$

12.10.1 Properties of the Associated Legendre's Functions

(i) $P_n^0(x) = P_n(x)$ (ii) $P_n^m(x) = 0$, *when* $m > n$

Proof: We know

$$P_n^m(x) = (1-x^2)^{\frac{m}{2}} \frac{d^m}{dx^m} P_n(x). \quad \text{(i)}$$

Putting $m = 0$, we get $P_n^0(x) = (1-x^2)^0 P_n(x) = P_n(x)$.

(ii) Since $P_n(x)$ is a polynomial of degree n, so when $m(>n)$

(*i*) gives [m times derivative must be vanish $\frac{d^m}{dx^m} P_n(x) = 0$].

Therefore, $P_n^m(x) = (1-x^2)^{\frac{m}{2}} \frac{d^m}{dx^m} P_n(x) = (1-x^2)^{\frac{m}{2}} \times 0 = 0.$

12.10.2 Orthogonality Relation for $P_n^m(x)$

Theorem 12.10.2. Prove that

$$\int_{-1}^{1} P_n^m(x) P_r^m(x) dx = \begin{cases} \dfrac{2(n+m)!}{(2n+r)(n-m)!} & if\ n \neq r \\ 0 & if\ n = r \end{cases}.$$

Proof: Case I: Let $n \neq r$.

Since $P_n^m(x)$ satisfies the Legendre's associated equation,

$$(1-x^2)\frac{d^2y}{dx^2} - 2x\frac{dy}{dx} + \left\{n(n+1) - \frac{m^2}{1-x^2}\right\} y = 0$$

or,
$$\frac{d}{dx}\left\{(1-x^2)\frac{dy}{dx}\right\} + \left\{n(n+1) - \frac{m^2}{1-x^2}\right\} y = 0$$

or,
$$\frac{d}{dx}\left\{(1-x^2)\frac{d}{dx} P_n^m(x)\right\} + \left\{n(n+1) - \frac{m^2}{1-x^2}\right\} P_n^m(x) = 0. \quad (12.72)$$

Similarly, for $P_r^m(x)$,

$$\frac{d}{dx}\left\{(1-x^2)\frac{d}{dx}P_r^m(x)\right\}+\left\{r(r+1)-\frac{m^2}{1-x^2}\right\}P_r^m(x)=0. \quad (12.73)$$

Multiplying Equation (i) by $P_r^m(x)$ and (ii) by $P_n^m(x)$ and subtracting, we get

$$P_r^m(x)\frac{d}{dx}\left\{(1-x^2)\frac{d}{dx}P_n^m(x)\right\}-P_n^m(x)\frac{d}{dx}\left\{(1-x^2)\frac{d}{dx}P_r^m(x)\right\}$$
$$+\left[n(n+1)-r(r+1)\right]P_n^m(x)P_r^m(x)=0.$$

Integrating, we get

$$\int_{-1}^{1}P_r^m(x)\frac{d}{dx}\left\{(1-x^2)\frac{d}{dx}P_n^m(x)\right\}dx$$
$$-\int_{-1}^{1}P_n^m(x)\frac{d}{dx}\left\{(1-x^2)\frac{d}{dx}P_r^m(x)\right\}dx$$
$$+\left[n(n+1)-r(r+1)\right]\int_{-1}^{1}P_n^m(x)P_r^m(x)dx=0.$$

Integrating by parts, we get

$$\left[P_r^m(x)\left\{(1-x^2)\frac{d}{dx}P_n^m(x)\right\}\right]_{-1}^{1}\int_{-1}^{1}\frac{d}{dx}P_r^m(x)\left\{(1-x^2)\frac{d}{dx}P_n^m(x)\right\}dx$$
$$-\left[P_n^m(x)\left\{(1-x^2)\frac{d}{dx}P_r^m(x)\right\}\right]_{-1}^{1}+\int_{-1}^{1}\frac{d}{dx}P_n^m(x)\left\{(1-x^2)\frac{d}{dx}P_r^m(x)\right\}dx$$
$$+[(n^2-r^2)+(n-r)\int_{-1}^{1}P_n^m(x)P_r^m(x)\,dx=0$$

or, $$[((n-r)(n+r+1)\int_{-1}^{1}P_n^m(x)P_r^m(x)dx=0.$$

$$\therefore \int_{-1}^{1} P_n^m(x) P_r^m(x) dx = 0 \text{ when } n \neq r. \tag{12.74}$$

Case II: When $n = r$ and $m > 0$,
we have

$$P_n^m(x) = (1-x^2)^{\frac{m}{2}} \frac{d^m}{dx^m} P_n(x). \tag{12.75}$$

$$\int_{-1}^{1} P_n^m(x) P_r^m(x) dx = \int_{-1}^{1} [P_n^m(x)]^2 dx = \int_{-1}^{1} (1-x^2)^m \frac{d^m}{dx^m} P_n(x) \frac{d^m}{dx^m} P_n(x) dx \text{ (using Equation 12.75)}$$

$$= \left[(1-x^2)^m \frac{d^m}{dx^m} P_n(x) \frac{d^{m-1}}{dx^{m-1}} P_n(x) \right]_{-1}^{1} - \int_{-1}^{1} \frac{d}{dx} \left\{ (1-x^2)^m \frac{d^m}{dx^m} P_n(x) \right\} \frac{d^{m-1}}{dx^{m-1}} P_n(x) dx$$

$$= -\int_{-1}^{1} \frac{d}{dx} \left\{ (1-x^2)^m \frac{d^m}{dx^m} P_n(x) \right\} \frac{d^{m-1}}{dx^{m-1}} P_n(x) dx. \tag{12.76}$$

Since $P_n(x)$ satisfies Legendre's equation, $(1-x^2)\frac{d^2 y}{dx^2} - 2x\frac{dy}{dx} + n(n+1)y = 0$.

$$\therefore (1-x^2)\frac{d^2 P_n(x)}{dx^2} - 2x\frac{dP_n(x)}{dx} + n(n+1)P_n(x) = 0. \tag{12.77}$$

Differentiating both sides of Equation (12.77) $(m-1)$ times with respect to x using the Leibnitz theorem, we get
or,

$$(1-x^2)\frac{d^{m+1}P_n(x)}{dx^{m+1}} - 2(m-1)x\frac{d^m P_n(x)}{dx^m} - \frac{2(m-1)(m-2)}{2}\frac{d^{m-1}P_n(x)}{dx^{m-1}}$$
$$-2x\frac{d^m P_n(x)}{dx^m} - 2(m-1)\frac{d^{m-1}P_n(x)}{dx^{m-1}}$$
$$+n(n+1)\frac{d^{m-1}P_n(x)}{dx^{m-1}} = 0$$

or,

$$(1-x^2)\frac{d^{m+1}P_n(x)}{dx^{m+1}} - 2mx\frac{d^m P_n(x)}{dx^m} + [n(n+1) - m(m+1)]\frac{d^{m-1}P_n(x)}{dx^{m-1}} = 0.$$

Multiplying by $(1-x^2)^{m-1}$,

$$(1-x^2)^m \frac{d^{m+1}P_n(x)}{dx^{m+1}} - 2mx(1-x^2)^{m-1}\frac{d^m P_n(x)}{dx^m} + [n(n+1) - m(m+1)](1-x^2)^{m-1}\frac{d^{m-1}P_n(x)}{dx^{m-1}} = 0$$

or,

$$\frac{d}{dx}\left\{(1-x^2)^m \frac{d^m P_n(x)}{dx^m}\right\} = -(n+m)(n-m+1)(1-x^2)^{m-1}\frac{d^{m-1}P_n(x)}{dx^{m-1}}. \tag{12.78}$$

Using Equation (12.78), Equation (12.76) becomes

$$\int_{-1}^{1}[P_n^m(x)]^2 dx = \int_{-1}^{1}\frac{d}{dx}\left\{(n+m)(n-m+1)(1-x^2)^{m-1}\frac{d^{m-1}P_n(x)}{dx^{m-1}}\right\}\frac{d^{m-1}}{dx^{m-1}}P_n(x)dx$$

$$=(n+m)(n-m+1)\int_{-1}^{1}\left\{(1-x^2)^{(m-1)/2}\frac{d^{(m-1)}P_n(x)}{dx^{m-1}}\right\}^2 dx$$

$=(n+m)(n-m+1)\int_{-1}^{1}\{P_n^{m-1}(x)\}^2 dx$ using Equation (12.75)

$$=\{(n+m)(n-m+1)\}\{(n+m-1)(n-m+2)\}..\{(n+1)n\}\int_{-1}^{1}\{P_n^0(x)\}^2 dx$$

[Similarly, it is repeating (m−1) times.]

$$=\{(n+m)(n+m-1)\ldots(n+1)\}\{n(n-1)..(n-m+1)\}\int_{-1}^{1}\{P_n(x)\}^2 dx$$

$$=\frac{(n+m)!}{(n-m)!}.\frac{2}{2n+1}. \tag{12.79}$$

Finally, let $m<0,$ *and if* $k>0,$ we get $m=-k.$ Then,

$$P_n^{-k}(x) = (-1)^k \frac{(n-k)!}{(n+k)!}P_n^k(x). \tag{12.80}$$

$$\therefore \int_{-1}^{1}[P_n^m(x)]^2 dx = \int_{-1}^{1}[P_n^{-k}(x)]^2 dx = = (-1)^{2k}\left[\frac{(n-k)!}{(n+k)!}\right]^2 \int_{-1}^{1}[P_n^k(x)]^2 dx$$

$$= \left[\frac{(n-k)!}{(n+k)!}\right]^2 \times \frac{(n+m)!}{(n-m)!}.\frac{2}{2n+1}$$

$$= \frac{(n+m)!}{(n-m)!}\frac{2}{2n+1}, \text{ using Equation (12.79) and } k=-m \qquad (12.81)$$

From Equations (12.79) and (12.81),

$$\int_{-1}^{1}[P_n^m(x)]^2 dx = \frac{(n+m)!}{(n-m)!}\frac{2}{2n+1} \qquad (12.82)$$

From Equations (12.74) and (12.82),

$$\therefore \int_{-1}^{1} P_n^m(x)P_r^m(x)dx = \frac{2(n+m)!}{(n-m)!}\frac{1}{(2n+1)!}\delta_r^n.$$

12.10.3 Recurrence Relations for $P_n^m(x)P_r^m$

Prove that

(a) $P_n^{m+1}(x) - \frac{2mx}{(1-x^2)^{1/2}}P_n^m(x) + \{n(n+1)-m(m-1)\}P_n^{m-1}(x) = 0$

(b) $(2n+1)P_n^m(x) = (n+m)P_{n-1}^m(x) + (n-m-1)P_{n+1}^m(x)$

(c) $(1\text{-}x^2)^{1/2}P_n^m(x) = \frac{1}{(2n+1)}\{P_{n+1}^{m+1}(x) - P_{n-1}^{m+1}(x)\}.$

Proof: (i) $P_n(x)$ satisfies Legendre's equation $(1-x^2)\frac{d^2y}{dx^2} - 2x\frac{dy}{dx} + n(n+1)y = 0.$

$$\therefore (1-x^2)\frac{d^2P_n(x)}{dx^2} - 2x\frac{dP_n(x)}{dx} + n(n+1)P_n(x) = 0. \qquad (12.82a)$$

Differentiating both sides of Equation (12.82a) $(m-1)$ times with respect to x using the Leibnitz theorem, we get
or,

$$(1-x^2)\frac{d^{m+1}P_n(x)}{dx^{m+1}}-2(m-1)x\frac{d^mP_n(x)}{dx^m}-\frac{2(m-1)(m-2)}{2}\frac{d^{m-1}P_n(x)}{dx^{m-1}}$$
$$-2x\frac{d^mP_n(x)}{dx^m}-2(m-1)\frac{d^{m-1}P_n(x)}{dx^{m-1}}+n(n+1)\frac{d^{m-1}P_n(x)}{dx^{m-1}}=0$$

or,

$$(1-x^2)\frac{d^{m+1}P_n(x)}{dx^{m+1}}-2mx\frac{d^mP_n(x)}{dx^m}+[n(n+1)-m(m+1)]\frac{d^{m-1}P_n(x)}{dx^{m-1}}=0.$$

Multiplying by $(1-x^2)^{(m-1)/2}$,

$$(1-x^2)^{(m+1)/2}\frac{d^{m+1}P_n(x)}{dx^{m+1}}-2mx(1-x^2)^{-1/2}(1-x^2)^{m/2}\frac{d^mP_n(x)}{dx^m}$$
$$+[n(n+1)-m(m+1)](1-x^2)^{(m-1)/2}\frac{d^{m-1}P_n(x)}{dx^{m-1}}=0.$$

Using $P_n^m(x)=(1-x^2)^{\frac{m}{2}}\frac{d^m}{dx^m}P_n(x)$ in the above equation, we get

$$P_n^{m+1}(x)-\frac{2mx}{(1-x^2)^{1/2}}P_n^m(x)++[n(n+1)-m(m+1)]P_n^{m-1}(x)=0.$$

(b) We know from Legendre's polynomial recurrence relations that

$$(2n+1)xP_n(x)=(n+1)P_{n+1}(x)+nP_{n-1}(x) \tag{12.83}$$

and,
$$(2n+1)P_n(x) = \frac{d}{dx}P_{n+1}(x) - \frac{d}{dx}P_{n-1}(x). \tag{12.84}$$

Differentiating both sides of Equation (12.83) m times with respect to x using the Leibnitz theorem, we get

$$(2n+1)\left\{x\frac{d^m}{dx^m}P_n(x) + \mathrm{m}\frac{d^{m-1}}{dx^{m-1}}P_n(x)\right\} =$$
$$(n+1)\frac{d^m}{dx^m}P_{n+1}(x) + n\frac{d^m}{dx^m}P_{n-1}(x), \tag{12.85}$$

and differentiating Equation (12.84) $(m-1)$ times with respect to x with the help of the Leibnitz theorem, we get

$$(2n+1)\frac{d^{m-1}}{dx^{m-1}}P_n(x) = \frac{d^m}{dx^m}P_{n+1}(x) - \frac{d^m}{dx^m}P_{n-1}(x). \tag{12.86}$$

Putting the value of $(2n+1)\frac{d^{m-1}}{dx^{m-1}}P_n(x)$ *in* Equation (12.85),

$$(2n+1)x\frac{d^m}{dx^m}P_n(x) + \mathrm{m}\left[\frac{d^m}{dx^m}P_{n+1}(x) - \frac{d^m}{dx^m}P_{n-1}(x)\right]$$
$$= (n+1)\frac{d^m}{dx^m}P_{n+1}(x) + n\frac{d^m}{dx^m}P_{n-1}(x)$$

or, $$(2n+1)x\frac{d^m}{dx^m}P_n(x) = -\mathrm{m}\left[\frac{d^m}{dx^m}P_{n+1}(x) - \frac{d^m}{dx^m}P_{n-1}(x)\right]$$
$$+ (n+1)\frac{d^m}{dx^m}P_{n+1}(x) + n\frac{d^m}{dx^m}P_{n-1}(x)$$
$$= (n+m)\frac{d^m}{dx^m}P_{n-1}(x) + (n-m+1)\frac{d^m}{dx^m}P_{n+1}(x).$$

Multiplying both sides by $(1-x^2)^{\frac{m}{2}}$, we get

$$(2n+1)x(1-x^2)^{\frac{m}{2}}\frac{d^m}{dx^m}P_n(x)=(n+m)(1-x^2)^{\frac{m}{2}}\frac{d^m}{dx^m}P_{n-1}(x)$$
$$+(n-m+1)(1-x^2)^{\frac{m}{2}}\frac{d^m}{dx^m}P_{n+1}(x).$$

Using $P_n^m(x)=(1-x^2)^{\frac{m}{2}}\frac{d^m}{dx^m}P_n(x)$ in the above equation, we get

$$(2n+1)xP_n^m(x)=(n+m)P_{n-1}^m(x)+(n-m+1)P_{n+1}^m(x).$$

(c) We know that from Legendre's polynomial recurrence relations,

$$(2n+1)P_n(x)=\frac{d}{dx}P_{n+1}(x)-\frac{d}{dx}P_{n-1}(x).$$

Differentiating m times with respect to x, we get

$$(2n+1)\frac{d^m}{dx^m}P_n(x)=\frac{d^{m+1}}{dx^{m+1}}P_{n+1}(x)-\frac{d^{m+1}}{dx^{m+1}}P_{n-1}(x).$$

Multiplying both sides by $(1-x^2)^{\frac{m+1}{2}}$, we get

$$(2n+1)(1-x^2)^{\frac{m+1}{2}}\frac{d^m}{dx^m}P_n(x)=(1-x^2)^{\frac{m+1}{2}}\frac{d^{m+1}}{dx^{m+1}}P_{n+1}(x)$$
$$-(1-x^2)^{\frac{m+1}{2}}\frac{d^{m+1}}{dx^{m+1}}P_{n-1}(x).$$

Using $P_n^m(x)=(1-x^2)^{\frac{m}{2}}\frac{d^m}{dx^m}P_n(x)$ in the above equation, we get

$$(2n+1)(1-x^2)^{\frac{1}{2}}P_n^m(x)=P_{n+1}^{m+1}(x)-P_{n-1}^{m+1}(x)$$

or, $$(1-x^2)^{\frac{1}{2}} P_n^m(x) = \frac{1}{2n+1}\{P_{n+1}^{m+1}(x) - P_{n-1}^{m+1}(x)\}.$$

12.11 Legendre's Functions of the Second Kind—$Q_n(x)$

12.11.1 Some Useful Results

$$\begin{aligned}
Q_n(x) &= \frac{n!}{[1.3.5.,,,,(2n+1)]}\left[x^{-n-1} + \frac{(n+1)(n+2)}{2(2n+3)}x^{-n-3} + \cdots\right] \\
&= \frac{n!(2.4.6.\ldots.2n)}{(2n+1)!}\left[x^{-n-1} + \frac{(n+1)(n+2)}{2(2n+3)}x^{-n-3} + \cdots\right] \\
&= \frac{2^n(n!)^2}{(2n+1)!}\left[x^{-n-1} + \frac{(n+1)(n+2)}{2(2n+3)}x^{-n-3} + \cdots\right],
\end{aligned} \tag{12.87}$$

which is a solution of Legendre's equation in descending powers of x being negative. Now,

$$\begin{aligned}
Q_n(x) &= \frac{2^n n!}{(2n+1)!}\left[n!x^{-n-1} + \frac{(n+2)!}{2(2n+3)}x^{-n-3} + \frac{(n+4)!}{2.4.(2n+3)(2n+5)}x^{-n-5} + \cdots\right] \\
&= \frac{2^n n!}{(2n+1)!}\sum_{r=0}^{\infty}\frac{(n+2r)!}{2.4.6\ldots 2r(2n+3)(2n+5)\ldots(2n+2r+1)}x^{-(n+2r+1)}.
\end{aligned}$$

$$\therefore Q_n(x) = \frac{2^n n!}{(2n+1)!}\sum_{r=0}^{\infty}\frac{(n+2r)!}{2.4.6\ldots 2r(2n+3)(2n+5)\ldots(2n+2r+1)}x^{-(n+2r+1)}. \tag{12.88}$$

Differentiating with respect to x, we get

$$\therefore Q_n'(x) = -\frac{2^n n!}{(2n+1)!}\sum_{r=0}^{\infty} \frac{(n+2r+1)!}{2^r.r!(2n+3)(2n+5)\ldots(2n+2r+1)} x^{-(n+2r+2)}. \tag{12.89}$$

Putting $(n-1)$ in place of n, *we get*

$$\begin{aligned} Q_{n-1}'(x) &= -\frac{2^{n-1}(n-1)!}{(2n-1)!}\sum_{r=0}^{\infty} \frac{(n+2r+1)!}{2^r.r!(2n+1)(2n+3)\ldots(2n+2r-1)} x^{-(n+2r+1)} \\ &= -\frac{2n.2^{n-1}(n-1)!}{2n.(2n-1)!}\sum_{r=0}^{\infty} \frac{(n+2r+1)!}{2^r.r!(2n+1)(2n+3)\ldots(2n+2r-1)} x^{-(n+2r+1)} \\ &= -\frac{2^n(n)!}{(2n)!}\sum_{r=0}^{\infty} \frac{(n+2r+1)!}{2^r.r!(2n+1)(2n+3)\ldots(2n+2r-1)} x^{-(n+2r+1)}. \end{aligned} \tag{12.90}$$

Again, putting $(n+1)$ *for* n in Equation (12.89),

$$\begin{aligned} Q_{n+1}'(x) &= -\frac{2^{n+1}(n+1)!}{(2n+3)!}\sum_{r=0}^{\infty} \frac{(n+2r+2)!}{2^r.r!(2n+5)(2n+7)\ldots(2n+2r+3)} x^{-(n+2r+3)} \\ &= -\frac{2^n n!(2n+2)}{(2n+3)(2n+2)(2n+1)(2n)!}\sum_{r=0}^{\infty} \frac{(n+2r+2)!}{2^r.r!(2n+5)(2n+7)\ldots(2n+2r+3)} x^{-(n+2r+3)} \\ &= -\frac{2^n n!}{(2n)!}\sum_{r=0}^{\infty} \frac{(n+2r+2)!}{2^r.r!(2n+3)(2n+1)(2n+5)(2n+7)\ldots(2n+2r+3)} x^{-(n+2r+3)}. \end{aligned} \tag{12.91}$$

Theorem 12.11.1. The associated Legendre's function of the second kind defined by

$$Q_n^m(x) = (1-x^2)^{\frac{m}{2}} \frac{d^m}{dx^m} Q_n(x)$$

satisfies Legendre's associated Equation $(1-x^2)\frac{d^2y}{dx^2} - 2x\frac{dy}{dx} + \left\{n(n+1) - \frac{m^2}{1-x^2}\right\} y = 0.$

Proof: Proceed similarly as Section (12.9).

Theorem 12.11.2. Assuming $P_n(x)$ as a solution of Legendre's equation, show that the complete solution of the equation is given by $aP_n(x)+bQ_n(x)$, where $Q_n(x)=cP_n(x)\int\frac{dx}{(1-x^2)P_n^2(x)}$, c being a constant.

12.11.2 Recurrence Relations for $Q_n(x)$

(i) $Q'_{n+1}(x)-Q'_{n-1}(x)=(2n+1)Q_n(x)$

(ii) $nQ'_{n+1}(x)+(n+1)Q'_{n-1}(x)=(2n+1)xQ'_n(x)$

(iii) $(2n+1)xQ_n(x)=(n+1)Q_{n+1}(x)+nQ_{n-1}(x)$

(iv) $(2n+1)(1-x^2)Q'_n(x)=n(n+1)[Q_{n-1}(x)-Q_{n+1}(x)]$

Proof. Here, $Q'_{n-1}(x)+(2n+1)Q_n(x)=$

$$-\frac{2^n(n)!}{(2n)!}\sum_{r=0}^{\infty}\frac{(n+2r+1)!}{2^r.r!(2n+1)(2n+3)\ldots(2n+2r-1)}x^{-(n+2r+1)}$$

$$+(2n+1)\frac{2^n n!}{(2n+1)!}\sum_{r=0}^{\infty}\frac{(n+2r)!}{2^r.r!(2n+3)(2n+5)\ldots(2n+2r+1)}x^{-(n+2r+1)}$$

$$=\frac{2^n n!}{(2n)!}\left[\sum_{r=0}^{\infty}\frac{(n+2r)!x^{-(n+2r+1)}}{2^r.r!(2n+3)(2n+5)\ldots(2n+2r+1)}-\sum_{r=0}^{\infty}\frac{(n+2r+1)!x^{-(n+2r+1)}}{2^r.r!(2n+1)(2n+3)\ldots(2n+2r-1)}-\right]$$

$$=\frac{2^n n!}{(2n)!}\left[\sum_{r=0}^{\infty}\frac{(n+2r)!x^{-(n+2r+1)}}{2^r.r!(2n+1)(2n+3)(2n+5)\ldots(2n+2r+1)}\times\{(2n+1)-(n+2r+1)\}\right]$$

$$=\frac{2^n n!}{(2n)!}\left[\sum_{r=0}^{\infty}\frac{-2r(n+2r)!x^{-(n+2r+1)}}{2^r.r!(2n+1)(2n+3)(2n+5)\ldots(2n+2r+1)}\right]$$

$$=-\frac{2^n n!}{(2n)!}\left[\sum_{r=0}^{\infty}\frac{(n+2r)!x^{-(n+2r+1)}}{2^{r-1}.(r-1)!(2n+1)(2n+3)(2n+5)\ldots(2n+2r+1)}\right]$$

Let $r=s+1$,, s=r-1 and therefore $\frac{1}{(-1)!}=0$

$$=-\frac{2^n n!}{(2n)!}\left[\sum_{s=0}^{\infty}\frac{(n-2s+2)!x^{-(n+2s+3)}}{2^s.(s)!(2n+1)(2n+3)(2n+5)\ldots(2n+2s+3)}\right]$$

$$=Q'_{n+1}(x).$$

(ii) $nQ'_{n+1}(x)+(n+1)Q'_{n-1}(x)=(2n+1)xQ'_n(x)$

Proof: We have $(2n+1)xQ'_n(x)-(n+1)Q'_{n-1}(x)$

$$=(2n+1)x\left[-\frac{2^n n!}{(2n+1)!}\sum_{r=0}^{\infty}\frac{(n+2r+1)!}{2^r.r!(2n+3)(2n+5)\ldots(2n+2r+1)}x^{-(n+2r+2)}\right]$$

$$-(n+1)\left[-\frac{2^n(n)!}{(2n)!}\sum_{r=0}^{\infty}\frac{(n+2r+1)!}{2^r.r!(2n+1)(2n+3)\ldots(2n+2r-1)}x^{-(n+2r+1)}\right]$$

$$=-\frac{2^n n!}{(2n)!}\sum_{r=0}^{\infty}\frac{(n+2r+1)!(2n+1)}{2^r.r!(2n+3)(2n+5)\ldots(2n+2r+1)}x^{-(n+2r+1)}$$

$$+\frac{-2^n(n)!}{(2n)!}\sum_{r=0}^{\infty}\frac{(n+2r)!(2n+2r+1)!(n+1)}{2^r.r!(2n+1)(2n+3)\ldots(2n+2r+1)}x^{-(n+2r+1)}$$

$$=\frac{-2^n n!}{(2n)!}\sum_{r=0}^{\infty}\frac{(n+2r)!x^{-(n+2r+1)}}{2^r.r!(2n+1)(2n+3)(2n+5)\ldots(2n+2r+1)}[(n+2r+1)(2n+1)-(n+1)((2n+2r+1)]$$

$$=\frac{-n.2^n n!}{(2n)!}\sum_{r=0}^{\infty}\frac{(n+2r)!x^{-(n+2r+1)}}{2^{r-1}(.r-1)!(2n+1)(2n+3)(2n+5)\ldots(2n+2r+1)}$$

$$=\frac{-n.2^n n!}{(2n)!}\left[\sum_{r=0}^{\infty}\left\{\frac{(n+2r)!x^{-(n+2r+1)}}{2^{r-1}(.r-1)!(2n+1)(2n+3)(2n+5)\ldots(2n+2r+1)}\right\}-0\right]$$

$$=\frac{-n.2^n n!}{(2n)!}\sum_{s=0}^{\infty}\frac{(n+2r)!x^{-(n+2s+3)}}{2^s(.s)!(2n+1)(2n+3)(2n+5)\ldots(2n+2s+3)}\quad \text{let } r=s+1$$

$$=nQ'_{n+1}(x).$$

(iii) $(2n+1)xQ_n(x)=(n+1)Q_{n+1}(x)+nQ_{n-1}(x)$

or, $$xQ_n(x)=\frac{(n+1)}{2n+1}Q_{n+1}(x)+\frac{n}{2n+1}Q_{n-1}(x)$$

or, $$(n+1)Q_{n+1}(x)-(2n+1)xQ_n(x)+\mathrm{n}Q_{n-1}(x)=0.$$

Proof: $nQ_{n-1}(x)-(2n+1)xQ_n(x)=$

$$n\frac{2^{n-1}(n-1)!}{(2n-1)!}\sum_{r=0}^{\infty}\frac{(n+2r-1)!}{2^r.r!(2n+3)(2n+5)\ldots(2n+2r-1)}x^{-(n+2r)}$$
$$-(2n+1)\frac{2^n n!}{(2n+1)!}\sum_{r=0}^{\infty}\frac{(n+2r)!}{2.4.6\ldots 2r(2n+3)(2n+5)\ldots(2n+2r-1)}x^{-(n+2r+1)}$$
$$=\frac{2^{n-1}(n)!2n}{(2n+1)!}\sum_{r=0}^{\infty}\frac{(n+2r-1)!(2n+2r+1)}{2^r.r!(2n+3)(2n+5)\ldots(2n+2r+1)}x^{-(n+2r)}$$
$$-\frac{2^n n!(2n+1)}{(2n+1)!}\sum_{r=0}^{\infty}\frac{(n+2r)!}{2.4.6\ldots 2r(2n+3)(2n+5)\ldots(2n+2r+1)}x^{-(n+2r+1)}$$
$$=\frac{2^n n!}{(2n+1)!}\left[\sum_{r=0}^{\infty}\frac{(n+2r-1)!}{2^r.r!(2n+3)(2n+5)\ldots(2n+2r+1)}x^{-(n+2r)}\{n((2n+2r+1)-(2n+1)(n+2r)\}\right]$$
$$=\frac{2^n n!}{(2n+1)!}\sum_{r=0}^{\infty}\frac{(n+2r-1)!}{2^r.r!(2n+3)(2n+5)\ldots(2n+2r+1)}x^{-(n+2r)}(-2r)(n+1)$$
$$=-\frac{2^n(n+1)!}{(2n+1)!}\left[\sum_{r=0}^{\infty\{}\frac{(n+2r-1)!}{2^r.r!(2n+3)(2n+5)\ldots(2n+2r+1)}x^{-(n+2r)}\right]$$
$$==-\frac{2^n(n+1)!n!}{(2n+1)!}\left[\sum_{r=0}^{\infty\{}\{\frac{(n+2r-1)!}{2^{r-1}.(r-1)!(2n+3)(2n+5)\ldots(2n+2r+1)}x^{-(n+2r)}\}+0\right]$$

Let $r=s+1$

$$=-\frac{2^n(n+1)!n!}{(2n+1)!}\left[\sum_{s=0}^{\infty\{}\frac{(n+2s+1)!}{2^s.s!(2n+3)(2n+5)\ldots(2n+2s+3)}x^{-(n+2s+2)}\right]$$
$$=-(n+1)\frac{2^n(2n+2)(n!)}{(2n+2)!}\left[\sum_{s=0}^{\infty\{}\frac{(n+2s+1)!}{2^s.s!(2n+3)(2n+5)\ldots(2n+2s+3)}x^{-(n+2s+2)}\}\right]$$
$$=-(n+1)\frac{2^n(n+1)!}{(2n+3)!}\left[\sum_{s=0}^{\infty\{}\frac{(n+2s+1)!}{2^s.s!(2n+5)\ldots(2n+2s+3)}x^{-(n+2s+2)}\}\right]$$

$$=-(n+1)Q_{n+1}(x).$$

(iv) $(2n+1)(1-x^2)Q_n'(x)=n(n+1)[Q_{n-1}(x)-Q_{n+1}(x)].$

Proof: Since $Q_n(x)$ is a solution of Legendre's equation, we have

$$(1-x^2)\frac{d^2y}{dx^2}-2x\frac{dy}{dx}+n(n+1)y=0,$$

or,
$$\frac{d}{dx}\left[(1-x^2)\frac{dy}{dx}\right]+n(n+1)y=0.$$

$$\therefore \frac{d}{dx}[(1-x^2)Q_n'(x)]=-n(n+1)Q_n(x). \quad (12.92)$$

Integrating both sides of Equation (12.92) from ∞ *to* x, we get

$$[(1-x^2)Q_n'(x)]_\infty^x=-n(n+1)\int_\infty^x Q_n(x)dx$$

or,
$$(1-x^2)Q_n'(x)=-n(n+1)\int_\infty^x Q_n(x)dx$$

as,
$$\left(1-x^2\right)Q_n'\left(x\right)=-n\left(n+1\right)\int_\infty^x Q_n\left(x\right)dx \quad (12.93)$$

We know the recurrence relation

$$Q_{n+1}'(x)-Q_{n-1}'(x)=(2n+1)Q_n(x). \quad (12.94)$$

Integrating both sides of Equation (12.94) between ∞ *to* x, we get

$$[Q_{n+1}(x)-Q_{n-1}(x)]_\infty^x=(2n+1)\int_\infty^x Q_n(x)dx$$

or,
$$Q_{n+1}(x)-Q_{n-1}(x)=(2n+1)\int_\infty^x Q_n(x)dx \quad (12.95)$$

as, $$Q_{n+1}(x)]_{x=\infty} = Q_{n-1}(x)]_{x=\infty} = 0.$$

From Equations (12.95) and (12.93),

$$(1-x^2)Q_n'(x) = -n(n+1)\frac{Q_{n+1}(x)-Q_{n-1}(x)}{2n+1}.$$

$$\therefore (2n+1)(1-x^2)Q_n'(x) = n(n+1)[Q_{n-1}(x)-Q_{n+1}(x)].$$

Other recurrence relations are as follows:

(a) $xQ_n'(x)-Q_{n-1}'(x)=nQ_n(x)$
(b) $Q_n'(x)-xQ_{n-1}'(x)=nQ_{n-1}(x)$
(c) $(x^2-1)Q_n'(x)=(n+1)Q_n(x)-(n+1)xQ_n(x)$.

12.11.3 Relation of $P_n(x)$ and $Q_n(x)$

$$\frac{1}{y-x} = \sum_{m=0}^{\infty}(2m+1)P_n(x)Q_n(x)$$

Proof: Let $f(x)=\dfrac{1}{y-x}$

or,

$$f(x)=\frac{1}{y-x}=\frac{1}{y(1-\frac{x}{y})}=\frac{1}{y}(1-\frac{x}{y})^{-1}=y^{-1}(1+\frac{x}{y}+\frac{x^2}{y^2}+\ldots\ldots+\frac{x^m}{y^m}+\cdots)$$
$$=(y^{-1}+y^{-2}x+y^{-3}x^2+\cdots+y^{-m-1}x^m+y^{-m-2}x^{m+1}+y^{-m-3}x^{m+2}+\cdots). \tag{12.96}$$

Let, $$f(x)=A_0+A_1x+A_2x^2+\cdots \tag{12.97}$$

Where the $A's$ are constants. Further suppose that $f(x)$ *is* also expressed as

$$f(x) = \sum_{m=0}^{\infty} B_m P_m(x). \tag{12.98}$$

We know that

$$B_m = \frac{1.2.3..m}{1.3.5...(2m-1)}\left[A_m + \frac{(m+1)(m+2)}{2(2m+3)}A_{m+2} + \frac{(m+1)(m+2)(m+3)(m+4)}{2.4(2m+3)(2m+5)}A_{m+4} + \cdots\right].$$

Comparing Equations (12.96) and (12.97),

$$A_0 = y^{-1}, A_1 = y^{-2} \ldots \ldots A_m = y^{-m-1}$$

$$B_m = \frac{m!}{1.3.5\ldots(2m-1)}\left[y^{-m-1} + \frac{(m+1)(m+2)}{2(2m+3)}y^{-m-3} + \frac{(m+1)(m+2)(m+3)(m+4)}{2.4(2m+3)(2m+5)}y^{-m-5} + \cdots\right]$$

$$== 2m+1)Qn(y) \text{ by definition.}$$

$$\therefore f(x) = \frac{1}{y-x} = \sum_{m=0}^{\infty} B_m P_m(x) = \sum_{m=0}^{\infty}(2m+1)P_m(x)Q_m(y).$$

12.12 Examples

Example 12.12.1. Prove that $\frac{d^{n+1}}{dx^{n+1}}[Q_n(x)] = -\frac{(-2)^n n!}{(x^2-1)^{n+1}}$.

Solution: We know that

$$Q_n(x) = \frac{n!}{[1.3.5.,,,,(2n+1)]}\left[x^{-n-1} + \frac{(n+1)(n+2)}{2(2n+3)}x^{-n-3} + \cdots\right]. \tag{i}$$

Differentiating both sides $(n+1)$ *times* with respect to x, *we get*

$$\frac{d^{n+1}}{dx^{n+1}}[Q_n(x)] = \frac{n!}{[1.3.5.,,,,(2n+1)]}\left[\frac{d^{n+1}}{dx^{n+1}}(x^{-n-1}) + \frac{(n+1)(n+2)}{2(2n+3)}\frac{d^{n+1}}{dx^{n+1}}(x^{-n-3}) + \cdots.\right]$$

$$= \frac{n!}{[1.3.5.,,,,(2n+1)]}\left[(-n-1)(-n-2)..(-n-1-n)x^{-n-1-n-1}\right.$$

$$\left.+\frac{(n+1)(n+2)}{2(2n+3)}(-n-3)(-n-4)..(-n-3-n)x^{-n-3-n-1} + \cdots\ldots\right]$$

$$= \frac{n!}{[1.3.5.,,,,(2n+1)]}\left[(-1)^{n+1}(n+1)(n+2)..(2n+1)x^{-2n-2}\right.$$

$$\left.+\frac{(n+1)(n+2)}{2(2n+3)}(-1)^{n+1}(n+3)(n+4)..(2n+3)x^{-2n-4} + \cdots\ldots\right]$$

$$= \frac{(-1)^{n+1}n!(n+1)(n+2)..(2n+1)}{[1.3.5.,,,,(2n+1)]}x^{-2n-2}\left[1 + \frac{(2n+2)(2n+3)}{2(2n+3)}x^{-2}\right.$$

$$\left.+\frac{(2n+2)(2n+3)(2n+4)(2n+5)}{2.4(2n+3)(2n+5)}x^{-4}\ldots\ldots\right]$$

$$= \frac{(-1)^{n+1}(2n+1)!}{[1.3.5.,,,,(2n+1)]}x^{-2n-2}\left[1 + (n+1)x^{-2} + \frac{(n+1)(n+2)}{2!}x^{-4}\ldots\ldots\right]$$

$$= (-1)^{n+1}2.4.6\ldots(2n)x^{-2n-2}\left(1 - \frac{1}{x^2}\right)^{-n-1} = (-1)^{n+1}2^n n! x^{-2n-2}\left(\frac{x^2-1}{x^2}\right)^{-n-1} = \frac{(-2)^n n!}{(x^2-1)^{n+1}}.$$

Example 12.12.2.
Prove that

(a) $(x^2-1)(Q_n P_n' - P_n Q_n')$

(b) $\dfrac{Q_n}{P_n} = \displaystyle\int_x^\infty \frac{dx}{(x^2-1)P_n^2}.$

From these, deduce that $Q_0(x) = \frac{1}{2}\log\frac{x+1}{x-1}$ and $Q_1(x) = \frac{x}{2}\log\frac{x+1}{x-1} - 1.$

Solution: Legendre's equation:

$$(1-x^2)\frac{d^2y}{dx^2} - 2x\frac{dy}{dx} + n(n+1)y = 0. \tag{i}$$

Putting P_n and Q_n in Equation (i) as a solution,

$$(1-x^2)\frac{d^2P_n}{dx^2} - 2x\frac{dP_n}{dx} + n(n+1)P_n = 0 \tag{ii}$$

$$(1-x^2)\frac{d^2Q_n}{dx^2}-2x\frac{dQ_n}{dx}+n(n+1)Q_n=0. \qquad \text{(iii)}$$

Multiplying Equation (ii) by Q_n and Equation (iii) by P_n, and then subtracting, we get

$$(1-x^2)(Q_nP_n''-P_nQ_n'')-2x(Q_nP_n'-P_nQ_n')=0$$

or,
$$(1-x^2)\frac{d}{dx}[Q_nP_n'-P_nQ_n']-2x(Q_nP_n'-P_nQ_n')=0$$

or,
$$\frac{d}{dx}[(1-x^2)(Q_nP_n'-P_nQ_n')]=0. \qquad \text{(iv)}$$

Integrating with respect to x, we get

$$(1-x^2)(Q_nP_n'-P_nQ_n')=-c,$$

or,
$$(x^2-1)(Q_nP_n'-P_nQ_n')=c \qquad \text{(v)}$$

where c is a constant.

(b) From Equation (a), $(Q_nP_n'-P_nQ_n')=\dfrac{c}{(x^2-1)}$

or,
$$(Q_nP_n'-P_nQ_n')=\frac{c}{x^2}(1-\frac{1}{x^2})^{-1}=\frac{c}{x^2}\left(1+\frac{1}{x^2}+\frac{1}{x^4}+\cdots\right). \qquad \text{(vi)}$$

We know that

$$Q_n=\frac{n!}{[1.3.5.,,,,(2n+1)]}\left[x^{-n-1}+\frac{(n+1)(n+2)}{2(2n+3)}x^{-n-3}+\cdots\right] \qquad \text{(vii)}$$

and, $$P_n(x) = \frac{1.3.5\ldots(2n-1)}{n!}\left[x^n - \frac{n(n-1)}{2(2n-1)}x^{n-2} + \cdots\right]. \tag{viii}$$

Using Equations (vii) and (viii) on the left-hand side of Equation (vi), we get

$$(Q_n P_n' - P_n Q_n') = \frac{n!}{[1.3.5.,,,,(2n+1)]}\left[x^{-n-1} + \frac{(n+1)(n+2)}{2(2n+3)}x^{-n-3} + \cdots\right]$$
$$\times \frac{1.3.5\ldots(2n-1)}{n!}\left[nx^{n-1} - \frac{n(n-1)}{2(2n-1)}(n-2)x^{n-3} + \cdots\right]$$
$$-\frac{1.3.5\ldots(2n-1)}{n!}\left[x^n - \frac{n(n-1)}{2(2n-1)}x^{n-2} + \cdots\right]$$
$$\times \frac{n!}{[1.3.5.,,,,(2n+1)]}\left[-(n+1)x^{-n-2} - \frac{(n+1)(n+2(n+3))}{2(2n+3)}x^{-n-4} + \cdots\right]$$

$\therefore$ The coefficient of $\frac{1}{x^2}$ on the left hand side of Equation (vi)

$$= \frac{n!}{[1.3.5.,,,,(2n+1)]}\frac{1.3.5\ldots(2n-1)}{n!}n - \frac{n!}{[1.3.5.,,,,(2n+1)]}\frac{1.3.5\ldots(2n-1)}{n!}(-n-1)$$
$$= \frac{n!}{[1.3.5.,,,,(2n+1)]}\frac{1.3.5\ldots(2n-1)}{n!}.(2n+1) = \frac{2n+1}{2n+1} = 1$$

Comparing its coefficient, we get $\therefore c = 1$.
From Equation (v), we get

$$(x^2-1)(Q_n P_n' - P_n Q_n') = 1$$

or, $$\frac{(P_n Q_n' - Q_n P_n')}{P_n^2} = -\frac{1}{(x^2-1)P_n^2}$$

or,
$$\frac{d}{dx}\left(\frac{Q_n}{P_n}\right)=-\frac{1}{(x^2-1)P_n^2}$$

Integrating with respect to x from ∞ *to x, we get*

$$\left[\frac{Q_n}{P_n}\right]_{\infty}^{x}=\int_x^{\infty}\frac{dx}{(x^2-1)P_n^2}$$

or,
$$\frac{Q_n}{P_n}-\lim_{x\to\infty}\frac{Q_n(x)}{P_n(x)}=\int_x^{\infty}\frac{dx}{(x^2-1)P_n^2}. \qquad \text{(ix)}$$

$$\lim_{x\to\infty}\frac{Q_n(x)}{P_n(x)}=\lim_{x\to\infty}\frac{\frac{d^n}{dx^n}Q_n(x)}{\frac{d^n}{dx^n}P_n(x)}$$ using L'Hospital's rule,

$$=\frac{\frac{n!}{[1.3.5.,,,,(2n+1)]}\left\{(-1)^n(n+1)(n+2)\ldots 2n\,x^{-(2n+1)}+\cdots\right\}}{\frac{1.3.5\ldots(2n-1)}{n!}n!}$$, using Equations (vii) and (viii)

$= 0$, taking the limit $x\to\infty$.

$$\therefore \text{Equation (ix) becomes}\frac{Q_n(x)}{P_n(x)}=\int_x^{\infty}\frac{dx}{(x^2-1)P_n^2}. \qquad \text{(x)}$$

Second Part: Replacing $n\,by\,0$ in Equation (x), we get

$$\frac{Q_0(x)}{P_0(x)} = \int_x^\infty \frac{dx}{(x^2-1)P_0^2}.$$

$$\therefore Q_0(x) = \int_x^\infty \frac{dx}{(x^2-1)P_0^2} \text{ as } P_0(x) = 1$$

$$= \frac{1}{2}\left[log\frac{x-1}{x+1}\right]_x^\infty = -\frac{1}{2}log\frac{x-1}{x+1} = \frac{1}{2}log\frac{x+1}{x-1}$$

$$\left\{\text{as } \lim_{x\to\infty} log\frac{x-1}{x+1} = \lim_{x\to\infty} log\frac{1-1/x}{1+1/x} = log 1 = 0\right].$$

Now, replacing n by 1 in Equation (x), we get $\frac{Q_1(x)}{P_1(x)} = \int_x^\infty \frac{dx}{(x^2-1)P_1^2}$.

$$\therefore Q_1(x) = P_1(x)\int_x^\infty \frac{dx}{(x^2-1)P_1^2} = x\int_x^\infty \frac{dx}{(x^2-1)x^2} \text{ as } P_1(x) = x$$

$$= x\int_x^\infty \left[\frac{1}{x^2-1} - \frac{1}{x^2}\right]dx = x\left[\frac{1}{2}log\frac{x-1}{x+1}\right]_x^\infty + x\left(\frac{1}{x}\right)_x^\infty = \frac{x}{2}Log\frac{x+1}{x-1} - 1.$$

Example 12.12.3. Prove that $Q_2(x) = \frac{1}{2}P_2(x)\log\frac{x+1}{x-1}x - \frac{3}{2}x.$

Solution: From recurrence relation,

$$(n+1)Q_{n-1}(x) = (2n+1)xQ_n(x) - nQ_{n-1}(x). \tag{i}$$

Putting $n = 1$, we get

$$2Q_2(x) = 3xQ_1(x) - Q_0(x)$$

$$= 3x\left[\frac{x}{2}log\frac{x+1}{x-1} - 1\right] - \frac{1}{2}log\frac{x+1}{x-1}$$

$$= \frac{3x^2-1}{2}log\frac{x+1}{x-1} - 3x = P_2(x)log\frac{x+1}{x-1} - 3x, \text{ as } P_2(x) = \frac{3x^2-1}{2}.$$

$$\therefore Q_2(x) = \frac{1}{2} P_2(x) \log \frac{x+1}{x-1} - \frac{3}{2} x.$$

12.13 Exercise

1. Show that (i) $P_n(-x) = (-1)^n P_n(x)$,

   (ii) $P'_{2n+1}(0) = \frac{(-1)^n (2n+1)!}{2^{2n} (n!)^2}$

   (iii) $P'_n(-1) = \frac{(-1)^n n(n+1)!}{2}$.

2. Express the following terms in Legendre's polynomials:

   (a) (i) $x^3 + 2x^2 - x - 3$ (ii) $x^4 + 3x^3 - x^2 + 5x - 2$

   (iii) $4x^3 + 6x^2 + 7x + 2$ (iv) $x^4 + 2x^3 - 2x^2 - x - 3$.

   (b) Expand $f(x) = x^2$ in a series of the form $\Sigma c_r P_r(x)$.

3. Show that $\int_{-1}^{1} (1 - x^2)[P'_n(x)]^2 \, dx = \frac{2n(n+1)}{2n+1}$.

4. Prove that the recurrence formula $Q'_{n+1}(x) - Q'_{n-1}(x) = (2n+1) Q_n(x)$.

5. Use Rodrigue's formula to derive an orthogonal property for $P_n(x)$, and show that

$$\int_{-1}^{1} P_n^2(x) dx = \frac{2}{2n+1}.$$

6. Define Legendre's function of the first and second kind, and obtain $P_4(x)$. Obtain the formula

$$(2n+1) P_n(x) = P'_{n+1}(x) - P'_{n-1}(x),$$

   and deduce that $P'_{n+1}(x) + P'_n(x) = P_0 + 3P_1 + \cdots . + (2n+1) P_n$.

7. Prove that $P_0^2 + 3P_1^2 + \cdots . + (2n+1) P_n^2 = (n+1)[P_n P'_{n+1} - P_{n+1} P'_n] = (n+1)^2 \{P_n(x)\}^2 + (1 - x^2)\{P'_n(x)\}^2$.

8. Evaluate (i) $\int_0^1 P_n(x) dx$, *when n is odd* and (ii) $\int_0^1 P_n(x) dx$, *when n is even.*

9. Prove that (i) $\int_{-1}^{1} P_n(x) dx = 2, \text{ if } n = 0$

   (ii) $\int_0^1 P_n(x) dx = 0, \text{ if } n \geq 1$.

10. (i) If $x > 1$, *show that* $P_n(x) < P_{n+1}(x)$, (ii) $|P_n(x)| \geq 1$, *if* $|x| \geq 1$

$$\int_{-1}^{1} x^2 P_n^2(x)dx = \frac{1}{8(2n-1)} + \frac{1}{4(2n+1)} + \frac{1}{8(2n+3)}.$$

11. Prove that $xP_n'(x) = nP_n + (2n-3)P_{n-2} + (2n-7)P_{n-4} + \cdots$.
and hence or otherwise, show that

(a) $\int_{-1}^{1} xP_n P_n' dx = \dfrac{2n}{(2n+1)}$

(b) $$\int_{-1}^{1} xP_n P_m' dx = \begin{cases} 0, \text{ when } n \neq m \\ \dfrac{2n}{(2n+1)}, \text{ when } n = m \\ 2, \text{ when } n = m-2, n \neq m-4, n \neq m-6, .. \end{cases}.$$

12. Show that

$$\int_{-1}^{1} x^m P_n(x)dx = \begin{cases} 0, \text{if } m < n \\ \dfrac{m!\Gamma\left(\dfrac{m}{2} - \dfrac{n}{2} + \dfrac{1}{2}\right)}{2^n (m-n)!\Gamma\left(\dfrac{m}{2} + \dfrac{n}{2} + \dfrac{3}{2}\right)} \text{ if } (m-n)(\geq 0) \text{ is even} \\ 0, \text{ if } (m-n)\ (>0) \text{ is odd} \end{cases}.$$

13. If n is a positive integer, show that $\int_{-1}^{1} P_n(x)(1-2xz+z^2)^{-\frac{1}{2}} dx = \dfrac{2z^n}{2n+1}$, and hence, using Rodrigue's formula, deduce that

$$\int_{-1}^{1} (1-x^2)^n (1-2xz+z^2)^{-n-\frac{1}{2}} dx = \frac{2^{2n+1}(n!)^2}{(2n+1)!}.$$

14. Show that $\sum_{n=0}^{\infty} P_{n+m}^{m}(x)t^{n} = \dfrac{(2m)!(1-x^{2})^{m+1/2}}{2^{m} m!(1-2xt+t^{2})^{m+1/2}}$.

15. Prove that $P_n^m(x) = \dfrac{(n+m)!(1-x^2)^{m/2}}{(n-m)!2^m m!} {}_2F_1(m-n, m+n+1;$

$m+1; \dfrac{1-x}{2})$.

13

Hermite Polynomials

Abstract

Hermite polynomials are a sequence of orthogonal polynomials that arise in various fields such as probability theory, physics, and combinatorics. They are solutions to a specific second-order differential equation and are closely connected to normal distributions.

This chapter explores their generating functions, recurrence relations, and orthogonality, as well as their applications in solving differential equations and relationships with other special functions.

***Keywords*:** Bilinear generating relation, orthogonality, Hermite equation, recurrence relation

13.1 Introduction

Hermite polynomials were defined by Pierre-Simon Laplace in 1810, though in scarcely recognizable form, and studied in detail by Pafnuty Chebyshev in 1859. Chebyshev's work was overlooked, and they were named after Charles Hermite, who wroked on the polynomials in 1864, describing them as new. They were consequently not new,although Hermite was the first to define the multidimensional polynomials.

Hermite polynomials are used to solve wave equations in quantum mechanics, creation of wavelets and Gaussian Quadrature, and in physics, it is used in heat equation. In this chapter we study its properties of orthogonality, generating function and recurrence relations.

13.2 Hermite Equation and Its Solution

A Hermite equation is

$$\frac{d^2y}{dx^2} - 2x\frac{dy}{dx} + 2ny = 0 \tag{13.1}$$

Dipankar De. Ordinary Differential Equations and Special Functions, (497–524) © 2026 Scrivener Publishing LLC

where n is a constant. By the method of Frobenius, we solve it.

Let the series solution of Equation (i) be

$$y = \sum_{m=0}^{\infty} c_m x^{k+m},\ c_m \neq 0. \tag{13.2}$$

Differentiating Equation (ii) and then putting the values of $\frac{d^2y}{dx^2}, \frac{dy}{dx}$ *and* y *in* Equation (i), we get

$$\sum_{m=0}^{\infty} c_m (k+m)(k+m-1)x^{k+m-2} - 2x\sum_{m=0}^{\infty} c_m (k+m)x^{k+m-1} + 2n\sum_{m=0}^{\infty} c_m x^{k+m} = 0$$

or, $$\sum_{m=0}^{\infty} c_m (k+m)(k+m-1)x^{k+m-2} - 2\sum_{m=0}^{\infty} c_m (k+m-n)x^{k+m} = 0. \tag{13.3}$$

To get indicial equation, equating coefficient of smallest power of, ie., x^{k-2} in Equation (13.3) to zero and obtain,

$$c_0(k)(k-1) = 0, \text{ since } c_0 \neq 0. \tag{13.4}$$

$\therefore k(k-1) = 0$ or $k = 0$ *or* $k = 1$.

Roots are distinct and differ by an integer.

The next smallest power is x^{k-1} and its coefficient is $c_1(k+1)(k)$.

Comparing it with the right side,

$$c_1(k+1)(k) = 0 \tag{13.5}$$

When we take $k = 0$ from an indicial equation, it implies that c_1 is indeterminate. Therefore, c_0, c_1 may be taken as arbitrary constants.

Now, equating the coefficient of $x^{k+m-2} = 0$, from Equation (13.3), we get

$$c_m (k+m)(k+m-1) - 2c_{m-2}(k+m-2-n) = 0.$$

$$\therefore c_m = \frac{2(k+m-2-n)}{(k+m)(k+m-1)} c_{m-2} \tag{13.6}$$

Putting

$$k = 0, c_m = \frac{2(m-2-n)}{(m)(m-1)} c_{m-2}. \tag{13.7}$$

For $m = 2, 4, 6, ..2m$ in Equation (13.7),

$$c_2 = \frac{-2n}{2.1} c_0 = -\frac{2n}{2!} c_0 = \frac{(-1)^1 2^1 .n}{2!} c_0$$

$$c_4 = \frac{2(2-n)}{4.3.} c_2 = \frac{2(2-n)}{4.3.} \times \frac{-2n}{2.1} c_0 = \frac{(-1)^2 2^2 n(n-2)}{4!} c_0$$

$$c_{2m} = \frac{(-1)^m 2^m n(n-2)\ldots.(n-2m+2)}{(2m)!} c_0.$$

Next, for $m = 3, 5, 7, ..2m+1$ in Equation (13.7),

$$c_3 = \frac{2(1-n)}{3.2.} c_1 = \frac{2(1-n)}{3!} c_1 = \frac{(-1)^1 2^1 (n-1)}{3!} c_1$$

$$c_5 = \frac{2(3-n)}{5.4} c_3 = \frac{2(3-n)}{5.4} \frac{(-1)^1 2^1 (n-1)}{3!} c_1 = \frac{(-1)^2 2^2 (n-3)}{} \frac{(n-1)}{5!} c_1$$

and, $$c_{2m+1} = \frac{(-1)^m 2^m n(n-2)\ldots.(n-2m+1)}{(2m+1)!} c_1.$$

Putting all the values of $c's$ in Equation (13.2) with $k=0$, *we get*

$$y=(c_0+c_2x^2+c_4x^4+....)+(c_1x+c_3x^3+c_5x^5...)$$
$$=c_0\left((1-\frac{2^1.n}{2!}x^2+\frac{2^2n(n-2)}{4!}x^4-\cdots+\frac{(-1)^m2^mn(n-2)...(n-2m+2)}{(2m)!}x^{2m}+\cdots.\right)$$
$$+c_1(x-\frac{2^1(n-1)}{3!}x^3+\frac{2^2(n-3)(n-1)}{5!}x^5$$
$$-\ldots+\frac{(-1)^m2^mn(n-2)....(n-2m+1)}{(2m+1)!}x^{2m+1}+.....) \tag{13.8}$$

or,
$$y=c_0u+c_1v. \tag{13.9}$$

Since $\frac{u}{v}$ is not a constant, u *and* v form a fundamental set. Hence, Equations (13.8) or (13.9) is the general solution of Equation (13.1) with c_0 *and* c_1 as arbitrary constants.

In practice, we require a solution of Equation (13.1) such that it is finite for all finite values of x, and as $x\to\infty, \exp\left(\frac{1}{2x^2}\right)\to 0.$

The solution Equation (13.8) in ascending power of x does not satisfy the condition $\exp\left(\frac{1}{2x^2}\right)\to 0\, as\, x\to\infty$. However, this requirement is easily seen to be satisfied provided the series is terminated.

Replacing m *by* $m+2$ *in* Equation (13.7), we have

$$c_{m+2}==\frac{2(m-n)}{(m+1)(m+2)}c_m. \tag{13.10}$$

Let n be a non-negative integer. Then, Equation (13.10) shows that c_{m+2} and the subsequent coefficients in Equation (13.2) will vanish, and so, the corresponding series terminates. Now, assume n to be a non-negative integer.

Rewriting Equation (13.2) for $k=0$, *we* have

$$y=c_nx^n+c_{n-2}x^{n-2}+c_{n-4}x^{n-4}+\cdots. \tag{13.11}$$

From Equation (13.10),

$$c_m = -\frac{(m+1)(m+2)}{2(m-n)} c_{m+2}. \tag{13.12}$$

Putting $m = n-2, n-4, \ldots\ldots$ in Equation (13.12), we get

$$c_{n-2} = -\frac{n(n-1)}{2.2} c_n,\ c_{n-4} = -\frac{(n-2)(n-3)}{2.4} c_{n-2} = \frac{(n-2)(n-3)}{2.4}\frac{n(n-1)}{2.2} c_n,$$

and so on.

Putting all these values in Equation (13.11), we have

$$\begin{aligned} y &= c_n x^n + c_{n-2} x^{n-2} + c_{n-4} x^{n-4} + \ldots \\ &= c_n \left\{ x^n - \frac{n(n-1)}{2.2} c_n x^{n-2} + \frac{(n-2)(n-3)}{2.4}\frac{n(n-1)}{2.2} x^{n-4} - \ldots \right\} \\ &= a_n \left\{ x^n - \frac{n(n-1)}{2.2} x^{n-2} + \frac{n(n-1)(n-2)(n-3)}{2^2.2.4} x^{n-4} - \ldots \right. \\ &\quad \left. +(-1)^r \frac{n(n-1)(n-2)(n-3)\ldots(n-2r+1)}{2.4.6..2r \quad 2^r} x^{n-2r} + \cdots \right\} \end{aligned}$$

or, $$y = a_n \sum_{r=0}^{\left[\frac{n}{2}\right]} (-1)^r \frac{n(n-1)(n-2)(n-3)\ldots(n-2r+1)}{2.4.6..2r \quad 2^r} x^{n-2r}$$

$$= a_n \sum_{r=0}^{[n/2]} (-1)^r \frac{n!}{r!\, 2^{2r} (n-2r)!} x^{n-2r}, \text{ where } [n/2]$$

$$= \begin{cases} \dfrac{n}{2}, \text{if } n \text{ is even} \\ \dfrac{n-1}{2}, \text{if } n \text{ is odd} \end{cases}.$$

Taking $a_n = 2^n$, we get the standard solution of Equation (13.1), known as *Hermite polynomial of order n*, which is denoted by $H_n(x)$:

$$H_n(x)=\sum_{r=0}^{[n/2]}(-1)^r\frac{n!}{r!(n-2r)!}(2x)^{n-2r}, \text{ where } [n/2]$$

$$=\begin{cases}\dfrac{n}{2}, if\ n\ is\ even\\ \dfrac{n-1}{2}, if\ n\ is\ odd\end{cases}.$$

13.3 Generating Function for Hermite Polynomials

Theorem 13.3.1. Prove that $e^{2xt-t^2}=\sum_{n=0}^{\infty}\frac{t^n}{n!}H_n(x)$.

Proof: Using the well-known exponential function, we get

$$e^{2xt-t^2}=e^{2xt}.e^{-t^2}=\left(\sum_{n=0}^{\infty}\frac{(2xt)^n}{n!}\right)\left(\sum_{m=0}^{\infty}\frac{(-t^2)^m}{m!}\right)$$

$$=\sum_{n=0}^{\infty}\sum_{m=0}^{\infty}\frac{(-1)^m(2x)^n t^{2m+n}}{n!m!}. \tag{13.13}$$

Let $2m+n=r\ or, n=r-2m$.

For a fixed value of m, the coefficient of t^r is given by

$$\frac{(-1)^m(2x)^{r-2m}}{(r-2m)!m!} \tag{13.14}$$

Now, $n\geq 0\ implies\ that\ r-2m\geq 0\Rightarrow r\geq 2m\Rightarrow m\leq\frac{r}{2}$, which gives all values of m for which Equation (ii) is the coefficient of t^r.

If r is even, $m\leq\frac{r}{2}$ shows that m varies from $0\ to\ \frac{r}{2}$. Again, if r is odd, $m\leq\frac{r}{2}$ shows that m varies from $0\ to\ \frac{r-1}{2}$. Here, m is an integer. Combining the results, we see that m varies from $0\ to [r/2]$

where $[r/2]=\begin{cases}\dfrac{r}{2}, if\ n\ is\ even\\ \dfrac{r-1}{2}, if\ n\ is\ odd\end{cases}.$

$\therefore$ The total coefficient of t^r in the expansion of e^{2xt-t^2} is given by

$$\sum_{m=0}^{[n/2]} \frac{(-1)^m (2x)^{r-2m}}{(r-2m)!m!} = \frac{1}{n!}\sum_{m=0}^{[n/2]} \frac{(-1)^m n!(2x)^{r-2m}}{(r-2m)!m!} = \frac{1}{n!}H_n(x),$$

i.e, the coefficient of t^n in $\sum_{n=0}^{\infty} \frac{t^n}{n!} H_n(x)$ is also $\frac{1}{n!}H_n(x)$.

Theorem 13.3.2. Prove that $H_n(x) = (-1)^n e^{x^2} \frac{d^n}{dx^n} e^{-x^2}$.

Proof: Using generating function, we have

$$\sum_{n=0}^{\infty} \frac{t^n}{n!} H_n(x) = e^{2xt-t^2}. \tag{13.15}$$

Expanding the function on the right-hand side by Taylor's theorem, Equation (13.15) gives

$$\sum_{n=0}^{\infty} \frac{t^n}{n!} H_n(x) = \sum_{n=0}^{\infty} \left[\frac{\partial^n}{\partial t^n} e^{2xt-t^2}\right]_{t=0} \frac{t^n}{n!}. \tag{13.16}$$

$$\therefore H_n(x) = \left[\frac{\partial^n}{\partial t^n} e^{2xt-t^2}\right]_{t=0} = \left[\frac{\partial^n}{\partial t^n} e^{x^2-(x-t)^2}\right]_{t=0} \left[\text{using} \frac{\partial^n}{\partial t^n} f(x-t) = (-1)^n \frac{\partial^n}{\partial x^n} f(x-t)\right]$$

$$= e^{x^2}\left[\frac{\partial^n}{\partial t^n} e^{-(x-t)^2}\right]_{t=0} = e^{x^2\left[(-1)^n \frac{\partial^n}{\partial x^n} e^{-(x-t)^2}\right]_{t=0}} = e^{x^2\left[(-1)^n \frac{d^n}{dx^n} e^{-x^2}\right]} = (-1)^n e^{x^2} \frac{d^n}{dx^n} e^{-x^2},$$

which is known as *Rodrigue's formula for* $H_n(x)$.

This formula gives the following values of $H_n(x)$ *for* $n = 0,1,..6,7$.

$$H_0(x) = (-1)^0 e^{x^2} \frac{d^0}{dx^0} e^{-x^2} = e^{x^2}.e^{-x^2} = 1$$

$$H_1(x) = (-1)^1 e^{x^2} \frac{d^1}{dx^1} e^{-x^2} = -e^{x^2}.[-2xe^{-x^2}] = 2x$$

$$H_2(x) = (-1)^2 e^{x^2} \frac{d^2}{dx^2} e^{-x^2} = e^{x^2}\left[(-2)e^{-x^2} + (-2x)(-2x)e^{-x^2}\right]$$

$$= -2 + 4x^2 = 4x^2 - 2$$

$$H_3(x) = (-1)^3 e^{x^2} \frac{d^3}{dx^3} e^{-x^2} = 8x^3 - 12x.$$

Using the formula $H_n(x) = \sum_{r=0}^{[n/2]} (-1)^r \frac{n!}{r!(n-2r)!}(2x)^{n-2r}$, where $\left[n/2\right] = \begin{cases} \frac{n}{2}, if\ n\ is\ even \\ \frac{n-1}{2}, if\ n\ is\ odd \end{cases}$,

$$H_4(x) = \sum_{r=0}^{2} (-1)^r \frac{4!}{r!(4-2r)!}(2x)^{4-2r} = \frac{4!}{(4)!}(2x)^4 - \frac{4!}{(2)!}(2x)^2 + \frac{4!}{2!}$$

$$= 16x^4 - 48x^2 + 12$$

$$H_5(x) = 32x^5 - 160x^3 + 120x$$

$$H_6(x) = 64x^6 - 480x^3 + 720x^2 - 120$$

$$H_7(x) = 128x^7 - 1344x^5 + 3360x^3 - 1680x.$$

Example 13.3.1 Prove that $H_n(x) = 2^n\left[exp\left(-\frac{1}{4}\frac{d^2}{dx^2}\right)x^n\right]$

Solution:

$$\exp\left(-\frac{1}{4}\frac{d^2}{dx^2}\right)e^{2tx} = \sum_{n=0}^{\infty}\frac{1}{n!}\left(-\frac{1}{4}\frac{d^2}{dx^2}\right)^n e^{2tx} = \sum_{n=0}^{\infty}\frac{(-1)^n}{n!}\left(\frac{1}{2}\frac{d}{dx}\right)^{2n} e^{2tx}$$

$$= \sum_{n=0}^{\infty}\frac{(-1)^n}{n!}\frac{1}{2^{2n}}\frac{d^{2n}}{dx^{2n}}e^{2tx} = \sum_{n=0}^{\infty}\frac{(-1)^n}{n!}\frac{1}{2^{2n}}(2t)^{2n}e^{2tx}$$

$$= e^{2tx}\sum_{n=0}^{\infty}\frac{(-1)^n}{n!}(t)^{2n} = e^{2tx}\sum_{n=0}^{\infty}\frac{(-t^2)^n}{n!} = e^{2tx}.e^{-t^2}. \tag{i}$$

We know the generating function $H_n(x)$ from Equation (i).

$$\sum_{n=0}^{\infty}\frac{t^n}{n!}H_n(x) = e^{2tx-t^2}. \tag{ii}$$

Also,

$$e^{2tx} = \sum_{n=0}^{\infty}\frac{(2tx)^n}{n!} = \sum_{n=0}^{\infty}\frac{2^n x^n}{n!}t^n. \tag{iii}$$

Using Equations (ii) and (iii), Equation (iii) becomes

$$e^{2tx-t^2} = \sum_{n=0}^{\infty}\exp\left(-\frac{1}{4}\frac{d^2}{dx^2}\right)\frac{2^n x^n}{n!}t^n = \sum_{n=0}^{\infty}\frac{t^n}{n!}H_n(x). \tag{iv}$$

Equating the coefficient of t^n on both sides of Equation (iv), we get

$$\frac{H_n(x)}{n!} = \exp\left(-\frac{1}{4}\frac{d^2}{dx^2}\right)\frac{2^n x^n}{n!}.$$

$$\therefore H_n(x) = 2^n\left[\exp\left(-\frac{1}{4}\frac{d^2}{dx^2}\right)x^n\right].$$

Example 13.3.2. Prove that $H_{2n}(0)=(-1)^n\dfrac{(2n)!}{n!}$ *and* $H_{2n-1}(0)=0.$

Solution: By the generating function, we have

$$\textstyle\sum_{n=0}^{\infty}\dfrac{t^n}{n!}H_n(x)=e^{2tx-t^2}. \qquad \text{(i)}$$

Now, $\sum_{n=0}^{\infty}\dfrac{t^n}{n!}H_n(0)=e^{-t^2}=\sum_{n=0}^{\infty}\dfrac{(-t^2)^n}{n!}=\sum_{n=0}^{\infty}\dfrac{(-1)^n}{n!}t^{2n}$

Or,
$$\textstyle\sum_{n=0}^{\infty}\dfrac{1}{(2n)!}H_{2n}(0)t^{2n}=\sum_{n=0}^{\infty}\dfrac{(-1)^n}{n!}t^{2n} \qquad \text{(ii)}$$

Comparing the coefficients, we get

$$\frac{1}{(2n)!}H_{2n}\left(0\right)=\frac{(-1)^n}{n!}.$$

$$\therefore H_{2n}\left(0\right)=\frac{\left(-1\right)^n\left(2n\right)!}{n!}$$

Since the right-hand side of Equation (ii) contains an even power of t, it means it has no coefficient of odd power of t^{2n+1}.

$$\therefore H_{2n+1}\left(0\right)=0.$$

13.4 Recurrence Relations

(i) $H_{n+1}(x)=2xH_n\left(x\right)-2nH_{n-1}(x).$

(ii) $H_n(x)=2xH_{n-1}\left(x\right)-H'_{n-1}(x).$

(iii) $H''_n\ (x)-2xH'_n\ \left(x\right)+2nH_n(x)=0.$

Solution: (i) We know that the generating function is

$$\sum_{n=0}^{\infty} \frac{t^n}{n!} H_n(x) = e^{2tx-t^2}. \tag{13.17}$$

Differentiating both sides of Equation (i) with respect to x, we get

$$\sum_{n=0}^{\infty} \frac{t^n}{n!} H_n'(x) = (2t)e^{2tx-t^2} = 2t\sum_{n=0}^{\infty} \frac{t^n}{n!} H_n(x)$$

or,
$$\sum_{n=0}^{\infty} \frac{t^n}{n!} H_n'(x) = 2\sum_{n=0}^{\infty} \frac{t^{n+1}}{n!} H_n(x). \tag{13.18}$$

Equating the coefficient of t^n : $\frac{1}{n!} H_n'(x) = \frac{2}{(n-1)!} H_{n-1}(x)$.

$$\therefore H_n'(x) = 2nH_{n-1}(x) \tag{13.19}$$

for $n = 0, H_0'(x) = 0$.

Differentiating both sides of Equation (13.17) with respect to t, we have

$$(2x-2t)e^{2tx-t^2} = \sum_{n=0}^{\infty} \frac{nt^{n-1}}{n!} H_n(x)$$

or,
$$2(x-t)\sum_{n=0}^{\infty} \frac{t^n}{n!} H_n(x) = 0 + \sum_{n=0}^{\infty} \frac{nt^{n-1}}{n!} H_n(x)$$

or,
$$2x\sum_{n=0}^{\infty} \frac{t^n}{n!} H_n(x) - 2\sum_{n=0}^{\infty} \frac{t^{n+1}}{n!} H_n(x) = \sum_{n=1}^{\infty} \frac{nt^{n-1}}{n!} H_n(x). \tag{13.20}$$

Equating the coefficient of t^n and $n=0$, we get $2xH_0(x)=H_1(x)$.
Equating the coefficient of t^n for $n\geq 1$:

$$2x\frac{1}{n!}H_n(x)-2\frac{1}{(n-1)!}H_{n-1}(x)=\frac{1}{n!}H_{n+1}(x). \tag{13.21}$$

Multiplying both sides by $n!$

$$2xH_n(x)-2nH_{n-1}(x)=H_{n+1}(x).$$

This relation is a pure recurrence relation that holds for $n\geq 1$.
(ii) Differentiate both sides with respect to x:

$$\sum_{n=0}^{\infty}\frac{t^n}{n!}H_n'(x)=(2t)e^{2tx-t^2}=2t\sum_{n=0}^{\infty}\frac{t^n}{n!}H_n(x). \tag{13.22}$$

Since $H_0'(x)=0$, we write $2\sum_{n=0}^{\infty}\frac{t^{n+1}}{n!}H_n(x)=\sum_{n=1}^{\infty}\frac{t^n}{n!}H_n'(x)$.
For equating the coefficient of t^n on both sides,

$$2\sum_{n=1}^{\infty}\frac{t^n}{(n-1)!}H_{n-1}(x)=\sum_{n=1}^{\infty}\frac{t^n}{n!}H_n'(x).$$

$$\therefore H_n'(x)=2nH_{n-1}(x). \tag{13.23}$$

Multiplying Equation (13.23) by t *and* Equation (13.22) *by* $(x-t)$ and subtracting, we get

$$\sum_{n=1}^{\infty}\frac{t^n}{(n-1)!}H_n(x)=(x-t)\sum_{n=1}^{\infty}\frac{t^n}{n!}H_n'(x),$$

$$\text{i.e.,}\sum_{n=1}^{\infty}\frac{t^n}{(n-1)!}H_n(x)+\sum_{n=1}^{\infty}\frac{t^{n+1}}{n!}H_n'(x)=x\sum_{n=1}^{\infty}\frac{t^n}{n!}H_n'(x)$$

$$\sum_{n=1}^{\infty} \frac{t^n}{(n-1)!} H_n(x) + \sum_{n=1}^{\infty} \frac{t^n}{(n-1)!} H'_{n-1}(x) = x \sum_{n=1}^{\infty} \frac{t^n}{n!} H'_n(x).$$

Comparing the coefficients of t^n on both sides:

$$\frac{1}{(n-1)!} H_n(x) + \frac{1}{n!} H'_{n-1}(x) = \frac{x}{n!} H'_n(x).$$

$$\therefore xH'_n(x) = nH_n(x) + nH'_{n-1}(x). \quad (13.24)$$

Combining Equations (vii) and (viii), we get

$$2xH_{n-1}(x) = H_n(x) + H'_{n-1}(x). \quad (13.25)$$

(iii) $H''_n(x) - 2xH'_n(x) + 2nH_n(x)$.

Since $H_n(x)$ is a solution of the Hermite differential equation $\frac{d^2y}{dx^2} - 2x \frac{dy}{dx} + 2ny = 0$,
therefore, it satisfies the equation

$$H''_n(x) - 2xH'_n(x) + 2nH_n(x) = 0.$$

13.4.1 Some Results

(i) $H_n(x) = (2x)^n \, {}_2F_0 \left[\begin{matrix} -\frac{n}{2}, -\frac{n}{2} + \frac{1}{2}; & -\frac{1}{x^2} \\ -; & \end{matrix} \right]$

Proof: We know that $H_n(x) = \sum_{m=0}^{[n/2]} \frac{(-1)^m n!(2x)^{n-2m}}{(n-2m)!m!}$

$$H_n(x) = (2x)^n \sum_{m=0}^{[n/2]} \frac{(-1)^m (-n)_{2m} (x)^{-2m}}{2^{2m} m!}$$

$$= (2x)^n \sum_{m=0}^{[n/2]} \frac{(-1)^m (-n/2)_m \left(-\frac{n}{2}+\frac{1}{2}\right)_m (x)^{-2m}}{m!}$$

$$\therefore H_n(x) = (2x)^n {}_2F_0\left[\begin{matrix} -\frac{n}{2}, -\frac{n}{2}+\frac{1}{2}; \\ -; \end{matrix} -\frac{1}{x^2}\right].$$

Here, ${}_2F_0$ is simply a terminating series, which is naturally convergent.

(ii) Integral representation for a Hermite polynomial:

We know, that $e^{2tx-t^2} = \sum_{n=0}^{\infty} \frac{t^n}{n!} H_n(x)$.

By Maclaurin's theorem, we have

$$f(t) = e^{2tx-t^2} = f(0) + tf'(0) + \cdots. + \frac{t^n}{n!} f^n(0) + \cdots = \sum_{n=0}^{\infty} f^n(0) \frac{t^n}{n!}$$

where $f^n(0)$ denotes the value of the *nth* partial derivative of e^{2tx-t^2} with respect to t *at* $t=0$,

$$\text{i.e., } H_n(x) = f^n(0) = \left[\frac{\partial^n}{\partial x^n} e^{2tx-t^2}\right]_{t=0}.$$

But by complex variable theory,

$$f^n(0) = \frac{n!}{2\pi i} \int_C u^{-n-1} f(u)du,$$

where C is the contour encircling the origin in the positive direction.

$$\therefore H_n(x) = \frac{n!}{2\pi i} \int_C u^{-n-1} f(u)du = \frac{n!}{2\pi i} \int_C u^{-n-1} e^{2ux-u^2} du.$$

If we choose the contour $u = e^{i\theta}$, the contour integral changes into a real integral, and we obtain

$$H_n(x) = \frac{n!}{\pi}\int_0^{\pi} \exp(2xcos\theta - cos2\theta) \times \cos(2xsin\theta - sin2\theta - n\theta)d\theta.$$

13.5 Orthogonal Property

Theorem 13.5.1. Prove that $\int_{-\infty}^{\infty} e^{-x^2} H_n(x)H_m(x)dx = \begin{cases} 0 \; if \; m \neq n \\ \sqrt{\pi}\, 2^n n! \end{cases}$.

Proof: The generating function of a Hermite polynomial is

$$\sum_{n=0}^{\infty} \frac{t^n}{n!} H_n(x) = e^{2tx-t^2} \text{ and } \sum_{m=0}^{\infty} \frac{s^m}{n!} H_m(x) = e^{2sx-s^2}.$$

Multiplying their corresponding sides, gives

$$\sum_{n=0}^{\infty} \sum_{m=0}^{\infty} \frac{H_n(x)H_m(x)}{n!m!} t^n s^m = e^{2tx-t^2} . e^{2sx-s^2} = e^{2tx-t^2+2sx-s^2}.$$

Multiplying both sides by e^{-x^2} and integrating with respect to x from $-\infty$ *to* ∞, we get

$$\begin{aligned}
&\sum_{n=0}^{\infty} \sum_{m=0}^{\infty} \left[\int_{-\infty}^{\infty} e^{-x^2} H_n(x)H_m(x)dx \right] \frac{t^n s^m}{n!m!} \\
&= \int_{-\infty}^{\infty} e^{-x^2} e^{2tx-t^2+2sx-s^2} dx \\
&= \int_{-\infty}^{\infty} e^{-x^2+2x(t+s)-\left(t^2+s^2\right)} dx \\
&= \int_{-\infty}^{\infty} e^{-x^2+2x(t+s)-(t+s)^2} e^{(t+s)^2-\left(t^2+s^2\right)} dx \\
&= e^{2ts} \int_{-\infty}^{\infty} e^{-\left[x-(t+s)\right]^2} dx = e^{2ts} \int_{-\infty}^{\infty} e^{-y^2} dy \\
&\quad [\text{where} = x-(t+s), dx = dy] \\
&= e^{2ts}\sqrt{\pi} \\
&= \sqrt{\pi} \sum_{n=0}^{\infty} \frac{(2ts)^n}{n!} = \sum_{n=0}^{\infty} \frac{2^n\sqrt{\pi}}{n!} t^n.s^n. \qquad (13.26)
\end{aligned}$$

We observe that the coefficients of t^n *and* s^n on the right-hand side of Equation (13.25) are always equal. Thus, when $m \neq n$ implies that there are no terms.

So, $\frac{1}{n!m!}\int_{-\infty}^{\infty} e^{-x^2} H_n(x)H_m(x)dx = 0,$

$$\text{i.e.,} \int_{-\infty}^{\infty} e^{-x^2} H_n(x)H_m(x)dx = 0 \; when \; m \neq n \tag{13.27}$$

Equating the coefficients of $t^n s^n$ on both sides of Equation (13.25), we have

$$\frac{1}{n!n!}\int_{-\infty}^{\infty} e^{-x^2} [H_n(x)]^2 dx = \frac{2^n\sqrt{\pi}}{n!}.$$

$$\therefore \int_{-\infty}^{\infty} e^{-x^2} [H_n(x)]^2 dx = 2^n n! \sqrt{\pi}. \tag{13.28}$$

Hence, by combining Equations (13.26) and (13.27), we get $\int_{-\infty}^{\infty} e^{-x^2} H_n(x)$ $H_m(x)dx = \begin{cases} 0 \; if \; m \neq n \\ \sqrt{\pi}\, 2^n n! \end{cases}.$

13.6 Expansion of Polynomials

Any polynomial, say $P(x)$, *of degree* n can be expanded as the linear combination of Hermite polynomials. Let it form

$$P(x) = \sum_{m=0}^{n} c_m H_m(x). \tag{13.29}$$

Multiplying both sides by $e^{-x^2} H_m(x)$, then integrating from $-\infty$ *to* ∞ w.r.t. x [using the property of orthogonality of Hermite polynomial] gives

$$c_m \int_{-\infty}^{\infty} e^{-x^2} [H_m(x)]^2 dx = \int_{-\infty}^{\infty} P(x)e^{-x^2} H_m(x)dx$$

or,
$$2^n m! \sqrt{\pi}\, c_m = \int_{-\infty}^{\infty} P(x)e^{-x^2} H_m(x)dx.$$

$$\therefore c_m = \frac{1}{2^n m! \sqrt{\pi}} \int_{-\infty}^{\infty} P(x) e^{-x^2} H_m(x) dx. \tag{13.30}$$

(a) We obtain, in particular, the expansion of x^n in terms of Hermite polynomials.

We know that $e^{2tx-t^2} = \sum_{n=0}^{\infty} \frac{t^n}{n!} H_n(x)$

or,

$$e^{2tx} = e^{t^2} \sum_{n=0}^{\infty} \frac{t^n}{n!} H_n(x)$$

or,

$$\sum_{n=0}^{\infty} \frac{(2x)^n t^n}{n!} = \sum_{m=0}^{\infty} \frac{t^{2m}}{m!} \sum_{n=0}^{\infty} \frac{t^n}{n!} H_n(x)$$

$$= \sum_{n=0}^{\infty} \sum_{m=0}^{\infty} \frac{t^{n+2m}}{m! n!} H_n(x) = \sum_{n=0}^{\infty} \sum_{m=0}^{\left[\frac{n}{2}\right]} \frac{t^n}{m!(n-2m)!} H_{n-2m}(x)$$

[replace n by $n-2m$]

$$\sum_{n=0}^{\infty} \frac{(2x)^n t^n}{n!} = \sum_{n=0}^{\infty} \sum_{m=0}^{\left[\frac{n}{2}\right]} \frac{t^n}{m!(n-2m)!} H_{n-2m}(x).$$

Comparing the coefficients of t^n, we get

$$\frac{(2x)^n}{n!} = \sum_{m=0}^{\left[\frac{n}{2}\right]} \frac{1}{m!(n-2m)!} H_{n-2m}(x).$$

$$\therefore x^n = \sum_{m=0}^{\left[\frac{n}{2}\right]} \frac{n!}{2^n m!(n-2m)!} H_{n-2m}(x). \tag{13.31}$$

(b) We can also expand Legendre polynomials in terms of Hermite polynomials.

Now, $$\sum_{n=0}^{\infty} P_n(x)t^n = \sum_{n=0}^{\infty} \sum_{m=0}^{\left[\frac{n}{2}\right]} \frac{(-1)^m (\frac{1}{2})_{n-m}}{m!(n-2m)!} t^n (2x)^{n-2m}$$

$$= \sum_{n=0}^{\infty} \sum_{m=0}^{\infty} \frac{(-1)^m (\frac{1}{2})_{n-m}}{m!(n)!} t^{n+2m} (2x)^n. \qquad (13.32)$$

From Equation (iii), $\frac{(2x)^n}{n!} = \sum_{s=0}^{\left[\frac{n}{2}\right]} \frac{1}{s!(n-2s)!} H_{n-2s}(x)$.

Hence, from Equation (iv), $\sum_{n=0}^{\infty} P_n(x)t^n = \sum_{n=0}^{\infty} \sum_{m=0}^{\infty} \sum_{s=0}^{\left[\frac{n}{2}\right]} \frac{1}{s!(n-2s)!} H_{n-2s}$

$(x) \frac{(-1)^m (\frac{1}{2})_{n-m}}{m!} t^{n+2m}$

$$or, \sum_{n=0}^{\infty} P_n(x)t^n = \sum_{n=0}^{\infty} \sum_{m=0}^{\infty} \sum_{s=0}^{\left[\frac{n}{2}\right]} \frac{1}{s!(n)!} H_n(x) \frac{(-1)^m \left(\frac{1}{2}\right)_{n+m+2s}}{m!} t^{n+2m+2s}$$

$$= \sum_{n=0}^{\infty} \sum_{m=0}^{\infty} \sum_{s=0}^{\left[\frac{n}{2}\right]} \frac{1}{s!(n)!} H_n(x) \frac{(-1)^{m-s} \left(\frac{1}{2}\right)_{n+m+s}}{(m-s)!} t^{n+2m}$$

$$= \sum_{n=0}^{\infty} \sum_{m=0}^{\infty} \sum_{s=0}^{\left[\frac{n}{2}\right]} H_n(x) \frac{(-1)^s (\frac{1}{2}+n+m)_s (-1)^m m! (\frac{1}{2})_{n+m}}{m!(m-s)! s! n!} t^{n+2m}$$

$$= \sum_{n=0}^{\infty} \sum_{m=0}^{\infty} \sum_{s=0}^{\left[\frac{n}{2}\right]} H_n(x) \frac{(-1)^s \left(\frac{1}{2}+n+m\right)_s (-1)^m m! \left(\frac{1}{2}\right)_{n+m} (-m)_s}{m!(-1)^s m! s! n!} t^{n+2m}$$

$$= \sum_{n=0}^{\infty} \sum_{m=0}^{\infty} H_n(x) \frac{{}_2F_0\left[-m, \left(\frac{1}{2}+n+m\right); -; 1\right] (-1)^m \left(\frac{1}{2}\right)_{n+m}}{m!n!} t^{n+2m}$$

$$= \sum_{n=0}^{\infty} \sum_{s=0}^{\left[\frac{n}{2}\right]} H_{n-2m}(x) \frac{{}_2F_0\left[-m, \left(\frac{1}{2}+n-m\right); -; 1\right] (-1)^m \left(\frac{1}{2}\right)_{n-m}}{m!(n-2m)!} t^n.$$

Thus, we get

$$P_n(x) = \sum_{m=0}^{\left[\frac{n}{2}\right]} \frac{{}_2F_0\left[-m,(\frac{1}{2}+n-m);-;1\right](-1)^m(\frac{1}{2})_{n-m}}{m!(n-2m)!} H_{n-2m}(x). \tag{13.33}$$

13.7 More Generating Functions

(a) $\sum_{n=0}^{\infty} \frac{1}{n!} H_{n+m}(x)t^n = e^{2tx-t^2} H_m(x-t)$

Solution: We have

$$\begin{aligned}\sum_{m=0}^{\infty}\sum_{n=0}^{\infty} \frac{1}{n!m!} H_{n+m}(x)t^n v^m &= \sum_{n=0}^{\infty}\sum_{m=0}^{\infty} \frac{1}{(n-m)!m!} H_n(x)t^{n-m}v^m \\ &= \sum_{n=0}^{\infty}(t+v)^n \frac{H_n(x)}{n!} \\ &= \exp\left[2x(t+v)-(t+v)^2\right] \\ &= e^{(2xt-t^2)}.e^{\left[2(x-t)v-v^2\right]} \\ &= e^{(2xt-t^2)}.\sum_{m=0}^{\infty} \frac{H_m(x-t)}{m!} v^m.\end{aligned}$$

Comparing the coefficient of v^m, we get

$$\sum_{n=0}^{\infty} \frac{1}{n!} H_{n+m}(x)t^n = e^{(2xt-t^2)} H_m(x-t). \tag{13.34}$$

(b) $\sum_{n=0}^{\infty} \frac{H_n(x)H_m(y)}{n!} t^n = (1-4t^2)^{-\frac{1}{2}} exp\left[\frac{-4t^2(y-2xt)^2}{1-4t^2}\right]$

Solution: Using Equation (i), we get,

$$\sum_{n=0}^{\infty}\frac{1}{n!}H_n(x)H_n(y)t^n=\sum_{n=0}^{\infty}\sum_{m=0}^{\left[\frac{n}{2}\right]}\frac{(-1)^m(2x)^{n-2m}}{m!(n-m)!}H_n(y)t^n$$

$$=\sum_{n=0}^{\infty}\sum_{m=0}^{\infty}\frac{(-1)^m(2x)^n}{m!(n)!}H_{n+2m}(y)t^{n+2m}$$

$$=\sum_{n=0}^{\infty}\sum_{m=0}^{\infty}\frac{(-1)^m(2xt)^n}{m!(n)!}H_{n+2m}(y)t^{2m}$$

$$=\sum_{n=0}^{\infty}\frac{\exp(4xyt-4x^2t^2)H_{2m}(y-2xt)(-1)^m}{\mathrm{m}!}t^{2m}$$

[applying Equation (i)].

Now, substituting,

$$H_{2m}(y-2xt)=\sum_{s=0}^{\infty}\frac{(-1)^s(2m)!(2y-4xt)^{2m-2s}}{s!(2m-2s)!}$$

and $(2m)!=(1)_{2m}=2^{2m}m!(\frac{1}{2})_m$.

We get $\sum_{n=0}^{\infty}\frac{1}{n!}H_n(x)H_n(y)t^n$

$$=\exp(4xyt-4x^2t^2)\sum_{m=0}^{\infty}\sum_{s=0}^{\infty}\frac{2^{2m}(\frac{1}{2})_m(2y-4xt)^{2m-2s}(-1)^{m+s}}{\mathrm{s}!(2\mathrm{m}-2\mathrm{s})!}t^{2m}$$

$$=\exp(4xyt-4x^2t^2)\sum_{m=0}^{\infty}\sum_{s=0}^{\infty}\frac{(-1)^m2^{2m+2s}(\frac{1}{2})_{m+s}\ (2y-4xt)^{2m}}{\mathrm{s}!(2\mathrm{m})!}t^{2m+2s}$$

$$=\exp(4xyt-4x^2t^2)\sum_{m=0}^{\infty}\sum_{s=0}^{\infty}\frac{(-1)^m2^{2m+2s}(\frac{1}{2})_{m+s}\ (2y-4xt)^{2m}}{\mathrm{s}!(2\mathrm{m})!}t^{2m+2s}$$

$$= \exp(4xyt - 4x^2t^2)\sum_{m=0}^{\infty}\sum_{s=0}^{\infty} \frac{(-1)^m 2^{2s}(\frac{1}{2}+m)_s\ (2y-4xt)^{2m}}{s!(m)!} t^{2m}$$

$$= \exp(4xyt - 4x^2t^2)\sum_{m=0}^{\infty} \frac{(-1)^m (2y-4xt)^{2m}}{m!(1-4t^2)^{\frac{1}{2}+m}} t^{2m}.$$

Hence, $\sum_{n=0}^{\infty} \frac{1}{n!} H_n(x)H_n(y)t^n == (1-4t^2)^{-1/2} exp\left[y^2 - \frac{(y-2xt)^2}{1-4t^2}\right],$

which is the *bilinear generating relation.*

13.8 Examples

Example 13.8.1. Prove that $\frac{d^m}{dx^m} H_n(x) = \frac{2^n n!}{(n-m)!} H_{n-m}(x), if\ m < n.$

Solution: We know that

$$\sum_{n=0}^{\infty} \frac{t^n}{n!} H_n(x) = e^{2tx-t^2}. \tag{i}$$

Differentiating both sides of Equation (i) with respect to $x\ m\ times,$ we get

$$\sum_{n=0}^{\infty} \frac{t^n}{n!} \frac{d^m}{dx^m} H_n(x) = \frac{d^m}{dx^m} e^{2tx-t^2} = (2t)^m e^{2tx-t^2} = 2^m t^m \sum_{n=0}^{\infty} \frac{t^n}{n!} H_n(x)$$

[by Equation (i)]

$$\sum_{n=0}^{\infty} \frac{t^n}{n!} \frac{d^m}{dx^m} H_n(x) = 2^m \sum_{n=0}^{\infty} \frac{t^{n+m}}{n!} H_n(x). \tag{ii}$$

Equating the coefficient of t^n of Equation (ii) on both sides, we get

$$\frac{1}{n!} \frac{d^m}{dx^m} H_n(x) = 2^m \frac{1}{(n-m)!} H_{n-m}(x)\ [\text{replacing}\ n\ by\ n-m].$$

$$\therefore \frac{d^m}{dx^m} H_n(x) = 2^m \frac{n!}{(n-m)!} H_{n-m}(x).$$

Example 13.8.2. $\int_{-\infty}^{\infty} x^2 e^{-x^2} [H_n(x)]^2 dx = 2^n (n+\frac{1}{2})\sqrt{\pi}n!$

Solution: We know that the recurrence relation of a Hermite polynomial is

$$H_{n+1}(x) = 2xH_n(x) - 2nH_{n-1}(x)$$

or, $$xH_n(x) = nH_{n-1}(x) + 1/2H_{n+1}(x) \tag{i}$$

or, $$x^2 H_n(x) = nxH_{n-1}(x) + \frac{1}{2} xH_{n+1}(x). \tag{ii}$$

Replacing $n \, by \, n-1 \, and \, n+1$ successively from Equation (i), we get

$$xH_{n-1}(x) = (n-1)H_{n-2}(x) + 1/2H_n(x) \tag{iii}$$

$$xH_{n+1}(x) = (n+1)H_n(x) + \frac{1}{2H_{n+2}(x)} \tag{iv}$$

Using Equations (iii) and (iv) in Equation (ii), we get

$$\begin{aligned} x^2 H_n(x) &= nxH_{n-1}(x) + \frac{1}{2} xH_{n+1}(x) \\ &= n\left[(n-1)H_{n-2}(x) + 1/2H_n(x)\right] \\ &+ \frac{1}{2}\left[(n-1)H_n(x) + 1/2H_{n+2}(x)\right] \\ &= n(n-1)H_{n-2}(x) + + \left(n + \frac{1}{2}\right) H_n(x) + \frac{1}{4} H_{n+2}(x). \end{aligned} \tag{v}$$

Multiplying both sides of Equation (v) by $e^{-x^2}H_n(x)$ and then integrating with respect to x from $-\infty$ *to* ∞, we get

$$\int_{-\infty}^{\infty} x^2 e^{-x^2} H_n(x)H_n(x)dx = n(n-1)\int_{-\infty}^{\infty} e^{-x^2} H_n(x)H_{n-2}(x)dx +$$

$$+(n+\frac{1}{2})\int_{-\infty}^{\infty} e^{-x^2} H_n(x)H_n(x)dx + \frac{1}{4}\int_{-\infty}^{\infty} e^{-x^2} H_n(x)H_{n+2}(x)dx$$

$$= 0 + 0 + (n+\frac{1}{2})\int_{-\infty}^{\infty} e^{-x^2} H_n(x)H_n(x)dx = (n+\frac{1}{2})\sqrt{\pi}\,2^n n!$$

$$[\text{since} .\int_{-\infty}^{\infty} x^2 e^{-x^2} H_m(x)H_n(x)dx = \begin{cases} 0, if\ m \neq n \\ \sqrt{\pi}\,2^n n! if\ m = n \end{cases}].$$

Example 13.8.3. Show that $P_n(x) = \frac{2}{n!\sqrt{\pi}}\int_0^{\infty} t^n e^{-t^2} H_n(xt).$

Solution: The Hermite polynomial is

$$H_n(xt) = \sum_{r=0}^{[n/2]} \frac{(-1)^r n!(2xt)^{n-2r}}{(n-2r)!r!}$$

$$= \frac{2}{n!\sqrt{\pi}}\int_0^{\infty} t^n e^{-t^2} H_n(xt)dt$$

$$= \frac{2}{n!\sqrt{\pi}}\int_0^{\infty} t^n e^{-t^2} \left[\sum_{r=0}^{[n/2]} \frac{(-1)^r n!(2xt)^{n-2r}}{(n-2r)!r!}\right]dt$$

$$= \sum_{r=0}^{[n/2]} \frac{(-1)^r (2x)^{n-2r}}{(n-2r)!r!}\frac{2}{\sqrt{\pi}}\int_0^{\infty} t^{2n-2r} e^{-t^2} dt$$

$$= \sum_{r=0}^{\left[\frac{n}{2}\right]} \frac{(-1)^r (2)^{n-2r+1} x^{n-2r}}{(n-2r)!r!}\frac{1}{\sqrt{\pi}}\frac{1}{2}\Gamma(n-r+\frac{1}{2})$$

$$[\therefore \int_0^{\infty} t^{2n-2r} e^{-t^2} dt = \int_0^{\infty} e^{-t^2} t^{2()\left(n-r+\frac{1}{2}\right)-1} dt$$

$$= \frac{1}{2}\Gamma\left(n-r+\frac{1}{2}\right) as\ \Gamma(x) = \int_0^{\infty} t^{2x-1} e^{-t^2} dt]$$

$$= \sum_{r=0}^{\left[\frac{n}{2}\right]} \frac{(-1)^r (2)^{n-2r} x^{n-2r}}{(n-2r)!r!} \frac{1}{\sqrt{\pi}} \frac{(2n-2r)!\sqrt{\pi}}{2^{2n-2r}(n-r)!}$$

$$\left[\because \Gamma\left(n+\frac{1}{2}\right) = \frac{(2n)!\sqrt{\pi}}{2^{2n} n!}, if\ n\ is\ a\ positive\ integer\right]$$

$$= \sum_{r=0}^{\left[\frac{n}{2}\right]} \frac{(-1)^r}{(n-2r)!r!} \frac{(2n-2r)!}{2^n (n-r)!} x^{n-2r}$$

$$= P_n(x). \quad \text{(i)}$$

Example 13.8.4. If $\varphi_n(x) = e^{-\frac{x^2}{2}} H_n(x)$, where $H_n(x)$ is a Hermite polynomial of degree n,

then, $$\int_{-\infty}^{\infty} \varphi_n'(x)\varphi_m(x)dx = \begin{cases} 0\ if\ m \neq n, if\ m \neq n \pm 1 \\ 2^{n-1} n!\sqrt{\pi}, if\ m = n-1 \\ -2^n (n+1)!\sqrt{\pi}, if\ m = n+1 \end{cases}.$$

Solution: Here,

$$\varphi_n(x) = e^{-\frac{x^2}{2}} H_n(x). \quad \text{(i)}$$

$$\therefore \varphi_n'(x) = -xe^{-\frac{x^2}{2}} H_n(x) + e^{-\frac{x^2}{2}} H_n'(x). \quad \text{(ii)}$$

From recurrence relations,

$$xH_n(x) = nH_{n-1}(x) + 1/2H_{n+1}(x). \quad \text{(iii)}$$

and, $$H_n'(x) = 2nH_{n-1}(x). \quad \text{(iv)}$$

Using Equations (iii) and (iv), Equation (ii) becomes

$$\varphi_n'(x) = -xe^{-\frac{x^2}{2}} H_n(x) + e^{-\frac{x^2}{2}} H_n'(x) = -e^{-\frac{x^2}{2}} \left[nH_{n-1}(x) + 1/2H_{n+1}(x) \right]$$

$$+e^{-\frac{x^2}{2}}.2nH_{n-1}(x)$$

$$= -e^{-\frac{x^2}{2}} \left[nH_{n-1}(x) + \frac{1}{2}H_{n+1}(x) - 2nH_{n-1}(x) \right]$$

$$= e^{-\frac{x^2}{2}} \left[nH_{n-1}(x) - \frac{1}{2}H_{n+1}(x) \right].$$

Now,

$$\int_{-\infty}^{\infty} \varphi_n'(x)\varphi_m(x)dx = \int_{-\infty}^{\infty} e^{-\frac{x^2}{2}} H_m(x) e^{-\frac{x^2}{2}} \left[nH_{n-1}(x) - \frac{1}{2}H_{n+1}(x) \right] dx$$

$$= n\int_{-\infty}^{\infty} e^{-x^2} H_m(x)H_{n-1}(x)dx$$

$$-\frac{1}{2}\int_{-\infty}^{\infty} e^{-x^2} H_m H_{n+1}(x)dx$$

$$= n2^{n-1}(n-1)!\sqrt{\pi}\delta_m^{n-1} - \frac{1}{2}2^{n+1}(n+1)!\sqrt{\pi}\delta_m^{n+1}$$

$$= \begin{cases} 0 \text{ if } m \neq n, \text{if } m \neq n \pm 1 \\ 2^{n-1}n!\sqrt{\pi}, \text{if } m = n-1 \\ -2^n(n+1)!\sqrt{\pi}, \text{if } m = n+1 \end{cases}.$$

Example 13.8.5. Express $f(x) = x^4 + 2x^3 + 2x^2 - x - 3$ in terms of a Hermite polynomial.

Solution: We know that $H_0(x) = 1, H_1(x) = 2x, H_2(x) = 4x^2 - 2, H_3(x) = 8x^3 - 12x, H_4(x) = 16x^4 - 48x^2 + 12.$

Now, we have

$$x^4 = \frac{1}{16}H_4(x) + 3x^2 - \frac{3}{4}$$
$$x^3 = \frac{1}{8}H_3(x) + \frac{3}{2}x$$
$$x^2 = \frac{1}{4}H_2(x) + \frac{1}{2},\ x = \frac{1}{2}H_1(x),\ H_0(x) = 1.$$

Now, $f(x) = x^4 + 2x^3 + 2x^2 - x - 3 = \left(\frac{1}{16}H_4(x) + 3x^2 - \frac{3}{4}\right)$

$$+2\left(\frac{1}{8}H_3(x) + \frac{3}{2}x\right) + 2\left(\frac{1}{4}H_2(x) + \frac{1}{2}\right) - \frac{1}{2}H_1(x) - 3$$
$$= \frac{1}{16}H_4(x) + \frac{1}{4}H_3(x) + \frac{1}{2}H_2(x) - \frac{1}{2}H_1(x) +$$
$$+3\left(\frac{1}{4}H_2(x) + \frac{1}{2}\right) + 3\left(\frac{1}{2}H_1(x)\right) + \left(-\frac{3}{4} + 1 - 3\right)$$
$$= \frac{1}{16}H_4(x) + \frac{1}{4}H_3(x) + \frac{5}{4}H_2(x) + H_1(x) - \frac{5}{4}.$$

13.9 Exercise

1. Prove that $H_n(-x) = (-1)^n H_n(x)$.
2. Show that $\frac{d^n}{dx^n}H_n(x) = 2^n n!$
3. Provethat $\int_0^\infty e^{-x^2}[H_{2k}(x)H_{2s+1}(x)dx = \frac{(-1)^{k+s}2^{2k+2s}\left(\frac{1}{2}\right)_k\left(\frac{3}{2}\right)_s}{(2s+1-2k)}$.

4. Establish that $H_n(x) = 2^n \left\{ exp\left(-\frac{1}{4}\frac{d^2}{dx^2} \right) \right\} x^n$, where $H_n(x)$ is a Hermite polynomial.
5. Use the relation

$$\exp(2xt - t^2) = \exp(2xt - x^2t^2)\exp\left[t^2(x^2 - 1)\right]$$

to prove that $H_n(x) = \sum_{k=0}^{\left[\frac{n}{2}\right]} \frac{n!H_n(1)}{(n-2k)!k!} \frac{(x^2-1)^k}{} x^{n-2k}$.
6. Show that $\int_0^\infty H_n(x)dx = \frac{1}{2(n+1)}\{H_{n+1}(x) - H_{n-1}(0)\}$.
7. Prove that $x^n = \sum_{k=0}^{\left[\frac{n}{2}\right]} \frac{n!H_{n-2k}(x)}{(n-2k)!k!\ 2^n}$, and express x^6 in terms of $H_n(x)$.

14
Laguerre Polynomials

Abstract

Laguerre polynomials are a family of orthogonal polynomials with numerous applications in mathematics, physics, and engineering. They are also used in signal processing and data analysis.

This chapter includes the study of the Rodrigues formula, orthogonality, recurrence relations, expansions of polynomials, and some special results involving Laguerre polynomials.

Keywords: Generalized Laguerre polynomial, Rodrigues formula, recurrence relations, orthogonality, associated Laguerre polynomials

14.1 Introduction

The Laguerre polynomials, named after French mathematician Edmond Laguerre (1834-1886) who is known for his work in infinite series and geometry, are nontrival solutions of l Laguerre differential equation: $xy''+(1-x)y'+ny=0$. The Laguerre polynomials arise in quantum mechanics, in the radial part of the solution of Schrodinger equation for a one-electron atom. Physicists sometimes use a definition for the Laguerre polynomials that is larger by a factor of $n!$ than the definition used here.

In this chapter, we shall discuss some properties of Laguerre polynomials (like generating function, its orthogonallity, Rodrigues fomula) and also discuss generalized laguerre polynomial and Associated Laguerre polynomials.

14.2 Laguerre's Equation and Its Solution

Laguerre's differential equation of order n is

$$x\frac{d^2y}{dx^2}+(1-x)\frac{dy}{dx}+ny=0 \tag{14.1}$$

Dipankar De. Ordinary Differential Equations and Special Functions, (525–560) © 2026 Scrivener Publishing LLC

where n is a positive integer. We obtain a solution of Equation (14.1), which is finite for all values of x and tends to infinity as $x \to \infty$. For the solution, we apply here the method of Frobenius.

Let the solution of Equation (14.1) be

$$y = \sum_{m=0}^{\infty} c_m x^{k+m} c_{0\neq 0}. \tag{14.2}$$

Differentiating Equation (14.2) and then putting the values of $\frac{d^2y}{dx^2}$, $\frac{dy}{dx}$ *and y in* Equation (14.1), we get

$$x\sum_{m=0}^{\infty} c_m(k+m)(k+m-1)x^{k+m-2} + (1-x)\sum_{m=0}^{\infty} c_m(k+m)x^{k+m-1}$$
$$+n\sum_{m=0}^{\infty} c_m x^{k+m} = 0$$

or, $$\sum_{m=0}^{\infty} c_m(k+m)(k+m-1)x^{k+m-1} + \sum_{m0}^{\infty} c_m(k+m)x^{k+m-1}$$
$$-\sum_{m=0}^{\infty} c_m(k+m)x^{k+m} + n\sum_{m=0}^{\infty} c_m x^{k+m} = 0$$

or,

$$\sum_{m=0}^{\infty} c_m(k+m)x^{k+m-1}\{(k+m-1)+1\} - \sum_{m=0}^{\infty} c_m x^{k+m}\{(k+m)-n\} = 0$$

or, $$\sum_{m=0}^{\infty} c_m(k+m)^2 x^{k+m-1} - \sum_{m=0}^{\infty} c_m\{k+m-n\}x^{k+m} = 0. \tag{14.3}$$

For the indicial equation, taking the coefficient of the smallest power of x, i.e., x^{k-1} in Equation (14.3),

$$c_0 k^2 = 0.$$

$\therefore k = 0$, since

$$c_0 \neq 0. \tag{14.4}$$

We see that the roots of the indicial equation are equal.

The next smallest power is x^{k+m-1}, and by equating its coefficient to zero, we have

$$c_m (k+m)^2 - c_{m-1}\{k+m-1-n\} = 0$$

$$c_m = \frac{k+m-1-n}{(k+m)^2} c_{m-1}. \tag{14.5}$$

The two independent solutions in the present case are $y_{k=0}$ and $\left(\frac{\partial y}{\partial x}\right)_{k=0}$. But $\left(\frac{\partial y}{\partial x}\right)_{k=0}$ involves the term of the form $logx$ and so is infinite when $x = 0$. Since we wish obtain a solution finite for all finite values of x, we consider only the former solution, i.e., $\left(\frac{\partial y}{\partial x}\right)_{k=0}$.

With $k = 0$, Equations (14.5) and (14.2) reduce to

$$c_m = \frac{m-1-n}{(m)^2} c_{m-1} \tag{14.6}$$

and, $$y = \sum_{m=0}^{\infty} c_m x^m = c_0 + c_1 x + c_2 x^2 + \cdots. \tag{14.7}$$

Putting $m = 1,2,3,\ldots..$ in Equation (14.6), we get

$$c_1 = (-n)/(1)^2 c_0 = \frac{(-1)}{(1!)^2} n c_0, \; c_2 = \frac{1-n}{(2!)^2} c_1 = (-1)^2 \frac{n(n-1)}{(2!)^2} c_0$$

$$c_3 = \frac{3-n}{(3)^2} c_2 = -\frac{n-2}{(3)^2} (-1)^2 \frac{n(n-1)}{(2!)^2} c_0 = (-1)^3 \frac{n(n-1)(n-2)}{(3!)^2} c_0$$

$$c_r = (-1)^r \frac{n(n-1)(n-3)\ldots(n-r+1)}{(r!)^2} c_0 \; for\, r \leq n.$$

Here, we observe that $c_{n+1} = 0$, so $c_{n+2} = c_{n+3} = \cdots = 0$.
Putting these values in Equation (14.7), we get

$$y = c_0 + c_1 x + c_2 x^2 + \cdots + .c_n x^n$$

$$= c_0 \left[1 - \frac{n}{(1!)^2} x + (-1)^2 \frac{n(n-1)}{(2!)^2} x^2 + \cdots . + (-1)^r \frac{n(n-1)(n-3)\ldots(n-r+1)}{(r!)^2} x^r + \cdots . \right]$$

$$= c_0 \sum_{r=0}^{n} (-1)^r \frac{n(n-1)(n-3)\ldots(n-r+1)}{(r!)^2} x^r$$

$$= c_0 \sum_{r=0}^{n} (-1)^r \frac{n(n-1)(n-3)\ldots(n-r+1)\ldots 3.2.1}{(n-r)!(r!)^2} x^r.$$

$$\therefore y = c_0 \sum_{r=0}^{n} (-1)^r \frac{n!}{(n-r)!(r!)^2} x^r.$$

Taking $c_0 = 1$,
we define the solution as the *Laguerre polynomial of order of* n and denoted it by $L_n(x)$. Thus,

$$L_n(x) = \Sigma_{r=0}^{n} (-1)^r \frac{n!}{(n-r)!(r!)^2} x^r.$$

Note: Some authors take $c_0 = n!$ and define $L_n(x) = \Sigma_{r=0}^{n} (-1)^r \frac{(n!)^2}{(n-r)!(r!)^2} x^r$.

14.3 Generating Function of Laguerre Polynomials

Theorem 14.3.1. Prove that if $L_n(x)$ is a Laguerre polynomial of order n, then

$$\frac{e^{\frac{-xt}{(1-t)}}}{(1-t)} = L_n(x) \frac{t^n}{n!}.$$

Proof: Here,

$$\frac{e^{\frac{-xt}{(1-t)}}}{(1-t)} = \frac{1}{(1-t)}\sum_{r=0}^{\infty}\frac{\left(\frac{-xt}{(1-t)}\right)^r}{r!} = \sum_{r=0}^{\infty}\frac{(-1)^r}{r!}x^r t^r (1-t)^{-(r+1)}$$

$$= \sum_{r=0}^{\infty}\frac{(-1)^r}{r!}x^r t^r \sum_{r=0}^{\infty}(1-t)^{-(r+1)}$$

$$\left[\begin{array}{l} (1-t)^{-n} = 1 + nt + \frac{n(n+1)}{2!}t^2 + \dots + \frac{n(n+1)\dots(n+r-1)}{r!}t^r.+\dots. \\ \therefore r-thterm = \frac{n(n+1)\dots(n+r-1)}{r!}t^r = \frac{(n+r-1)!}{(n+1)!r!}t^r \end{array}\right]$$

$$\frac{e^{\frac{-xt}{(1-t)}}}{(1-t)} = \sum_{r=0}^{\infty}\frac{(-1)^r}{r!}x^r t^r \sum_{r=0}^{\infty}(1-t)^{-(r+1)} = \sum_{r=0}^{\infty}\sum_{s=0}^{\infty}\frac{(-1)^r}{r!}x^r t^r \frac{(r+s)!}{s!r!}t^s$$

(here, for simplicity, $n = r+1$, and for summation, r is replaced by s, i.e., $n = s+1$)

$$\frac{e^{\frac{-xt}{(1-t)}}}{(1-t)} = \sum_{r=0}^{\infty}\sum_{s=0}^{\infty}\frac{(-1)^r}{r!}x^r t^r \frac{(r+s)!}{s!r!}t^s = \sum_{r=0}^{\infty}\sum_{s=0}^{\infty}\frac{(-1)^r}{(r!)^2}\frac{(r+s)!}{s!}x^r t^{r+s}.$$

Let r be fixed. Then, the coefficient of t^n can be obtained by setting $r+s=n$, i.e., $s=n-r$.

Hence, for the chosen fixed value of r, the coefficient of t^n is

$$t^n = (-1)^r \frac{n!}{(n-r)!(r!)^2}x^r.$$

Now, $s \geq 0 \rightarrow n \geq r$, which gives all allowed values of r for finding the coefficient of t^n.

Thus, the total coefficient of t^n is given by

$$\sum_{s=0}^{\infty} \frac{(-1)^r}{(r!)^2} \frac{n!}{(n-r)!} x^r \equiv L_n(x).$$

If we take $L_n(x) = \sum_{r=0}^{\infty} \frac{(-1)^r}{(r!)^2} \frac{(n!)^2}{(n-r)!} x^r$, then $\frac{e^{\frac{-xt}{(1-t)}}}{(1-t)} = \sum_{n=0}^{\infty} L_n(x) \frac{t^n}{n!}$.

14.3.1 Alternative Expression of Laguerre Polynomials

Theorem 14.3.2. Prove that $L_n(x) = \frac{e^x}{n!} \frac{d^n}{dx^n}(x^n e^{-x})$.

Proof: By the Leibnitz's theorem, we have

$$D^n(uv) = \sum_{k=0}^{n} \binom{n}{k} D^{n-k} \mathrm{u} D^k \mathrm{v}. \tag{14.8}$$

$$\therefore \frac{e^x}{n!} \frac{d^n}{dx^n}(x^n e^{-x}) = \frac{e^x}{n!} \sum_{k=0}^{n} \binom{n}{k} D^{n-k}(\mathrm{x}^{\mathrm{n}}) D^k(e^{-x}) \text{ by (i)} \tag{14.9}$$

$$= \frac{e^x}{n!} \sum_{k=0}^{n} \binom{n}{k} \frac{\mathrm{n!}}{\mathrm{k!}} x^k (-1)^k e^{-x}$$

$$= \sum_{k=0}^{n} \frac{e^x}{n!} \frac{n!}{(n-k!)k!} \frac{\mathrm{n!}}{\mathrm{k!}} x^k (-1)^k e^{-x} = \sum_{k=0}^{n} (-1)^k \frac{n!}{(n-k!)(k!)^2} x^k = L_n(x)$$

$$\left[\begin{array}{c} \because D^n x^m = m(m-1)(m-2)\ldots(m-n+1)x^{m-n}. = \dfrac{m!}{(m-n)!} x^{m-n} \\ D^n e^{ax} = a^n e^{ax} \end{array} \right].$$

Remark: If we use $L_n(x) = \sum_{k=0}^{n} (-1)^k \frac{(n!)^2}{(n-k!)(k!)^2} x^k$, then $L_n(x) = e^x \frac{d^n}{dx^n}(x^n e^{-x})$.

This formula is known as *Rodrigue's formula of Laguerre polynomials.*

14.3.2 First Few Laguerre Polynomials

Let us find L_0, L_1, L_2, L_3, L_4.

We know that $L_n(x) = \frac{e^x}{n!}\frac{d^n}{dx^n}(x^n e^{-x})$.

Putting $n = 0,1,2,3,4.....$ in succession in Equation (14.8), we have

$$L_0(x) = \frac{e^x}{0!}\frac{d^0}{dx^0}(x^0 e^{-x}) = 1$$

$$L_1(x) = \frac{e^x}{1!}\frac{d}{dx}(xe^{-x}) = e^x[e^{-x} - xe^{-x}] = 1 - x$$

$$L_2(x) = \frac{e^x}{2!}\frac{d^2}{dx^2}(x^2 e^{-x}) = \frac{e^x}{2}\frac{d}{dx}[2xe^{-x} - x^2 e^{-x}]$$

$$= \frac{e^x}{2}[2e^{-x} - 2xe^{-x} - 2xe^{-x} + x^2 e^{-x}] = 1 - 2x + \frac{x^2}{2}$$

$$L_3(x) = \frac{e^x}{3!}\frac{d^3}{dx^3}(x^3 e^{-x}) = \frac{e^x}{6}\frac{d^2}{dx^2}\{e^{-x}(3x^2 - x^3)\} = \frac{e^x}{6}\frac{d}{dx}$$

$$e^{-x}[(6x - 3x^2) - (3x^2 - x^3)] = \frac{e^x}{6}\frac{d}{dx}e^{-x}(x^3 - 6x^2 + 6x)$$

$$= \frac{e^x}{6}e^{-x}[(3x^2 - 12x + 6) - (x^3 - 6x^2 + 6x)]$$

$$= \frac{1}{6}(-x^3 + 9x^2 - 18x + 6)$$

$$L_4(x) = \frac{e^x}{4!}\frac{d^4}{dx^4}(x^4 e^{-x}) = \frac{1}{4!}(24 - 96x + 72x^2 - 16x^3 + x^4).$$

14.4 Orthogonality Properties of Laguerre Polynomials

Theorem 14.4.1. Prove that $\int_0^\infty e^{-x} L_n(x) L_n(x) dx = \delta_n^m = \begin{cases} 0, if\ m \neq n \\ 1, if\ m = n \end{cases}$.

Proof: Using the generating function, we get

$$\sum_{n=0}^{\infty} L_n(x) t^n = \frac{e^{\frac{-xt}{(1-t)}}}{(1-t)} \text{ and } \sum_{m=0}^{\infty} L_m(x) s^m = \frac{e^{\frac{-xs}{(1-s)}}}{(1-s)}.$$

Multiplying corresponding sides, we have

$$\sum_{n=0}^{\infty}\sum_{m=0}^{\infty} L_n(x)L_m(x)t^n s^m = \frac{e^{\frac{-xt}{(1-t)}}}{(1-t)}\frac{e^{\frac{-xs}{(1-s)}}}{(1-s)} = \frac{e^{-x[1+\frac{t}{(1-t)}+\frac{s}{(1-s)}]}}{(1-t)(1-s)}. \tag{14.10}$$

Multiplying both sides by e^{-x} and then integrating both sides with respect to x from $0\ to\ \infty$,

$$\sum_{n=0}^{\infty}\sum_{m=0}^{\infty}[\int_0^{\infty} e^{-x}L_n(x)L_m(x)dx]t^n s^m dx = \frac{1}{(1-t)(1-s)}\int_0^{\infty} e^{-x}e^{\frac{-xt}{(1-t)}}e^{\frac{-xs}{(1-s)}}dx$$

or,

$$\sum_{n=0}^{\infty}\sum_{m=0}^{\infty}[\int_0^{\infty} e^{-x}L_n(x)L_m(x)dx]t^n s^m dx = \frac{1}{(1-t)(1-s)}\int_0^{\infty} e^{-x[1+\frac{t}{(1-t)}+\frac{s}{(1-s)}]}dx$$

$$\sum_{n=0}^{\infty}\sum_{m=0}^{\infty}[\int_0^{\infty} e^{-x}L_n(x)L_m(x)dx]t^n s^m = \frac{1}{(1-t)(1-s)}\left[\frac{e^{-x[1+\frac{t}{(1-t)}+\frac{s}{(1-s)}]}}{-[1+\frac{t}{(1-t)}+\frac{s}{(1-s)}]}\right]_0^{\infty}$$

$$= \frac{1}{(1-t)(1-s)}\frac{1}{[1+\frac{t}{(1-t)}+\frac{s}{(1-s)}]}$$

$$= \frac{1}{(1-t)(1-s)+t(1-s)+s(1-t)}$$

$$= \frac{1}{1-st} = (1-st)^{-1}$$

$$= \sum_{n=0}^{\infty}(st)^n.$$

$$\therefore \sum_{n=0}^{\infty}\sum_{m=0}^{\infty}[\int_0^{\infty} e^{-x}L_n(x)L_m(x)dx]t^n s^m = \sum_{n=0}^{\infty}(st)^n \tag{14.11}$$

Here, we observe that the indices of $t\, and\, s$ of the right-hand side of Equation (14.11) are always equal.

Thus, we conclude that when the indices $m \neq n\, of\, t\, and\, s\, on\, the\, left-hand\, side\, of$ Equation (14.11), the right-hand side has no such coefficient, which has non-equal indices.

Hence,

$$\int_0^\infty e^{-x} L_n(x) L_m(x) dx = 0, if\ m \neq n \tag{14.12}$$

$$=$$

$$\int_0^\infty e^{-x} L_n(x) L_m(x) dx = 1, if\ m = n. \tag{14.13}$$

By Equations (14.12) and (14.13), we get

$$\int_0^\infty e^{-x} L_n(x) L_n(x) dx = \delta_n^m = \begin{cases} 0, if\ m \neq n \\ 1, if\ m = n \end{cases}.$$

Remark: If we use the Laguerre polynomial $\frac{e^{\frac{-xt}{(1-t)}}}{(1-t)} = \sum_{n=0}^\infty L_n(x) \frac{t^n}{n!}$, then

$$\int_0^\infty e^{-x} L_n(x) L_n(x) dx = n! m! \delta_n^m.$$

14.5 Recurrence Relations

I. $(n+1)L_{n+1}(x) = (2n+1-x)L_n(x) - nL_{n-1}(x)$
II. $xL_n'(x) = nL_n(x) - nL_{n-1}(x)$
III. $L_n'(x) = -\sum_{r=0}^{n-1} L_r(x)$

I. Proof: We know the generating function, and we get

$$\sum_{n=0}^\infty L_n(x) t^n = \frac{e^{\frac{-xt}{(1-t)}}}{(1-t)}. \tag{14.14}$$

Differentiating with respect to t, we get

$$\sum_{n=0}^{\infty} L_n(x)nt^{n-1} = \frac{e^{\frac{-xt}{(1-t)}}}{(1-t)^2}\times[\left\{\frac{[-x(1-t)-xt\}(1-t)}{(1-t)^2}+1\right\}$$

$$= \frac{1}{(1-t)^2}e^{\frac{-xt}{(1-t)}}\left[\frac{-x}{(1-t)}+1\right]$$

$$= \frac{1}{1-t}\sum_{n=0}^{\infty} L_n(x)t^n\left[\frac{-x}{(1-t)}+1\right]$$

$$= \frac{1}{1-t}\sum_{n=0}^{\infty} L_n(x)t^n - \frac{x}{(1-t)^2}\sum_{n=0}^{\infty} L_n(x)t^n.$$

Multiplying both sides by $(1-t)^2$:

$$(1-t)^2\sum_{n=0}^{\infty} L_n(x)nt^{n-1} = (1-t)\sum_{n=0}^{\infty} L_n(x)t^n - x\sum_{n=0}^{\infty} L_n(x)t^n$$

or,
$$\sum_{n=0}^{\infty} L_n(x)nt^{n-1} - 2t\sum_{n=0}^{\infty} L_n(x)nt^{n-1} + t^2\sum_{n=0}^{\infty} L_n(x)nt^{n-1}$$
$$= \sum_{n=0}^{\infty} L_n(x)t^n - \sum_{n=0}^{\infty} L_n(x)t^{n+1} - x\sum_{n=0}^{\infty} L_n(x)t^n. \tag{14.15}$$

Equating the coefficients of t^n on both sides, we get

$$(n+1)L_{n+1}(x) - 2nL_n(x) + (n+1)L_{n-1}(x) = L_n(x) - L_{n-1}(x) - xL_n(x)$$

or,
$$(n+1)L_{n+1}(x) = (2n+1-x)L_n(x) - nL_{n-1}(x). \tag{14.16}$$

[If we put $\sum_{n=0}^{\infty} L_n(x)\frac{t^n}{n!} = \frac{e^{\frac{-xt}{(1-t)}}}{(1-t)}$, we proceed as above and get

$$L_{n+1}(x) = (2n+1-x)L_n(x) - n^2L_{n-1}(x).]$$

II. We know the generating function, and we get

$$\sum_{n=0}^{\infty} L_n(x)t^n = \frac{e^{\frac{-xt}{(1-t)}}}{(1-t)}.$$

Differentiating with respect to x, we get

$$\sum_{n=0}^{\infty} L_n'(x)t^n = \frac{e^{\frac{-xt}{(1-t)}}}{(1-t)}\left(\frac{-t}{(1-t)}\right)$$

$$= \frac{-t}{1-t}\sum_{n=0}^{\infty} L_n(x)t^n \qquad (14.17)$$

or, $(1-t)\sum_{n=0}^{\infty} L_n'(x)t^n = -t\sum_{n=0}^{\infty} L_n(x)t^n$ [multiplying both sides by $(1-t)$]

or, $$\sum_{n=0}^{\infty} L_n'(x)t^n - \sum_{n=0}^{\infty} L_n'(x)t^{n+1} = -\sum_{n=0}^{\infty} L_n(x)t^{n+1} \qquad (14.17a)$$

Equating the coefficient of t^n on both sides of Equation (14.17a), we get

$$L_n'(x)L_{n-1}'(x) = -L_{n-1}(x). \qquad (14.18)$$

From recurrence relation I,

$$(n+1)L_{n+1}(x) = (2n+1-x)L_n(x) - nL_{n-1}(x). \qquad (14.19)$$

Differentiating with respect to x, we get

$$(n+1)L_{n+1}'(x) = (2n+1-x)L_n'(x) - nL_{n-1}'(x) - L_n(x). \qquad (14.20)$$

From Equation (14.18),

$$L'_{n-1}(x) = L'_n(x) + L'_{n-1}(x). \tag{14.21}$$

Replacing n *by* $n+1$ in Equation (14.18), we get

$$L'_{n+1}(x) - L'_n(x) = -L_n(x).$$

or,

$$L'_{n+1}(x) = L'_n(x) - L_n(x). \tag{14.22}$$

Putting the values of $L'_{n-1}(x)$ *and* $L'_{n+1}(x)$ from Equations (14.21) and (14.22) in Equation (14.20), we have

$$(n+1)[L'_n(x) - L_n(x)] = (2n+1-x)L'_n(x) - n[L'_n(x) + L'_{n-1}(x)] - L_n(x).$$

By simplifying the above relation, we get $xL'_n(x) = nL_n(x) - nL_{n-1}(x)$.

[if we put $\sum_{n=0}^{\infty} L_n(x)\frac{t^n}{n!} = \frac{e^{\frac{-xt}{(1-t)}}}{(1-t)}$, we proceed as above and get

$$L'_n(x) - xL'_{n-1}(x) + nL_{n-1}(x) = 0].$$

III. We know the generating function, and we get

$$\sum_{n=0}^{\infty} L_n(x)t^n = \frac{e^{\frac{-xt}{(1-t)}}}{(1-t)}. \tag{14.23}$$

Differentiating with respect to x, we get

$$\begin{aligned}
\sum_{n=0}^{\infty} L'_n(x)t^n &= \frac{e^{\frac{-xt}{(1-t)}}}{(1-t)}\left(\frac{-t}{(1-t)}\right) \\
&= \frac{-t}{1-t}\sum_{r=0}^{\infty} L_r(x)t^r = -t(1-t)^{-1}\sum_{r=0}^{\infty} L_r(x)t^r \\
&= -t\sum_{s=0}^{\infty} t^s \sum_{r=0}^{\infty} L_r(x)t^r.
\end{aligned}$$

or, $$\sum_{n=0}^{\infty} L_n'(x)t^n = -\sum_{s=0}^{\infty}\sum_{r=0}^{\infty} L_r(x)t^{r+s+1}. \tag{14.24}$$

For equating the coefficient of t^n on both sides, we take $n = r+s+1$ on the right-hand side of Equation (14.17a), so that $s = n-r-1$. Hence, for a fixed value of r, *the coefficient of* t^n on the right-hand side of Equation (14.17a) is $-L_r(x)$.

But $s \geq 0 \rightarrow n-r-1 \geq 0 \rightarrow r \leq n-1$, which gives all the values of r for which $-L_r(x)$ is the coefficient of t^n.

Hence, the total coefficient of t^n on the right-hand side of Equation (14.17a) is given by $-\sum_{r=0}^{\infty} L_r(x)$.

Hence, $L_n'(x) = -\sum_{r=0}^{\infty} L_r(x)$.

14.6 Expansion of Laguerre Polynomials

Theorem 14.6.1. If $f(x)$ is a polynomial of degree m, show that $f(x)$ may be expressed in the form

$$f(x) = \sum c_r L_r(x), \text{ where } c_r = \int_0^{\infty} e^{-x} L_r(x) f(x)dx.$$

Proof: Since $f(x)$ is a polynomial of degree m, *we have*

$$f(x) = a_m x^m + a_{m-1}x^{m-1} + \cdots\ldots\ldots + a_1 x + a_0, \tag{14.25}$$

and $L_m(x)$ is a polynomial of degree m in the form

$$L_m(x) = b_m x^m + b_{m-1}x^{m-1} + \cdots\ldots\ldots + b_1 x + b_0. \tag{14.26}$$

Consider $f(x) - \frac{a_m}{b_m} L_m(x)$.

There are two cases as follows:

Case (i): Let $f(x) - \frac{a_m}{b_m} L_m(x) = 0$ so that $f(x) = \frac{a_m}{b_m} L_m(x)$, which proves the required result.

Case (ii): Let $f(x) - \frac{a_m}{b_m} L_m(x) = g_{m-1}(x)$, $g_{m-1}(x)$ being the polynomial of degree $m-1$.

Taking $c_m = \dfrac{a_m}{b_m}$, we may write

$$f(x) = c_m L_m(x) + g_{m-1}(x). \tag{14.27}$$

Taking $g_{m-1}(x)$ in place of $f(x)$ and proceeding as above,

$$g_{m-1}(x) = c_{m-1} L_{m-1}(x) + g_{m-2}(x). \tag{14.28}$$

Using Equation (14.28) in Equation (14.27), we get

$$f(x) = c_m L_m(x) + c_{m-1} L_{m-1}(x) + g_{m-2}(x) \tag{14.29}$$

$$= c_m L_m(x) + +c_{m-1} L_{m-1}(x) + c_{m-2} L_{m-2}(x) + \ldots + c_0$$
$$\left[\because L_0(x) = 1\right]$$

$$f(x) = \textstyle\sum_{r=0}^{m} c_r L_r(x) \tag{14.30}$$

or,

$$f(x) = \textstyle\sum_{s=0}^{m} c_s L_s(x). \tag{14.31}$$

Multiplying both sides by $e^{-x} L_r(x)$ and then integrating with respect to x from $0\ to\ \infty$, we get

$$\textstyle\int_0^\infty e^{-x} L_r(x) f(x) dx = \sum_{s=0}^{m} c_s \{\int_0^\infty e^{-x} L_r(x) L_s(x) dx\}. \tag{14.32}$$

We know that

$$\textstyle\int_0^\infty e^{-x} L_r(x) L_s(x) dx = \begin{bmatrix} 0, if\ r = s \\ 1, if r \neq s \end{bmatrix}. \tag{14.33}$$

Using Equation (14.33), Equation (14.32) gives

$$\therefore c_r = \textstyle\int_0^\infty e^{-x} L_r(x) f(x) dx.$$

14.7 Properties of Laguerre Polynomials

14.7.1 Differential Equation of $L_n(x)$

$$xL_n''(x)+(1-x)L_n'(x)+nL_n(x)=0.$$

Proof: We know from recurrence relation that

$$xL_n'(x)=nL_n(x)-nL_{n-1}(x). \tag{14.34}$$

Differentiating it with respect to x,

$$L_n'(x)+xL_n''(x)=n[L_n'(x)-L_{n-1}'(x)]$$

$$=-nL_{n-1}(x)$$

$$=xL_n'(x)-nL_n(x)\ \text{[from Equation (14.34)]}$$

$$\therefore xL_n''(x)+(1-x)L_n'(x)+nL_n(x)=0.$$

14.7.2 Integral Properties

(i) $(m-n)\int_0^\infty e^{-x}L_{m,}(x)L_n(x)dx=\left[xe^{-x}\left\{L_{n,}'(x)L_m(x)-L_{m,}'(x)L_n(x)\right\}\right]$.

(ii) $\int_0^\infty e^{-x}x^kL_n(x)dx=\begin{cases}0, & k=0,1,2\ldots,n-1\\ (-1)^n n! & ,k=n\end{cases}$.

Proof: (i) We know the Laguerre differential equation

$$xL_n''(x)+(1-x)L_n'(x)+nL_n(x)=0.$$

Multiplying both sides by e^{-x}: $xe^{-x}L_n''(x)+(1-x)e^{-x}L_n'(x)+ne^{-x}L_n(x)=0$

or,
$$\frac{d}{dx}\left[xe^{-x}L_n'(x)\right]+ne^{-x}L_n(x)=0$$

and,
$$\frac{d}{dx}\left[xe^{-x}L_m'(x)\right]+me^{-x}L_m(x)=0.$$

Multiply the first equation by $L_m(x)$ and the second equation by and $L_n(x)$ subtracting, we get

$$(m-n)e^{-x}L_m(x)L_n(x) = \frac{d}{dx}[xe^{-x}L_n'(x)]L_m(x) - \frac{d}{dx}[xe^{-x}L_n'(x)]L_n(x)$$
$$= \frac{d}{dx}[L_m(x)xe^{-x}L_n'(x)] - \frac{d}{dx}[L_n(x)xe^{-x}L_n'(x)]$$
$$= \frac{d}{dx}[xe^{-x}\{L_m(x)L_n'(x) - L_n(x)L_m'(x)\}].$$

By integrating with respect to x on both sides from a *to* b,

$$(m-n)\int_a^b e^{-x}L_m(x)L_n(x)dx = [xe^{-x}\{L_m(x)L_n'(x) - L_n(x)L_m'(x)\}]_a^b.$$

(ii) Since $L_r(x)$ is a polynomial of degree r, we can express x^k, where k is an integer, as a linear combination of polynomials $L_0(x), L_1(x), \ldots\ldots L_k(x)$. So let

$$x^k = \sum_{r=0}^{k} c_r L_r(x).$$

Multiplying both sides by $e^{-x}L_n(x)$ and integrating with respect to x on both sides from 0 *to* ∞, we get

$$\int_0^\infty x^k e^{-x} L_n(x)dx = \sum_{r=0}^{k} c_r \int_0^\infty e^{-x}L_n(x)L_r(x)dx. \qquad (14.35)$$

Case I: The right-hand side of the above Equation (14.35) becomes 0 because r, ranges from 0 *to* k, is never equal to n; i.e., $L_n(x) \neq L_r(x)$.

$$\therefore \int_0^\infty x^k e^{-x} L_n(x)dx = 0.$$

Case II: $k = n$, in Orthogonality of Laguerre polynomial Theoreom 14.4.1

$$\int_0^\infty e^{-x}L_n(x)L_n(x)dx = \delta_n^m = \begin{cases} 0, if\ m \neq n \\ 1, if\ m = n \end{cases}$$

Right hand side of above equation are only non-zero term:

$$c_n \int_0^\infty e^{-x}\left[L_n(x)\right]^2 dx.$$

Now, $x^k = x^n = \sum_{r=0}^{n} c_r L_r(x) = c_0 L_0(x) + c_1 L_1(x) + \cdots + c_n L_n(x)$.
But x^n occurs on the right-hand side only in the last term, which is

$$c_n \left\{ \frac{(-1)^n}{n!} x^n + \text{a polynomial of } \textit{degree } (n-1) \right\}.$$

Comparing with the left-hand side of the equation, we have

$$1 = c_n \frac{(-1)^n}{n!}.$$

$$\therefore c_n = (-1)^n n!$$

Thus, in this case,

$$\int_0^\infty x^k e^{-x} L_n(x) dx = (-1)^n n!$$

14.8 Generalized Laguerre Polynomial

14.8.1 Introduction

The generalized Laguerre polynomials $L_n^{(\alpha)}(x)$ are classical orthogonal polynomials that are orthogonal with respect to the gamma distribution $e^{-x}x^{\alpha}$ on the interval $(0,\infty)$. The generalized Laguerre polynomial $L_n^{(\alpha)}(x)$ associated with the non-negative integer n and real number $\alpha > -1$ are widely used in many problems of quantum mechanics, mathematical physics, and engineering. $L_n^{(\alpha)}(x)$ is in connection with the solution of the second-order linear differential equation

$$x\frac{d^2y}{dx^2} + (\alpha + 1 - x)\frac{dy}{dx} - ny = 0. \tag{14.39a}$$

These polynomials satify some recurrence relations. One very useful, when extracting properties of the wave functions of the hydrogen atom, is the following:

$$L_{n+1}^{(\alpha)}(x) = \frac{2n+1+\alpha-x}{n+1} L_n^{(\alpha)}(x) - \frac{n+\alpha}{n+1} L_{n-1}^{(\alpha)}(x), (n \geq 1).$$

We define the Laguerre polynomial $L_n^{(\alpha)}(x)$ by means of the generating relation

$$e^t{}_0F_1[-;1+\alpha;-xt]=\sum_{n=0}^{\infty}\frac{L_n^{(\alpha)}(x)}{(1+\alpha)_n}t^n. \tag{14.36}$$

Here, ${}_0F_1$ is a Bessel function. We may write the above equation in the form of

$$(xt)^{-\alpha}\Gamma(1+\alpha)e^tJ_n(2\sqrt{xt})=\sum_{n=0}^{\infty}\frac{L_n^{(\alpha)}(x)t^n}{(1+\alpha)_n}. \tag{14.37}$$

From Equation (14.36),

$$\begin{aligned}\therefore \sum_{n=0}^{\infty}\frac{L_n^{(\alpha)}(x)}{(1+\alpha)_n}t^n &=\sum_{n=0}^{\infty}\frac{t^n}{n!}\sum_{m=0}^{\infty}\frac{(-1)^m x^m t^m}{(1+\alpha)_m m!}\\ &=\sum_{n=0}^{\infty}\sum_{m=0}^{\infty}\frac{(-1)^m x^m t^{m+n}}{(1+\alpha)_m m!n!}.\\ &=\sum_{n=0}^{\infty}\sum_{m=0}^{\infty}\frac{(-1)^m x^m}{(1+\alpha)_m m!(n-m)!}t^n.\end{aligned}$$

Comparing the coefficient of t^n, we get

$$L_n^{(\alpha)}(x)=\sum_{m=0}^{n}\frac{(-1)^m x^m(1+\alpha)_n}{(1+\alpha)_m m!(n-m)!}, \tag{14.38}$$

which shows that $L_n^{(\alpha)}(x)$ is a polynomial of degree n.

Again, Equation (14.38) can be in the form

$$L_n^{(\alpha)}(x)=\frac{(1+\alpha)_n}{n!}\sum_{m=0}^{n}\frac{(-1)^m x^m}{(1+\alpha)_m m!}=\frac{(1+\alpha)_n}{n!}\,{}_1F_1\left[-n;1+\alpha;x\right], \tag{14.39}$$

whence it follows that the Laguerre polynomial is a terminating confluent hypergeometric series.

For the case $\alpha = 0$, we get

$$L_n^{(0)}(x) = L_n(x) = {}_1F_1[-n;1;x]. \tag{14.40}$$

$L_n^{(0)}(x)$ is called a simple *Laguerre polynomial.*

Some authors call $L_n^{(0)}(x)$ as a *generalized Laguerre polynomial.*

If $\alpha = k$ (a positive integer), then $L_n^{(0)}(x)$ is called the *associated Laguerre polynomial.*

14.8.2 Recurrence Relations

We derive the recurrence relations with the help of generating relations. For establishing the generating relations, we have from Equation (14.38)

$$\begin{aligned}
\sum_{n=0}^{\infty}\frac{(c_n)L_n^{(\alpha)}(x)}{(1+\alpha)_n}t^n &= \sum_{n=0}^{\infty}\sum_{m=0}^{n}\frac{(c)_n(-x)^m}{(1+\alpha)_m m!(n-m)!}t^n\\
&= \sum_{n=0}^{\infty}\sum_{m=0}^{\infty}\frac{(c)_{n+m}(-x)^m}{(1+\alpha)_m m!n!}t^{n+m}\\
&= \sum_{n=0}^{\infty}\sum_{m=0}^{\infty}\frac{(c+m)_n(-x)^m}{n!}t^n.\frac{(c)_m}{(1+\alpha)_m m!}(-xt)^m\\
&= \sum_{m=0}^{\infty}\frac{(c)_m}{(1+\alpha)_m m!(1-t)^{c+m}}(-xt)^m\\
&= (1-t)^{-c}{}_1F_1\left[\begin{matrix}c; & -\dfrac{xt}{}\\ 1+\alpha; & 1-t\end{matrix}\right].
\end{aligned}$$

Hence, we arrive at the family of generating relations for arbitrary c:

$$(1-t)^{-c}{}_1F_1\left[\begin{matrix}c; & -\dfrac{xt}{1-t}\\ 1+\alpha; & \end{matrix}\right] = \sum_{n=0}^{\infty}\frac{(c_n)L_n^{(\alpha)}(x)}{(1+\alpha)_n}t^n. \tag{14.41}$$

If we put $c = 1+\alpha$, we get

$$(1-t)^{-(1+\alpha)}\exp\left(-\frac{xt}{1-t}\right) = \sum_{n=0}^{\infty}L_n^{(\alpha)}(x)t^n. \tag{14.42}$$

(a) Differentiating both sides of Equation (14.42) with respect to t, we get

$$\frac{1+\alpha}{(1-t)^{(2+\alpha)}}\exp\left(-\frac{xt}{1-t}\right)+\frac{1}{(1-t)^{(1+\alpha)}}\exp\left(-\frac{xt}{1-t}\right)\left\{-\frac{x}{(1-t)^2}\right\}=\sum_{n=0}^{\infty} L_n^{(\alpha)}(x)nt^{n-1}$$

or, $$\frac{1+\alpha}{(1-t)}\sum_{n=0}^{\infty} L_n^{(\alpha)}(x)t^n-\frac{x}{(1-t)^2}\sum_{n=0}^{\infty} L_n^{(\alpha)}(x)t^n=\sum_{n=0}^{\infty} L_n^{(\alpha)}(x)nt^{n-1}$$

or,

$$(1+\alpha)(1-t)\sum_{n=0}^{\infty} L_n^{(\alpha)}(x)t^n-x\sum_{n=0}^{\infty} L_n^{(\alpha)}(x)t^n=(1-t)^2\sum_{n=0}^{\infty} L_n^{(\alpha)}(x)nt^{n-1}$$

or, $$(1+\alpha)\sum_{n=0}^{\infty} L_n^{(\alpha)}(x)t^n-(1+\alpha)\sum_{n=0}^{\infty} L_n^{(\alpha)}(x)t^{n+1}-x\sum_{n=0}^{\infty} L_n^{(\alpha)}(x)t^n$$

$$=\sum_{n=0}^{\infty} L_n^{(\alpha)}(x)nt^{n-1}-2\sum_{n=0}^{\infty} L_n^{(\alpha)}(x)nt^n+\sum_{n=0}^{\infty} L_n^{(\alpha)}(x)nt^{n+1}.$$

Equating the coefficients of t^n from both sides, we get

$$(1+\alpha)L_n^{(\alpha)}(x)-(1+\alpha)L_{n-1}^{(\alpha)}(x)-xL_n^{(\alpha)}(x)=(n+1)L_{n+1}^{(\alpha)}(x)-2nL_n^{(\alpha)}(x)+(n-1)L_{n-1}^{(\alpha)}(x).$$

Now, shifting n *to* $n-1$ in the above equation, we get

$$nL_n^{(\alpha)}(x)=(2n-x-1+\alpha)L_{n-1}^{(\alpha)}(x)-(n-1+\alpha)L_{n-2}^{(\alpha)}(x). \quad (14.43)$$

This relation Equation (14.43) is a pure in terms of Laguerre polynomials.

(b) Differentiating Equation (14.42) with respect to x and writing $D \equiv \frac{d}{dx}$, we get

$$(1-t)^{-(1+\alpha)} \exp\left(-\frac{xt}{1-t}\right)\left(\frac{-t}{1-t}\right) = \sum_{n=0}^{\infty} D\left\{L_n^{(\alpha)}(x)\right\} t^n$$

or,
$$-\frac{t}{1-t} \sum_{n=0}^{\infty} L_n^{(\alpha)}(x) t^n = \sum_{n=0}^{\infty} D\left\{L_n^{(\alpha)}(x)\right\} t^n \tag{14.44}$$

or,
$$-\sum_{n=0}^{\infty} L_n^{(\alpha)}(x) t^{n+1} = \sum_{n=0}^{\infty} D\left\{L_n^{(\alpha)}(x)\right\} t^n - \sum_{n=0}^{\infty} D\left\{L_n^{(\alpha)}(x)\right\} t^{n+1}.$$

Equating the coefficient of t^n on both sides,
we get the *first differential recurring relation*

$$D\left\{L_n^{(\alpha)}(x)\right\} = D\left\{L_{n-1}^{(\alpha)}(x)\right\} - L_{n-1}^{(\alpha)}(x). \tag{14.45}$$

Differentiating Equation (14.43) with respect to x, we get

$$\begin{aligned} nD\{L_n^{(\alpha)}(x)\} &= -L_{n-1}^{(\alpha)}(x) + (2n - x - 1 + \alpha) D\left\{L_{n-1}^{(\alpha)}(x)\right\} \\ &\quad - (n - 1 + \alpha) D\left\{L_{n-2}^{(\alpha)}(x)\right\} \end{aligned} \tag{14.46}$$

From Equation (14.45), it follows that

$$D\{L_{n-2}^{(\alpha)}(x)\} = D\{L_{n-1}^{(\alpha)}(x)\} + L_{n-2}^{(\alpha)}(x). \tag{14.47}$$

Using the values of $D\{L_n^{(\alpha)}(x)\}$ *and* $D\{L_{n-2}^{(\alpha)}(x)\}$ from Equations (14.45) and (14.47) in (14.46), we get

$$nD\{L_{n-1}^{(\alpha)}(x)\}-nL_{n-1}^{(\alpha)}(x)=-L_{n-1}^{(\alpha)}(x)+\left(2n-x-1+\alpha\right)D\{L_{n-1}^{(\alpha)}(x)\}$$
$$-(n-1+\alpha)D\{L_{n-1}^{(\alpha)}(x)\}-\left(n-1+\alpha\right)D\{L_{n-2}^{(\alpha)}(x)\}$$

or, $$xD\{L_{n-1}^{(\alpha)}(x)\}=(n-1)L_{n-1}^{(\alpha)}(x)-(n-1+\alpha)L_{n-2}^{(\alpha)}(x).$$

On shifting n *to* $(n+1)$, we write it

$$xD\{L_n^{(\alpha)}(x)\}=nL_n^{(\alpha)}(x)-(n+\alpha)L_{n-1}^{(\alpha)}(x), \tag{14.48}$$

which is *the second differential recurrence relation.*

14.8.3 Rodrigues Formula

We know that

$$L_n^{(\alpha)}(x)=\sum_{m=0}^{n}\frac{(-1)^m x^m(1+\alpha)_n}{(1+\alpha)_m m!(n-m)!}$$
$$=x^{-\alpha}\sum_{m=0}^{n}\frac{(-1)^m(1+\alpha)_n}{(1+\alpha)_m m!(n-m)!}x^{m+\alpha}. \tag{14.49}$$

When k is a non-negative integer, we have

$$\frac{d^k}{dx^k}(x^{n+\alpha})=(n+\alpha)(n+\alpha-1)\ldots(n+\alpha-k+1)x^{n+\alpha-k}.$$

If we put $k=n-m$, we get

$$D^{n-m}(x^{n+\alpha})=\frac{d^{n-m}}{dx^{n-m}}(x^{n+\alpha})$$
$$=(n+\alpha)(n+\alpha-1)\ldots(\alpha+m+1)x^{\alpha+m}=\frac{(1+\alpha)_n}{(1+\alpha)_m}x^{\alpha+m}.$$

Hence, Equation (14.49) becomes

$$L_n^{(\alpha)}(x) = x^{-\alpha} \sum_{m-0}^{n} \frac{(-1)^m}{(m!(n-m)!} D^{n-m}(x^{n+\alpha})$$

$$= x^{-\alpha} e^x \sum_{m-0}^{n} \frac{1}{(m!(n-m)!} D^{n-m}(x^{n+\alpha})(-1)^m e^{-x}$$

$$[\therefore \frac{d^k}{dx^k}(e^{-x}) = (-1)(^k e^{-x}].$$

We may write $L_n^{(\alpha)}(x)$

$$= \frac{x^{-\alpha} e^x}{n!} \sum_{m=0}^{n} \frac{n!}{(m!(n-m)!} D^{n-m}(x^{n+\alpha}) D^m(e^{-x}).$$

Using the Leibnitz's theorem, we get $L_n^{(\alpha)}(x) = \dfrac{x^{-\alpha} e^x}{n!} D^n[(x^{n+\alpha})e^{-x}]$, which is known as the *Rodrigue's formula for Laguerre polynomials.*

14.8.4 Expansions

We can expand $x^n, H_n(x), \textit{and } P_n(x)$.

(a) We know that ${}_0F_1\left[-;1+\alpha;-xt\right] = e^{-t} \sum_{n=0}^{\infty} \dfrac{L_n^{(\alpha)}(x)}{(1+\alpha)_n} t^n$

or,
$$\sum_{n=0}^{\infty} \frac{(-x)^n}{(1+\alpha)_n n!} t^n = \sum_{n=0}^{\infty} \frac{(-1)^n t^n}{n!} \sum_{m=0}^{\infty} \frac{L_m^{(\alpha)}(x)}{(1+\alpha)_m} t^m$$

$$= \sum_{n=0}^{\infty} \sum_{m=0}^{\infty} \frac{(-1)^n}{n!} \frac{L_m^{(\alpha)}(x)}{(1+\alpha)_m} t^{n+m}$$

$$= \sum_{n=0}^{\infty} \sum_{m=0}^{\infty} \frac{(-1)^{n-m}}{(n-m)!} \frac{L_m^{(\alpha)}(x)}{(1+\alpha)_m} t^n.$$

Comparing the coefficient of t^n

$$x^n = \sum_{m=0}^{\infty} \frac{(-1)^m n!(1+\alpha)_n}{(n-m)!} \frac{L_m^{(\alpha)}(x)}{(1+\alpha)_m}. \tag{14.50}$$

(b) We know that

$$\begin{aligned}\sum_{n=0}^{\infty} \frac{H_n(x)}{n!} t^n &= e^{(2xt-t^2)} = e^{(2xt)} e^{(-t^2)} \\ &= \sum_{n=0}^{\infty} \frac{(2x)^n t^n}{n!} \sum_{m=0}^{\infty} \frac{(-1)^m t^{2m}}{m!} \\ &= \sum_{n=0}^{\infty} \sum_{m=0}^{\infty} \frac{(2x)^n (-1)^m t^{n+2m}}{n!m!}.\end{aligned}$$

Putting the value of x^n and simplifying, we finally get

$$H_n(x) = 2^n (1+\alpha)_n \sum_{m=0}^{\infty} {}_2F_2 \left[\begin{matrix} -\frac{1}{2}(n-m), -\frac{1}{2}(n-m-1); \\ -\frac{1}{2}(\alpha+n), -\frac{1}{2}(\alpha+n-1); \end{matrix} \; -\frac{1}{4} \right] \times \frac{(-n)_m L_m^{(\alpha)}(x)}{(1+\alpha)_m} \tag{14.51}$$

and the Legendre polynomial

$$P_n(x) = \frac{2^n (\frac{1}{2})_n (1+\alpha)_n}{n!} \sum_{m=0}^{n} {}_2F_3 \left[\begin{matrix} -\frac{1}{2}(n-m), -\frac{1}{2}(n-m-1); \\ \frac{1}{2}-n, -\frac{1}{2}(\alpha+n), -\frac{1}{2}(\alpha+n-1); \end{matrix} \; -\frac{1}{4} \right] \times \frac{(-n)_m L_m^{(\alpha)}(x)}{(1+\alpha)_m}. \tag{14.52}$$

14.8.5 Some Special Results

Here, we derive some special results of the Laguerre polynomials.
(a) Generating function of a Laguerre polynomial

$$\sum_{n=0}^{\infty} L_n^{(\alpha)}(x)t^n = (1-t)^{-(1+\alpha)} \exp\left(-\frac{xt}{1-t}\right)$$
$$= (1-t)^{(\alpha-\beta)}\left[(1-t)^{-(1-\beta)} e^{\left(-\frac{xt}{1-t}\right)}\right]$$
$$= \left(\sum_{m=0}^{\infty} \frac{(\alpha-\beta)_m}{m!} t^m\right)\left(\sum_{n=0}^{\infty} L_n^{(\beta)}(x)t^n\right)$$
$$= \sum_{n=0}^{\infty} \sum_{m=0}^{\infty} \frac{(\alpha-\beta)_m}{m!} L_{n-m}^{(\beta)}(x)t^m. \tag{14.53}$$

Comparing the coefficient of t^n, we get

$$L_n^{(\alpha)}(x) = \sum_{m=0}^{\infty} \frac{(\alpha-\beta)_m}{m!} L_{n-m}^{(\beta)}(x). \tag{14.54}$$

(b) Replacing $x \; by \; x+y \; and \; \alpha \; by \; \alpha+\beta+1$ in Equation (14.53), we get

$$\sum_{n=0}^{\infty} L_n^{(\alpha+\beta+1)}(x+y)t^n = (1-t)^{-1-(1+\alpha+\beta)} \exp\left(-\frac{(x+)t}{1-t}\right)$$
$$= \left\{(1-t)^{-1-\alpha} \exp\left(-\frac{xt}{1-t}\right)\right\}\left\{(1-t)^{-1-\beta} \exp\left(-\frac{yt}{1-t}\right)\right\}$$
$$= \sum_{n=0}^{\infty} \sum_{m=0}^{\infty} L_n^{(\alpha)}(x) L_{n-m}^{(\beta)}(y)t^n.$$

Comparing both sides of the coefficients of t^n, we get

$$L_n^{(\alpha+\beta+1)}(x+y) = \sum_{m=0}^{\infty} L_n^{(\alpha)}(x) L_{n-m}^{(\beta)}(y). \tag{14.55}$$

(c) The generating function

We know that $e^t{}_0F_1\left[-;1+\alpha;-xt\right]=\sum_{n=0}^{\infty}\frac{L_n^{(\alpha)}(x)}{(1+\alpha)_n}t^n.$

Replacing x by xy, it becomes

$$e^{(1-y)t}\{e^{yt}{}_0F_1[-;1+\alpha;-xt]\}=\sum_{n=0}^{\infty}\frac{L_n^{(\alpha)}(xy)}{(1+\alpha)_n}t^n$$

or, $$\left(\sum_{n=0}^{\infty}\frac{(1-y)^n t^n}{n!}\right)\left(\sum_{m=0}^{\infty}\frac{L_n^{(\alpha)}(x)(y)^m t^m}{(1+\alpha)_m}\right)=\sum_{n=0}^{\infty}\frac{L_n^{(\alpha)}(xy)}{(1+\alpha)_n}t^n$$

or, $$\sum_{n=0}^{\infty}\sum_{m=0}^{\infty}\frac{(1-y)^{n-m}}{(n-m)!}\frac{L_m^{(\alpha)}(x)(y)^m t^n}{(1+\alpha)_m}=\sum_{n=0}^{\infty}\frac{L_n^{(\alpha)}(xy)}{(1+\alpha)_n}t^n.$$

Comparing the coefficients of t^n, we get

$$L_n^{(\alpha)}(xy)=\sum_{m=0}^{\infty}\frac{(1+\alpha)_n(1-y)^{n-m}}{(n-m)!}\frac{L_m^{(\alpha)}(x)y^m}{(1+\alpha)_m}.$$

(d) The confluent hypergeometric function ${}_1F_1(a;b;x)$ satisfies the differential equation

$$xL_n^{(k)}(x)+(b-x)\frac{dy}{dx}-ay=0.$$

From $e^t{}_0F_1\left[-;1+\alpha;-xt\right]=\sum_{n=0}^{\infty}\frac{L_n^{(\alpha)}(x)}{(1+\alpha)_n}t^n$, we see that the differential equation satisfied by the Laguerre polynomial $L_n^{(\alpha)}(x)$ is

$$x\frac{d^2y}{dx^2}+(1+\alpha-x)\frac{dy}{dx}-ny=0. \tag{14.56}$$

If α is a non-negative integer, say k, then $L_n^k(x)$ stands for a polynomial of more practical importance. We can establish a relation between $L_n^k(x)$ and the Laguerre polynomial $L_n^k(x)$:

$$L_n^k(x) = \frac{d^k}{dx^k}\{L_{n+k}(x)\}$$

where $L_n^k(x)$ is the associated Laguerre polynomial.

The explicit relation for the associated Laguerre polynomial may be written as

$$L_n^k(x) = \Sigma_{m=0}^{n}(-1)^m \frac{(n+k)!x^m}{(k-m)!(k+m)!m!}.$$

For $\alpha = k$, the differential Equation (14.35a) is called the *Laguerre's associated differential equation.*

14.9 Examples

Example 14.9.1. Prove that (i) $L_n'(0) = -n$ (ii) $L_n''(0) = \frac{n(n-1)}{2}$.

Solution: (i) We know that $L_n(x)$ is the solution of $x\frac{d^2y}{dx^2} + (1-x)\frac{dy}{dx} + ny = 0$

$$\therefore x\frac{d^2L_n(x)}{dx^2} + (1-x)\frac{dL_n(x)}{dx} + nL_n(x) = 0.$$

Putting $x = 0$ and using $L_n(0) = 1$, the above equation gives

$$L_n'(0) + nL_n(0) = 0 \text{ or, } L_n'(0) + n = 0 \therefore L_n'(0) = -n.$$

(ii) Take

$$\Sigma_{n=0}^{\infty} t^n L_n(x) = \frac{1}{1-t}e^{-\frac{xt}{1-t}}. \quad \text{(ii)}$$

Differentiating two times with respect to x, we get

$$\sum_{n=0}^{\infty} \frac{t^n}{n!} L_n''(x) = \frac{1}{1-t} e^{-\frac{xt}{1-t}} \left(-\frac{t}{1-t}\right)^2. \qquad \text{(iii)}$$

Putting $x = 0$ in Equation (iii), we get

$$\sum_{n=0}^{\infty} t^n L_n''(0) = \frac{1}{1-t}\left(-\frac{t}{1-t}\right)^2 = t^2(1-t)^{-3}. \qquad \text{(iv)}$$

Equating the coefficient of t^n on both sides of Equation (iv), we get

$$L_n''(0) = coefficient\ of\ t^{n-2} \text{ of } (1-t)^{-3}$$

$$= \binom{-3}{n-2} = \frac{(-3)(-3-1)\ldots.(-n)}{(n-2)!}(-1)^{n-2}$$

$$= \frac{3.4.5\ldots.n}{(n-2)!}(-1)^{n-2}(-1)^{n-2} = \frac{n!}{2(n-2)!}$$

$$= \frac{n(n-1)}{2}[\binom{-3}{n-2} = \frac{(-3)!}{(n-2)![-3-(n-2)]!}].$$

Example 14.9.2. Taking $\sum_{n=0}^{\infty} \frac{t^n}{n!} L_n(x) = \frac{1}{1-t} e^{-\frac{xt}{1-t}}$,

(a) Prove that $\int_0^{\pi/2} \frac{e^{-tan\theta}}{cos^2\theta} L_n(tan\theta) L_m(tan\theta) d\theta = \delta_n^m$.

(b) (i) $L_n'(x) = n[L_{n-1}'(x) - L_{n-1}(x)]$.

(ii) $xL_n'(x) = nL_n(x) - n^2 L_{n-1}(x)$.

Solution: (a) $\int_0^{\pi/2} \frac{e^{-tan\theta}}{cos^2\theta} L_n(tan\theta) L_m(tan\theta) d\theta$.

Putting $x = tan\theta, dx = sec^2\theta d\theta$, and limit $\theta = 0 \rightarrow x = 0 \, and \, \theta = \pi/2 \rightarrow$ $x = \infty$

$$\therefore I = \int_0^{\pi/2} \frac{e^{-tan\theta}}{cos^2\theta} L_n(tan\theta)L_m(tan\theta)d\theta = \int_0^{\infty 2} e^{-x} L_n(x)L_m(x)dx = \delta_n^m.$$

(b) (i) We have

$$\sum_{n=0}^{\infty} \frac{t^n}{n!} L_n(x) = \frac{1}{1-t} e^{-\frac{xt}{1-t}}. \qquad \text{(i)}$$

Differentiating two times with respect to x, we get

$$\sum_{n=0}^{\infty} \frac{t^n}{n!} L_n'(x) = \frac{1}{1-t} e^{-\frac{xt}{1-t}} \left(-\frac{t}{1-t}\right)$$

$$= -\frac{t}{1-t} \sum_{n=0}^{\infty} \frac{t^n}{n!} L_n(x)$$

or,
$$(1-t)\sum_{n=0}^{\infty} \frac{t^n}{n!} L_n'(x) = -t\sum_{n=0}^{\infty} \frac{t^n}{n!} L_n(x)$$

or,
$$\sum_{n=0}^{\infty} \frac{t^n}{n!} L_n'(x) = \sum_{n=0}^{\infty} \frac{t^{n+1}}{n!} L_n'(x) - \sum_{n=0}^{\infty} \frac{t^{n+1}}{n!} L_n(x). \qquad \text{(ii)}$$

Equating the coefficient of t^n on both sides, we get

$$\frac{1}{n!} L_n'(x) = \frac{1}{(n-1)!} L_{n-1}'(x) - \frac{1}{(n-1)!} L_{n-1}(x).$$

$$\therefore L_n'(x) = n[L_{n-1}'(x) - L_{n-1}(x)]. \qquad \text{(iii)}$$

(ii) From recurrence relations I [for $\sum_{n=0}^{\infty} \frac{t^n}{n!} L_n(x) = \frac{1}{1-t} e^{-\frac{xt}{1-t}}$],

$$L_{n+1}(x) = (2n+1-x)L_n(x) - n^2 L_{n-1}(x).$$

Differentiating both sides with respect to, we get

$$L'_{n+1}(x) = (2n+1-x)L'_n(x) - L_n(x) - n^2 L'_{n-1}(x). \quad \text{(iv)}$$

From Equation (iii),

$$L'_n(x) = n[L'_{n-1}(x) - L_{n-1}(x)]. \quad \text{(v)}$$

Replacing $n\,by$ n+1 in the above Equation (v),

$$L'_{n+1}(x) = n[L'_n(x) - L_n(x)]. \quad \text{(vi)}$$

Again, from Equation (v), we get

$$nL'_{n-1}(x) = nL_{n-1}(x) + L'_n(x). \quad \text{(vii)}$$

Using Equations (vi) and (vii) in Equation (v), it becomes

$$n[L'_n(x) - L_n(x)] = \left(2n+1-x\right)L'_n(x) - L_n(x) - n[nL_{n-1}(x) + L'_n(x)].$$

$$\therefore xL'_n(x) = nL_n(x) - n^2 L_{n-1}(x).$$

Example 14.9.3. Prove that $\int_0^\infty e^{-st} L_n(t)dt = \left(\frac{1}{s}\right) \times (1-\frac{1}{s})^n$.

Solution: We know that $L_n(x) = \sum_{r=0}^\infty \frac{(-1)^r}{(r!)^2} \frac{n!}{(n-r)!} t^r$.

Now,

$$\int_0^\infty e^{-st} L_n(t)dt = \sum_{r=0}^\infty \frac{(-1)^r}{(r!)^2} \frac{n!}{(n-r)!} \int_0^\infty e^{-st} t^{(r+1)-1} dt$$

$$= \sum_{r=0}^\infty \frac{(-1)^r}{(r!)^2} \frac{n!}{(n-r)!} \frac{\Gamma(r+1)}{s^{r+1}} = \frac{1}{s} \sum_{r=0}^n \frac{(-1)^r}{r!} \frac{n!}{(n-r)!} \left(\frac{1}{s}\right)^r$$

$$= \frac{1}{s} \sum_{\mathrm{r}=0}^{\mathrm{n}} \binom{n}{k} \left(\frac{1}{s}\right)^r = \frac{1}{s}\left(1-\frac{1}{s}\right)^r.$$

Example 14.9.4. Prove that $\int_0^\infty e^{-y} x^k L_n(x)dx = \begin{cases} 0, if\ k < n \\ (-1)^n n!, if\ k = n \end{cases}$.

Solution: If $f(x)$ is a polynomial of degree m, $f(x)$ may be expressed in the form

$$f(x) = \sum_{r=0}^{\infty} c_r L_r(x), \tag{i}$$

where, $c_r = \int_0^\infty e^{-x} L_r(x) f(x)dx$ [from Art. (14.6)] (ii)

Let $f(x) = x^k$, then

$$x^k = \sum_{r-0}^{k} c_r L_r(x) \tag{iii}$$

where, $$c_r = \int_0^\infty e^{-x} x^k L_r(x)dx. \tag{iv}$$

Case (i): Let $k < n$, then $c_r = 0$.
Hence, from Equation (iv), we get (with $r = n$)

$$\int_0^\infty e^{-x} x^k L_n(x)dx = 0 \tag{v}$$

Case (ii): Let $k = n$. Then, Equations (iii) and (iv) give

$$x^n = \sum_{r=0}^{n} c_r L_r(x) = c_0 L_0(x) + c_1 L_1(x) + \cdots. + c_n L_n(x) \tag{vi}$$

where, $$c_r = \int_0^\infty e^{-x} x^n L_n(x)dx. \tag{vii}$$

We know by definition that

$$L_n(x) = \sum_{r=0}^{n} \frac{(-1)^n}{(r!)^2} \frac{n!}{(n-r)!} x^r. \tag{viii}$$

From Equation (viii), the coefficient of

$$x^n in L_n(x) = \frac{n!}{(0)!} \frac{(-1)^n}{(n!)^2} = \frac{(-1)^n}{n!}. \tag{ix}$$

Equating the coefficient of x^n from both side of Equation (vi) and using Equation (ix), we get

$$1 = c_n \frac{(-1)^n}{n!} \text{implies } c_n = (-1)^n n! \tag{x}$$

From Equations (v) and (x), we get $\int_0^\infty e^{-x} x^k L_n(x) dx = \begin{cases} 0, if\ n < k \\ (-1)^n n!, if\ n = k \end{cases}.$

Example 14.9.5. Show that (i) $H_{2n}(x) = (-1)^n 2^{2n} n! L_n^{\left(-\frac{1}{2}\right)}(x^2)$.

(ii) $H_{2n+1}(x) = (-1)^n 2^{2n+1} n! x L_n^{\left(\frac{1}{2}\right)}(x^2)$.

Solution: (i) We have

$$H_{2n}(x) = \sum_{m=0}^{n} \frac{(-1)^m (2n)!(2x)^{2n-2m}}{m!(2n-2m)!}$$

$$= \sum_{m=0}^{n} \frac{(-1)^m n! 2^{2n} (\frac{1}{2})_n (2)^{2n-2m} (x^2)^{n-m}}{m!(2)^{2n-2m} (\frac{1}{2})_{n-m} (1)_{n-m}}$$

$$= n! 2^{2n} \sum_{m=0}^{n} \frac{(-1)^m (\frac{1}{2})_n (x^2)^{n-m}}{m!(n-m)!(\frac{1}{2})_{n-m}}$$

$$= n! 2^{2n} (-1)^n \sum_{m=0}^{n} \frac{(-1)^m (\frac{1}{2})_n (x^2)^m}{m!(n-m)!(\frac{1}{2})_m} = 2^{2n} n! (-1)^n L_n^{\left(-\frac{1}{2}\right)}(x^2).$$

(ii) Also,

$$H_{2n+1}(x) = \sum_{m=0}^{n} \frac{(-1)^m (2n+1)!(2x)^{2n+1-2m}}{m!(2n+1-2m)!}$$

$$= \sum_{m=0}^{n} \frac{(-1)^m (2)_{2n} (2x)^{2n+1-2m}}{m!(2n+1-2m)!}$$

$$= \sum_{m=0}^{n} \frac{(-1)^m 2^{2n} (1)_n (\frac{3}{2})_n (2)^{2n+1-2m} (x)^{2n+1-2m}}{m!(1)_{n-m} (\frac{3}{2})_{n-m} 2^{2n-2m}}$$

$$= 2^{2n+1} n! x \sum_{m=0}^{n} \frac{(-1)^m (\frac{3}{2})_n (x^2)^{n-m}}{m!(n-m)!(\frac{3}{2})_{n-m}}$$

$$= (-1)^n 2^{2n+1} n! x \sum_{m=0}^{n} \frac{(-1)^m (\frac{3}{2})_n (x^2)^m}{m!(n-m)!(\frac{3}{2})_m}$$

$$= (-1)^n 2^{2n+1} n! x L_n^{\left(\frac{1}{2}\right)}(x^2).$$

Example 14.9.6. Show that $\int_0^t x^\alpha (t-x)^{\beta-1} L_n^{(\alpha)}(x)dx = \frac{\Gamma(1+\alpha)\Gamma(\beta)(1+\alpha)_n}{\Gamma(1+\alpha+\beta)(1+\alpha+\beta)_n}$ $t^{(\alpha+\beta)} L_n^{(\alpha+\beta)}(t)$.

Solution:

$$\int_0^t x^\alpha (t-x)^{\beta-1} L_n^{(\alpha)}(x)dx$$

$$= \int_0^t x^\alpha (t-x)^{\beta-1} \sum_{m=0}^{n} \frac{(-1)^m}{m!} \frac{(1+\alpha)_n x^m}{(n-m)!(1+\alpha)_m} x^r dx$$

$$= \sum_{m=0}^{n} \frac{(-1)^m}{m!} \frac{(1+\alpha)_n x^m}{(n-m)!(1+\alpha)_m} \int_0^t (t-x)^{\beta-1} x^{\alpha+m} dx$$

$$= \sum_{m=0}^{n} \frac{(-1)^m}{m!} \frac{(1+\alpha)_n t^{\alpha+\beta+m}}{(n-m)!(1+\alpha)_m} \int_0^t (1-v)^{\beta-1} v^{\alpha+m} dv$$

$$= \sum_{m=0}^{n} \frac{(-1)^m}{m!} \frac{(1+\alpha)_n t^{\alpha+\beta+m}}{(n-m)!(1+\alpha)_m} \frac{\Gamma(1+\alpha+m)\Gamma(\beta)}{\Gamma(1+\alpha+m+\beta)}$$

$$= \frac{\Gamma(1+\alpha)\Gamma(\beta)(1+\alpha)_n t^{\alpha+\beta}}{\Gamma(1+\alpha+\beta)(1+\alpha+\beta)_n} \sum_{m=0}^{n} \frac{(-1)^m}{m!} \frac{(1+\alpha+\beta)_n}{(n-m)!(1+\alpha+\beta)_m} t^m$$

$$= \frac{\Gamma(1+\alpha)\Gamma(\beta)(1+\alpha)_n t^{\alpha+\beta}}{\Gamma(1+\alpha+\beta)(1+\alpha+\beta)_n} L_n^{(\alpha+\beta)}(t).$$

14.10 Exercise

1. Obtain the recurrence relation $xL_n''(x) + (1-x)L_n'(x) + nL_n(x) = 0.$
2. If $e^{-x} = \sum c_n L_n(x)$, then show that $c_n = \frac{1}{2^{n+1} n!}$.
3. Show that $L_n(x)$ is defined by $e^{\frac{-xt}{1-t}} = (1-t)\sum_{n=0}^{\infty} \frac{L_n(x)}{n!} t^n$.
4. Show that (i) $L_n^{(n)}(0) = (-1)^n$ and (ii) $L_n^{(n)}(0) = (-1)^n n!$, where $L_n^{(n)}(x)$ stands for $\frac{d^n}{dx^n} L_n(x)$.
5. Prove that $\int_x^{\infty} e^{-t} L_n(t) dt = e^{-x}\left[L_n(x) - L_{n-1}(x)\right]$.
6. Find the value of (i) $\int_0^{\infty} e^{-t} L_3(t) L_5(t) dt$ (ii) $\int_x^{\infty} e^{-t} [L_4(t)]^2 dt$.
7. Express $10 - 23x + 10x^2 - x^3$ in terms of Laguerre polynomials.
8. If $L_n(x)$ is defined by $e^{\frac{-xt}{1-t}} = (1-t)\sum_{n=0}^{\infty} \frac{L_n(x)}{n!} t^n, t < 1,$

9. Show that (i) $L_n'(x) - nL_{n-1}'(x) = -nL_n(x)$

   (ii) $L_{n+1}(x) = (2n+1-x)L_n(x) - n^2 L_{n-1}(x)$.

10. Show that (i) ${}_1F_1(-n;1;x) = L_n(x)$

    (ii) $e^t {}_0F_1(-;1;nx) = \sum_{n=0}^{\infty} \frac{L_n(x)}{n!} t^n$.

15
Jacobi Polynomials

Abstract
Jacobi polynomials are powerful tools with broad applications, ranging from solving differential equations and modeling physical phenomena to developing numerical methods and exploring advanced mathematical theories.

They are a set of orthogonal polynomials defined by a specific weight function and serve as generalizations of classical polynomials like Legendre and Chebyshev polynomials. In this chapter, we study their properties- orthogonality, generating functions, recurrence relations, and the Rodrigues formula.

Keywords: Ultraspherical polynomial, Bateman's generating functions, Rodrigues formula, orthogonality, recurrence relations, zeros

15.1 Introduction

Jacobi polynomials $P_n^{(\alpha,\beta)}(x)$ (occasionally called hypergeometric polynomials) are a class of classical orthogonal polynomials. They are orthogonal with respect to the weight $(1-x)^{\alpha}(1+x)^{\beta}$ on the interval $[-1,1]$. The Legendre, Zernike, and Chebyshev polynomials are special cases of the Jacobi polynomials. The Jacobi polynomials were introduced by Carl Gustav Jacob Jacobi (1804–1851) who made significant contributions to partial differential equations, astronomy, elliptic functions, and calculus of variation.

The Jacobi polynomials are also known as hypergeometric polynomials, occur in the study of rotation groups and in the solution to the equations of motion of the symmetric top. They are solutions to the Jacobi differential equation and yield some other specially named polynomials as special cases. Here, we discuss its orthogonality, generating functions, Rodrigues formulas, and recurrence relations.

Dipankar De. Ordinary Differential Equations and Special Functions, (561–586) © 2026 Scrivener Publishing LLC

15.2 Jacobi Polynomial

Jacobi Polynomial, an orthogonal polynomial $P_n^{(\alpha,\beta)}(x)$ is defined by the relation

$$P_n^{(\alpha,\beta)}(x) = \frac{(1+\alpha)_n}{n!}\, {}_2F_1\left[\begin{matrix} -n, 1+\alpha+\beta+n; & \frac{1-x}{2} \\ 1+\alpha; & \end{matrix}\right] \tag{15.1}$$

where $(1+\alpha)_n$ is a factorial function (Pochhammer's symbol). In this case, the series for the hypergeometric function is finite; therefore, one obtains the equivalent expression

$$P_n^{(\alpha,\beta)}(x) = \frac{\Gamma(\alpha+n+1)}{n!\,\Gamma(\alpha+\beta+n+1)} \sum_{m=0}^{n} \binom{n}{m} \frac{\Gamma(\alpha+\beta+n+m+1)}{\Gamma(\alpha+m+1)} \left(\frac{x-1}{2}\right)^m \tag{15.2}$$

Equation (15.2) can be written (expansion form) as

$$P_n^{(\alpha,\beta)}(x) = \sum_{m=0}^{n} \frac{(1+\alpha)_n (1+\alpha+\beta)_{n+m}}{m!(n-m)!(1+\alpha)_m (1+\alpha+\beta)_n} \left(\frac{x-1}{2}\right)^m, \tag{15.3}$$

which shows that $P_n^{(\alpha,\beta)}(x)$ is a polynomial of degree n.

Classical orthogonal polynomials are solutions of the second-order linear homogeneous differential equation of the form

$$a(x)y_n''(x) + b(x)y_n'(x) + \lambda_n y_n(x) = 0$$

where $a(x)$ is a polynomial of the degree at most 2, $b(x)$ is a polynomial of the degree 1, and λ_n does not depend on x. For the classical Jacobi polynomials, this equation has become

$$(1-x^2)y_n''(x) + [\beta - \alpha - \{\alpha+\beta+2\}x]y_n'(x) + n(n+\alpha+\beta+1)y_n(x) = 0,$$

which, in the case of the classical Legendre polynomials, is reduced to the equation

$$(1-x^2)y_n''(x) + 2xy_n'(x) + n(n+1)y_n(x) = 0 \text{ (here, } \alpha = 0, \beta = 0)$$

Clearly, $P_n^{(0,0)}(x)$ is a *Legendre polynomial.* A special case of Jacobi polynomial, i.e., $P_n^{(\alpha,\alpha)}(x)$, is known as *ultraspherical polynomial.* Special ultraspherical polynomials are Chebyshev polynomials $T_n(x)$ and $U_n(x)$, which are defined by

$$T_n(x) = \frac{n!}{(\frac{1}{2})_n} P_n^{\left(-\frac{1}{2},-\frac{1}{2}\right)}(x) \tag{15.4}$$

and,

$$U_n(x) = \frac{(n+1)!}{(\frac{1}{2})_n} P_n^{\left(\frac{1}{2},\frac{1}{2}\right)}(x). \tag{15.5}$$

If we replace the order of Equation (15.3) $(n-m)$ in place of m, we get

$$\begin{aligned}P_n^{(\alpha,\beta)}(x) &= \sum_{m=0}^{n} \frac{(1+\alpha)_n (1+\alpha+\beta)_{2n-m}}{m!(n-m)!(1+\alpha)_{n-m}(1+\alpha+\beta)_n} \left(\frac{x-1}{2}\right)^{n-m} \\ &= \frac{(1+\alpha+\beta)_{2n}}{n!(1+\alpha+\beta)_n} \left(\frac{x-1}{2}\right)^n \sum_{m=0}^{n} \frac{(-n)_m(-\alpha-n)_n(1+\alpha+\beta)_m}{m!(-\alpha-\beta-2n)_m} \left(\frac{2}{x-1}\right)^m \\ &= \frac{(1+\alpha+\beta)_{2n}}{n!(1+\alpha+\beta)_n} \left(\frac{x-1}{2}\right)^n {}_2F_1\left[\begin{matrix}-n,-\alpha-n; & \frac{2}{1-x} \\ -\alpha-\beta-2n; & \end{matrix}\right].\end{aligned}$$

Applying the transformation to the function ${}_2F_1$ of Equation (15.1),

$$\left\{\text{using } (1-x)^{-\alpha}\left[\begin{matrix}a,c-b; & -\frac{x}{1-x} \\ c & \end{matrix}\right] = F[a,b;c;x]\right\},$$

we have

$$P_n^{(\alpha,\beta)}(x) = \frac{(1+\alpha)_n}{n!}\left(\frac{x+1}{2}\right)^n {}_2F_1\left[\begin{matrix}-n,-\beta-n; & \frac{x-1}{x+1} \\ 1+\alpha; & \end{matrix}\right]. \tag{15.6}$$

In the expansion form, it can be written as

$$P_n^{(\alpha,\beta)}(x)=\sum\nolimits_{m=0}^{n}\frac{(1+\alpha)_n(1+\beta)_n}{m!(n-m)!(1+\alpha)_m(1+\beta)_{n-m}}\left(\frac{x-1}{2}\right)^m\left(\frac{x+1}{2}\right)^{n-m} \tag{15.7}$$

If we reverse the order of Equation (15.7), we get

$$P_n^{(\alpha,\beta)}(x)=\sum\nolimits_{m=0}^{n}\frac{(1+\alpha)_n(1+\beta)_n}{(n-m)!m!(1+\alpha)_{n-m}(1+\beta)_m}\left(\frac{x-1}{2}\right)^{n-m}\left(\frac{x+1}{2}\right)^m$$

$$=\frac{(1+\beta)_n}{n!}\left(\frac{x-1}{2}\right)^n {}_2F_1\left[\begin{matrix}-n,-\alpha-n; & \frac{x+1}{x-1}\\ 1+\beta; & \end{matrix}\right]. \tag{15.8}$$

15.3 Generating Functions

(a) The Jacobi polynomial is

$$P_n^{(\alpha,\beta)}(x)=\sum\nolimits_{m=0}^{n}\frac{(1+\alpha)_n(1+\alpha+\beta)_{n+m}}{m!(n-m)!(1+\alpha)_m(1+\alpha+\beta)_n}\left(\frac{x-1}{2}\right)^m$$

or,

$$\sum\nolimits_{n=0}^{\infty}\frac{(1+\alpha+\beta)_n P_n^{(\alpha,\beta)}(x)}{(1+\alpha)_n}t^n=\sum\nolimits_{n=0}^{\infty}\sum\nolimits_{m=0}^{n}\frac{(1+\alpha+\beta)_{n+m}}{m!(n-m)!(1+\alpha)_m}\frac{(x-1)^m}{2^m}t^n$$

$$=\sum\nolimits_{n=0}^{\infty}\sum\nolimits_{m=0}^{n}\frac{(1+\alpha+\beta)_{n+2m}}{m!n!(1+\alpha)_m}\frac{(x-1)^m}{2^m}t^{n+m} \text{ (replacing } n \text{ by } n+m)$$

$$=\sum\nolimits_{n=0}^{\infty}\sum\nolimits_{n=0}^{n}\frac{(1+\alpha+\beta+2m)_n}{n!}t^n\frac{(1+\alpha+\beta)_{2m}}{m!(1+\alpha)_m}\frac{(x-1)^m}{2^m}t^m$$

$$=\sum\nolimits_{m=0}^{\infty}(1-t)^{-1-\alpha-\beta-2m}2^{2m}\frac{\left(\frac{1+\alpha+\beta}{2}\right)_m\left(\frac{2+\alpha+\beta}{2}\right)_m}{m!(1+\alpha)_m}\frac{(x-1)^m}{2^m}t^m$$

$$=(1-t)^{-1-\alpha-\beta}\sum\nolimits_{m=0}^{\infty}\frac{\left(\frac{1+\alpha+\beta}{2}\right)_m\left(\frac{2+\alpha+\beta}{2}\right)_m}{m!(1+\alpha)_m}\left\{\frac{2t(x-1)}{(1-t)^2}\right\}^m$$

$$=(1-t)^{-1-\alpha-\beta}{}_2F_1\left[\begin{matrix}\left(\frac{1+\alpha+\beta}{2}\right),\left(\frac{2+\alpha+\beta}{2}\right); & \frac{2t(x-1)}{(1-t)^2}\\ 1+\alpha; & \end{matrix}\right]. \tag{15.9}$$

This is the generating relation of Jacobi polynomial.

(b) The expansion form $P_n^{(\alpha,\beta)}(x)$ may be written as

$$P_n^{(\alpha,\beta)}(x)=\sum_{m=0}^{n}\frac{(1+\alpha)_n(1+\beta)_n}{m!(n-m)!(1+\alpha)_m(1+\beta)_{n-m}}\left(\frac{x-1}{2}\right)^m\left(\frac{x+1}{2}\right)^{n-m}$$

$$\text{or, } \sum_{n=0}^{\infty}\frac{P_n^{(\alpha,\beta)}(x)}{(1+\alpha)_n(1+\beta)_n}t^n$$

$$=\sum_{n=0}^{\infty}\sum_{m=0}^{\infty}\frac{1}{m!(n-m)!(1+\alpha)_m(1+\beta)_{n-m}}\left(\frac{x-1}{2}\right)^m\left(\frac{x+1}{2}\right)^{n-m}t^n$$

$$=\sum_{n=0}^{\infty}\sum_{m=0}^{\infty}\frac{1}{m!n!(1+\alpha)_m(1+\beta)_n}\left(\frac{x-1}{2}\right)^m\left(\frac{x+1}{2}\right)^n t^{n+m}\ \text{(replacing } n \text{ by } n+m)$$

$$=\left(\sum_{n=0}^{\infty}\frac{\left(\frac{x-1}{2}t\right)^m}{m!(1+\alpha)_m}\right)\left(\sum_{n=0}^{\infty}\frac{\left(\frac{x+1}{2}t\right)^n}{n!(1+\beta)_n}\right)\sum_{n=0}^{\infty}\sum_{m=0}^{\infty}\frac{1}{m!(1+\beta)_n}t^{n+m}$$

$$={}_0F_1\left[\begin{matrix}-;\\1+\alpha;\end{matrix}\left(\frac{x-1}{2}t\right)\right]{}_0F_1\left[\begin{matrix}-;\\1+\beta;\end{matrix}\left(\frac{x+1}{2}t\right)\right]$$

$$\therefore {}_2F_1\left[\begin{matrix}-;\\1+\alpha;\end{matrix}\left(\frac{x-1}{2}t\right)\right]{}_2F_1\left[\begin{matrix}-;\\1+\beta;\end{matrix}\left(\frac{x+1}{2}t\right)\right]$$

$$={}_2F_1\left[\begin{matrix}-;\\1+\alpha;\end{matrix}\left(\frac{x-1}{2}t\right)\right]{}_2F_1\left[\begin{matrix}-;\\1+\beta;\end{matrix}\left(\frac{x+1}{2}t\right)\right]. \tag{15.10}$$

(c) If we replace $x\ by -x\ and\ t\ by -t$ in the above equation, we get

$$\sum_{n=0}^{\infty}\frac{(-1)^n P_n^{(\alpha,\beta)}(-x)}{(1+\alpha)_n(1+\beta)_n}t^n={}_0F_1\left[\begin{matrix}-;\\1+\alpha;\end{matrix}\left(\frac{x+1}{2}t\right)\right]{}_0F_1\left[\begin{matrix}-;\\1+\beta;\end{matrix}\left(\frac{x-1}{2}t\right)\right]$$

$$=\sum_{n=0}^{\infty}\frac{P_n^{(\beta,\alpha)}(x)}{(1+\alpha)_n(1+\beta)_n}t^n.$$

By comparing the coefficients of t^n, we get

$$P_n^{(\alpha,\beta)}(-x)=(-1)^n P_n^{(\beta,\alpha)}(x). \tag{15.11}$$

Applying Equation (15.11) in

$P_n^{(\alpha,\beta)}(x)=\frac{(1+\alpha)_n}{n!}\,{}_2F_1\left[\begin{matrix}-n,1+\alpha+\beta+n;\frac{1-x}{2}\\1+\alpha;\end{matrix}\right]$, we get

$$P_n^{(\alpha,\beta)}(x)=(-1)^n\frac{(1+\beta)_n}{n!}\,{}_2F_1\left[\begin{matrix}-n,1+\alpha+\beta+n;\frac{1+x}{2}\\1+\beta;\end{matrix}\right] \tag{15.12}$$

or, $$P_n^{(\alpha,\beta)}(x)=\sum_{m=0}^{n}(-1)^{n-m}\frac{(1+\beta)_n(1+\alpha+\beta)_{n+m}}{m!(n-m)!(1+\beta)_m(1+\alpha+\beta)_n}\left(\frac{1+x}{2}\right)^m. \tag{15.13}$$

By reversing the order, it becomes

$$\begin{aligned}P_n^{(\alpha,\beta)}(x)&=\sum_{m=0}^{n}(-1)^{m}\frac{(1+\beta)_n(1+\alpha+\beta)_{2n-m}}{m!(n-m)!(1+\beta)_{n-m}(1+\alpha+\beta)_n}\left(\frac{1+x}{2}\right)^{n-m}\\&=\frac{(1+\alpha+\beta)_{2n}}{n!(1+\alpha+\beta)_n}\left(\frac{1+x}{2}\right)^n{}_2F_1\left[\begin{matrix}-n,-\beta-n;\ \frac{2}{1+x}\\-\alpha-\beta-2n;\end{matrix}\right].\end{aligned} \tag{15.14}$$

From Equation (15.11), we get $P_n^{(\alpha,\beta)}(-1)=(-1)^n P_n^{(\beta,\alpha)}(1)$

or, $$P_n^{(\alpha,\beta)}(-1)=(-1)^n\frac{(1+\beta)_n}{n!}. \tag{15.15}$$

It follows that

$$P_n^{(\alpha,\beta)}(1)=\frac{(1+\alpha)_n}{n!}. \tag{15.16}$$

15.4 Rodrigues' Formula

The Rodrigues' formula for the Jacobi polynomial is stated below.

Theorem 15.4.1

$$(1-x)^{\alpha}(1+x)^{\beta}P_n^{(\alpha,\beta)}(x)=\frac{(-1)^n}{2n!}\frac{d^n}{dx^n}\left[(1-x)^{n+\alpha}(1+x)^{n+\beta}\right]$$

Proof: We know that

$$P_n^{(\alpha,\beta)}(x)=\sum_{m=0}^{n}\frac{(1+\alpha)_n(1+\beta)_n}{m!(n-m)!(1+\alpha)_m(1+\beta)_{n-m}}\left(\frac{x-1}{2}\right)^m\left(\frac{x+1}{2}\right)^{n-m}$$

$$=\frac{(x-1)^{-\alpha}(x+1)^{-\beta}}{2^n}\sum_{m=0}^{n}\frac{(1+\alpha)_n(1+\beta)_n}{m!(n-m)!(1+\alpha)_m(1+\beta)_{n-m}}$$

$$(x-1)^{m+\alpha}(x+1)^{n-m+\beta}. \tag{15.17}$$

If s *and* k are non-negative integers, we have

$$\frac{d^s}{dx^s}(x^{k+\alpha})\equiv D^s(x^{k+\alpha})$$

$$=(k+\alpha)(k+\alpha-1)\ldots.(k+\alpha-s+1)x^{k+\alpha-s}$$

$$=\frac{(1+\alpha)_k}{(1+\alpha)_{k-s}}x^{k+\alpha-s}.$$

Hence, we obtain $D^m\left[(x+1)^{n+\beta}\right]=\dfrac{(1+\beta)_n}{(1+\beta)_{n-m}}(x+1)^{n+\beta-m}$

and,

$$D^{m-n}[(x-1)^{n+\alpha}]=\frac{(1+\alpha)_n}{(1+\alpha)_m}(x-1)^{m+\alpha}. \tag{15.18}$$

Using these results, we can write Equation (15.17) in the form

$$P_n^{(\alpha,\beta)}(x) = \frac{(x-1)^{-\alpha}(x+1)^{-\beta}}{n!2^n} \sum_{m=0}^{n} \frac{1}{\mathrm{m!(n-m)!}} D^m\left[(x+1)^{n+\beta}\right]$$
$$D^{m-n}[(x-1)^{n+\alpha}].$$

Using the Leibnitz theorem, we get

$$P_n^{(\alpha,\beta)}(x) = \frac{(x-1)^{-\alpha}(x+1)^{-\beta}}{n!2^n} D^n\left[(x-1)^{n+\alpha}(x+1)^{n+\beta}\right]$$

$$\therefore (1-x)^{\alpha}(1+x)^{\beta} P_n^{(\alpha,\beta)}(x) = \frac{(-1)^n}{2n!} \frac{d^n}{dx^n}\left[(1-x)^{n+\alpha}(1+x)^{n+\beta}\right]. \tag{15.19}$$

15.5 Orthogonality of Jacobi Polynomial

The Jacobi polynomial is an orthogonal polynomial in the interval (-1,1) with respect to the weight function $(1-x)^{\alpha}(1+x)^{\beta}$. Now, using the Rodrigues formula $(1-x)^{\alpha}(1+x)^{\beta} P_n^{(\alpha,\beta)}(x) = \frac{(-1)^n}{2n!} \frac{d^n}{dx^n}\left[(1-x)^{n+\alpha}(1+x)^{n+\beta}\right]$, we have

$$\varphi = \int_{-1}^{1} (1-x)^{\alpha}(1+x)^{\beta} P_n^{(\alpha,\beta)}(x) P_m^{(\alpha,\beta)}(x)dx$$
$$= \frac{(-1)^n}{2n!} \int_{-1}^{1} \frac{d^n}{dx^n}\left[(1-x)^{n+\alpha}(1+x)^{n+\beta}\right] P_m^{(\alpha,\beta)}(x)dx.$$

For integration, let us apply integration by parts taking $P_m^{(\alpha,\beta)}(x)$ as first function and $\frac{d^n}{dx^n}\left[(1-x)^{n+\alpha}(1+x)^{n+\beta}\right]$ as the second function; we get

$$\varphi = \frac{(-1)^n}{2^n n!}\left[\frac{d^{n-1}}{dx^{n-1}}\left\{(1-x)^{n+\alpha}(1+x)^{n+\beta}\right\} P_m^{(\alpha,\beta)}(x)\right]_{-1}^{1}$$
$$-\frac{(-1)^n}{2n!} \int_{-1}^{1} \frac{d^{n-1}}{dx^{n-1}}\left[(1-x)^{n+\alpha}(1+x)^{n+\beta}\right]\left\{\frac{d}{dx} P_m^{(\alpha,\beta)}(x)\right\}dx.$$

Under this condition, $Re(\alpha)>-1, Re(\beta)>-1$, the first term vanishes, and we get

$$\varphi=-\frac{(-1)^n}{2^n n!}\int_{-1}^{1}\frac{d^{n-1}}{dx^{n-1}}\left[(1-x)^{n+\alpha}(1+x)^{n+\beta}\right]\left\{\frac{d}{dx}P_m^{(\alpha,\beta)}(x)\right\}dx.$$

In this process, repeating integration *n times*, we have

$$\varphi=\int_{-1}^{1}(1-x)^{\alpha}(1+x)^{\beta}P_n^{(\alpha,\beta)}(x)P_m^{(\alpha,\beta)}(x)dx$$

$$=\frac{(-1)^n}{2n!}\int_{-1}^{1}\frac{d^n}{dx^n}\left[(1-x)^{n+\alpha}(1+x)^{n+\beta}\right]P_m^{(\alpha,\beta)}(x)dx. \quad (15.19a)$$

Case I. If $n>m$, since $P_m^{(\alpha,\beta)}(x)$ is a polynomial of *mth* degree, $\frac{d^n}{dx^n}P_m^{(\alpha,\beta)}(x)$ must be vanished.

$$\int_{-1}^{1}(1-x)^{\alpha}(1+x)^{\beta}P_n^{(\alpha,\beta)}(x)P_m^{(\alpha,\beta)}(x)dx=0$$

Case II. If $n<m$, instead of substituting the value of $(1-x)^{\alpha}(1+x)^{\beta}P_n^{(\alpha,\beta)}$, we substitute

$$(1-x)^{\alpha}(1+x)^{\beta}P_m^{(\alpha,\beta)}.$$

Repeating the previous same process of integration, we find that φ is again zero.

Hence, we arrive at the orthogonality relation

$$\int_{-1}^{1}(1-x)^{\alpha}(1+x)^{\beta}P_n^{(\alpha,\beta)}(x)P_m^{(\alpha,\beta)}(x)dx=0 \ \textit{for}\ m\neq n$$

under the condition $Re(\alpha)>-1, Re(\beta)>-1$.

15.5.1 Properties of Orthogonality

Show that

$$\int_{-1}^{1}(1-x)^{\alpha}(1+x)^{\beta}\left\{P_n^{(\alpha,\beta)}(x)\right\}^2(x)dx=\frac{2^{\alpha+\beta+1}\Gamma(1+\alpha+n)\Gamma(1+\beta+n)}{n!(1+\alpha+\beta+2n)\Gamma(1+\alpha+\beta+n)}.$$

Solution: When $m = n$, from Equation (15.19a),

$$\varphi_1 = -\frac{1}{2^n n!}\int_{-1}^{1}\left[(1-x)^{n+\alpha}(1+x)^{n+\beta}\right]\left\{\frac{d^n}{dx^n}P_n^{(\alpha,\beta)}(x)\right\}dx. \quad (15.20)$$

But $\frac{d^n}{dx^n}P_n^{(\alpha,\beta)}(x) = \frac{(1+\alpha)_n}{n!}D^n\left\{{}_2F_1\left[\begin{matrix}-n, 1+\alpha+\beta+n; \frac{1-x}{2}\\ 1+\alpha\end{matrix}\right]\right\}$.

Using $\frac{d}{dx}F\left[\begin{matrix}a,b; & x\\ c\end{matrix}\right] = \frac{ab}{c}F\left[\begin{matrix}a+1, b+1; & x\\ c+1\end{matrix}\right]$ and n times differentiation, we have

$$D^n\left\{{}_2F_1\left[\begin{matrix}-n, 1+\alpha+\beta+n; \frac{1-x}{2}\\ 1+\alpha\end{matrix}\right]\right\}$$

$$= \frac{(\frac{1}{2})_n(-n)_n(1+\alpha+\beta+n)_n}{(1+\alpha)_n}\,{}_2F_1\left[\begin{matrix}0, 1+\alpha+\beta+2n; \frac{1-x}{2}\\ 1+\alpha+n\end{matrix}\right]$$

$$= \frac{(-1)^n n!(1+\alpha+\beta)_{2n}}{2^n(1+\alpha)_n(1+\alpha+\beta)_n}.$$

Thus,

$$\frac{d^n}{dx^n}P_n^{(\alpha,\beta)}(x) = \frac{(1+\alpha+\beta)_{2n}}{2^n(1+\alpha+\beta)_n}. \quad (15.21)$$

From Equation (15.19a), $\varphi_1 = \frac{1}{2^n n!}\int_{-1}^{1}\left[(1-x)^{n+\alpha}(1+x)^{n+\beta}\right]\left\{\frac{d^n}{dx^n}P_n^{(\alpha,\beta)}(x)\right\}$ dx.

$$= \frac{1}{2^{2n}n!}\frac{(1+\alpha+\beta)_{2n}}{2^n(1+\alpha+\beta)_n}\int_{-1}^{1}\left[(1-x)^{n+\alpha}(1+x)^{n+\beta}\right]dx$$

For evaluation, substitute $1+x=2y$, then

$$\varphi_1 = \frac{1}{2^{2n} n!} \frac{(1+\alpha+\beta)_{2n} 2^{1+\alpha+\beta+2n}}{2^n (1+\alpha+\beta)_n} \int_0^1 \left[(1-y)^{n+\alpha} (y)^{n+\beta}\right] dy$$

$$= \frac{2^{1+\alpha+\beta}}{n!} \frac{(1+\alpha+\beta)_{2n}}{2^n (1+\alpha+\beta)_n} \frac{\Gamma(1+\alpha+n)\Gamma(1+\beta+n)}{\Gamma(2+\alpha+\beta+2n)}$$

$$= \frac{2^{1+\alpha+\beta}}{n!} \frac{(1+\alpha+\beta)_{2n}}{2^n (1+\alpha+\beta+2n)} \frac{\Gamma(1+\alpha+n)\Gamma(1+\beta+n)}{\Gamma(2+\alpha+\beta+n)}.$$

Thus, we obtain

$$\int_{-1}^{1} (1-x)^{\alpha} (1+x)^{\beta} \left\{P_n^{(\alpha,\beta)}(x)\right\}^2 (x) dx = \frac{2^{\alpha+\beta+1} \Gamma(1+\alpha+n)\Gamma(1+\beta+n)}{n!(1+\alpha+\beta+2n)\Gamma(1+\alpha+\beta+n)}.$$

Remark 15.5.1. For the orthogonality of Jacobi polynomials, it also follows that

$$\int_{-1}^{1} (1-x)^{\alpha} (1+x)^{\beta} x^m P_n^{(\alpha,\beta)}(x)\, dx = 0, \text{ when } m = 0,1,\ldots.(n-1).$$

Remark 15.5.2. We know that $\frac{d^n}{dx^n} P_n^{(\alpha,\beta)}(x) = \frac{(1+\alpha+\beta)_{2n}}{2^n (1+\alpha+\beta)_n}$.

Using this relation, we get $P_n^{(\alpha,\beta)}(x) = \frac{(1+\alpha+\beta)_{2n}}{2^n n!(1+\alpha+\beta)_n} x^n$ + a polynomial of the $(n-1)th$ degree.

Hence, $x^n = \frac{2^n n!(1+\alpha+\beta)_n}{(1+\alpha+\beta)_{2n}} P_n^{(\alpha,\beta)}(x)$ + a polynomial of the $(n-1)th$ degree.

Any polynomial of degree m can be expressed as a linear combination of polynomials $P_1^{(\alpha,\beta)}(x), P_2^{(\alpha,\beta)}(x), \ldots.P_m^{(\alpha,\beta)}(x)$.

Using the above two relations, we get

$$\int_{-1}^{1} (1-x)^{\alpha} (1+x)^{\beta} x^n P_n^{(\alpha,\beta)}(x)(x) dx = \frac{2^{\alpha+\beta+1+n} \Gamma(1+\alpha+n)\Gamma(1+\beta+n)}{\Gamma(2+\alpha+\beta+2n)}.$$

15.6 Recurrence Relations

We will find two types of recurrence relations as follows: (a) relations in which a shift of index n only is involved and (b) relations in which a shift in α and β is also involved.

15.6.1 (a) Relations in which shift n is involved

We have a generating function

$$\sum_{n=0}^{\infty} \frac{(1+\alpha+\beta)_n P_n^{(\alpha,\beta)}(x)}{(1+\alpha)_n} t^n = (1-t)^{-1-\alpha-\beta} {}_2F_1\left[\begin{matrix}\left(\frac{1+\alpha+\beta}{2}\right),\left(\frac{2+\alpha+\beta}{2}\right); \\ 1+\alpha;\end{matrix} \frac{2t(x-1)}{(1-t)^2}\right]. \tag{15.22}$$

If we write

$$f(x,t) = (1-t)^{-1-\alpha-\beta} {}_2F_1\left[\begin{matrix}\left(\frac{1+\alpha+\beta}{2}\right),\left(\frac{2+\alpha+\beta}{2}\right); \\ 1+\alpha;\end{matrix} \frac{2t(x-1)}{(1-t)^2}\right]$$

or,

$$f(y,t) = (1-t)^{-c}\psi\left\{\frac{-4ty}{(1-t^2)}\right\},$$

where $c = 1+\alpha+\beta$ and ψ is ${}_2F_1$ with $y = \frac{1}{2}(1-x)$, then we have

$$\frac{\partial f}{\partial y} = -4t(1-t)^{-c-2}\psi'$$

and $\frac{\partial f}{\partial t} = c(1-t)^{-c-1}\psi - 4y(1+t)(1-t)^{-c-3}\psi'.$

Therefore, $y(1+t)\frac{\partial f}{\partial y}-t(1-t)\frac{\partial f}{\partial t}=-ctf$,

which may be written in the forms

$$y\frac{\partial f}{\partial y}-t\frac{\partial f}{\partial t}=-ctf-t^2\frac{\partial f}{\partial t}-yt\frac{\partial f}{\partial y} \tag{15.23}$$

$$y\frac{\partial f}{\partial y}-t\frac{\partial f}{\partial t}=\frac{-ct}{1-t}f-\frac{2yt}{1+t}\frac{\partial f}{\partial y} \tag{15.24}$$

$$y\frac{\partial f}{\partial y}-t\frac{\partial f}{\partial t}=\frac{-ct}{1+t}f-\frac{2t^2}{1+t}\frac{\partial f}{\partial t}. \tag{15.25}$$

(i) Since $f=\sum_{n=0}^{\infty}\frac{(c)_n P_n^{(\alpha,\beta)}(1-2y)}{(1+\alpha)_n}t^n$

(ii) becomes

$$\sum_{n=0}^{\infty}\left[\frac{y(c)_n}{(1+\alpha)_n}\frac{\partial}{\partial y}\{P_n^{(\alpha,\beta)}(1-2y)\}-n\frac{(c)_n}{(1+\alpha)_n}\{P_n^{(\alpha,\beta)}(1-2y)\}\right]t^n$$

$$=-\sum_{n=0}^{\infty}\frac{(c)_n t}{(1+\alpha)_n}P_n^{(\alpha,\beta)}(1-2y)t^{n+1}-\sum_{n=0}^{\infty}\frac{n(c)_n}{(1+\alpha)_n}\{P_n^{(\alpha,\beta)}(1-2y)\}t^{n+1}$$

$$-\sum_{n=0}^{\infty}\frac{y(c)_n}{(1+\alpha)_n}\frac{d}{dy}\{P_n^{(\alpha,\beta)}(1-2y)\}t^{n+1}$$

$$=-\sum_{n=0}^{\infty}\frac{(c+n-1)\,(c)_{n-1}}{(1+\alpha)_{n-1}}P_n^{(\alpha,\beta)}(1-2y)t^n$$

$$-\sum_{n=1}^{\infty}\frac{y(1+\alpha+\beta)_{n-1}}{(1+\alpha)_{n-1}}\frac{d}{dy}\{P_{n-1}^{(\alpha,\beta)}(1-2y)\}t^n.$$

Comparing the coefficients of t^n on both sides, we get

$$\frac{y\,(c)_n}{(1+\alpha)_n}\frac{d}{dy}\left\{P_n^{(\alpha,\beta)}(1-2y)\right\}-n\frac{(c)_n}{(1+\alpha)_n}\left\{P_n^{(\alpha,\beta)}(1-2y)\right\}$$

$$=-\frac{(1+\alpha=n)_{n-1}\,(c)_{n-1}}{(1+\alpha)_{n-1}}P_n^{(\alpha,\beta)}(1-2y)$$

$$-\frac{y\,(c)_{n-1}}{(1+\alpha)_{n-1}}\frac{d}{dy}\left\{P_{n-1}^{(\alpha,\beta)}(1-2y)\right\}$$

or, $$y(\alpha+\beta+n)\frac{d}{dy}\left\{P_n^{(\alpha,\beta)}(1-2y)\right\}+y(n+\alpha)\frac{d}{dy}\left\{P_{n-1}^{(\alpha,\beta)}(1-2y)\right\}$$

$$=n(\alpha+\beta+n)\frac{d}{dy}\left\{P_n^{(\alpha,\beta)}(1-2y)\right\}$$

$$-(\alpha+\beta+n)(\alpha+n)\frac{d}{dy}\left\{P_{n-1}^{(\alpha,\beta)}(1-2y)\right\}.$$

Therefore,

$$(x-1)\left[(\alpha+\beta+n)\frac{d}{dx}\left\{P_n^{(\alpha,\beta)}(x)\right\}+(n+\alpha)\frac{d}{dx}\left\{P_{n-1}^{(\alpha,\beta)}(x)\right\}\right]$$

$$=(\alpha+\beta+n)\left[nP_n^{(\alpha,\beta)}(x)-(\alpha+n)P_{n-1}^{(\alpha,\beta)}(x)\right], \tag{15.26}$$

which is one recurrence relation.

(ii) The relation $y\frac{\partial f}{\partial y}-t\frac{\partial f}{\partial t}=\frac{-ct}{1-t}f-\frac{2yt}{1+t}\frac{\partial f}{\partial y}$ gives

$$\sum_{n=0}^{\infty}\left[y\frac{(c)_n\frac{d}{dy}\{P_n^{(\alpha,\beta)}(1-2y)\}}{(1+\alpha)_n}-\frac{n(c)_n}{(1+\alpha)_n}\{P_n^{(\alpha,\beta)}(1-2y)\}\right]t^n$$

$$=-\frac{ct}{1-t}\sum_{n=0}^{\infty}\frac{(c)_n\frac{d}{dy}\{P_n^{(\alpha,\beta)}(1-2y)\}}{(1+\alpha)_n}t^n$$

$$-\frac{2yt}{1+t}\sum_{n=0}^{\infty}\frac{(c)_n}{(1+\alpha)_n}\frac{d}{dy}\{P_n^{(\alpha,\beta)}(1-2y)\}t^n$$

$$=-c\sum_{n=0}^{\infty}\sum_{m=0}^{\infty}\frac{(c)_n\{P_n^{(\alpha,\beta)}(1-2y)\}}{(1+\alpha)_n}t^{n+m+1}$$

$$-2y\sum_{n=0}^{\infty}\sum_{m=0}^{\infty}\frac{(c)_n\frac{d}{dy}\{P_n^{(\alpha,\beta)}(1-2y)\}}{(1+\alpha)_n}t^{n+m+1}$$

$$=-c\sum_{n=0}^{\infty}\sum_{m=0}^{\infty}\frac{(c)_{n-m}\{P_{n-m}^{(\alpha,\beta)}(1-2y)\}}{(1+\alpha)_{n-m}}t^{n+1}$$

$$-2y\sum_{n=0}^{\infty}\sum_{m=0}^{\infty}\frac{(c)_{n-m}\frac{d}{dy}\{P_{n-m}^{(\alpha,\beta)}(1-2y)\}}{(1+\alpha)_{n-m}}t^{n+1}$$

$$=-c\sum_{n=0}^{\infty}\sum_{m=0}^{n}\frac{(c)_m\{P_m^{(\alpha,\beta)}(1-2y)\}}{(1+\alpha)_m}t^{n+1}$$

$$-2y\sum_{n=0}^{\infty}\sum_{m=0}^{n}\frac{(c)_m\frac{d}{dy}\{P_m^{(\alpha,\beta)}(1-2y)\}}{(1+\alpha)_m}t^{n+1}.$$

Comparing the coefficients of t^n from both sides, we get

$$(x-1)\frac{d}{dx}P_n^{(\alpha,\beta)}(x)-nP_n^{(\alpha,\beta)}(x)$$

$$=\frac{(1+\alpha)_n}{(1+\alpha+\beta)_n}\sum_{m=0}^{n=1}\frac{(1+\alpha+\beta)_m}{(1+\alpha)_m}\left[(1+\alpha+\beta)P_m^{(\alpha,\beta)}(x)+2(x-1)\frac{d}{dx}P_n^{(\alpha,\beta)}(x)\right],\tag{15.27}$$

which is another recurrence relation.

(iii) Similarly, from $y\frac{\partial f}{\partial y}-t\frac{\partial f}{\partial t}=\frac{-ct}{1+t}f-\frac{2t^2}{1+t}\frac{\partial f}{\partial t}$, we get

$$\sum_{n=0}^{\infty}\left[y\frac{(c)_n}{(1+\alpha)_n}\frac{d}{dy}\left\{P_n^{(\alpha,\beta)}(1-2y)\right\}-\frac{n(c)_n}{(1+\alpha)_n}\left\{P_n^{(\alpha,\beta)}(1-2y)\right\}\right]t^n$$

$$=-\frac{ct}{1-t}\sum_{n=0}^{\infty}\frac{(c)_n\left\{P_n^{(\alpha,\beta)}(1-2y)\right\}}{(1+\alpha)_n}t^n-\frac{2t^2}{1+t}\sum\frac{n(c)_n}{(1+\alpha)_n}\left\{P_n^{(\alpha,\beta)}(1-2y)\right\}t^{n+1}$$

$$=-c\sum_{n=0}^{\infty}\sum_{m=0}^{\infty}\frac{(-1)^m(c)_n\left\{P_n^{(\alpha,\beta)}(1-2y)\right\}}{(1+\alpha)_n}t^{n+m+1}$$

$$-2\sum_{n=1}^{\infty}\sum_{m=0}^{\infty}\frac{(-1)^m n(c)_n}{(1+\alpha)_n}\left\{P_n^{(\alpha,\beta)}(1-2y)\right\}t^{n+m+1}$$

$$=-c\sum_{n=0}^{\infty}\sum_{m=0}^{n}\frac{(-1)^m(c)_{n-m}\left\{P_{n-m}^{(\alpha,\beta)}(1-2y)\right\}}{(1+\alpha)_{n-m}}t^{n+1}$$

$$-2\sum_{n=1}^{\infty}\sum_{m=0}^{\infty}\frac{(-1)^m(n-m)(c)_{n-m}}{(1+\alpha)_{n-m}}\left\{P_{n-m}^{(\alpha,\beta)}(1-2y)\right\}t^{n+1}$$

$$=-c\sum_{n=0}^{\infty}\sum_{m=0}^{n}\frac{(-1)^{n-m}(c)_m\left\{P_m^{(\alpha,\beta)}(1-2y)\right\}}{(1+\alpha)_m}t^{n+1}$$

$$-2\sum_{n=1}^{\infty}\sum_{m=0}^{n}\frac{(-1)^{n-m}(m)(c)_m}{(1+\alpha)_m}\left\{P_m^{(\alpha,\beta)}(1-2y)\right\}t^{n+1},$$

which gives the third recurrence relation.

(iv) In the relation Equation (v), if we replace x *by* $-x$ and interchange α *and* β and use the relation

$$P_n^{(\alpha,\beta)}(-x)=(-1)^n P_n^{(\alpha,\beta)}(x),$$

we obtain the recurrence relation

$$(x+1)\left[(\alpha+\beta+n)\frac{d}{dx}P_n^{(\alpha,\beta)}(x)-(\beta+n)\frac{d}{dx}P_{n-1}^{(\alpha,\beta)}(x)\right]$$
$$=(\alpha+\beta+n)\left[nP_n^{(\alpha,\beta)}(x)+(\beta+n)P_{n-1}^{(\alpha,\beta)}(x)\right]. \quad (15.28)$$

(v) From the relation Equations (v) and (vii), if we eliminate $\frac{d}{dx}P_{n-1}^{(\alpha,\beta)}(x)$, we get another recurrence relation

$$(\alpha+\beta+2n)(x^2-1)\frac{d}{dx}P_n^{(\alpha,\beta)}(x)=n[\beta-\alpha+(\alpha\beta+2n)x]P_n^{(\alpha,\beta)}(x)$$
$$-2(\alpha+n)(\beta+n)P_{n-1}^{(\alpha,\beta)}(x). \quad (15.29)$$

15.6.2 (b) Relations in which shifts in α, β, and n are involved

The definition of Jacobi polynomial is

$$P_n^{(\alpha,\beta)}(x)=\frac{(1+\alpha)_n}{n!}\,{}_2F_1\left[\begin{matrix}-n,1+\alpha+\beta+n;\frac{1-x}{2}\\ 1+\alpha;\end{matrix}\right].$$

Now,

$$\frac{d}{dx}P_n^{(\alpha,\beta)}(x)=\frac{n(1+\alpha)_n(1+\alpha+\beta+n)}{n!2(1+\alpha)}$$
$${}_2F_1\left[\begin{matrix}-n+2+\alpha+\beta+n;\frac{1-x}{2}\\ 2+\alpha;\end{matrix}\right]$$
$$=\frac{(2+\alpha)_{n-1}(1+\alpha+\beta+n)}{2(n-1)!}$$
$${}_2F_1\left[\begin{matrix}-(n-1),1+(\alpha+1)(1+\beta)+n-1;\frac{1-x}{2}\\ 1+\alpha+1;\end{matrix}\right]$$
$$=\frac{1}{2}(1+\alpha+\beta+n)P_{n-1}^{(\alpha+1,\beta+1)}(x).$$

Thus,

$$\frac{d}{dx}P_n^{(\alpha,\beta)}(x)=\frac{1}{2}(1+\alpha+\beta+n)P_{n-1}^{(\alpha+1,\beta+1)}(x). \tag{15.30}$$

It follows that

$$\begin{aligned}\frac{d^2}{dx^2}P_n^{(\alpha,\beta)}(x)&=\frac{1}{2}(1+\alpha+\beta+n)\frac{d}{dx}P_{n-1}^{(\alpha+1,\beta+1)}(x)\\&=\frac{1}{2^2}(1+\alpha+\beta+n)(2+\alpha+\beta+n)P_{n-2}^{(\alpha+2,\beta+2)}(x).\end{aligned}$$

Repeating the process of differentiation n times,

$$\begin{aligned}\frac{d^m}{dx^m}P_n^{(\alpha,\beta)}(x)=\frac{1}{2^m}&(1+\alpha+\beta+n)(2+\alpha+\beta+n)\\&\ldots(m+\alpha+\beta+n)P_{n-m}^{(\alpha+m,\beta+m)}(x)\end{aligned} \tag{15.31}$$

where $0<m\leq n$.

(ii) We recall

$$P_n^{(\alpha,\beta)}(x)=\frac{(1+\alpha)_n}{n!}\left(\frac{x+1}{2}\right)^n {}_2F_1\left[\begin{matrix}-n,-\beta-n; & \dfrac{x-1}{x+1}\\ 1+\alpha; & \end{matrix}\right].$$

Differentiating both sides of this and using

$\frac{d}{dx}F\left[\begin{matrix}a,b; & x\\ c; & \end{matrix}\right]=\frac{ab}{c}F\left[\begin{matrix}a+1,b+1; & x\\ c+1 & \end{matrix}\right]$, we get

$$\begin{aligned}&\frac{d}{dx}P_n^{(\alpha,\beta)}(x)\\&=\frac{(1+\alpha)_n}{2(n-1)!}\left(\frac{x+1}{2}\right)^{n-1}{}_2F_1\left[\begin{matrix}-n,-\beta-n; & \dfrac{x-1}{x+1}\\ 1+\alpha; & \end{matrix}\right]\\&+\frac{(2+\alpha)_{n-1}(\beta+n)}{(n-1)!(x+1)}\left(\frac{x+1}{2}\right)^{n-2}{}_2F_1\left[\begin{matrix}-(n-1),-\beta-n+1; & \dfrac{x-1}{x+1}\\ 1+(1+\alpha); & \end{matrix}\right]\frac{d}{dx}P_n^{(\alpha,\beta)}(x)\\&=\frac{n}{x+1}P_n^{(\alpha,\beta)}(x)+\frac{\beta+n}{x+1}P_{n-1}^{(\alpha+1,\beta)}(x)(x+1)\frac{d}{dx}P_n^{(\alpha,\beta)}(x)\\&=nP_n^{(\alpha,\beta)}(x)+(\beta+n)P_{n-1}^{(\alpha+1,\beta)}(x)\end{aligned} \tag{15.32}$$

(iii) We know [from equation (15.8)]

$$P_n^{(\alpha,\beta)}(x) = \frac{(1+\beta)_n}{n!}\left(\frac{x-1}{2}\right)^n {}_2F_1\left[\begin{matrix}-n, -\alpha-n; & \dfrac{x+1}{x-1} \\ 1+\beta; & \end{matrix}\right]$$

and we get by differentiating it

$$\frac{d}{dx}P_n^{(\alpha,\beta)}(x) = n(x-1)^{-1}P_n^{(\alpha,\beta)}(x) - (x-1)^{-1}$$

$$(\alpha+n)\frac{(1+\beta)_{n-1}}{(n-1)!}\left(\frac{x-1}{2}\right)^{n-1} {}_2F_1\left[\begin{matrix}-(n-1), -\alpha-n+1; & \dfrac{x+1}{x-1} \\ 1+(1+\beta); & \end{matrix}\right]$$

$$= n(x-1)^{-1}P_n^{(\alpha,\beta)}(x) - (x-1)^{-1}(\alpha+n)P_{n-1}^{(\alpha,\beta+1)}(x)$$

$$(x-1)\frac{d}{dx}P_n^{(\alpha,\beta)}(x) = nP_n^{(\alpha,\beta)}(x) - (\alpha+n)P_{n-1}^{(\alpha,\beta+1)}(x) \quad (15.33)$$

(iv) Subtracting Equation (15.33) from Equation (15.32), we get

$$2\frac{d}{dx}P_n^{(\alpha,\beta)}(x) = (\beta+n)P_{n-1}^{(\alpha+1,\beta)}(x) + (\alpha+n)P_{n-1}^{(\alpha,\beta+1)}(x). \quad (15.34)$$

(v) Substituting the value of $\frac{d}{dx}P_n^{(\alpha,\beta)}(x)$ from Equation (15.30) in Equation (15.34), we get

$$(1+\alpha+\beta+n)P_{n-1}^{(\alpha+1,\beta+1)}(x) = (\beta+n)P_{n-1}^{(\alpha+1,\beta)}(x) + (\alpha+n)P_{n-1}^{(\alpha,\beta+1)}(x),$$

which gives, on shifting n *to* $n+1, \alpha$ *to* $\alpha-1,$ *and* β *to* $\beta-1,$

$$(\alpha+\beta+n)P_n^{(\alpha,\beta)}(x) = (\beta+n)P_n^{(\alpha,\beta-1)}(x) + (\alpha+n)P_{n-1}^{(\alpha-1,\beta)}(x). \quad (15.35)$$

Other relations are as follows:

(vi) $\frac{1}{2}(2+\alpha+\beta+2n)(x-1)P_n^{(\alpha+1,\beta)}(x) = (n+1)P_{n+1}^{(\alpha,\beta)}(x)$
$-(1+\alpha+n)P_n^{(\alpha,\beta)}(x)$

(vii) $\frac{1}{2}(2+\alpha+\beta+2n)(x+1)P_n^{(\alpha,\beta+1)}(x)=(n+1)P_{n+1}^{(\alpha,\beta)}(x)$
$+(1+\alpha+n)P_n^{(\alpha,\beta)}(x)$

(viii) $(\alpha+\beta+2n)P_n^{(\alpha,\beta-1)}(x)=(\alpha+n)P_{n-1}^{(\alpha,\beta)}(x)+(\alpha+\beta+n)$
$P_n^{(\alpha,\beta)}(x)$

(ix) $P_n^{(\alpha,\beta-1)}(x)-P_n^{(\alpha-1,\beta)}(x)=P_{n-1}^{(\alpha,\beta)}(x)$

(x) $(\alpha+\beta+2n)P_n^{(\alpha-1,\beta)}(x)=(\alpha+\beta+n)P_n^{(\alpha,\beta)}(x)-(\beta+n)$
$P_{n-1}^{(\alpha,\beta)}(x)$

(xi) $2P_n^{(\alpha,\beta)}(x)=(1+x)P_n^{(\alpha,\beta+1)}(x)+(1-x)P_n^{(\alpha+1,\beta)}(x).$

15.7 Expansions

Let us take the generating relation

$$\sum_{n=0}^{\infty}\frac{(1+\alpha+\beta)_n P_n^{(\alpha,\beta)}(x)}{(1+\alpha)_n}t^n$$

$$=(1-t)^{-1-!\alpha-\beta}\ {}_2F_1\left[\begin{matrix}\frac{1}{2}(1+\alpha+\beta),\frac{1}{2}(2+\alpha+\beta); & \dfrac{2t(x-1)}{(1-t)^2}\\ 1+\alpha; & \end{matrix}\right],$$

with $c=1+\alpha+\beta, y=\frac{1}{2}(1-x)$ and

$$\psi(z)={}_2F_1\left[\begin{matrix}\frac{1}{2}(1+\alpha+\beta),\frac{1}{2}(2+\alpha+\beta); & z\\ 1+\alpha; & \end{matrix}\right].$$

It takes the form

$$(1-t)^{-c}\psi\left(\frac{-4ty}{(1-t)^2}\right)=\sum_{n=0}^{\infty}\frac{(c)_n P_n^{(\alpha,\beta)}(1-2y)}{(1+\alpha)_n}t^n. \tag{15.36}$$

If we let $\nu=\frac{-4t}{(1-t)^2}$,

then, on solving this for t and taking the positive root of the discriminant of the quadratic equation, we find $t=1-\frac{2}{1+\sqrt{(1-\nu)}}=\frac{-\nu}{\left[1+\sqrt{(1-\nu)}\right]^2}$.

Hence, Equation (15.36) becomes

$$\psi(yv) = \left(\frac{2}{1+\sqrt{(1-v)}}\right)^{c} \sum_{m=0}^{\infty} \frac{(c)_m P_m^{(\alpha,\beta)}(1-2y)}{(1+\alpha)_m} \frac{(-1)^m v^m}{\left[1+\sqrt{(1-v)}\right]^{2m}}$$

$$= \sum_{m=0}^{\infty} \frac{(c)_m P_m^{(\alpha,\beta)}(1-2y)}{(1+\alpha)_m} \frac{(-1)^m v^m}{2^{2m}[1+\sqrt{(1-v)}]^{2m}} \left(\frac{2}{1+\sqrt{(1-v)}}\right)^{c+2m}.$$

We know that $\left(\frac{2}{1+\sqrt{(1-v)}}\right)^{c+2m} = {}_2F_1\left[\begin{matrix}\frac{1}{2}(1+c+2m), \frac{1}{2}(c+2m); \\ 1+c+2m;\end{matrix} v\right].$

Then,

$$\psi(yv) = \sum_{m=0}^{\infty} {}_2F_1\left[\begin{matrix}\frac{1}{2}(1+c+2m), \frac{1}{2}(c+2m); \\ 1+c+2m;\end{matrix} v\right] \times \frac{(c)_m P_m^{(\alpha,\beta)}(1-2y)}{(1+\alpha)_m} \frac{(-1)^m v^m}{2^{2m}}$$

$$= \sum_{n=0}^{\infty} \sum_{m=0}^{\infty} \frac{\left(\frac{1}{2}(1+c+2m)\right)_n \left(\frac{1}{2}(c+2m)\right)_n}{(1+c+2m)_n} \frac{(c)_m P_m^{(\alpha,\beta)}(1-2y)}{n!(1+\alpha)_m} \frac{(-1)^m v^{m+n}}{2^{2m}}$$

$$= \sum_{n=0}^{\infty} \sum_{m=0}^{\infty} \frac{(c+2m)_n}{2^{2n}(1+c+2m)_n} \frac{(c)_m P_m^{(\alpha,\beta)}(1-2y)}{n!(1+\alpha)_m} \frac{(-1)^m v^{m+n}}{2^{2m}}$$

$$= \sum_{n=0}^{\infty} \sum_{m=0}^{\infty} \frac{(c)_{2m+1}(c)_m}{2^{2n}(1+c+2m)_n} \frac{(c)_{2n+2m} P_m^{(\alpha,\beta)}(1-2y)}{n!(1+\alpha)_m} \frac{(-1)^m v^{m+n}}{2^{2m}}$$

$$= \sum_{n=0}^{\infty} \sum_{m=0}^{\infty} \frac{(c)_{2n}(c+2m)}{(c)_{n+1+m}} \frac{(c)_m P_m^{(\alpha,\beta)}(1-2y)}{(n-m)!(1+\alpha)_m} \frac{(-1)^m v^m}{2^{2n}}.$$

Thus, we have

$$\psi(yv) = {}_2F_1\left[\begin{matrix}\frac{1}{2}(1+\alpha+\beta), \frac{1}{2}(2+\alpha+\beta); \\ 1+\alpha;\end{matrix} yv\right]$$

$$= \sum_{n=0}^{\infty} \sum_{m=0}^{\infty} \frac{(c)_{2n}(c+2m)}{2^{2n}(c)_{n+1+m}} \frac{(c)_m P_m^{(\alpha,\beta)}(1-2y)}{(n-m)!(1+\alpha)_m} \frac{(-1)^m v^n}{}$$

$$\therefore \sum_{n=0}^{\infty} \frac{\left(\frac{1}{2}(1+\alpha+\beta)\right)_n \left(\frac{1}{2}(2+\alpha+\beta)\right)_n}{(1+\alpha)_n n!} y^n v^n$$

$$= \sum_{n=0}^{\infty} \sum_{m=0}^{\infty} \frac{(c)_{2n}(c+2m)}{2^{2n}(c)_{n+1+m}} \frac{(c)_m P_m^{(\alpha,\beta)}(1-2y)}{(n-m)!(1+\alpha)_m} \frac{(-1)^m v^n}{}.$$

On comparing the coefficients of v^n,

$$y^n = \frac{n!(1+\alpha)_n}{2^{2n}\left(\frac{1}{2}(1+\alpha+\beta)\right)_n \left(\frac{1}{2}(2+\alpha+\beta)\right)_n} \times \sum_{m=0}^{\infty} \frac{(c)_{2n}\,(c+2m)}{2^{2n}(c)_{n+1+m}} \frac{(c)_m P_m^{(\alpha,\beta)}(1-2y)}{(n-m)!(1+\alpha)_m} \frac{(-1)^m}{}$$

$$= (1+\alpha)_n \sum_{n=0}^{\infty} \frac{(-n)_m (1+\alpha+\beta+2m)(1+\alpha+\beta)_m P_m^{(\alpha,\beta)}(1-2y)}{(1+\alpha+\beta)_{n+1+m}(1+\alpha)_m}.$$

$$\therefore (1-x)^n = 2^n (1+\alpha)_n \sum_{n=0}^{\infty} \frac{(-n)_m (1+\alpha+\beta+2m)(1+\alpha+\beta)_m P_m^{(\alpha,\beta)}(x)}{(1+\alpha+\beta)_{n+1+m}(1+\alpha)_m}. \tag{15.37}$$

This result is made use of in expressing the Jacobi polynomial in terms of other polynomials.

15.8 Examples

Example 15.8.1. Prove that

$$\sum_{n=0}^{\infty} \frac{(1+x)^n P_n^{(\alpha,\beta)}\left(\frac{1-x}{1+x}\right)}{(1+\alpha)_n (1+\beta)_n} t^n = {}_0F_1(-;1+\alpha;-xt)\,{}_0F(-;1+\beta;t).$$

Solution: Bateman's generating relation for $P_n^{(\alpha,\beta)}(x)$ is

$$\sum_{n=0}^{\infty} \frac{(P_n^{(\alpha,\beta)}(x)}{(1+\alpha)_n(1+\beta)_n} t^n = {}_0F_1\left[\begin{matrix} -; & \frac{t(x-1)}{2} \\ 1+\alpha; & \end{matrix}\right] {}_0F_1\left[\begin{matrix} -; & \frac{t(x-1)}{2} \\ 1+\beta; & \end{matrix}\right].$$

If we replace in this relation x *by* $\frac{1-x}{1+x}$ and t *by* $t(x+1)$, we get

$$\sum_{n=0}^{\infty} \frac{(1+x)^n P_n^{(\alpha,\beta)}\left(\frac{1-x}{1+x}\right)}{(1+\alpha)_n(1+\beta)_n} t^n = {}_0F_1\left[\begin{matrix} -; & \\ 1+\alpha; & -xt \end{matrix}\right] {}_0F_1\left[\begin{matrix} -; & \\ 1+\beta; & t \end{matrix}\right].$$

Example 15.8.2. Prove that

$$P_n^{(\alpha,\beta)}(x) = \frac{(1+\alpha)_n}{(1+\alpha+\beta)_n} \times \sum \frac{(-1)^{n-m}(\beta)_{n-m}(1+\alpha+\beta)_{n+m}(1+\alpha+2m)P_m^{(\alpha,\beta)}(x)}{(n-m)!(1+\alpha)_{n+m+1}}.$$

Solution: Using the expansion of Jacobi polynomial [Article 15.7, Equation (15.37)]

$$\left(\frac{1-x}{2}\right)^s = (1+\alpha)_s \sum_{m=0}^{\infty} \frac{(-s)_m(1+\alpha+\beta+2m)(1+\alpha+\beta)_m P_m^{(\alpha,\beta)}(x)}{(1+\alpha+\beta)_{s+1+m}(1+\alpha)_m}.$$

Putting $\beta=0$, it becomes

$$\left(\frac{1-x}{2}\right)^s = (1+\alpha)_s \sum_{m=0}^{\infty} \frac{(-s)_m(1+\alpha+2m)P_m^{(\alpha,0)}(x)}{(1+\alpha)_{s+1+m}}.$$

Using Equation (15.1), we get,

$$\sum_{n=0}^{\infty)} \frac{(1+\alpha+\beta)_n P_n^{(\alpha,\beta)}(x)}{(1+\alpha)_m} = \sum_{n=0}^{\infty}\sum_{s=0}^{n} \frac{(-1)^s(1+\alpha+\beta)_{n+s}}{s!(n-s)!(1+\alpha)_s}\left(\frac{x-1}{2}\right)^s t^n$$

$$= \sum_{n=0}^{\infty}\sum_{s=0}^{n} \frac{(-1)^s(1+\alpha+\beta)_{n+2s}}{s!n!(1+\alpha)_s}\left(\frac{x-1}{2}\right)^s t^{n+s}$$

$$= \sum_{n=0}^{\infty}\sum_{s=0}^{\infty}\sum_{m=0}^{s} \frac{(-1)^{s+m}(1+\alpha+\beta)_{n+2s}(1+\alpha+2m)}{(s-m)!n!(1+\alpha)_{s+1+m}} P_m^{(\alpha,0)}(x)t^{n+s}$$

$$= \sum_{n,s,m=0}^{\infty} \frac{(-1)^{s+2m}(1+\alpha+\beta)_{n+2s+2m}(1+\alpha+2m)}{s!n!(1+\alpha)_{s+1+2m}} P_m^{(\alpha,0)}(x)t^{n+s+m}$$

$$= \sum_{n,m=0}^{\infty}\sum_{s=0}^{n} \frac{(-1)^s(1+\alpha+\beta)_{n+2s+2m}(1+\alpha+2m)}{s!(n-s)!(1+\alpha)_{s+1+2m}} P_m^{(\alpha,0)}(x)t^{n+m}$$

$$= \sum_{n,m=0}^{\infty} {}_2F_1\left[\begin{matrix}-n,(1+\alpha+\beta+n+2m); \\ 2+\alpha+2m;\end{matrix}\; 1\right]$$

$$\times \frac{(1+\alpha+\beta)_{n+2m}(1+\alpha+2m)}{n!(1+\alpha)_{1+2m}} P_m^{(\alpha,0)}(x)t^{n+m}$$

$$= \sum_{n,m=0}^{\infty} \frac{\Gamma(2+\alpha+2m)\Gamma 1-\div\beta)}{\Gamma(2+\alpha+n+2m)\Gamma(1-\beta-n)} \frac{(1+\alpha+\beta)_{n+2m}(1+\alpha+2m)}{n!(1+\alpha)_{1+2m}} P_m^{(\alpha,0)}(x)\, t^{n+m}$$

$$= \sum_{n=0}^{\infty}\sum_{m=0}^{n} \frac{\Gamma(1-\div\beta)}{\Gamma(1-\beta-n+m)(2+\alpha+2m)_{n-m}} \frac{(1+\alpha+\beta)_{n+2m}(1+\alpha+2m)}{(n-m)!(1+\alpha)_{1+2m}} P_m^{(\alpha,0)}(x)\, t^n.$$

$$\because \frac{\Gamma(1-\div\beta)}{\Gamma(1-\beta-n+m)} = \frac{\Gamma(1-\div\beta)\Gamma(1-\beta-n)}{\Gamma(1-\beta-n)\Gamma(1-\beta-n+m)} = \frac{(-1)^n(\beta)_n}{(1-\beta-n)_m} = (-1)^{n-m}(\beta)_{n-m}$$

and,

$$(2+\alpha+2m)_{n-m}(1+\alpha)_{2m+1} = \frac{\Gamma(2+\alpha+n+m)\Gamma(2+\alpha+2m)}{\Gamma(2+\alpha+2m)\Gamma(1+\alpha)} = (1+\alpha)_{n+m+1}.$$

We obtain

$$\frac{(1+\alpha+\beta)_n}{(1+\alpha)_n} P_n^{(\alpha,\beta)}(x)t^n = \sum_{n=0}^{\infty}\sum_{m=0}^{n} \frac{(-1)^{n-m}}{} \frac{(\beta)_{n-m(1+\alpha+\beta)_{n+m}}(1+\alpha+2m)}{(n-m)!(1+\alpha)_{1+m+n}}$$

$$P_m^{(\alpha,0)}(x)\, t^n.$$

Comparing the coefficient of t^n on both sides, we get

$$P_n^{(\alpha,\beta)}(x) = \frac{(1+\alpha)_n}{(1+\alpha+\beta)_n} \frac{(-1)^{n-m}}{} \frac{(\beta)_{n-m}(1+\alpha+\beta)_{n+m}(1+\alpha+2m)}{(n-m)!(1+\alpha)_{1+m+n}} P_m^{(\alpha,0)}(x).$$

Example 15.8.3. Evaluate the integral

$$\int_{-1}^{1}(1-x)^{\alpha} P_n^{(\alpha,\beta)}(x)P_k^{(\alpha,0)}(x)dx.$$

Solution: We have proved above that

$$P_n^{(\alpha,\beta)}(x) = \frac{(1+\alpha)_n}{(1+\alpha+\beta)_n} \sum_{m=0}^{n} \frac{(-1)^{n-m}}{} \frac{(\beta)_{n-m}(1+\alpha+\beta)_{n+m}(1+\alpha+2m)}{(n-m)!\,(1+\alpha)_{1+m+n}} P_m^{(\alpha,0)}(x).$$

Using this result, we get

$$\int_{-1}^{1}(1-x)^{\alpha} P_n^{(\alpha,\beta)}(x)P_k^{(\alpha,0)}(x)dx = \frac{(1+\alpha)_n}{(1+\alpha+\beta)_n} \sum \frac{(-1)^{n-m}}{} \frac{(\beta)_{n-m}(1+\alpha+\beta)_{n+m}(1+\alpha+2m)}{(n-m)!\,(1+\alpha)_{1+m+n}}$$
$$\times \int_{-1}^{1}(1-x)^{\alpha} P_m^{(\alpha,0)}(x)P_k^{(\alpha,0)}(x)dx.$$

Using the orthogonality property for Jacobi polynomials with $\beta = 0$, since the only non-vanishing term is given by $m = k$, we find that if $k > n$, the given integral is zero, and if $0 \le k \le n$, we get

$$\int_{-1}^{1}(1-x)^{\alpha} P_n^{(\alpha,\beta)}(x)P_k^{(\alpha,0)}(x)$$
$$= \frac{(1+\alpha)_n(-1)^{n-k}(\beta)_{n-k}(1+\alpha+\beta)_{n+k}(1+\alpha+2k)2^{1+\alpha}\Gamma(1+\alpha+k)k!}{(1+\alpha+\beta)_n(n-k)!\,(1+\alpha)_{1+k+n}k!\,(1+\alpha+2k)\Gamma(1+\alpha+k)}$$
$$= \frac{2^{1+\alpha}(-1)^{n-k}(\beta)_{n-k}(1+\alpha+\beta+n)_k}{(n-k)!(1+\alpha+n)_{k+1}}.$$

15.9 Exercise

1. Prove that

$$\int_{-1}^{1}(1-x^2)^{\alpha} P_n^{(\alpha,\beta)}(x)P_k^{(\alpha,\alpha)}(x)dx = 0, \text{ if } k > n$$
$$= \frac{(-1)^{n-k}(\beta-\alpha)_{n-k}2^{1+2\alpha}\Gamma(1+\alpha+\beta+n+k)\Gamma(1+\alpha+k)}{k!(n-k)!\Gamma(2+2\alpha+n+k)\Gamma(1+\alpha+\beta+n)}$$

if $0 \le k \le n$.

2. Prove that

$$
{}_2F_1\left[\begin{matrix} a, b; \\ c; \end{matrix} v\right] {}_2F_1\left[\begin{matrix} a, b; \\ 1+a+b-c; \end{matrix} v\right]
$$

$$
= {}_4F_3\left[\begin{matrix} a, b, \frac{1}{2}(1+b), \frac{1}{2}(a+b+1); \\ \alpha+b, c, 1+a+b-c;; \end{matrix} 4v(1-v)\right].
$$

3. Is the Jacobi polynomial orthogonal in the interval $(-1,1)$? If yes, prove it, and write the weight function.
4. Prove that

$$
P_n^{(\alpha,0)}(x) = \frac{(\pi+\alpha)_n}{(1+\alpha+\beta)_n}
$$

$$
\times\sum_{k=0}^{n} \frac{(-1)^{n-k}(\beta-\alpha)_{n-k}(1+\alpha+\beta)_{n+k}(1+\alpha+2k)P_k^{(\alpha,0)}(x)}{(n-k)!(1+\alpha+\beta)_{n+k+1}}.
$$

Hence, deduce that

$$
\int_{-\frac{\pi}{2}}^{\frac{\pi}{2}} (1-sin\theta)^{\alpha} cos\theta P_n^{(\alpha,\beta)}(sin\theta)P_k^{(\alpha,0)}(sin\theta)d\theta
$$

$$
= \frac{(2)^{1+\alpha}(-1)^{n-k}(\beta)_{n-k}(1+\alpha+\beta+n)_k}{(n-k)!(1+\alpha+n)_{k+1}}.
$$

16
Chebyshev Polynomials

Abstract
Chebyshev polynomials are a sequence of orthogonal polynomials known for their significance in numerical analysis, approximation theory, and signal processing. They come in two forms—first kind and second kind—and are closely related to trigonometric functions. Their orthogonality and extremal properties make them especially useful in applications. This chapter discusses their trigonometric identities, generating functions, Rodrigues formula, and expansions.

Keywords: Chebyshev polynomial of the first kind, Chebyshev polynomial of the second kind, recurrence relations, orthogonality, zeros, identities

16.1 Introduction

Chebyshev polynomials are sequence of orthogonal polynomials, widely used in approximation theory and is a powerful tool in numerical analysis and differential equations due to their optimal convergence theory and connection to the Fourier transform. They are defined in terms of trigonometric functions and have two main types: Chebyshev polynomial of first kind by $T_n(x)$ and Chebyshev polonomial of second kind by $U_n(x)$.

This polynomial is named after the Russian mathematician Pafnuti Chebyshev (1821-1894) who is known as for his work to theory of numbers and approximation theory. In this chapter we study its properties- orthogonality, recurrence relations, expansion, Rodrigues formula and identities of chebyshev polynomials.

16.2 Chebyshev Polynomials

A special case of Jacobi polynomial, i.e., $P_n^{(\alpha,\alpha)}(x)$, is known as the *ultraspherical polynomial.* Special ultraspherical polynomials are Chebyshev polynomials $T_n(x)$ and $U_n(x)$, which are defined by

Dipankar De. Ordinary Differential Equations and Special Functions, (587–602) © 2026 Scrivener Publishing LLC

$$T_n(x) = \frac{n!}{(\frac{1}{2})_n} P_n^{\left(-\frac{1}{2},-\frac{1}{2}\right)}(x) \tag{16.1}$$

and,
$$U_n(x) = \frac{(n+1)!}{(\frac{1}{2})_n} P_n^{\left(\frac{1}{2},\frac{1}{2}\right)}(x). \tag{16.2}$$

Now, with the Jacobi polynomial, the Chebyshev polynomial $T_n(x)$ is

$$T_n(cos\theta) = {}_2F_1\left[\begin{matrix} -n, n; \\ 1/2 \end{matrix} sin^2\theta\right] \tag{16.3}$$

We know that

$$cosmz = {}_2F_1\left[\begin{matrix} \frac{m}{2}, -\frac{m}{2}; \\ 1/2 \end{matrix} sin^2 z\right]. \tag{16.4}$$

Putting $z = \theta/2$ and $m = 2n$, Equation (16.3) becomes

$$cosn\theta = {}_2F_1\left[\begin{matrix} n, -n; \\ 1/2 \end{matrix} sin^2\theta/2\right].$$

Then, Equation (16.2) becomes

$$T_n(cos\theta) = cosn\theta \tag{16.5}$$

and,
$$T_n(x) = \cos(ncos^{-1}x). \tag{16.6}$$

Similarly, with the Jacobi polynomial, Equation (16.2) becomes

$$U_n(cos\theta) = (n+1)\,{}_2F_1\left[\begin{matrix} -n, 2+n; \\ \frac{3}{2} \end{matrix}\; sin^2\frac{\theta}{2}\right]. \tag{16.7}$$

Using

$$\sin mz = m\sin z\,{}_2F_1\left[\begin{matrix} \frac{1}{2}-\frac{m}{2}, \frac{1}{2}+\frac{m}{2}; \\ \frac{3}{2} \end{matrix}\; sin^2 z\right],$$

then,

$$U_n(cos\theta) = \frac{\sin(n+1)\theta}{sin\theta}, \tag{16.8}$$

it could also be written as

$$U_n^*(cos\theta) = \frac{\sin(n+1)\theta}{sin\theta}$$

$$\therefore U_n(x) = \frac{\sin(n+1)cos^{-1}x}{\sqrt{(1-x^2)}}. \tag{16.9}$$

But in a restricted manner, the Chebyshev polynomials $T_n(x)$ *and* $U_n(x)$ are defined by

$$T_n(x) = \cos\,(ncos^{-1}x)$$

$$U_n(x) = \sin(ncos^{-1}x). \tag{16.10}$$

Writing $x = cos\theta$, we obtain

$$T_n(cos\theta) = \cos\,(n\theta)$$

$$U_n(cos\theta) = \mathrm{sinn}\theta. \tag{16.11}$$

From the hypergeometric differential equation

$$z(1-z)y''+[c-(a+b+1)z]y'-aby=0,$$

it follows that the differential equation satisfied by Jacobi polynomials is

$$(1-x^2)D^2P_n^{(\alpha,\beta)}(x)+[\beta-\alpha-(2+\alpha+\beta)x]DP_n^{(\alpha,\beta)}(x)$$
$$+n(1+\alpha+\beta+n)\,P_n^{(\alpha,\beta)}(x)=0. \tag{16.12}$$

Hence, the differential equations satisfied by $T_n(x)$ *and* $U_n(x)$ are

$$(1-x^2)D^2T_n(x)-xDT_n(x)+n^2T_n(x)=0 \tag{16.13}$$

and, $$(1-x^2)D^2U_n(x)-3xDU_n(x)+n(2+n)U_n(x)=0. \tag{16.14}$$

Theorem 16.2.1. The differential equation satisfied by $T_n(x)$ *and* $U_n(x)$ is

$$(1-x^2)D^2y-xDy+n^2y=0. \tag{16.15}$$

Proof: From Equation (16.5), we have the differentiations

$$\frac{dT_n(x)}{dx}=\frac{n}{\sqrt{1-x^2}}\sin\,(n\cos^{-1}x)\text{ and}$$
$$\frac{d^2T_n(x)}{dx^2}=\frac{nx}{(1-x^2)^{3/2}}\sin(n\cos^{-1}x)-\frac{n^2}{1-x^2}\cos\,(n\cos^{-1}x)$$

or, $$(1-x^2)\frac{d^2T_n(x)}{dx^2}=\frac{nx}{\sqrt{1-x^2}}\sin(n\cos^{-1}x)-n^2\cos\,(n\cos^{-1}x)$$
$$=x\frac{dT_n(x)}{dx}-n^2T_n(x).$$

Therefore, $(1-x^2)D^2T_n(x)-xDT_n(x)+n^2T_n(x)=0.$
Hence, $T_n(x)$ is the solution of

$$(1-x^2)D^2y-xDy+n^2y=0.$$

We show similarly that $U_n(x)$ is a solution of Equation (16.15). Further, $T_n(x)$ *and* $U_n(x)$ are independent solutions of Equation (16.15) for $T_n(1)=1$, $T_n(1)=1$, so that $T_n(1)$ cannot be expressed as a multiple of $U_n(1)$.

16.3 Orthogonality Property

Theorem 16.3.1. Show that

(i) $$\int_{-1}^{1}\frac{T_m(x)T_n(x)}{\sqrt{1-x^2}}dx=\begin{cases}0, & if m\neq n\\ \frac{\pi}{2}, & if m=n\neq 0\\ \pi, & if\ m=n=0\end{cases}$$

and (ii) $$\int_{-1}^{1}\frac{U_m(x)U_n(x)}{\sqrt{1-x^2}}dx=\begin{cases}0, & if m\neq n\\ \frac{\pi}{2}, & if m=n\neq 0.\\ 0, & if\ m=n=0\end{cases}$$

Proof: We know that $T_m(cos\theta)=cosm\theta$ *and* $T_n(cos\theta)=cosn\theta.$
Therefore,

$$I=\int_{-1}^{1}\frac{T_m(x)T_n(x)}{\sqrt{1-x^2}}dx=-\int_{\pi}^{0}\frac{T_m(cos\theta)T_n(cos\theta)}{sin\theta}sin\theta\, d\theta\ (\text{putting } x=cos\theta)$$

$$=\int_{0}^{\pi}cosm\theta\, d\theta\, cosn\theta\, d\theta.$$

Case I. When $m\neq n$, we have

$$I=\frac{1}{2}\int_{0}^{\pi}\cos(m-n)\theta+\cos(m+n)\theta d\theta\, d\theta$$

$$=\frac{1}{2}\left[\frac{\sin(m-n)\theta}{(m-n)}+\frac{\sin(m+n)\theta}{(m+n)}\right]_0^{\pi}=0.$$

Case II. When $m = n \neq 0,$ we have

$$I = \frac{1}{2}\int_0^\pi \cos 0.\theta + \cos\,(2n)\theta d\theta\; d\theta = \frac{1}{2}\int_0^\pi 1 + \cos\,(2n)\theta d\theta\; d\theta$$

$$= \frac{1}{2}\left[\theta + \frac{\sin 2n\theta}{2n}\right]_0^\pi = \frac{1}{2}[\pi + 0] = \frac{\pi}{2}.$$

Case III. When $m = n = 0,$ we have

$$I = \frac{1}{2}\int_0^\pi (1+1)\; d\theta = \pi.$$

(ii) We know from Equation (x) that $U_m(cos\theta) = \text{sinm}\theta$ and $U_n(cos\theta) = \text{sinn}\theta$.

Therefore, $I = \int_{-1}^{1} \frac{U_m(x)U_n(x)}{\sqrt{1-x^2}} dx = -\int_\pi^0 \frac{\text{sinm}\theta\text{sinn}\theta}{sin\theta}\text{sin}\theta d\theta$

$$= \int_0^\pi \text{sinm}\theta\text{sinn}\theta d\theta.$$

Case I. When $m \neq n$

$$I = \frac{1}{2}\int_0^\pi \cos(m-n)\theta - \cos(m+n)\theta d\theta\; d\theta$$

$$= \frac{1}{2}\left[\frac{\sin(m-n)\theta}{(m-n)} - \frac{\sin(m+n)\theta}{(m+n)}\right]_0^\pi = 0.$$

Case II. When $m = n \neq 0$

$$I = \frac{1}{2}\int_0^\pi \cos(m-n)\,\theta - \cos\,(m+n)\theta d\theta\; d\theta = \frac{1}{2}\int_0^\pi 1 - \cos\,(2n)\theta d\theta\; d\theta$$

$$= \frac{1}{2}\left[\theta - \frac{sin2n\theta}{2n}\right]_0^\pi = \frac{1}{2}(\pi - 0) = \frac{\pi}{2}.$$

Case III. When $m = n = 0\; I = \frac{1}{2}\int_0^\pi (1-1) d\theta\; d\theta = 0$

It shows that the Chebyshev polynomials are orthogonal with respect to the weight function $(1-x^2)^{-\frac{1}{2}}$.

16.4 Recurrence Relations

The recurrence relations between the Chebyshev polynomials are as follows:

I. $T_{n+1}(x) = 2xT_n(x) - T_{n-1}(x)$
II. $(1-x^2)T_n'(x) = n[T_{n-1}(x) - xT_n(x)]$
III. $U_{n+1}(x) = 2xU_n(x) - U_{n-1}(x)$
IV. $(1-x^2)U_n'(x) = n[U_{n-1}(x) - xU_n(x)]$.

Proof: We know that $T_n(cos\theta) = cosn\theta$.
Therefore,

$$\begin{aligned} &T_{n+1}(cos\theta) + T_{n-1}(cos\theta) \\ &= \cos(n+1)\,\theta + \cos\,(n-1)\theta \\ &= 2cosn\theta cos\theta \\ &= 2T_n(cos\theta)cos\theta \end{aligned}$$

or,
$$T_{n+1}(x) + T_{n-1}(x) = 2xT_n(x).$$

II. We know that $T_n(x) = \cos\,(ncos^{-1}x)$.

$$\therefore \frac{dT_n(x)}{dx} = \frac{n}{\sqrt{1-x^2}}\sin\,(ncos^{-1}x)$$

so that

$$\frac{d}{dx}T_n(cos\theta) = \frac{n}{sin\theta}\sin\,(n\theta)$$

or,
$$sin^2\theta\frac{d}{dx}T_n(cos\theta) = nsin\theta sinn\theta = \frac{1}{2}n[\cos(n-1)\,\theta - \cos(n+1)\theta].$$

$$\therefore (1-x^2)\frac{d}{dx}T_n(cos\theta) = \frac{1}{2}n[T_{n-1}(x) - T_{n+1}(x)]$$
$$= \frac{1}{2}n[T_{n-1}(x) - \{2xT_n(x) - T_{n-1}(x)\}]$$
$$= \frac{1}{2}n[2T_{n-1}(x) - 2xT_n(x)] = n\{T_{n-1}(x) - xT_n(x)\}$$

$$(1-x^2)\frac{d}{dx}T_n(x) = n\{T_{n-1}(x) - xT_n(x)\}.$$

III and IV. These results are exactly similar to I and II, respectively, and are obtained in the exact the same manner.

16.5 Identities of Chebyshev Polynomials

The Chebyshev polynomials are orthogonal polynomial as well as trigonometric cosine functions, so it satisfies a large number of useful relationships.

Now, we have $T_{n+1}(x) = \cos[(n+1)\, cos^{-1}x]$

and $T_{n-1}(x) = \cos[(n-1)\, cos^{-1}x]$.

Differentiating the above relations, we get

$$\frac{1}{(n+1)}\frac{d(T_{n+1}(x))}{dx} = \frac{-\sin\,(n+1)}{-\sqrt{1-x^2}}cos^{-1}x \tag{16.16}$$

and,
$$\frac{1}{(n-1)}\frac{d(T_{n-1}(x))}{dx} = \frac{-\sin\,(n-1)}{-\sqrt{1-x^2}}cos^{-1}x. \tag{16.17}$$

Subtracting Equation (16.17) from Equation (16.16), we get

$$\frac{1}{(n+1)}\frac{d(T_{n+1}(x))}{dx} - \frac{1}{(n-1)}\frac{d(T_{n-1}(x))}{dx} = \frac{\sin\,(n+1)}{\sqrt{1-x^2}}cos^{-1}x - \frac{\sin\,(n-1)}{\sqrt{1-x^2}}$$

$$cos^{-1}x = \frac{\sin\,(n+1)\theta}{sin\theta} - \frac{\sin\,(n-1)\theta}{sin\theta} \quad [\text{let}\, cos^{-1}x = \theta\,]$$

$$\frac{1}{(n+1)}\frac{d(T_{n+1}(x))}{dx} - \frac{1}{(n-1)}\frac{d(T_{n-1}(x))}{dx} = \frac{\sin(n-1)\theta - \sin(n-1)\theta}{sin\theta}$$

$$= \frac{2cosn\theta sin\theta}{sin\theta} = 2cosn\theta . = 2T_n(x)$$

$$\therefore T_n(x) = \frac{1}{2}\left[\frac{T'_{n+1}(x)}{(n+1)} - \frac{T'_{n-1}(x)}{(n-1)}\right], n \geq 2.$$

Here, we get $T_2'(x) = 4T_1(x), T_1'(x) = T_0(x)$ and $T_0'(x) = 0$.
Here the figure [Figure 16.1] of $T_0(x), T_0(x), T_0(x)$ *and* $T_0(x)$ *are given below:*

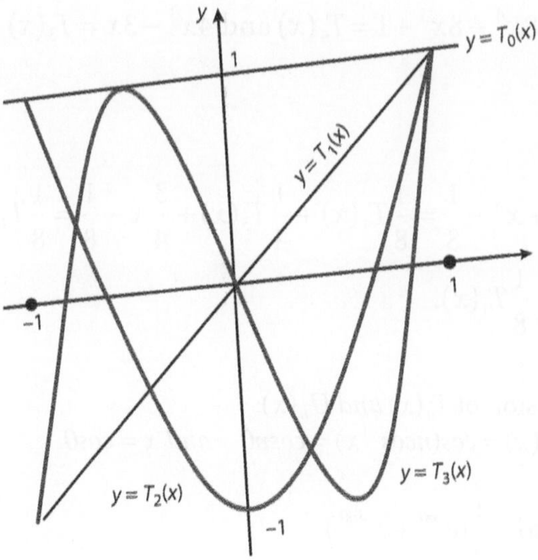

Figure 16.1 The Chebyshev polynomial.

16.6 Expansions

(i) The first few Chebyshev polynomials are obtained from definitions and recurrence relations.

We have $T_n(x) = cos(ncos^{-1}x)$.
For $n = 0,1$,

$$T_0(x) = cos(0.\ cos^{-1}x) = 1$$

$$T_1(x) = cos(1.\ cos^{-1}x) = x.$$

From the recurrence relation, $T_{n+1}(x) = 2xT_n(x) - T_{n-1}(x)$.

$$\therefore\ T_2(x) = 2xT_1(x) - T_0(x) = 2x.x - 1 = 2x^2 - 1$$

$$T_3(x) = 2xT_2(x) - T_1(x) = 2x(2x^2 - 1) - x = 4x^3 - 3x$$

$$T_4(x) = 2xT_3(x) - T_2(x) = 8x^4 - 8x^3 + 1\ ; T_5(x) = 2xT_4(x) - T_3(x)$$

$$= 6x^5 - 20x^3 + 5x, \text{ and so on.}$$

(ii) Expansion of x^4 in terms of $T_n(x)$:

$$8x^4 - 8x^3 + 1 = T_4(x) \text{ and } 4x^3 - 3x = T_3(x)$$

or,

$$x^4 = \frac{1}{8}T_4(x) + x^3 - \frac{1}{8} = \frac{1}{8}T_4(x) + \frac{1}{4}T_3(x) + \frac{3}{4}x - \frac{1}{8} = \frac{1}{8}T_4(x) + \frac{1}{4}T_3(x)$$

$$+ \frac{3}{4}T_1(x) - \frac{1}{8}T_0(x).$$

(iii) The expansion of $T_n(x)$ *and* $U_n(x)$:

We have $T_n(x) = cos(ncos^{-1}x) = cosn\theta \quad and\ x = cos\theta$

$$T_n(x) = cosn\theta = \frac{1}{2}(e^{in\theta} + e^{-in\theta})$$

$$= \frac{1}{2}\left[(e^{i\theta})^n + (e^{-i\theta})^n\right] = \frac{1}{2}[(cos\theta + isin\theta)^n + (cos\theta - isin\theta)^n]$$

$$= \frac{1}{2}\left[(x + i\sqrt{1-x^2})^n + (x - i\sqrt{1-x^2})^n\right]$$

$$= \frac{1}{2}\left[\sum_{r=0}^{n}\binom{n}{r}(i)^r(1-x^2)^{\frac{r}{2}}(x)^{n-r} + \sum_{r=0}^{n}\binom{n}{r}(-i)^r(1-x^2)^{\frac{r}{2}}(x)^{n-r}\right]$$

$$= \frac{1}{2}\left[\sum_{r=0}^{n}\binom{n}{r}\{(i)^r + (-i)^r\}(1-x^2)^{\frac{r}{2}}(x)^{n-r}\right]$$

$$\{(i)^r + (-i)^r\} = \begin{cases} 0\ if\ r\ is\ odd \\ 2i^{2k}, if\ r\ is\ even \end{cases}$$

$$T_n(x) = \sum_{k=0}^{\left[\frac{n}{2}\right]}\binom{n}{2k}i^{2k}(1-x^2)^k(x)^{n-2k}\ (let\ r = 2k)$$

where $\left[\frac{n}{2}\right] = \begin{cases} \frac{n}{2}, \textit{when n is even} \\ \frac{n+1}{2}, \textit{when n is odd} \end{cases}$.

Therefore,

$$T_n(x) = \sum_{k=0}^{\left[\frac{n}{2}\right]} \frac{n!}{(2k)!(n-2k)!}(-1)^k(1-x^2)^k(x)^{n-2k}$$

$$= \sum_{k=0}^{\left[\frac{n}{2}\right]} \frac{n!}{(2k)!(n-2k)!}(x^2-1)^k(x)^{n-2k}$$

Similarly, we have, $U_n(x) = \frac{1}{2i}\left[(x+i\sqrt{1-x^2})^n - (x-i\sqrt{1-x^2})^n\right]$

$$= -\frac{i}{2}\left[\sum_{r=0}^{n}\binom{n}{r}\{(i)^r-(-i)^r\}(1-x^2)^{\frac{r}{2}}(x)^{n-r}\right].$$

Since $\{(i)^r-(-i)^r\} = \begin{cases} 0, \textit{when n is even} \\ 2i^{2k+1}, \textit{when n is odd } (r=2k+1) \end{cases}$

and $\left[\frac{n-1}{2}\right] = \begin{cases} \frac{n-1}{2} \textit{ when n is odd} \\ \frac{n-2}{2} \textit{ when n is even} \end{cases}$,

$$U_n(x) = -i\sum_{r=0}^{n} i^{2k+1}\binom{n}{2k+1}(1-x^2)^{k+\frac{1}{2}}(x)^{n-2k-1}$$

$$= \sum_{r=0}^{n} i^{2k}\binom{n}{2k+1}(1-x^2)^{k+\frac{1}{2}}(x)^{n-2k-1}.$$

$$\therefore U_n(x) = \left[\sum_{k=0}^{\left[\frac{n-1}{2}\right]}(-1)^k\frac{n!}{(2k+1)!(n-2k-1)!}(1-x^2)^{k+\frac{1}{2}}(x)^{n-2k-1}\right].$$

$T_n(x)$ can be written as

$$T_n(x)=\frac{n}{2}\sum_{r=0}^{N}(-1)^r\frac{(n-r-1)!}{r!(n-2r)!}(2x)^{n-2r}, \text{where } N=\begin{cases}\frac{n}{2}, if\ n\,is\,even\\ \frac{1}{2}(n-1), if\ n\,is\,odd\end{cases}.$$

16.7 Generating Function

Theorem 16.7.1.

(i) $\frac{1-xt}{1-2xt+t^2}=\sum_{n=0}^{\infty}T_n(x)t^n$

(ii) $\frac{\sqrt{1-x^2}}{1-2xt+t^2}=\sum_{n=0}^{\infty}U_{n+1}(x)t^n$

Proof: $\frac{1-xt}{1-2xt+t^2}=\frac{1-tcos\theta}{1-2tcos\theta+t^2}, let\ x=cos\theta$

$$=\frac{1}{2}\left[\frac{2-t(e^{i\theta}+e^{-i\theta})}{1-t(e^{i\theta}+e^{-i\theta})+t^2}\right]$$

$$=\frac{1}{2}\left[\frac{2-t(e^{i\theta}+e^{-i\theta})}{(1-te^{i\theta})(1-te^{-i\theta})}\right]$$

$$=\frac{1}{2}\left[\frac{1}{(1-te^{-i\theta})}+\frac{1}{(1-te^{i\theta})}\right]$$

$$=\frac{1}{2}\left[(1-te^{-i\theta})^{-1}+(1-te^{i\theta})^{-1}\right]$$

$$=\frac{1}{2}\left[\sum_{n=0}^{\infty}(te^{-i\theta})^n+\sum_{n=0}^{\infty}(te^{i\theta})^n\right]$$

$$=\frac{1}{2}\left[\sum_{n=0}^{\infty}t^n(e^{-in\theta}+e^{-in\theta})\right]=\sum_{n=0}^{\infty}t^n cosn\theta=\sum_{n=0}^{\infty}T_n(x)\,t^n.$$

(iii) $$\frac{\sqrt{1-x^2}}{1-2xt+t^2}=\frac{sin\theta}{1-2tcos\theta+t^2}$$

$$=\frac{1}{2i}\left[\frac{(e^{i\theta}-e^{-i\theta})}{(1-te^{i\theta})(1-te^{-i\theta})}\right]$$

$$=\frac{1}{2it}\left[\frac{1}{(1-te^{i\theta})}-\frac{1}{(1-te^{-i\theta})}\right]$$

$$=\frac{1}{2it}\left[(1-te^{i\theta})^{-1}-(1-te^{-i\theta})^{-1}\right]$$

$$=\frac{1}{2it}\left[\sum_{n=0}^{\infty}t^n e^{in\theta}-\sum_{n=0}^{\infty}t^n e^{-in\theta}\right]$$

$$=\frac{1}{2it}\left[\sum_{n=0}^{\infty}t^n(e^{in\theta}-e^{-in\theta})\right]$$

$$\frac{\sqrt{1-x^2}}{1-2xt+t^2}=\sum_{n=0}^{\infty}t^{n-1}sinn\,\theta$$

$$=\sum_{n=1}^{\infty}\sin(n+1)\theta t^n=\sum_{n=1}^{\infty}U_{n+1}(x)t^n.$$

Also, we write it as $\frac{1}{1-2xt+t^2}=\sum_{n=1}^{\infty}\frac{U_{n+1}(x)}{\sqrt{1-x^2}}t^n=\sum_{n=1}^{\infty}U_n^*(x)t^n.$

16.8 Rodrigues Formula of Chebyshev Polynomials

Theorem 16.8.1. Show that

$$T_n(x)=\frac{(-2)^n n!}{(2n)!}\sqrt{1-x^2}\,\frac{d^n}{dx^n}(1-x^2)^{n-1/2}.$$

Proof: The weight function of the Chebyshev polynomial is $W(x)=\frac{1}{\sqrt{1-x^2}}$

According to the Rodrigues formula for classical orthogonal polynomials, $T_n(x)$ is proportional to

$$\frac{1}{W(x)}\frac{d^n}{dx^n}[Q(x)^n W(x)] \tag{16.18}$$

where $Q(x) = (1 - x^2)$.

[$Q(x)$ is the coefficient of y'' in the Chebyshev differential equation $(1 - x^2)y'' - xy' + n^2 y = 0$].

Here, it is enough to prove that the proportionality constant is $(-1)^n \frac{2^n n!}{(2n)!}$.

Now, after taking the derivative n times and dividing by $W(x)$, we get

$$\frac{1}{W(x)} \frac{d^n}{dx^n}[Q(x)^n W(x)]$$
$$= \sum_{k=0}^{n} \binom{n}{k} (-1)^k (n - \frac{1}{2})_k (n - \frac{1}{2})_{(n-k)} (1 - x)^{n-k} (1 + x)^k.$$

The above expression has one non-zero term at $x = 1$, and it is

$$\binom{n}{n} (-1)^n (n - \frac{1}{2})_n (2)^n = (-1)^n \frac{(2n)!}{2^n n!}.$$

We know that $T_n(1) = 1$.
Therefore,

$$T_n(x) = (-1)^n \frac{(2)^n n!}{(2n)!} \frac{1}{W(x)} \frac{d^n}{dx^n}[Q(x)^n W(x)]$$
$$= (-1)^n \frac{(2)^n n!}{(2n)!} \frac{1}{\frac{1}{\sqrt{1 - x^2}}} \frac{d^n}{dx^n}\left[(1 - x^2)^n \frac{1}{\sqrt{1 - x^2}}\right]$$
$$= \frac{(-2)^n n!}{(2n)!} \sqrt{1 - x^2} \frac{d^n}{dx^n}\left[(1 - x^2)^{n - 1/2}\right].$$

16.8.1 Special Values of $T_n(x)$

The special values and properties of $T_n(x)$ are often useful:

(i) $T_n(-x) = (-1)^n T_n(x)$

(ii) $T_{2n}(0) = (-1)^n$

(iii) $T_n(1) = 1$

(iv) $T_{2n+1}(0) = 0$

(v) $T_n(-1) = (-1)^n$

16.9 Exercise

1. Show that (i) $T_n(1) = 1$ *and* $T_n(-1) = (-1)^n$.

   (ii) $T_n(0) = 0$, *if n is odd and* $T_n(0) = (-1)^{\frac{n}{2}}$, *if n is even.*

2. Show that $T_n(x) = \frac{1}{2}\left[(x + i\sqrt{1-x^2})^n + (x - i\sqrt{1-x^2})^n\right]$,

   setting $x = \cos\theta, i = \sqrt{-1}$.

3. Obtain the relations

$$2[T_n(x)]^2 = 1 + T_{2n}(x)$$
$$[T_n(x)]^2 - T_{n+1}(x)T_{n+1}(x) = 1 - x^2$$
$$T_{m+n}(x) + T_{m-n}(x) = 2T_m(x)T_n(x).$$

4. Show that the Chebyshev polynomials are orthogonal polynomials with respect to the weight function.

5. Prove that

   (i) $\sqrt{(1-x^2)}T_n(x) = U_{n+1}(x) - xU_n(x)$

   (ii) $[T_n(x)]^2 - T_{(n+1)}(x)T_{(n-1)}(x) = 1 - x^2$

Appendix A: Answers to Even-Numbered Excercises

Exercise 1.9

1. i) $x = \log\left[1 + tan\frac{1}{2}(x+y)\right] + c$, *ii)* $4x + y + 1 = 2\tan(2x + \frac{\pi}{4})$
2. i) $x + ye^{\frac{x}{y}} = c$ 2. ii) $\log\left(x + y + \frac{1}{3}\right) + \frac{3}{2}(y - x) = c,$
3. *i)* $(x^2 + y^2)y' = 2xy$
4. $e^x(x^2 + y^2) = 2e,$
5. $xy = c(x + y + 1)^3,$
6. $2x = (1 + 2cx^2)logy$
7. $(y - x)^2(ce^{x^2} - x^2 - 1) = 1$
8. $p - eliminate\ of\ y = p^2x + p\ and\ (p-1)^2x = logp - p + c$
9. $x^2 + y^2 = c(x + y) - \frac{1}{4}c^2$
10. $y = Acos\,nx + B\sin nx + \frac{cosnx}{n^2}logcosnx$
11. $y = c_1 + c_2e^x + c_3e^{-x} + c_4cosx + c_5sinx + \frac{1}{4}xe^x + \frac{1}{4}xsinx + \frac{1}{2}x^2$
12. $y = (c_1 + c_2x)e^x + \frac{1}{6}x^3e^x$
13. $y = x^{-1}\left(c_1 + c_2logx + log\frac{x}{x-1}\right),$
14. $y = c_1(1 - xcotx) + c_2cotx,$
15. $(1 + x + x^2)y = \frac{c_1}{2}x^2 + c_2x + c_3$
16. $y = x^2 + x + 1 + c_1 + c_2e^x$
17. $y = \frac{c_1}{2}sin^2xlogtan\frac{x}{2} - \frac{c_1}{2}cosx + c_2sin^2x$
18. $logy = c_2e^x + c_3e^{-x}$
19. $y = (c_1 + c_2x)cosx + (c_3 + c_4x)sinx + \frac{1}{48}[4x^3sinx - x^2(x^2 - 9)$ $cosx]$
20. $y = x^2\left(c_1x^{\sqrt{3}} + c_2x^{-\sqrt{3}}\right) + \frac{1}{x}[\frac{1}{6}(logx + 1) + \frac{logx}{61}\{5\sin(logx)$ $+ 6\cos(logx)\} + \frac{2}{3721}\{27\sin(logx) + 191\cos(logx)\}$

21. $y=\frac{1}{2}\left(c\frac{x^2}{2}-\frac{1}{c}logx\right)+c'$,23 $y=\left(c_1-\frac{1}{2}cosx-\frac{1}{5}c_2e^{-2x}\{cosx\right.$
$\left.+2sinx\}\right)e^x$

Exercise 2.8

1. $x^3, \frac{1}{3}x^3-\frac{1}{12}x^4, \frac{1}{3}x^3-\frac{1}{12}x^4+\frac{x^5}{60}$
2. (i) $y_1=1+x^2, y_2=1+x^2+\frac{x^4}{2}, y_3=1+x^2+\frac{x^4}{2}+\frac{x^6}{6}$
   (ii) $y_1=3(e^x-1), y_2=9e^x-6x-9, y_3=21e^x-6x^2-18x-21$
   (iii) $y_1=2+x+x^2, y_2=2+x+x^2+\frac{x^3}{3}+\frac{x^4}{4}, y_3=2+x+x^2$
   $+\frac{x^3}{3}+\frac{x^4}{4}+\frac{x^5}{15}+\frac{x^6}{24}$
3. $\frac{1}{2}+\frac{3x^4}{8}+\frac{x^5}{10}+\frac{3x^8}{64}+\frac{7x^9}{360}+\frac{x^{12}}{256}$
4. $y_3=1+\frac{x}{2}+\frac{3x^5}{40}+\frac{x^6}{60}+\frac{x^9}{192}$
5. $a+1$

Exercise 3.8

1. $\phi(t)=\begin{pmatrix} e^t & e^{2t} & e^{2t} \\ e^t & -e^{2t} & 0 \\ 3e^t & 0 & e^{2t} \end{pmatrix}$,

$$e^{At}=\begin{pmatrix} -e^t+2e^{2t} & -e^t+e^{2t} & -e^t-e^{2t} \\ -e^t+e^{2t} & -e^t+2e^{2t} & -e^t-e^{2t} \\ -3e^t+9te^{2t} & -3e^t+3e^{2t} & -3e^t-2e^{2t} \end{pmatrix}$$

2. $x(t)=e^{4t}\begin{pmatrix} 3c-b \\ 6c-2b \\ 3c-b \end{pmatrix}+e^t[\{E+t(A-E)u_2\}$ and

$$e^{At}=\begin{pmatrix} e^t & -e^{4t}+e^t & 4e^{4t}-3e^t \\ 0 & -3e^{4t}+2e^t & 6e^{4t}-6e^t \\ 0 & -e^{4t}+e^t & 3e^{4t}-3e^t \end{pmatrix}$$

3. (i) $\det\phi(t)=e^{4t}$ and (ii) $det\phi(t)=e^{-8t}$

4. (i) $\phi(t)=\begin{pmatrix} \frac{1}{2} & -4e^{3t} & 2e^{15t} \\ 1 & -2e^{3t} & -2e^{15t} \\ 1 & 4e^{3t} & e^{15t} \end{pmatrix}$,

(ii) $\phi(t)=\begin{pmatrix} e^{3t} & e^{3t} & e^{-2t} \\ 0 & -e^{3t} & -e^{-2t} \\ -e^{2t} & -e^{3t} & 4e^{2t} \end{pmatrix}$

(iii) $e^{At}=\frac{1}{8}\begin{pmatrix} 8e^{t} & -3e^{t}+e^{t}+4te^{t} & e^{3t}-e^{t}+12te^{t} \\ 0 & 6e^{3t}+2e^{t} & -6e^{3t}+6e^{t} \\ 0 & -2e^{3t}+2e^{t} & 2e^{3t}+6e^{t} \end{pmatrix}$,

$\phi(t)=\begin{pmatrix} e-^{3t} & 0 & 2e^{t} \\ -6e^{-3t} & e^{t} & 0 \\ 2e^{-3t} & e^{t} & 0 \end{pmatrix}$

5. $\phi(t)=\begin{pmatrix} e^{t} & e^{2t} & e^{2t} \\ e^{t} & -e^{2t} & 0 \\ 3e^{t} & 0 & e^{2t} \end{pmatrix}$

6. $\begin{pmatrix} x \\ y \end{pmatrix}'=\begin{pmatrix} 1 & 2 \\ 4 & 3 \end{pmatrix}\begin{pmatrix} x \\ y \end{pmatrix}$

9. $\phi(t)=\begin{pmatrix} e^{2t} & e^{3t} & 3e^{5t} \\ -e^{2t} & -2e^{3t} & -6e^{5t} \\ -e^{2t} & -e^{3t} & -2e^{5t} \end{pmatrix}$

10. $\begin{pmatrix} e^{5t} \\ 2e^{5t} \end{pmatrix},\begin{pmatrix} e^{t} \\ -2e^{t} \end{pmatrix} \textit{and} \begin{pmatrix} e^{5t} & e^{t} \\ 2e^{5t} & -2e^{t} \end{pmatrix}$

Exercise 4.7

2. $c_1x+c_2x^2-xsinx$
3. (i) $(c_1+c_2x+c_3x^2+c_4x^3)e^x+\frac{1}{24}x^4e^x$

   (ii) $c_1e^{i(\sqrt{2}-1)x}$

   (iii) $e^{2x}(c_1sinx+c_2cosx)$
4. $y_1=x^2, y_2=x^3, y_3=\frac{1}{x^2}; W(y_1,y_2,y_3)(x)=20$

5. $y(x) = 1 + \frac{1}{2}(sinx - cosx - e^{-x})$
6. (i) $y_p(x) = -\frac{1}{4}cos2xlog(sec2x + tan2x)$
   (ii) $y_p(x) = 30x^2e^{-2x}\text{-}60x^2e^{-2x}logx$
7. (b) $W(\phi_1, \phi_2)(x) = 0$ for all x
8. $W' = \alpha_1 W$
9. (i) $\phi(x) = c_1cos3x + c_2sin3x - \frac{1}{6}xcos3x$
   (ii) $\phi(x) = c_1e^{x/2} + c_2e^{-x/2} + \frac{1}{3}e^x.$
   (iii) $\phi(x) = c_1e^{2x/3} + c_2e^{-3x/2} - \frac{1}{6}x - \frac{5}{36}$
10. (i) $\phi(x) = c_1e^{2x} + c_2\exp\left\{(-1+i\sqrt{3})x\right\} + c_3\exp\left\{\left(-1-i\sqrt{3}\right)x\right\}$
    (ii) $\phi(x) = e^{\sqrt{2}x}(c_1cos\sqrt{2}x + c_2sin\sqrt{2}x) + e^{-\sqrt{2}x}(c_3cos\sqrt{2}x +$
    $+ c_4sin\sqrt{2}x)$
    (iii) $\phi(x) = \sum_{k=1}^{100} c_ke^{r_kx}$, $r_k = (100)^{\frac{1}{100}}exp\left[\frac{(2k-1)\pi}{100}\right]$, $k =$
    $1, 2..100$
    (iv) $\phi(x) = (c_1 + c_2x + c_3x^2)e^{ix}, c_1, c_2, c_3,$ are any constants
11. (b) $\phi(x) = cos\sqrt{2}x\ cosh\sqrt{2}x)$
12. (a) $\phi_1(x) = 1, \phi_2(x) = x, \phi_3(x) = x^2, \phi_4(x) = x^3,$
    (b) $\phi_1(x) = \cosh kx\ coskx, \phi_2(x) = \cosh kx\ sinkx, \phi_3(x) = \sin$
    $h\ kx\ coskx,$
    $\phi_4(x) = \sin h\ kx\ sinkx$; where $\lambda = 4k^4$
    (c) $\phi_1(x) = \cosh kx, \phi_2(x) = sinhkx;\ \phi_3(x) = coskx, \phi_4(x)$
    $= sinkx$; where $\lambda = -k^4$
13. (a) $\phi_1(x) = e^x, \phi_2(x) = xe^x, \phi_3(x) = e^{-x}, \phi_4(x) = cosx,\ \ \phi_5(x) =$
    $sinx$
    (b) $W(\phi_1, \phi_2, \phi_3, \phi_4, \phi_5) = 32e^x$
    (c) $\phi(x) = \frac{5}{8}e^x + \frac{1}{4}xe^x + \frac{1}{8}e^{-x} - \frac{1}{4}cosx - \frac{1}{4}sinx$
14. (i) $\phi(x) = (c_1 + c_2e^{x,} + c_3e^{-x,}) - \frac{x^2}{2}$ $c_1, c_2, c_3,$ are any constants
    (ii) $\phi(x) = \left(c_1e^{\sqrt{2}x,}cosx + c_2e^{\sqrt{2}x,}sinx + c_3e^{-\sqrt{2}x,}cosx + c_4e^{-\sqrt{2}x,}\right.$
    $\left. sinx\right) + \frac{1}{17}cosx$
    (iii) $\phi(x) = c_1e^{x,} + c_2e^{-x,} + c_3cosx + c_4sinx - \frac{x}{4}sinx$
15. $B = \dfrac{A}{p(\alpha)}$
16. (i) $\psi(x) = -\frac{1}{2}x^2 + \frac{1}{2}x - \frac{3}{4} - \frac{3}{10}cosx - \frac{1}{10}sinx$

(ii) $\psi(x)=-\frac{1}{60}x^5+\frac{1}{4}e^{-x}sinx-\frac{1}{4}e^{-x}cosx$

(iii) $\psi(x)=-\frac{1}{60}x^5\frac{1}{4}e^{-x}$

Exercise 5.5

4(i). $t^2\frac{d^2x}{dt^2}-(2t^3+3t)\frac{dx}{dt}+(2t^2+3)x=0$

5. $\frac{d}{dx}\left[cosx\frac{dy}{dx}\right]+(cosx)y=0$

Exercise 6.4

1. $y(x)==\int_0^1 G(x,t)ty(t)dt+\frac{1}{2}x(x-1), where\ G(x,t)=x(1-t), (x<t)\ and=t(1-x), (x>t)$
2. $y(x)==\int_0^1 G(x,t)y(t)dt+\frac{1}{6}(x^3-3x+6), where\ G(x,t)=x, (x<t)\ and=t,(x>t)$
3. $u(x)==\int_0^1 G(x,t)e^t u(t)dt+\frac{1}{6}(x^3+x), where\ G(x,t)= x(1-2t),(x<t)\ and=t(1-2x),(x>t)$
4. $y_p(x)=\frac{3}{4}+\frac{3}{4}cos2x$
5. $y_p(x)=12x-15x^3+3x^5$
6. $G(x,t)=\begin{cases}\frac{sinhxsinh(t-1)}{sinh1}, x<t,\\ \frac{sinhtsinh(x-1)}{\mu sinh1}, x>t\end{cases}$
7. For second order equation

$$G=\begin{cases}\frac{y_2(t)y_1(x)}{a_0(t)W(t)}, if\ c\le x\le t,\\ \frac{y_2(x)y_1(t)}{a_0(t)W(t)}, if\ t\le x\le d,\end{cases}$$

Exercise 7.8

1. $y_n=\frac{cosn\pi x}{l};\ \lambda_n=\frac{n^2\pi^2}{l^2}$;n=0,1,2,...
2. $y_n=sinnx;\ \lambda_n=n^2$;n=1,2,3,...

3. $y_n = \sin(2n-1)x\,;\ \lambda_n = (2n-1)^2$; n=1,2,3,...
4. $y_n = \cos n\pi x/l\,;\ \lambda_n = n\pi/l$; n=1,2,3,...
5. $(i)\, y_{n'}(x) = cos\left(\dfrac{n\pi x}{l}\right), \mu_n = n^2\pi^2/l^2$

   $(ii)\, y_{n'}(x) = \sin n\pi, \mu_n = n^2, n = 1,2,3,\ldots$

Exercise 8.6

1. (i) oscillatory
   (ii) convergent
4. $-1 < x \leq 1$

Exercise 9.6

2. $y = ax^3\left(1 - \frac{4}{2.7}x + \frac{4.5}{2.1.7.8}x^2 + \cdots\right) + bx^{-3}\left(1 - \frac{2}{5}x + \frac{1}{20}x^2 + \cdots.\right)$
3. $y = x^{-1}(acosx + bsinx)$
4. $y = a\left(1 - 2x + \frac{3}{2!}x^2 - \frac{4}{3!}x^3 + \cdots\right) + b\left(y_1 logx + (3x - \frac{13}{4}x^2)\right)$
5. $y = ax^{-\frac{1}{2}}\left(1 + \frac{x}{2} + \frac{x^2}{40} + \cdots\right) + bx^{\frac{1}{4}}\left(1 + \frac{x}{14} + \frac{x^2}{616} + \cdots\right)$
6. $y = e^{mx}$
7. $y = c_1 x^{-\frac{1}{2}}\left(1 + \dfrac{x}{2} + \dfrac{x^2}{40} + \ldots\right) + c_2 x^{\frac{1}{4}}(1 + \dfrac{x}{14} + \dfrac{x^2}{616} + \ldots)$
8. $y = c_1\left(1 + x + \dfrac{x^2}{4.7} + \ldots\right) + c_2 x^{\frac{2}{3}}\left(1 + \dfrac{x}{3} + \dfrac{x^2}{3.6} + \dfrac{x^3}{3.6.9} + \ldots\right)$
9. $y = ax^3\left(1 - \dfrac{3x^2}{5} + \ldots\right) + \dfrac{b}{x^4}\left(1 + \dfrac{10}{9x^2} + \dfrac{35}{33x^4} + \ldots\right)$
10. $y = a(1 + 2^2x + 3^2x^2 + 4^2x^3 + \ldots..) + b[ulogx - 2(1.2x + 2.3x^2 + \cdots.)]$,
    $where\ u = (1 + 2^2x + 3^2x^2 + 4^2x^3 + \ \ldots..)$
11. $y = a\left(1 + x + \dfrac{2x^2}{4} + \dfrac{2.5x^3}{4.9} + \ldots\right) + b\left[logx\left(1 + x + \dfrac{2x^2}{4} + \dfrac{2.5x^3}{4.9} + \ldots\right)\right.$
    $\left. + \left(-2x - x^2 - .\right)\right]$

Exercise 10.13

7. $y = a\,{}_2F_1\left[\dfrac{n+1}{2}, -\dfrac{n}{2}; \dfrac{1}{2}; x^2\right] + b\,{}_2F_1\left[\dfrac{n+1}{2}, -\dfrac{n}{2}; \dfrac{1}{2}; x^2\right]$

Exercise 12.13

2. (a) (i) $\frac{2}{5}P_3+\frac{4}{3}P_2-\frac{2}{5}P_1-\frac{7}{3}P_0$

   (ii) $\frac{8}{35}P_4+\frac{6}{5}P_3-\frac{2}{21}P_2+\frac{34}{5}P_1-\frac{224}{105}P_0$

   (iii) $\frac{8}{5}P_3+\frac{32}{5}P_2+7P_1+4P_0$

   (iv) $\frac{8}{35}P_4+\frac{4}{5}P_3-\frac{4}{21}P_2-\frac{34}{25}P_1-\frac{336}{105}P_0$

   (b) $\frac{1}{3}P_0(x)-\frac{2}{3}P_2(x)$

8. (i) $\dfrac{(-1)^{\frac{n-1}{2}}(n-1)!}{2^n(\frac{n}{2}+\frac{1}{2})!(\frac{n}{2}-\frac{1}{2})!}$

   (ii) 0

Exercise 14.10

6. (i) 0, (ii) 1.
7. $L_0(x)-L_1(x)+2L_2(x)-6L_3(x)$

References

Bell, W.W., *Special Functions for Scientists and Engineers*, D Van Nostrand co. Ttd., London, 1968.

E.A., Coddington, Levinson, N., *Theory of Ordinary Differential Equations*, McGraw-Hill, 1955.

E.A. and Coddington, *A introduction to Ordinary Differential Equations*, Printice Hall of India Private Limited, New Delhi, India, 1991.

Rai, B. and Choudhury, D.P., *Ordinary Differential Equations*, Narosa Publishing House, 2011.

Birkhoff, G. and Rota, G.-C., *Ordinary Differential Equations*, John Wiley & Sons, Singapore, 1989.

Ince, E.L., *Ordinary Differential Equations*, Dover Publications, 1956.

Miller, R.K. and Michel, A.N., *Ordinary Differential Equations*, Academic Press, 1982.

Piaggio, H.T.H., *An Elementary Treatise of Differential Equations and their Applications*, G. Bell and Sons, 1956.

Raisinghania, M.D., *Advanced Differential Equations*, S. Chand & Company Pvt. Ltd, New Delhi, India, 2014.

Rainville, E.D., *Special Functions*, Macmillan, New York, 1960.

Senddon, I.N., *Special Functions of Mathematical Physics and Chemistry*, Oliver and Boyd, Edinburg.

Simmons, G.R., *Differential with Applications and Historical Notes*, McGraw-Hill, 1972.

Somasundram, D., *Ordinary Differential Equations, A First Course*, Narosa Publishing House, 2016.

Dipankar De. Ordinary Differential Equations and Special Functions, (611–612) © 2026 Scrivener Publishing LLC

Index

Also of Interest

Books in the series, "Mathematics and Computer Science"

Graph Theory for Computer Science, Edited by Manikandan Rajagopal, Ramkumar Sivasakthivel, Joseph Varghese Kureethara, Niranjanamurthy M., and Biswadip Basu Mallik, ISBN: 9781394302598. This book is a vital resource for anyone looking to understand the essential role of graph theory as the unifying thread that connects and provides innovative solutions across a wide spectrum of modern computer science disciplines.

Mathematics and Computer Science for Real-World Applications, Edited by Biswadip Basu Mallik, M. Niranjanamurthy, Sharmistha Ghosh, Krishanu Deyasi, and Santanu Dasc ISBN: 9781394275328. *Mathematics and Computer Science for Real-World Applications* gives invaluable insights into how mathematical and computer sciences drive essential modern innovations that enhance everyday life, making it a must-read for anyone interested in the intersection of mathematics and technology and their real-world applications.

MATHEMATICS AND COMPUTER SCIENCE VOLUME 1: Concepts and Applications, Edited by Sharmistha Ghosh, M. Niranjanamurthy, Krishanu Deyasi, Biswadip Basu Mallik, and Santanu Das, ISBN: 9781119879671. This first volume in a new multi-volume set gives readers the basic concepts and applications for diverse ideas and innovations in the field of computing together with its growing interactions with mathematics.

MATHEMATICS AND COMPUTER SCIENCE VOLUME 2, Edited by Sharmistha Ghosh, M. Niranjanamurthy, Krishanu Deyasi, Biswadip Basu Mallik, and Santanu Das, ISBN: 9781119896326. This second volume in a new multi-volume set builds on the basic concepts and fundamentals laid out in the previous volume, presenting the reader with more advanced and cutting-edge topics being developed in this exciting field.

Mathematical Modeling for Computer Applications, Edited by Biswadip Basu Mallik, M. Niranjanamurthy, Sharmistha Ghosh, Valentina Emilia Balas, Krishanu Deyasi, and Santanu Das. Series: Mathematics and Computer Science, ISBN: 9781394248407. This new volume in this groundbreaking series on mathematics and computer science builds on the basic concepts and fundamentals laid out in previous volumes, presenting the reader with more advanced and cutting-edge topics being developed in this exciting field.

Check Out These Other Related Titles from Scrivener Publishing

Mathematics for Engineers, By Ritu Shrivastava, Ramakant Bhardwaj, and Satyendra Narayan, ISBN: 9781394214588. Mathematics for Engineers serves as a comprehensive guide on the basics of mathematics and their applications in engineering for students and seasoned professionals alike.

Mathematical Models Using Artificial Intelligence for Surveillance Systems, Edited by Padmesh Tripathi, Mritunjay Rai, Nitendra Kumar, and Santosh Kumar, ISBN: 9781394200733. This book gives comprehensive insights into the application of AI, machine learning, and deep learning in developing efficient and optimal surveillance systems for both indoor and outdoor environments, addressing the evolving security challenges in public and private spaces.

MATHEMATICS IN COMPUTATIONAL SCIENCE AND ENGINEERING, Edited by Ramakant Bhardwaj, Jyoti Mishra, and Satyendra Narayan, ISBN: 9781119777151. This groundbreaking new volume, written by industry experts, is a must-have for engineers, scientists, and students across all engineering disciplines working in mathematics and computational science who want to stay abreast with the most current and provocative new trends in the industry.

Introduction to Differential Geometry with Tensor Applications, By Dipankar De, ISBN: 9781119795629. This is the only volume of its kind to explain, in precise and easy-to-understand language, the fundamentals of tensors and their applications in differential geometry and analytical mechanics with examples for practical applications and questions for use in a course setting.

DATA WRANGLING: Concepts, Applications, and Tools, Edited by M. Niranjanamurthy, Kavita Sheoran, Geetika Dhand, and Prabhjot Kaurk, ISBN: 9781119879688. Written and edited by some of the world's top experts in the field, this exciting new volume provides state-of-the-art research and latest technological breakthroughs in next-data wrangling, its theoretical concepts, practical applications, and tools for solving everyday problems.

ADVANCES IN DATA SCIENCE AND ANALYTICS, Edited by M. Niranjanamurthy, Hemant Kumar Gianey, and Amir H. Gandomi, ISBN: 9781119791881. Presenting the concepts and advances of data science and analytics, this volume, written and edited by a global team of experts, also goes into the practical applications that can be utilized across multiple disciplines and industries, for both the engineer and the student, focusing on machining learning, big data, business intelligence, and analytics.

CONVERGENCE OF DEEP LEARNING IN CYBER-IOT SYSTEMS AND SECURITY, Edited by Rajdeep Chakraborty, Anupam Ghosh, Jyotsna Kumar Mandal and S. Balamurugan, ISBN: 9781119857211. In-depth analysis of Deep Learning-based cyber-IoT systems and security which will be the industry leader for the next ten years.

MACHINE INTELLIGENCE, BIG DATA ANALYTICS, AND IOT IN IMAGE PROCESSING: Practical Applications, Edited by Ashok Kumar, Megha Bhushan, José A. Galindo, Lalit Garg and Yu-Chen Hu, ISBN: 9781119865049. Discusses both theoretical and practical aspects of how to harness advanced technologies to develop practical applications such as drone-based surveillance, smart transportation, healthcare, farming solutions, and robotics used in automation.

Printed and bound by CPI Group (UK) Ltd, Croydon, CR0 4YY
15/09/2026
14962464-0004